Linear Algebra for Large Scale and Real-Time Applications

NATO ASI Series

Advanced Science Institutes Series

A Series presenting the results of activities sponsored by the NATO Science Committee, which aims at the dissemination of advanced scientific and technological knowledge, with a view to strengthening links between scientific communities.

The Series is published by an international board of publishers in conjunction with the NATO Scientific Affairs Division

A **Life Sciences**	Plenum Publishing Corporation
B **Physics**	London and New York
C **Mathematical**	Kluwer Academic Publishers
and Physical Sciences	Dordrecht, Boston and London
D **Behavioural and Social Sciences**	
E **Applied Sciences**	
F **Computer and Systems Sciences**	Springer-Verlag
G **Ecological Sciences**	Berlin, Heidelberg, New York, London,
H **Cell Biology**	Paris and Tokyo
I **Global Environmental Change**	

NATO-PCO-DATA BASE

The electronic index to the NATO ASI Series provides full bibliographical references (with keywords and/or abstracts) to more than 30000 contributions from international scientists published in all sections of the NATO ASI Series.
Access to the NATO-PCO-DATA BASE is possible in two ways:

– via online FILE 128 (NATO-PCO-DATA BASE) hosted by ESRIN,
Via Galileo Galilei, I-00044 Frascati, Italy.

– via CD-ROM "NATO-PCO-DATA BASE" with user-friendly retrieval software in English, French and German (© WTV GmbH and DATAWARE Technologies Inc. 1989).

The CD-ROM can be ordered through any member of the Board of Publishers or through NATO-PCO, Overijse, Belgium.

Linear Algebra for Large Scale and Real-Time Applications

edited by

Marc S. Moonen
Katholieke Universiteit Leuven,
ESAT Laboratory,
Leuven, Belgium

Gene H. Golub
Stanford University,
Computer Science Department,
Stanford, California, U.S.A.

and

Bart L. R. De Moor
Katholieke Universiteit Leuven,
ESAT Laboratory,
Leuven, Belgium

Kluwer Academic Publishers

Dordrecht / Boston / London

Published in cooperation with NATO Scientific Affairs Division

Proceedings of the NATO Advanced Study Institute on
Linear Algebra for Large Scale and Real-Time Applications
Leuven, Belgium
August 3–14, 1992

Library of Congress Cataloging-in-Publication Data

```
Linear algebra for large scale and real-time applications / edited by
  Marc S. Moonen, Gene H. Golub, Bart L.R. de Moor.
      p.    cm. -- (NATO ASI series. Series E, Applied sciences ; vol.
232)
    "Proceedings of the NATO advanced study institute, Linear algebra
for large scale and real-time applications, Leuven, Belgium, August
3-14, 1992."
    Includes index.

    1. Algebras, Linear--Data processing--Congresses.   I. Moonen,
Marc S., 1963-   . II. Golub, Gene H. (Gene Howard), 1932-   .
III. Moor, Bart L. R. de, 1960-   . IV. Series: NATO ASI series.
Series E, Applied sciences ; no. 232.
QA185.D37L56   1993
512'.5--dc20                                          92-46135
```

ISBN 978-90-481-4246-0

Published by Kluwer Academic Publishers,
P.O. Box 17, 3300 AA Dordrecht, The Netherlands.

Kluwer Academic Publishers incorporates the publishing programmes of
D. Reidel, Martinus Nijhoff, Dr W. Junk and MTP Press.

Sold and distributed in the U.S.A. and Canada
by Kluwer Academic Publishers,
101 Philip Drive, Norwell, MA 02061, U.S.A.

In all other countries, sold and distributed
by Kluwer Academic Publishers Group,
P.O. Box 322, 3300 AH Dordrecht, The Netherlands.

Printed on acid-free paper

TABLE OF CONTENTS

viii

PREFACE

In recent years, there has been a great interest in large-scale and real-time matrix computations; these computations arise in a variety of fields, such as computer graphics, imaging, speech and image processing, telecommunication, biomedical signal processing, optimization and so on. This volume gives an account of recent research advances in numerical techniques used in large-scale and real-time computations and their implementation on high performance computers. For anyone interested in any of the aforementioned disciplines, this collection of papers is of value and provides state-of-the-art expositions as well as new and important trends and directions for the future, motivated and illustrated by a wealth of scientific and engineering applications.

The volume is an outgrowth of the NATO Advanced Study Institute "Linear Algebra for Large-Scale and Real-Time Applications," held at Leuven, Belgium, August 1992. We were quite fortunate to be able to gather such an excellent group of researchers to participate in this Institute. We are indebted to all the participants who enriched the meeting through their many contributions. Special thanks are due to the invited speakers at the Institute, P. Bjørstad, Universitetet i Bergen -Norway, S. Boyd, Stanford University -U.S.A., G. Cybenko, Dartmouth College -U.S.A., J. Demmel, University of California Berkeley -U.S.A, E. Deprettere, Technische Universiteit Delft -The Netherlands, P. Dewilde, Technische Universiteit Delft -The Netherlands, R. Freund, AT & T -U.S.A., M. Gentleman, National Research Council -Canada, S. Haykin, McMaster University -Canada, B. Kågström, University of Umeå -Sweden, N. Kalouptsidis, University of Athens -Greece, F. Luk, Rensselaer Polytechnic Institute -U.S.A., G.W. Stewart, University of Maryland -U.S.A, P. Toint, Université de Namur -Belgium, C. Van Loan, Cornell University -U.S.A., and M. Wright, AT & T -U.S.A. Finally, we also wish to thank the other members of the organizing committee, A. Bunse-Gerstner, Universität Bremen -Germany, S. Hammarling, Numerical Algorithms Group Ltd, Oxford -U.K., J. Meinguet, Université Catholique de Louvain -Belgium, J. Vandewalle, Katholieke Universiteit Leuven -Belgium.

The generous support of the NATO Scientific Affairs Division, which made this Institute possible, is gratefully acknowledged. We are also indebted to the co-sponsors of the Institute, the *Belgian National Fund for Scientific Research (NFWO)*, the *European Office of Aerospace Research and Development*, and the *Office of Naval Research Europe, London*, *BARCO*, and *BELGACOM*.

Marc Moonen
Gene Golub
Bart De Moor

INVITED LECTURES

LARGE SCALE STRUCTURAL ANALYSIS ON MASSIVELY PARALLEL COMPUTERS

P.E. BJØRSTAD and J. COOK
Para//ab, Institutt for Informatikk
University of Bergen
N-5020 Bergen, Norway
na.bjorstad@na-net.ornl.gov, Jeremy.Cook@ii.uib.no

ABSTRACT. We discuss the use of a massively parallel computer in large scale structural analysis computations. In particular, we consider the necessary modifications to a standard industrial code and a strategy where the code may be allowed to evolve gradually from a pure F77 version to a form more suitable for a range of parallel computers. We present preliminary computational results from a DECmpp computer, showing that we can achieve good price performance when compared to traditional vector supercomputers.

Introduction

Structural analysis provided the problems, motivation and pioneering work that led to the development of the powerful finite element method. This field is still the most important application area of finite element based analysis. There exist a number of highly successful commercial codes that can be used to analyze the behavior of almost any kind of structure under rather general conditions. In almost all of these codes the technique called substructuring [1] is used in order to simplify modeling and post processing. The physical domain is divided into several disjoint pieces called substructures, and each substructure can be assembled separately. The global stiffness matrix is then usually formed in order to solve the equations.

One commercial code, SESAM, [3, 11] pioneered the further use of this concept, by also taking advantage of the substructures (in particular identical ones) in the solution algorithm. In this code the solution procedure is carried out by the elimination of interior unknowns from each substructure, followed by the calculation of Schur complements corresponding to the unknowns on interior interfaces between substructures.

This strategy is implemented in a recursive fashion, thus making a multilevel substructure implementation. At any given level in this procedure, the unknowns can be uniquely divided into two disjoint sets, the internal variables and the retained or external variables. The

3

M. S. Moonen et al. (eds.), Linear Algebra for Large Scale and Real-Time Applications, 3–11.
© 1993 *Kluwer Academic Publishers.*

4

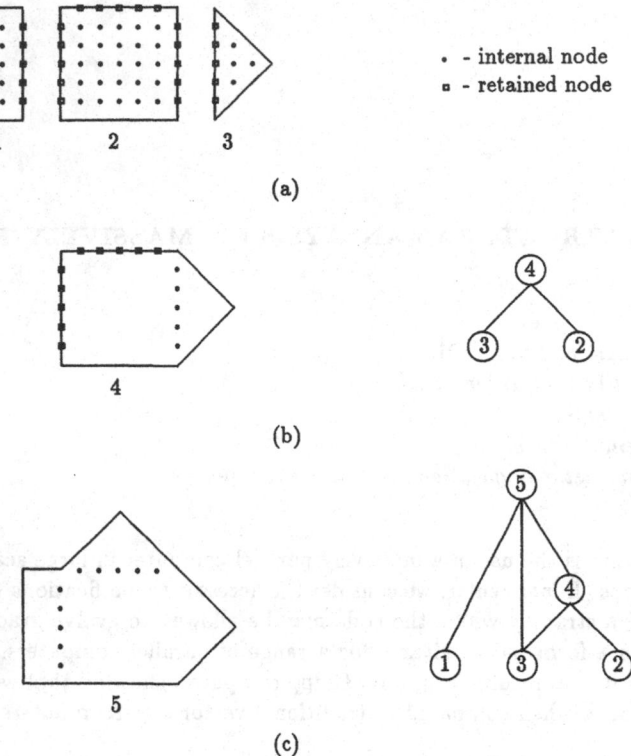

Figure 1: *The multilevel substructure technique.*

latter group of variables can be interpreted as boundary values or fixed degrees of freedom at this level of the algorithm. Before the algorithm proceeds to the next level, all internal variables are eliminated. At the next level the retained variables from the previous level are again split into two sets and this process repeats until one reaches the highest level where all remaining variables (or unknowns) in the problem will be in the internal set.

The organization of a finite element model, as well as the computational procedure outlined above, can be represented as in Figure 1. The dots represent *internal* nodes, these are nodes which cannot be shared with other substructures [1]. The small squares are *retained* nodes, which are nodes retained for the purpose to be coupled together with retained nodes from other substructures. Retained nodes at a given level may become internal or specified nodes at the next (upward) level. Figure 1(b) shows an intermediate level in the assembly process where substructure 2 and one instance of substructure 3 are coupled together to form the new substructure 4. The retained nodes on the interface between these two substructures become internal nodes in substructure 4, while the remaining retained nodes

---

[1]This figure is taken from [9]. The description is taken from [2].

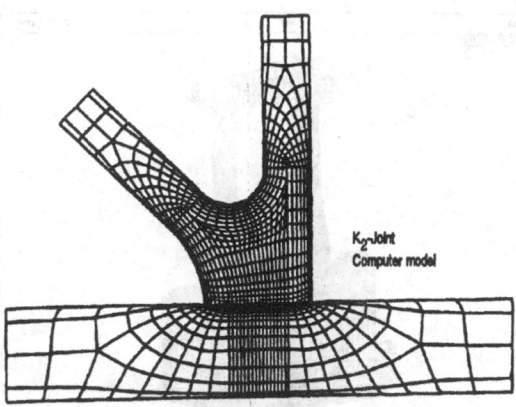

Figure 2: *Finite element model of a tubular joint*

from substructure 2 keep their status. The substructure tree of this intermediate level is shown to the right in 1(b). Finally, Figure 1(c) shows the entire computational model where substructure 1 and a rotated instance of substructure 3 are matched with the remaining retained nodes of substructure 4 to form substructure 5, the *structure level*. On the structure level there are only internal variables. A small, but realistic model of a tubular joint assembled from 9 substructures is pictured in Figure 2.

1 Block algorithms for MPP architecture

The first proposal for a parallel, direct solver in the SESAM system, was described in [4]. At the end of 1989, many of these ideas had been carefully investigated and implemented by Anders Hvidsten as described in his thesis [9]. We note that the superelement technique can be interpreted as a multi-frontal code (Ch. 10 in [8]) with respect to the finite elements, while most general sparse codes normally consider the nodes as the fundamental quantity. Experiments with iterative methods using domain decomposition preconditioners, that have direct relevance for this application code, have been reported in [5] and more recently in [10]. An overview of both direct and iterative solvers applied to this application is given in [2]. The SESAM system can perform the reduction of different superelements in parallel on a heterogeneous, loosely coupled computing system.

Our focus in this paper, is the very large, dense superelements that typically arise towards the end of a multilevel, superelement analysis. We show that this part of the overall process can be separated and successfully computed by a massively parallel processing system.

The design of the SESAM system, with emphasis on hierarchical storage and a design requirement to handle arbitrary large problems, makes the relative speed of computation and data movement the most important consideration when porting the application to a new platform. The application is therefore I/O-bound on most of today's computers unless there is enough RAM that can be used to cache the disk access. Computations are always localized in data to a substructure at a time. A reasonable requirement is therefore to

6

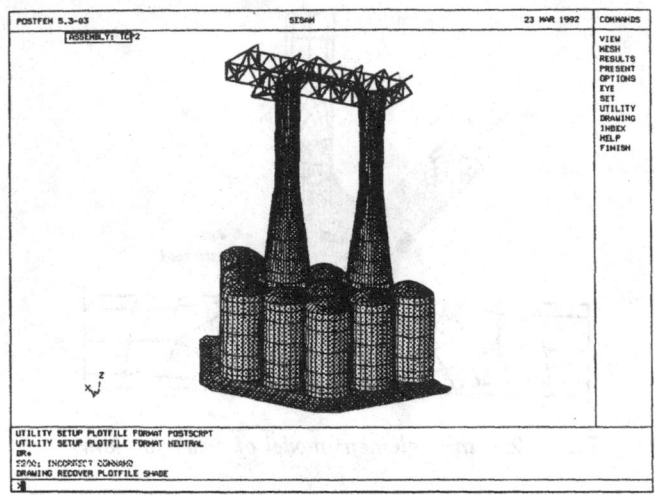

Figure 3: *Finite element model of TCP-2*

assume that the size of substructures and the size of memory that can be reserved as buffer space for the disk, are of similar size. For most substructures being modeled today, a working set of a few hundred Mbytes will be sufficient.

In a strategy to migrate very large software systems (the SESAM system consists of close to a million lines of Fortran code), the attention should first be directed to the most time consuming parts. When SESAM is used to perform static analysis, this has traditionally been the direct linear solver. This part of the code is the only place where the computational work grows faster than linearly with the size of the substructure. Each superelement is represented by a few large matrices as described in the introduction. The three fundamental operations matrix multiplication, solution of triangular systems, and Cholesky factorization, are implemented using block-sparse algorithms, where the blocksize is typically in the range from 32 to 100. This blocksize used in the computation also determines the smallest unit of disk storage.

It was realized that the coupling of computational blocks with the size of a disk storage block would pose severe restrictions on an efficient MPP implementation. We therefore proceeded to define 'superblocks' consisting of one or more diskblocks, as a new basic unit of I/O for the MPP system. The superblock concept allows us, for example, to increase the amount of work that is scheduled for computation on the MPP system by a factor of 4, 16 or 64. This increases the computational efficiency and significantly reduces the overall volume of data that is read from and written to the disk. The implementation required only four logical modules on the MPP system, effectively replacing corresponding modules in the standard code. In addition to the three computational kernels referred to earlier, a module to read a superblock from the disk into the MPP, or write it back from the MPP was needed. The computational modules on the MPP, were written as block algorithms based on BLAS kernels developed earlier [6, 7]. In this way we achieved independence between the I/O blocksize, the physical nonzero matrix-blocks stored on disk (and accessed by many other parts of the software as well), and the optimal computational blocksize. The latter is typically machine dependent and may match the number of processing elements on the MPP system.

All matrix computations in SESAM operate with matrix (disk) blocks as the fundamental computational unit. (Unlike many sparse codes where a single number may be the computational unit.) Every matrix calculation is therefore preceded by a symbolic phase where the exact order of calculations and an appropriate sparse data structure are determined. It was possible to make this preprocessing apply to the new concept of superblocks with very few modifications in the original code. We could therefore start using the MPP unit as an attached matrix processor without porting the entire code immediately. This opens the way for a more gradual port, where one can focus on the compute intensive parts of the code and at the same time allow the F90 compiler technology to mature before taking on the full code. All logic, including the processing of the sparse data structures, are performed on the front end computer. It executes the entire code, but communicates with the MPP system requesting it to fill a (distributed) memory buffer from the disk, write a buffer back, or carry out a matrix operation involving one or more buffers.

The size of a superblock can be changed at run time, this facilitates debugging and testing in an early phase and may be used to enhance performance at a later stage.

2 Large scale performance tests

In this section, we report on two different calculations of realistic problems, using our implementation. The first example shown in Figure 2, is a tubular joint calculation. The particular design was tested by industrial parties about ten years ago, at a time where this represented a substantial computational effort on a Vax class computer. The model has slightly more than 20000 degrees of freedom. The entire computation can be performed using approximately 200 Mbyte of secondary storage.

This case is included to highlight the strong coupling between the front end system (a DEC 5000-240) and the MPP system (A DECmpp 12000/Sx with 16384 processors, this machine is identical to a Maspar MP-1.) We will also report on a MasPar MP-1 equipped with 256 Mbytes of IO-RAM. Unfortunately, the front end computer on this system was a DEC 5000-200. (When doing large scale tests of this kind it is virtually impossible to perform experiments where only one system component changes at a time.) All times reported here are elapsed wall clock time, on machines that had very light loads. The times would be expected to increase in a production environment. The measurements obtained on the Cray Y-MP are all from running the industrial production version of the software. (A version that has been optimized for Cray computers.)

The results are summarized in Figure 4. We make several observations: The most cost effective platform for this problem is the workstation, provided that time to solution is not critical. We further see that the use of the Y-MP disk and the DECmpp diskarray (used as secondary storage), results in fairly similar total times. The superblock size was set to

8

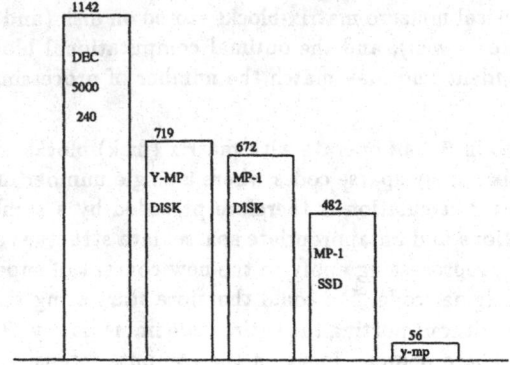

Figure 4: *Elapsed time to perform a complete reduction of the tubular joint model.*

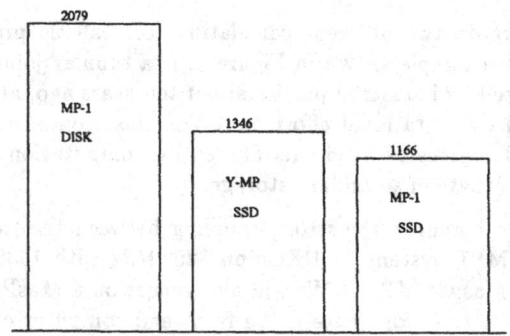

Figure 5: *Elapsed time to reduce the three largest substructures of the TCP-2 model.*

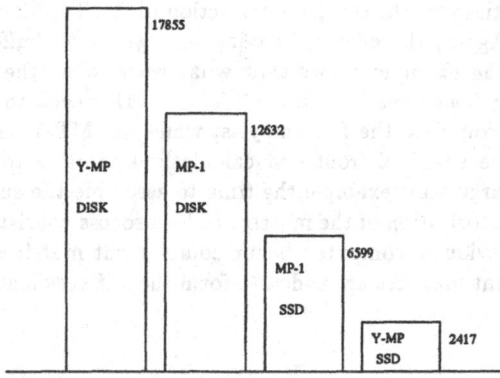

Figure 6: *Elapsed time to perform a complete reduction of the TCP-2 model.*

256 in this run. Most striking is the performance of the Cray equipped with a RAM disk[2]. The figure reveals the savings in IO-time by noting the difference between the times for the MP-1 system with and without SSD. The remaining difference reflects the difference in processing speed between the DEC workstation and the Y-MP (on scalar and short loop code). This example is too small to show the benefits of an MPP system. If this example was an 'inner loop' problem in an interactive design process, then the preferred machine would likely be the Y-MP.

Our second example is the global analysis of an oil platform designed by Elf-Aquitaine, see Figure 3. The physical structure can be represented by a computer model of half the structure due to symmetry. The model has almost ten times as many unknowns as our first example. The required disk space is almost 3 Gigabytes. This analysis represents a significant amount of computation even on a state of art supercomputer.

Figure 5 shows the elapsed time for reducing the three largest, higher level substructures. The computation is floating point intensive and requires about 20 hours on a SUN Sparc-2. We see that the time is reduced to less than 20 minutes on an MP-1 with an SSD. We note the effect of the SSD, the time for the MP-1 is almost halved compared to a configuration with only a diskarray. The Y-MP performs about as well as the MP-1 in this phase of the calculation. In this example we adjusted the size of a superblock up to 1024. Clearly, there is a tradeoff between the effiency of large dense matrix calculations and the ability to exploit the sparse structure of the matrix calculations. This is most important on the lower level superelements, since the Shur complements tend to produce quite dense matrices at the higher levels of the multifrontal elimination tree.

---

[2]For convenience we will use the terms IO-RAM and SSD, without regard to vendor, to mean a large RAM area used to cache disk storage.

Figure 6 shows the time for the complete reduction of the TCP-2 oil production platform structural analysis. Again, the effect of having a large cache buffer to disk is the most significant finding. The example shows that what used to be the dominant part of the computation has been been greatly reduced. The Y-MP needs to double the time from Figure 5, in order to complete the full analysis, while the MP-1 needs six times as much time. This reflects the speed of front end calculations relative to the Y-MP processing speed. Even for this large scale example the time to assemble the superelements now takes more time than the factorization of the matrix. (This process consists of adding the various contributions from previously computed Schur complement matrices together, in precisely the same way as element matrices are added to form the stiffness matrix in a standard finite element calculation.)

3 Conclusions

We have shown how the most time consuming part of a large scale structural analysis code can be reduced by using a massively parallel machine on this part of the problem. The use of a large RAM partition as a cache buffer to disk, reduces the running time on both traditional vector supercomputers like the Y-MP as well as on MPP systems significantly.

Two lines of development are likely to further reduce the elapsed time for the MPP system. The front end part of these systems should follow the performance curve of workstations and move considerably closer to the Y-MP in scalar performance over the next few years. At the same time, we expect considerable improvements in the Fortran 90 compiler technology for MPP architectures with distributed memory. This will make it feasible to compile more of the code for MPP execution and reduce the amount of rewrite that would be needed today. The development will be encouraged even more as soon as Fortran 90 compilers become generally available on workstation platforms.

Today, structural analysis calculations involving millions of unknowns can be performed on high performance computers. The algorithm of choice remains sophisticated implementations of the Cholesky factorization. It remains an open question whether iterative methods will gain more prominence in the future. The problem size that can be solved using a direct method, keeps increasing with time and new computer generations. The iterative methods are indeed facing a moving target.

Acknowledgments

The authors work would not have been possible without assistance and support from many sources. DnV Sesam Systems provided their software and generous support. Special thanks are due to Jon Brækhus and Anders Hvidsten for many extra hours. Pål Hellesnes wrote part of the required I/O module, while Tor Sørevik assisted with several of the computational kernels. We thank MasPar Computer Corporation for the use of a machine with appropriate IO-RAM. The use of the TCP-2 model was granted by Elf-Aquitaine. Financial support from Digital Equipment Corporation is also gratefully acknowledged.

References

[1] K. Bell, B. Hatlestad, O. E. Hansteen, and P. O. Araldsen. *NORSAM, a programming system for the finite element method. Users manual, Part 1, General description.*. NTH, Trondheim, 1973.

[2] P. Bjørstad, J. Brækhus, and A. Hvidsten. Parallel substructuring algorithms in structural analysis, direct and iterative methods. In : R. Glowinski, Y. A. Kuznetsov, G. Meurant, J. Périaux and O. B. Widlund (Eds), *Fourth International Symposium on Domain Decomposition Methods for Partial Differential Equations*, SIAM, Philadelphia, pp 321–340, 1991.

[3] P. E. Bjørstad. SESAM'80: A modular finite element system for analysis of structures.. In : B. Enquist and T. Smedsaas (Eds.), *PDE Software: Modules, Interfaces and Systems*, North Holland, Amsterdam, pp 19–27, 1984.

[4] ——. A large scale, sparse, secondary storage, direct linear equation solver for structural analysis and its implementation on vector and parallel architectures. *Parallel Computing*, pp 3–12, 1987.

[5] P. E. Bjørstad and A. Hvidsten, Iterative methods for substructured elasticity problems in structural analysis. In : R. Glowinski, G. H. Golub, G. A. Meurant and J. Périaux (Eds.), *Domain Decomposition Methods for Partial Differential Equations*, SIAM, Philadelphia, 1988.

[6] P. E. Bjørstad and T. Sørevik, Data-parallel BLAS as a basis for LAPACK on massively parallel computers. To appear in *Linear Algebra for Large Scale and Real-Time Applications*, 1993.

[7] ——, Two different data-parallel implementations of the BLAS. In : J. S. Kowalik (Ed.), *Software for Parallel Computation*, To appear 1993.

[8] I. Duff, A. Erisman, and J. Reid, *Direct Methods for Sparse Matrices*, Oxford Science Publications, 1986.

[9] A. Hvidsten, *A parallel implementation of the finite element program SESTRA*, PhD thesis, Department of Informatics, University of Bergen, Norway, 1990.

[10] B. F. Smith, *Domain decomposition algorithms for the partial differential equations of linear elasticity*, PhD thesis, Courant Institute of Mathematical Sciences, New York, September 1990.

[11] Veritas Sesam Systems, P.O. box 300, N-1322 Høvik, NORWAY, *SESAM technical description*, April 1989.

DATA-PARALLEL BLAS AS A BASIS FOR LAPACK ON MASSIVELY PARALLEL COMPUTERS

P.E. BJØRSTAD and T. SØREVIK
Para//ab, Institutt for Informatikk
University of Bergen
N-5020 Bergen, Norway
na.bjorstad@na-net.ornl.gov

ABSTRACT. We consider a data-parallel implementation of LU-factorization based on the LAPACK routine DGETRF. We analyze the performance of the required BLAS routines and show that high performance is inhibited by current compiler limitations. In particular, we show that optimal data movement when performing rank-1 updates is crucial. The rank-1 update is available as a BLAS-2 routine and can also easily be expressed using the intrinsic SPREAD in Fortran 90. However, in order to minimize processor communication, this operation should be explicitly inlined in the computational kernels. Using this observation we identify the need for an explicit LU-factorization applied to a single block. With the freedom to adjust the block-size to hardware, this is a much simpler task than writing the full code in a low level machine language. With this extension, we show that high performance is achievable without modifying the block structure of the LAPACK routine. We expect similar observations to hold for other modules in LAPACK.

Introduction

Parallel computers, in particular massively parallel processors (MPPs), claim a potential price/performance advantage as well as a desirable scalability over traditional high performance computers. This claim, unfortunately, depends heavily on the availability of software that is able to use the underlying hardware efficiently. The shortage of such software and the long development time for good compilers for this class of machines, have increased the importance of mathematical libraries that can be used by application programs. Modular software, efficient libraries, and a restructuring of many application codes to take advantage of data parallelism are necessary in order to foster widespread and diverse use of MPP computers.

A major obstacle to widespread use of parallel computers is the many machine specific language constructions to obtain high performance. Tuning a program for a specific application on any particular computer is often non-trivial. One way to alleviate this problem

M. S. Moonen et al. (eds.), Linear Algebra for Large Scale and Real-Time Applications, 13–20.

is to identify computational kernels on which we can build our algorithms. With code for such kernels available the porting of code to new architectures should ideally only require a recompilation and linking with the local library. The best known and most important collection of computational kernels is the BLAS [6, 7, 10], proposed originally in 1973 [8]. As this strategy promoted efficient and well structured code, and as new high performance computers depended more on data locality, higher level and more complex computational kernels were added to the BLAS standard. [6, 7]. After having reduced the effort of writing efficient code for many linear algebra problems to that of writing efficient BLAS routines, we must stress that this is by no means a trivial task, and may account for several man-years of effort. A complementary effort must therefore be directed to produce a new generation of compilers for high level languages. The Fortran 90 standard (F90) [1] and the recent work on the HPF (High Performance Fortran) definition [9], makes it likely that better and more portable software will emerge across different MPP platforms.

The newly released linear algebra library, LAPACK, [2] is entirely built upon the BLAS. It is designed to run efficiently on a wide range of high performance computers, including vector machines and parallel, shared memory computers. We consider the use of LAPACK on massively parallel computers with **local** memory. We study the question of achieving high performance from linear algebra subprograms keeping the general algorithmic framework adopted by the LAPACK software. In order to develop models for a data-parallel realization, we devote Section 1 to the performance of the relevant BLAS modules. In particular, we study the rank-1 update (DGER in BLAS-2), as well as DTRSM and DGEMM. In Section 2, we consider the use of this modules in the context of LU-factorization. It is of interest to determine if the current BLAS kernels are sufficient to achieve good performance or if additional kernels are needed. A model of the execution behavior is derived and discussed. In the last section, we compare the model with actual computer implementations on two MasPar computers.

1 Data-parallel BLAS modules

The main structure of the LAPACK LU-factorization routine DGETRF is:

```
DO I = 1,N,NB
   M = N-J+1
   K = J+NB
   CALL DGETF2(M,NB,A(J:N,J:K-1) ... )

check for singularity and perform row interchanges

   CALL DTRSM(M-NB,NB,A(J:K-1,J:K-1),A(J:K-1,K:N), ... )
   IF ( K < N ) THEN
      CALL DGEMM(M-NB,M-NB,NB,A(K:N,J:K-1),A(J:K-1,K:N),A(K:N,K:N),.)
   END IF
END DO.
```

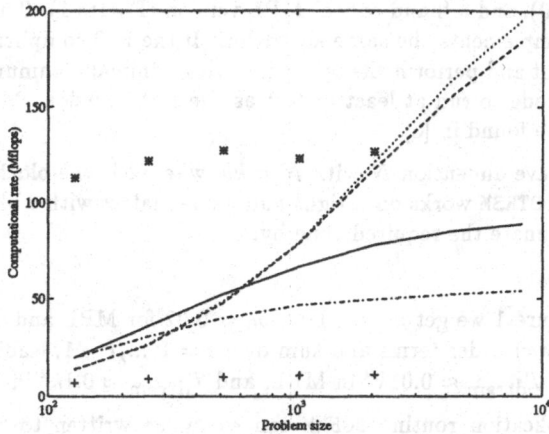

Figure 1: *Performance of* DGER *on an MP-1 with 16K PE. Solid curve is F90 with N=128, variable M, dashed-dot curve is F90 with M=128, variable N. Dashed curve is MPL with N=128, variable M, and dotted curve is MPL with M=128, variable N. Performance of* DTRSM *in F90 is marked with a '+', while an MPL version of* DTRSM *is marked with '*'.*

The BLAS routines called by DGETF2 are IDAMAX, DSCAL, DSWAP (BLAS-1), and DGER (BLAS-2). In a data-parallel context, many BLAS-1 routines are trivial in the sense that no communication is required. Data reduction operations like IDAMAX are also normally well supported. Moreover, these operations contribute very little to the overall time. A more general discussion of data-parallel BLAS is reported in [3]. The rank-1 update DGER is important, we show its performance in Figure 1. DGER is always called with one (short) vector of block size length b, while the other vector is often long. The figure shows a significant difference in performance between an F90 based implementation using the intrinsic function SPREAD and a more machine specific language MPL (A low level data-parallel C language). We note that there is a marked difference in F90 performance depending on which vector is the longer one, while the MPL routine is less sensitive.

An algorithm for DTRSM as called above, consists of an outer loop with a call to DGER inside. If we let B be the right hand side matrix to be updated, then DTRSM takes the form:

```
DO I = 1,N
   B(I,:)  = B(I,:)  / A(I,I)
   B(I+1:N,1:M) = B(I+1:N,1:M)
              - SPREAD(A(I+1:N,I),DIM=2,NCOPIES=M)
              * SPREAD(B(I,1:M),DIM=1,NCOPIES=N-I)
ENDDO.
```

We now observe an enormous difference in performance between an implementation as outlined above (in F90), and a 'hand coded' MPL version (Figure 1). This is disappointing since the MPL code implements the same algorithm. If the F90 compiler had been able to analyze the data layout and perform the operation with minimal communication, we would expect the resulting code to run at least as fast as the MPL version. More details on this implementation can be found in [3].

Let the matrix A have dimension N with $N = kb$, where b is the block size called NB in the code segments. If DTRSM works on a right hand side matrix with i block columns (that is ib columns), we estimate the required time by:

$$t_{\text{dtrsm}} = a_0 + b_0 i \tag{1}$$

From the data in Figure 1 we get $a_0 \approx 0.002$, $b_0 \approx 0.02$ for MPL and $a_0 \approx 0.1$, $b_0 \approx 0.2$ for F90. We ignore lower order terms and sum over $i = 1, ..., k - 1$, leading to a total time estimate for DTRSM of $T_{\text{dtrsm}} \approx 0.01k^2$ in MPL, and $T_{\text{dtrsm}} \approx 0.1k^2$ in F90.

The matrix multiplication routine DGEMM can easily be written to run near the peak performance of the machine as soon as the number of elements in the matrix is reasonable relative to the number of processors. This kernel is so important that F90 supports an intrinsic function MATMUL. The routine is therefore never written in a high level language. The time model is here:

$$t_{\text{dgemm}} = a_2 + b_2 i^2. \tag{2}$$

Proceeding as above, we find $a_2 \approx 0.005$ and $b_2 \approx 0.01$ to obtain $T_{\text{dgemm}} \approx 0.003k^3$.

2 Data-parallel LAPACK

It remains to consider DGETF2. The important part of DGETF2 (considering floating point operations) is a loop with the structure:

```
DO I = 1,N)

    find pivot and test for singularity

        A(I+1:M,I) = A(I+1:M,I) / A(I,I)
        A(I+1:M,I+1:N) = A(I+1:M,I+1:N)
                       - SPREAD(A(I+1:M,I),DIM=2,NCOPIES=N-I)
                       * SPREAD(A(I,I+1:N),DIM=1,NCOPIES=M-I)
    ENDDO
```

Note that this is essentially the same code segment as for DTRSM in the previous section. We can therefore expect the same qualitatively behavior. The performance is plotted in Figure 2. But here there is no BLAS routine that can be used to effectively hide this loop. It is likely that a better F90 compiler will be able to generate near optimal code for this loop some time in the future. In the meantime, a new computational kernel is needed. It may appear that defining an LU-factorization routine as a kernel for the LU routine DGETRF is solving the original problem outside of LAPACK. However, by restricting the routine to

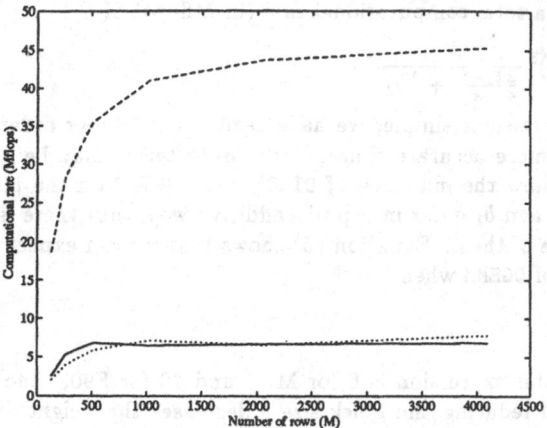

Figure 2: *Performance of* DGETF2 *with* N *fixed at 128 and variable* M *on an MP-1 with 16K PE. Solid line is F90 with* SPREAD, *dashed line is MPL implementation, dotted line is F90 with* DGER.

handle a single column block, with a machine dependent maximum block size, one can often reduce the programming effort and keep the code more modular.

By identifying such modules in LAPACK one may achieve two goals: Establish the package on the MPP system with reasonable performance and minimal effort, and at the same time direct the attention to areas where the compiler technology should improve by establishing performance goals for an optimizing F90 based compiler.

The time model for DGETF2 when working on a column of length ib is:

$$t_{\text{dgetf2}} = a_1 + b_1 i \tag{3}$$

Proceeding as before, we estimate $a_1 \approx 0.03$, $b_1 \approx 0.04$ for MPL and $a_1 \approx 0.1$, $b_1 \approx 0.2$ for F90. This gives a total DGETF2 time of $T_{\text{dgetf2}} \approx 0.02k^2$ in MPL, and $T_{\text{dgetf2}} \approx 0.1k^2$ in F90. The performance of the single block column version of DGETF2 written in MPL (Figure 2), is much better than the purely BLAS based version. It should be stressed that this routine is not yet written optimally. Our models for timing holds for any implementation of the computational kernels. In a tuned implementation the performance of this kernel should be much closer to that of DTRSM.

Putting it all together, we find approximately

$$T_{\text{dgetrf}} \approx T_{\text{dtrsm}} + T_{\text{dgetf2}} + T_{\text{dgemm}}, \tag{4}$$

or in terms of our model

$$T_{\text{dgetrf}} \approx \frac{1}{2}(b_0 + b_1)k^2 (1 + \frac{k}{\frac{3}{2}\frac{b_0 + b_1}{b_2}}). \tag{5}$$

18

This corresponds to a total computational rate (in Mflops) of

$$\text{Mflops} \approx \left(\frac{b}{100}\right)^3 \frac{1}{\frac{3}{4}\frac{b_0+b_1}{k} + \frac{1}{2}b_2} \tag{6}$$

In order to keep expressions simple, we have neglected all lower order terms. Obviously, in order to provide more accurate times, lower order terms may be included. The main purpose here is to show the influence of DTRSM and DGETF2 on the block algorithm. We see from (5) that b_0 and b_1 enter in a purly additive way, thus there is little to be gained by only reducing one of them. Equation (5) shows that we can expect performance in the same range as that of DGEMM when

$$k \gg \frac{b_0 + b_1}{b_2}. \tag{7}$$

In our experiment, this expression is 6 for MPL and 40 for F90. One can also note from (6) that the effect of reducing the block size b decreases the weight of $b_0 + b_1$ relative to b_2, by the same factor (k will increase). On an MPP like the MasPar this implies that the optimal block size is equal to the size of the processor array. (Any smaller block will lose processor utilization).

3 Performance of DGETRF from LAPACK

We ran several versions of DGETRF on both a MasPar MP-1 and a MasPar MP-2. Both machines had 16384 processors. The MP-2 has a floating point speed that is 3 to 4 times as fast as the MP-1, while the speed of communication is the same for both machines. Figure 3 summarizes the results for both machines. We observe that the performance on the MP-1 is in good agreement with the predictions made in the previous section. The difference between a fully optimized 'hand-coded' routine and our LAPACK structured routine is a little more pronounced for the MP-2, reflecting that communication is relatively more expensive on this machine. The DGETF2 kernel has not been optimized for these experiments, and may make the code even more satisfactory in the future. Still, it should be noticed that we do sustain a computational rate in excess of one Gflops for LU-factorization on this low cost MPP system.

4 Conclusions

We have investigated the performance of the LAPACK routine DGETRF performing LU-factorization on massively parallel computers using a data-parallel programming model. The time complexity model of the routine shows, and the computational experiments confirm, that the performance of the level-2 factorization module DGETF2 is critical for the overall performance of the algorithm. Asymptotically, the performance will only depend on the performance of the BLAS-3 matrix multiplication kernel DGEMM, but this range is beyond the practical limitations set by the memory size.

This difficulty can be overcome in two different ways. In the short term, the definition of a computational kernel in addition to the BLAS, that does LU-factorization on a column block, provides a way out. It should be noted that such a routine written in a lower level

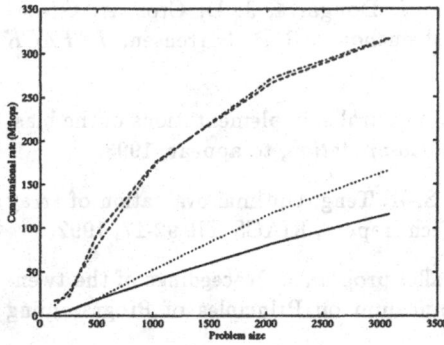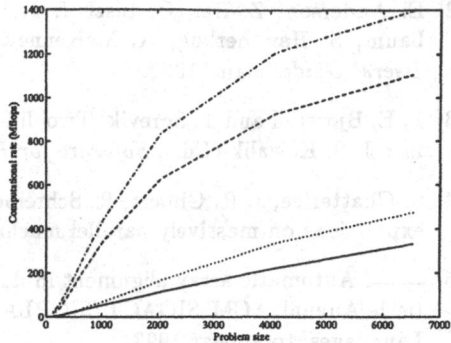

Figure 3: *Performance of DGETRF on MP-1 and MP-2 with 16K PE. Solid curves are F90 kernels, dashed curves are MPL kernels. Dash-dot curves are 'best effort' and dotted curves are MPL kernels without DGETF2.*

language, may be considerably easier to write and maintain than a general LU-routine. The task of explicit processor virtualization can for example be simplified. Such a kernel can be used within the framework of LAPACK, just like the BLAS. We expect to identify a small set of similar 'new' kernels when implementing other LAPACK routines.

In the longer term, we expect many of these problems to be handled by better compiler technology. The problem at hand, has been traced to the inefficient execution of rank-1 updates. A much better implementation of the intrinsic function SPREAD when it is being used in F90 expressions, is needed. Important work on compiler algorithms for optimal data alignment and evaluation of array expressions, is starting to emerge [4, 5]. We hope that our work will contribute to making efficient rank-1 updates a high priority. As we have shown, the underlying programming model and the parallel hardware provide a very satisfactory performance of the LU-factorization within the algorithmic structure of LAPACK.

Acknowledgments

We thank MasPar Computer Corporation for providing an MP-2 for numerical experiments. The first author thanks Dr. Robert Schreiber for several discussions on this subject. This research was partially supported by NAVF grant D.01.08.054 and by NTNF grant IT2.28.28484

References

[1] American National Standards Institute. *Fortran 90: X3J3 internal document S8.118.* Submitted as Text for ANSI X3.198-1991, and ISO/IEC JTC1/SC22/WG5 internal document N692. Submitted as Text for ISO/IEC 1539:1991, May 1991, ANSI, 1991.

[2] E. Anderson, Z. Bai, C. Bischof, J. Demmel, J. Dongarra, J. D. Croz, A. Greenbaum, S. Hammerling, A. McKenney, S. Ostrouchov and D. Sorrensen. *LAPACK Users' Guide*, Siam, 1992.

[3] P. E. Bjørstad and T. Sørevik. Two different data-parallel implementations of the blas. In : J. S. Kowalik (Ed.), *Software for Parallel Computation*, to appear, 1993.

[4] S. Chatterjee, J. R. Gilbert, R. Schreiber and S.-H. Teng. Optimal evaluation of array expressions on massively parallel machines. Tech. report, RIACS TR 92-17, 1992.

[5] ——. Automatic array alignment in data-parallel programs. Proceedings of the twentieth Annual ACM SIGACT/SIGPLAN Symposium on Principles of Programming Languages, to appear 1993.

[6] J. Dongarra, J. Du Croz, I. Duff and S. Hammarling. A set of level 3 basic linear algebra subprograms: Model implementation and test programs. *ACM Trans. Math. Soft.*, **16**, 1990.

[7] J. Dongarra, J. Du Croz, S. Hammarling and R. Hanson. An extended set of fortran basic linear algebra subprograms. *ACM Trans. Math. Soft.*, **14**, 1988.

[8] R. Hanson, F. Krogh and C. Lawson. A proposal for standard linear algebra subprograms. *SIGNUM*, **8**, 1973.

[9] High Performance Fortran Forum, *High Performance Fortran language specification, version 0.4. draft, Nov. 1992*, Rice University, 1992.

[10] C. L. Lawson, R. J. Hanson, D. R. Kincaid and F. T. Krogh. Basic Linear Algebra Subprograms for fortran usage. *ACM Trans. Math. Soft.*, **5**, 1979.

LARGE-SCALE NONLINEAR CONSTRAINED OPTIMIZATION

A.R. CONN
IBM T.J. Watson Research Center
Yorktown Heights, U.S.A.

N. GOULD
Rutherford Appleton Laboratory
Chilton, Oxfordshire, England

Ph.L. TOINT
Department of Mathematics
Facultés Universitaires ND de la Paix
Namur, Belgium
na.toint@na-net.ornl.gov

ABSTRACT. During the past ten years, much progress has been made in the theory and practice of constrained nonlinear optimization. However, considerable obstacles appear when these ideas are applied to large-scale problems. This is important as many real applications require the solution of problems in thousands of unknowns. In some areas, in particular linear programming, considerable progress has been made. But even modest departures into nonlinearity, for example the solution of large, general quadratic programs, present considerable challenges. This is apparent when one views the paucity of software for solving such problems. Unsurprisingly, the position does not improve as more drastic forms of nonlinearity are encountered.

In this paper, we will try to explain why the difficulties arise, how attempts are being made to overcome them and what the problems are that still remain.

KEYWORDS. Large-scale, constraints, nonlinear optimization, overview.

Introduction

Our purpose in this paper is to present an overview of the state-of-the art in large-scale nonlinear optimization. This article is a personal response to the questions, why we are interested in large-scale nonlinear optimization, what are the difficulties and what kind

M. S. Moonen et al. (eds.), Linear Algebra for Large Scale and Real-Time Applications, 21–48.
© 1993 *Kluwer Academic Publishers.*

of progress has been made. Although we have made some effort to be complete in our references and thus hope to provide a useful bibliography, we have not attempted to be as complete in our overview. Rather, we have tried to include enough details of the general issues to indicate the nature and reasons for some of the current research in the field. Moreover, the length of treatment is frequently an indicator that the work is new and less well-known, and a brief mention does not mean that the work is relatively less important.

It seems appropriate to first state the most general form of the problem that we are addressing, namely

$$\text{minimize} \quad f(x) \qquad (1)$$
$$x \in \Re^n$$

subject to the general (possibly nonlinear) inequality constraints

$$c_j(x) \leq 0, \quad 1 \leq j \leq l, \qquad (2)$$

to the (possibly nonlinear) equality constraints

$$c_j(x) = 0, \quad l+1 \leq j \leq m, \qquad (3)$$

and the simple bounds

$$l_i \leq x_i \leq u_i, \quad 1 \leq i \leq n. \qquad (4)$$

Here, f and the c_j are all assumed to be twice-continuously differentiable and any of the bounds in (4) may be infinite.

At the outset we should make it clear that we only expect to obtain local minimizers. This is in marked contrast to combinatorial optimizers who, typically, are only interested in global solutions. This presents no problems in convex programming, where all local minima are indeed global (for example, in linear programming), but even for small, general nonlinear programming problems it is usually extremely difficult to verify globality. For large problems, it is practically impossible. Fortunately, in many situations, an algorithm that determines local optima suffices.

Our primary interest here is in problems that involve a large number of variables and/or constraints. Consequently, it seems worthwhile to elaborate as to what we mean by large.

Firstly, this notion is clearly *computer dependent*. What is large on an Apple Macintosh is significantly different from what is large on an IBM 3090 or a Cray 2. The first machine has a substantially smaller memory and storage than the other two, and therefore has more difficulty handling problems involving a large amount of data. Secondly, a highly nonlinear problem in one hundred variables could be considered large, whereas in linear programming it is possible to solve problems in five million variables. The notion of size is thus *problem dependent*. It also depends upon the *structure of the problem*. Many large-scale nonlinear problems arise from the modelling of very complicated systems that may be subdivided into loosely connected subsystems. This structure may often be reflected in the mathematical formulation of the problem and exploiting it is often crucial if one wants to obtain an answer efficiently. The complexity of the structure is often a key factor in assessing the size of a problem. Lastly, the notion of a large problem depends upon the *frequency* with which one expects to solve a particular instance or closely related problem. When one anticipates solving the same class of problems many times, one can afford to expend a significant amount of energy analyzing and exploiting the underlying structure. Thus, although it is not possible to say categorically that a problem in say seven hundred variables is large, suffice it to say that, today, a problem in fifty variables is small and a generally nonlinear problem in five thousand variables and one thousand nonlinear constraints is large.

One might suppose that intellectual curiosity alone is sufficient reason to be interested in large-scale nonlinear optimization. However, although we readily admit to the fact that this is an important element of our interest (and indeed if our research had been confined to the publication of theoretical articles, arguably the main one), much of our joint effort has been devoted to the time consuming and often tedious task of writing software, preparing input and testing. The salient point is that there is a need for algorithms to solve large-scale nonlinear optimization problems. The accurate modelling of physical and scientific phenomena frequently leads to such problems. Nature loves to optimize: minimum energy, minimum potential difference, shortest paths. Moreover, the universe certainly is not linear. If the model is to be accurate (for example, if it is derived from a discretization of a continuous process), the number of variables is necessarily large. Another area where large nonlinear problem arise naturally is in economics, where one often wishes to maximize profit (or minimize losses) in complex situations involving many parameters. The proliferation of large linear models, rather than nonlinear ones, is sometimes a consequence of our lack of knowledge concerning the phenomena being modelled, in which case assuming linearity is about the simplest assumption one can make. As our knowledge improves, often the models are refined and nonlinearity should be introduced. In our opinion, the frequent use of linear models is not an indication that nonlinear problems do not abound. Rather, it is a statement of the desire to use an algorithm (the simplex method) that is readily understood and is well-known to be suitable for large problems. In particular, it is one of our tasks to convince you that you should consider solving nonlinear programs, when they are more appropriate. As a necessary corollary, it should be emphasized that solutions to large nonlinear problems on moderate workstations in a reasonable amount of time are currently quite possible. Furthermore, in practice one is often only seeking marked improvement rather than assured optimality (another reason why globality is not necessarily an issue). This fact makes even problems that at first sight seem impossible (for example, control problems that one wishes to solve in something like real-time), tractable.

Without a doubt, the ubiquity of powerful workstations and the availability of supermachines (both parallel and sequential) have encouraged research in algorithms for large-scale problems. However, we concur with a remark that Martin Beale once made that he would 'much rather work with today's algorithms on yesterday's computers than with yesterday's algorithms on today's computers' [128].

1 Examples of applications

As we already stated, our interest in developing algorithms for large-scale optimization was created out of necessity. There is an increasing demand for such software as the size and nonlinearity of the problems that practitioners are interested in solving steadily grows. The same evolution that leads to larger and larger nonlinear optimization models for physical phenomena is observed in data fitting, econometrics and operations research models. In particular, it is perhaps worth listing some examples:

- Discretizations of variational calculations and optimal control problems involving both state and control variables.

  These arise, for example, in quantum physics, tidal flow analysis, design (of aircraft, journal bearings and other mechanical devices), structural optimization, ceramics manufacturing, chemical process control and satellite piloting.

- Nonlinear equations arising both in their own right and in the solution of ordinary and partial differential equations.

  These occur, for instance, in elasticity, semiconductor simulation, chemical reaction modelling and radiative transfer.

- Nonlinear least squares or regression.

  Some examples include fluid dynamic calculations, tomography (both seismic and medical), combustion, isomerization and metal coating thickness assessment.

- Nonlinear approximation.

  These include antenna design, power transmission, maximum likelihood and robust regression.

- Nonlinear networks.

  Examples occur in traffic modelling, energy and water distribution/management systems, and neural networks. These problems can have hundreds of thousands of variables but they are tractable because of their very special structure.

- Other interesting problems occur in macro- and micro-economics, equilibrium calculations, production planning, energy scenario appraisal and portfolio analysis. These problems often give rise to quadratic programs, particularly in portfolio analysis.

The increasing interest in the solution of large-scale nonlinear optimization problems is also related to the realisation by users that today's advances in computer technology are making the solution of such problems possible.

Recent articles and books devoted primarily to large-scale optimization include [19], [20], [24], [33], [35], and [133]. Some examples of applications are given in [59], [87], [95], [96], [123] and [131]. Background material on nonlinear optimization is given in [54], [60], [61], [70] and [108]. An overview of what is involved in a mathematical programming system, especially with respect to linear problems, is given in [129].

2 What are the difficulties?

Efficient algorithms for small-scale problems do not necessarily translate into efficient algorithms for large-scale problems. This is unfortunate, since in the past twenty years rather sophisticated and reliable techniques for small-scale problems have been developed (see [108] for good surveys). As a consequence, it is just not adequate to take existing optimization software for small problems and apply it to large ones, hoping that the increased capacity in computing will take care of the growth in problem size. By contrast, we could expect that an efficient method for large-scale problems be at least moderately efficient for small-scale problems.

Perhaps the most important difficulty in the large-scale context is that of *exploiting structure*. The fact that we are able to solve large problems at all is because they are structured. Even in linear programming, a problem in one thousand variables devoid of structure (happily, usually an indication of a bad formulation) severely taxes codes. Thus, it is absolutely essential for efficient algorithms to exploit structure. Moreover, this means exploiting more than just sparsity. Unfortunately, this exploitation often complicates the question of stability, that is, the ability of an algorithm to guarantee that small perturbations in the data will only result in small perturbations to the solution for 'satisfactorily conditioned' problems. By contrast, algorithms for small problems have the possibility to ignore structure.

Another significant difficulty is that of *scaling*. Perhaps the main reason algorithms for small-scale problems do not necessarily translate into efficient algorithms for large-scale problems is that in order to be able to handle large problems the algorithms have to necessarily be as simple as possible. Consequently, relative to many of the more successful algorithms for small, dense problems, the amount of information available at any given iteration may be severely restricted. This makes designing algorithms that are scale invariant (in the sense that, assuming infinite precision arithmetic, quasi-Newton methods for unconstrained optimization are invariant under linear transformations) more difficult for large-scale problems.

One consequence of the necessity to keep the algorithms for large-scale nonlinear optimization '*simple*' is that typically one has rather incomplete information available at every iteration. Thus it becomes difficult to successfully merging two distinct (and in many ways, conflicting) aspects of any nonlinear programming algorithm. The first aspect is that which guarantees global convergence. By global convergence (not to be confused with convergence to global optima), we mean convergence to a stationary point from any starting point. Essentially, this is a weak requirement (the method of steepest descent with a suitable line search condition will suffice) which combined with the desire to stress simplicity encourages us to use a steepest-descent-like method. On the other hand, ultimately we want faster convergence. This means using Newton's method or quasi-Newton methods, although in the case of large problems a fast linear rate might suffice.

From a more mundane point of view there are very real difficulties *inputting* large problems. In particular, the amount of information present in the structure of a large-scale optimization problem, although crucial for the acceptable performance of algorithms, is also very difficult to specify in a complete and understandable format. A standard input that is a simple formal language in which these structural concepts could be expressed unambiguously, has been rather well-established and successful in the more restricted domain of linear programming [47]. We have extended this input format to the nonlinear case (see [36]). However, particularly because of the necessity of exploiting structure in a rather general sense, the resulting standard is not as simple as it was in the linear case. Other approaches are based upon using a high-level modelling language. The high-level aspect makes these rather more user friendly but they require an interpreter and are thus not normally in the public domain. Moreover, they do not, currently, exploit structure as generally as we would like. Well-known examples of this approach are GAMS ([14]) and AMPL ([65]).

Obviously, a primary requirement in evaluating the quality of an algorithm is to have a good set of *test problems*. In the first instance, it is by no means obvious what constitutes such a set for large-scale optimization, in the main because of the complexity of the problems and the lack of experience in solving them. Nevertheless, important starts have been made, and although we do not yet have collections as readily available as those for linear programming and sparse linear algebra (see [57] and [66]), we currently have over nine hundred problem instances in (our) 'standard input format' [40]. We are also asking for more test problems from the community ([125]). Moreover, as a part of the MINPACK-2 project 'a collection of significant optimization problems' is being made, see [4]. Earlier collections for unconstrained problems include [15] and [102].

Just as important is the ability to *evaluate results*. Given the number of variables (in both senses of the word) at hand, it is a complex task to interpret the results of testing. It is fair to say that, at present, we require more established test problems and a broader experience of the behaviour with various algorithms.

3 Current approaches

Having considered the difficulties, we now examine how they are addressed. It is convenient to consider three broad classes, namely, approaches based upon classical large linear programming, approaches based upon small-scale nonlinear programming and approaches based upon a mixed linear programming/interior point method, even though the ideas in each approach are not mutually exclusive.

3.1 APPROACH BASED UPON CLASSICAL LARGE LINEAR PROGRAMMING

We will begin with a terse and somewhat eccentric summary of the simplex method. For those who need further details, an excellent recent survey article, that includes the interior point method that is relevant to the third approach below, is given in [73]. Consider the linear programming problem in the form

$$\min_{x\in\Re^n} \; c^T x \tag{5}$$

subject to the m $(\leq n)$ general linear constraints

$$Ax = b \tag{6}$$

and to the simple bound constraints

$$x \geq 0. \tag{7}$$

The solution to the above problem normally occurs at a vertex of the feasible region, that is a point defined by the equations (6) and $n - m$ of the variables lying on their bounds (7). Without loss of generality, we can assume that the last $n - m$ components of x are at their bounds. We call such variables *non-basic* — the remaining m variables are termed *basic*. We may thus consider the activities (i.e. those constraints satisfied as equalities) to be given by

$$Cx = \begin{pmatrix} B & N \\ & I \end{pmatrix} x = \begin{pmatrix} b \\ 0 \end{pmatrix}. \tag{8}$$

Now

$$C^{-1} = \begin{pmatrix} B^{-1} & -B^{-1}N \\ & I \end{pmatrix}. \tag{9}$$

The fact that the kth column of C^{-1} is orthogonal to the other $n - 1$ rows of C, along with the fact that we start at a vertex of the feasible polytope and insist on following a path of objective-improving feasible vertices to optimality is really the heart of the simplex method. The first statement means that by moving along this k^{th} column the remaining equations corresponding to the other $n - 1$ rows of C stay active. Being at a vertex ensures that we can refer to C^{-1}. Objective-improving is just a matter of sign and maintaining feasibility requires that we move to an adjacent vertex.

More importantly, from the point of view of this article, the method is efficient because it exploits heavily the structure of B, making use of techniques such as the Markowitz strategy, [93], and sparse Bartels-Golub, [6], updating of LU factors (see [117] for further details and the first implementation in Fortran). Another important feature of linear programming software is the ability to have crash starts, i.e. a relatively simple method for finding a good starting basis (see, for example [76]).

The highly successful package MINOS, [106] can be viewed as an extension of the simplex method as a reduced gradient technique. Its origins come from [118] and [119]. One should also note that for practitioners who are used to linear programming approaches, MINOS serves as an *extremely* useful bridge to nonlinear programming. In particular, MINOS replaces

$$\begin{array}{ll} \text{minimize} & F(x) + c^T x + d^T y \\ x \in \mathbf{R}^n, y \in \mathbf{R}^m & \\ \text{subject to} & f(x) + A_1 y = b_1 \\ & A_2 x + A_3 y = b_2 \\ \text{and} & \\ & l_x \leq x \leq u_x \\ & l_y \leq y \leq u_y \end{array} \tag{10}$$

with

$$\text{minimize} \quad F(x) + c^T x + d^T y + \lambda_k^T(f(x) - \tilde{f}(x)) + \tfrac{1}{2}\rho(f(x) - \tilde{f}(x))^T(f(x) - \tilde{f}(x))$$
$$x \in \mathbf{R}^n, y \in \mathbf{R}^m$$

subject to

$$\tilde{f}(x) + A_1 y = b_1$$
$$A_2 x + A_3 y = b_2 \tag{11}$$

and

$$l_x \le x \le u_x$$
$$l_y \le y \le u_y,$$

where

$$\tilde{f}(x) = f(x_k) + J_k(x - x_k),$$

and J_k denotes the Jacobian of f evaluated at x_k. In other words, the nonlinear contribution to the constraints is linearized so that we can exploit linear programming technology. It then formulates a quadratic model for the artficial objective function. A reduced gradient technique is used, that is one determines a search direction that maintains the current activities to first-order (i.e. the linearized approximations that were active stay active). Writing the activities that are determined by the general linear constraints as

$$\hat{A}x = \left(\begin{array}{ccc} B & S & N \end{array} \right) x = b, \tag{12}$$

this means that our search direction is given by

$$h = Zd, \tag{13}$$

where

$$Z^T = \left(\begin{array}{ccc} -[B^{-1}S]^T & I & 0 \end{array} \right). \tag{14}$$

This follows directly from the fact that

$$\hat{A}Z = 0 \quad \text{and} \quad (0 \quad 0 \quad I)Z = 0. \tag{15}$$

Analogously to the simplex method, the columns of B correspond to basic variables and the columns of N correspond to non-basic variables. However, because of the nonlinearity of the objective function, we are no longer able to ensure that optima lie at vertices (the number of columns of B and N may not add up to the dimension of the space). The ensuing deficiences are made up by the columns of S, the so-called superbasic columns. Because of the similarities in the resulting linear algebra the exploitation of structure is much the same as that in the simplex method. It is worth pointing out that exploitation of the structure of Z and the simple bounds is especially attractive in the context of network problems (see, for example, [51], [82], [127] and [126]).

It should be clear that the fewer superbasic columns, the closer the problem is to a linear programming problem. MINOS works particularly well when there are relatively few superbasics.

A related approach that was one of the earliest successful pieces of software that could handle large nonlinear problems was an implementation of the generalised reduced gradient method of Abadie ([1]) by Lasdon ([90]). A quadratic programming algorithm that uses similar ideas to MINOS and is for large-scale problems is given in [75].

Not surprisingly, the earliest approach to large-scale nonlinear optimization was a successive linear programming technique (see [81]). A more recent successive linear programming technique that uses an exact l_1 penalty function and incorporates trust region constraints is given in [64], although numerical results are given for small problems only.

3.2 APPROACHES BASED UPON SMALL-SCALE NONLINEAR PROGRAMMING

3.2.1 Sequential quadratic programming. One of the best known techniques for nonlinear programming is the so-called sequential quadratic programming approach (see, for example, [61], Chapter 12 and [70], Chapter 6). Recent work by Eldersveld [58] and colleagues uses the augmented Lagrangian. The vector s represents slack or surplus variables (see below for some motivation for this function and the introduction of slack or surplus variables)

$$f(x) - \lambda^T[c(x) - s] + [c(x) - s]^T[c(x) - s]/\mu, \tag{16}$$

with the quadratic programming search-direction subproblem

$$\underset{p,q}{\text{minimize}} \qquad \tfrac{1}{2}p^T H p + g^T p$$

$$Ap - q = -[c(x_k) - s_k]$$

$$\hat{l} \le \begin{bmatrix} p \\ q \end{bmatrix} \le \hat{u}, \tag{17}$$

using a suitable symmetric matrix H, to solve (1) to (4). A prototype implementation has been developed that uses a modification of the MINOS code. They use for the active set

$$\hat{A} = \begin{pmatrix} B & S & N \\ 0 & 0 & I \end{pmatrix} \tag{18}$$

and solve

$$Z^T H Z y_z = -Z^T(g + Hp),$$

$$y = Z y_z,$$

where

$$Z^T = \begin{pmatrix} -[B^{-1}S]^T & I & 0 \end{pmatrix}$$

and $Z^T H Z$ is small. Noting that one needs H to evaluate the gradient of the quadratic objective function, they make use of the fact that if we define $Q = [Z\ Y]$, choosing Z and Y so that $Z^T H Y = 0$ and $Y^T H Y = I$, then we can write $H = Q^{-T}(Q^T H Q)Q^{-1}$, where

$$Q^T H Q = \begin{bmatrix} Z^T H Z & 0 \\ 0 & I \end{bmatrix}.$$

For example, taking

$$Y = \begin{bmatrix} B^{-1} & 0 \\ 0 & 0 \\ 0 & I \end{bmatrix}$$

then

$$Q^{-1} = \begin{bmatrix} 0 & I & 0 \\ B & S & 0 \\ 0 & 0 & I \end{bmatrix}.$$

Details are given in [58]. An approach that also uses sequential quadratic programming, but 'solves' the quadratic program using an interior point method (see below) is given by [13]. Although these methods hold promise, computational experience to date has been insufficient to make definitive statements as to their effectiveness. Other sequential quadratic programming based methods include [92] and [105].

3.2.2 The LANCELOT project.

The approach to which we wish to devote much of the rest of this article is based upon the adaptation of trust region methods to the problem with simple bounds. The method is extended to general constraints by using an augmented Lagrangian function and the bounds are handled directly via projections that are easy to compute. We use group partial separability (a generalisation of sparsity, introduced in [79]) to allow efficient storage and updating of matrices in matrix-vector product form. This approach has the further advantage that accurate approximations to the second derivatives of the element functions, normally being of low rank, are easier to obtain than for the assembled matrices. This structure is extremely general. Indeed, any sufficiently differentiable function with a sparse Hessian matrix may be written in this form. An introduction to group partial separability is given by [34]. The entire project has resulted in a substantial amount of software that is available at nominal cost for research purposes. There is also a book ([42]) to accompany the software. Returning to the underlying concepts of LANCELOT, we will now give some details.

Trust region methods in the context of unconstrained optimization have been able to combine a rather intuitive framework and robust numerical implementations with a powerful and elegant theoretical foundation. An excellent reference is [101]. The basic idea is to model the objective function (by a quadratic given by the first three terms of a Taylor's series expansion about the current point x^k, for example). One then 'trusts' this model in a neighbourhood (called the trust region) of x^k. The next step is to approximately minimize the model in the trust region, thereby obtaining a point $x^k + s^k$, say. One now determines how well the model actually predicted the change in the true objective function. If good descent is obtained, the next iterate, x^{k+1}, is set to $x^k + s^k$ and the trust region is expanded. If moderate descent is obtained, the next iterate, x^{k+1} is set to $x^k + s^k$ and the trust region remains unchanged. Otherwise, x^{k+1} is set to x^k and the trust region is contracted. The beauty of such an approach is that, when the trust region is small enough and the problem smooth, the approximation is good, provided the model gradient is sufficiently accurate. Moreover, assuming one does at least as well as the minimum along the steepest descent direction of the model within the trust region (that determines the so-called Cauchy point), one can ensure convergence to a stationary point ([18]). In addition, eventually the trust region is expanded sufficiently that it does not interfere with the subsequent iterates, and thus, assuming that in this situation the underlying algorithm is sufficiently sophisticated, one can ensure fast asymptotic convergence. Details are given in [101].

The algorithm that is at the heart of LANCELOT is a method for which all the constraints are just simple bounds. The extension of the above ideas are relatively straightforward in this case. Essentially, one generalizes the Cauchy point to the minimum along the *projected* gradient path within the trust region, where the projection is with respect to the simple bounds. It is important to note that it is trivial computationally to compute such a projection: components of x that hit a bound just remain fixed. This approach was first carried out by McCormick in [94], and independently by Bertsekas [10] and Levitin and Polyak [91]. More recently it has been exploited extensively in the context of large-scale optimization by many authors, see for example [32], [52], [103], and [104]. As in the unconstrained case, global convergence can be guaranteed, provided one does at least as well as the generalized Cauchy point. One obtains better convergence, and ultimately a satisfactory asymptotic convergence rate, by further reducing the model function. In the context of LANCELOT, this is achieved by fixing the activities determined by the generalized Cauchy point and further reducing the model within the feasible region and trust region using just the remaining free variables. Updating of the trust region size is handled in exactly the same way as it is in the unconstrained case. The basic algorithm can be summarised as follows:

- Find the generalized Cauchy point based upon a local (quadratic) model.

- Fix activities to those at the generalized Cauchy point.

- Using the free variables further reduce the model within the feasible region and the trust region. Of course this may, and typically does, introduce new activities in addition to those determined by the generalized Cauchy point.

- Determine whether the current point is acceptable and update the trust region radius accordingly.

Provided the quadratic model is reasonable, we are able to prove that we converge to a Kuhn-Tucker point. Moreover, we identify the correct active constraints (activities) after a finite number of iterations assuming that strict complementarity is satisfied and the activities determined by the generalised Cauchy point are kept active when the model is further reduced. Details are given in [31].

The extension to general constraints is carried out by means of an augmented Lagrangian function. In order to understand this extension we need to motivate this function. We have known for nearly two hundred years ([89], part 1, section 4, article 2), that a solution to

$$\begin{aligned} \text{minimize} \quad & f(x) \\ x \in \mathbf{R}^n \\ \text{subject to} \quad & c(x) = 0 \end{aligned} \tag{19}$$

is a feasible stationary point of the Lagrangian

$$f(x) - \lambda^T c(x). \tag{20}$$

It is only since the Second World War, that we have recognised ([48]) that we can solve

(19) using the quadratic penalty function

$$\underset{x\in\mathbf{R}^n}{\text{minimize}} \quad f(x) + c^2(x)/\mu \tag{21}$$

as μ tends to zero from above. The idea here is that as μ becomes small the 'penalty term' $c^2(x)/\mu$ forces one to become feasible. This intuitive idea was not accorded a sound theoretical basis until the work of Fiacco and McCormick (see for example [60]). Augmented Lagrangians combine both ideas, thereby convexifying the Lagrangian and circumventing the necessity of requiring small μ by, instead, approximating the Lagrange multipliers, λ. Thus we use the problem

$$\underset{x\in\mathbf{R}^n}{\text{minimize}} \quad f(x) - \lambda^T c(x) + c^2(x)/\mu. \tag{22}$$

This approach was first suggested by K. J. Arrow and R. M. Solow in [3] but is better known through the work of Hestenes and Powell in [85] and [115].

In LANCELOT one thus solves the general problem by first introducing slack or surplus variables, if necessary, to change inequalities to equalities. Subsequently one minimizes the augmented Lagrangian

$$\Phi(x, \lambda, S, \mu) = f(x) + \sum_{i=1}^{m} \lambda_i c_i(x) + \frac{1}{2\mu} \sum_{i=1}^{m} s_{ii} c_i(x)^2 \tag{23}$$

(where the diagonal matrix S is introduced to incorporate scalings) subject to the explicit bounds, using the earlier algorithm[1]. This approach can be summarised as follows:

1. Test for convergence using the two following conditions.
   Sufficient stationarity — the projected gradient of the augmented Lagrangian with respect to the simple bounds is sufficiently small;
   Sufficient feasibility — the norm of the constraint violations is sufficiently small .

2. Use the simple bounds algorithm to find a sufficiently stationary approximate minimizer of Φ (considered as a function of x only) subject to simple bounds.

3. If sufficiently feasible (both the 'local convergence' values here and in 2) are greater than the test for convergence, in general) update the multipliers and decrease the tolerances for stationarity and feasibility.

4. Otherwise, decrease the penalty parameter and reset tolerances for stationarity and feasibility.

We are able to show, under suitable conditions, that we converge to a first-order stationary point for the nonlinear programming problem. Furthermore, if we have a single limit point, we eventually stop reducing the penalty parameter, μ. Under somewhat stronger conditions we are able to show that one requires only a single iteration of the simple bounds algorithm

---

[1] It is worth noting that MINOS uses for its objective function an augmented Lagrangian function with corresponding constraints $f - \tilde{f} = 0$, whose relaxation can be considered as *trusting* the linear approximation to f.

to satisfy the conditions of the third item above. Details of these important properties are given in [38] and [43].

As we have already seen, a significant (and often dominant) cost in optimization is solving a linear system. Typically these arise from the necessity to determine an approximate stationary point for a quadratic function — equivalently, the necessity to solve a linear system whose coefficient matrix is a symmetric matrix. If the system is large there are two possible approaches. The first is to use direct methods based upon multifrontal techniques. These use partial assembly and dense matrix technology on sparse matrices. General details are given in [56] and an application in the context of LANCELOT is given in [44]. Our experience to date, however, has been that an iterative approach is more robust. The most popular such approach is preconditioned conjugate gradients. For ease of motivation we first consider conjugate gradients without a preconditioner. Directions, d_i, are called conjugate with respect to a positive definite matrix A if $d_1^T A d_2 = \langle d_1, d_2 \rangle_A = 0$. In other words, they are orthogonal in the A-metric. The best known conjugate set of vectors are the set of orthonormal eigenvectors. If one considers minimizing a strictly convex quadratic form $\frac{1}{2}x^T A x - b^T x + c$, it is easy to see from the geometry that if one minimizes along the eigenvectors of A, then at each stage one determines the minimum of the quadratic on the space spanned by the eigenvectors used, and thus, after at most n steps, if A is n by n, the quadratic function's (unique) minimum is determined.

The appeal of conjugate-gradient methods is that this finite termination result for quadratics is true for general conjugate directions. Moreover, the attraction for large-scale optimization is that such an orthogonalisation can be determined via a three term recurrence and thus the method is particularly simple and only requires that we store three vectors.

However, if n is large, performing n steps may be prohibitively expensive. Moreover, a quadratic is only being used to model a nonlinear problem and so what we have is really a 'moving quadratic'. What makes this technique remain attractive is the use of preconditioners. The essential result is that whenever one has multiple eigenvalues, the conjugate gradient method *minimizes the quadratic* in the space spanned by the corresponding eigenvectors. If we can cluster eigenvectors (i.e. approximately have multiple eigenvectors) we can reduce the number of iterations for good approximations to minimizers, from n to the number of clusters. The perfect way to do this in the quadratic case is to precondition with A^{-1} — but then this is equivalent to doing Newton's method. Surprisingly one can often do very well by using very crude approximations to A^{-1} (diagonal matrices, for instance). A good reference is [67].

It is fortunate that most optimization problems in thousands of variables are structured; fortunate but demanding of respect. If one considers the arrow-head matrix ($a_{i,1}, a_{1,i}, a_{i,i}$ non-zero, $i = 1, 2, \ldots n$, all other entries zero), it is clear that one neither wants to input or nor wants to store this matrix as a dense matrix, for large n. A little less obvious is the fact that if one does Gaussian elimination without pivoting, after the first column is updated the remaining $n - 1$ by $n - 1$ block will be full — in other words, fill-in is disastrous. On the other hand, if we first reverse the ordering of the rows and columns (which amounts to changing the orderings of the equations and the labellings of the variables) there is no fill-in at all. Since optimal numerical stability typically dictates row and column orderings and limitations on storage motivate one to minimize fill-in, we are immediately aware of a major conflict in numerical linear algebra when one wants to account for structure. One form of compromise is known as threshold pivoting (see, for example, [56]).

In LANCELOT we take the point of view that invariant subspaces are more important than sparsity. For example, consider $f(x) = x_{50}^4$, and $F(x) = \left(\sum_{i=1}^{5,000,000} x_i\right)^4$, $x \in \mathbf{R}^{5,000,000}$. In the first case the Hessian is sparse, while in the second case the Hessian is dense. But they both have an invariant subspace of dimension $n-1$. In the first case, it is the orthogonal complement of e_{50} (the vector with a single non-zero entry, one, in the fiftieth component), and in the second case it is the orthogonal complement of e (the vector of all ones). We exploit invariant subspaces by writing our functions as $f(x) = \sum_{i=1}^m g_i \left(l_i(x) + \sum_{j \in I_i} w_j f_j(x)\right)$, where g_i is a scalar function, the l_i are linear functions and the w_j are weights for the nonlinear functions $f_j(x)$. The essential point is that the rank of $\nabla_{xx} f_j$) is much smaller than n and the null space of $\nabla_{xx} f_j$ is fixed. This, and the use of linear transformations to consider expressions like $e^{(x+y)}$ as $e^u, u = x + y$, enables us to use very compact representations of the problems we are optimizing. In particular we can store these as *dense* matrices. We note, however, that it is no longer reasonable to expect these matrices to be positive definite. This has led to a revival of interest in rank one secant methods. For an introduction to these considerations, with many more details, the reader is urged to read [34]. A related means of exploiting structure is given in [96]. For recent work on rank one updating see [37], [55], [88], [111], and [132].

Although we are unaware of the details, an augmented Lagrangian approach that is designed for large-scale optimization has been developed by Contesse, [46]. There are some similarities between the approach of [38], [45] and [64]. Fletcher and Sainz de la Maza use a piecewise linear model and an l_1 penalty function for the merit function but improvements over the Cauchy point involve projected Hessian approximations.

Current work in LANCELOT has been to consider the special case of convex constraints. The motivation is that, in particular, linear constraints are very common, are too simple to handle effectively in the same manner as we handle general constraints, but are nevertheless too complicated to handle the projections easily in the context of large problems. We use a combination of a simple piecewise linear line search (with only two pieces) and the trust region approach. The projections are handled approximately, but the approximation has to be 'good enough'. Details are given in [45]. In particular, a special case gives a convergence proof for sequential linear programming in the case of convex constraints. In the case of nonlinear networks, Sartenaer, in [122], has obtained some very encouraging numerical results.

In addition we have a test-bed of around nine hundred problems written in our standard data format. We have solved almost all these examples, many of which are of a substantial size. We expect to have them available electronically via netlib or something similar some time this year (1992). Some typical examples of applications we have solved using LANCELOT are tabulated above, in Figure 1. Detailed analysis of these results and our interpretation will be reported separately.

What about other current work on LANCELOT and the future? As an alternative to

Problem	n	m
1-d nonlinear boundary value problem	5002	0
Economic model from Thailand	2230	1112
1-d variational problem from ODEs	1001	0
2-d variational problem from PDEs	5184	0
3-d variational problem from PDEs	4913	0
Nonlinear network gas flow problem	2734	2727
Chemical reaction problem	5000	5000
Oscillation problem in structural mechanics	5041	0
Nonlinear optimal control problem	9006	7000
Maximum pivot growth in Gaussian Elimination	3946	3690
Nonlinear network problem on a square grid	13284	6724
Nonlinear optimal control problem	10001	5000
Hydro-electric reservoir management	2017	1008
Minimum surface problem with nonlinear boundary conditions	15625	0
Economic equilibrium	1825	730
Nonlinear optimal control problem	7011	5005
Orthogonal regression problem	8197	4096
Analysis of semiconductors	1002	1000
Elastic-plastic torsion problem	14884	0

Figure 1: *Some typical examples.*

the augmented Lagrangian we are implementing a Lagrangian barrier method [41], that is closely related to the modified and shifted barrier function method (see, for example, [68], and [114]). We use the Lagrangian barrier function

$$\Psi(x, \lambda, s) = f(x) - \sum_{i=1}^{m} \lambda_i s_i \log(s_i - c_i(x)),$$

which we can then optimize with respect to the simple bounds. A possible choice is $s_i = \mu \lambda_i^{\alpha_\lambda}$, where μ is a penalty parameter and $0 < \alpha_\lambda \leq 1$.

We also would like to exploit group partial separability at a more fundamental level in our trust region algorithm. We have been able to use the structure of the problem explicitly in the definition of the trust region and have suitable convergence properties [39] and we are currently investigating the computational implications. In order to exploit linear constraints more successfully we are investigating interior point methods. Finally, we always need to perform more testing.

3.3 APPROACH BASED UPON A MIXED LINEAR PROGRAMMING/INTERIOR POINT METHOD

Although not strictly speaking interior point methods, we have already mentioned modified barrier methods above. More generally, interior point methods, although originally developed for nonlinear programming, have had a spectacular success in the context of linear programming and have thereby generated new interest in their use in nonlinear optimization. See [134], for example, for background material. Encouraging results have already been obtained in the context of linear-like problems (see [23], [25], [26], [100] and [120]) and quadratic programming ([2], [9], [12], [22] [72], [83], [84], [99], [112], and [137]). Nash and Sofer have computational experience with a barrier method applied to a thousand variable nonlinear problem with bound constraints, [107]. Work in convex programming includes [5], [53], [86], [97], [98], [135] and [136]. It seems reasonable to expect further developments within nonlinear programming, especially since these techniques appear to be especially appropriate for large-scale problems.

3.4 OTHER ISSUES

There is obviously a number of issues that are relevant to our subject, but that fall slightly out of the context of the current paper. We now briefly mention some of the most important ones.

Both primal and dual degeneracy are intrinsic difficulties in that they are often a manifestation of the problem that can be troublesome to the method of solution. By primal degeneracy, we mean that the dual variables are not uniquely defined. By dual degeneracy, we mean that some of the dual variables are zero. Moreover degeneracy, especially primal degeneracy, is not a rare occurrence. Revelent work includes [17], [50], [62], [63], [69] and [121]. This is an important area in which there is a need for further research.

The primary level of formulation is clearly important. Augmented Lagrangians are rather different from barrier functions. If one exploits the structure using group partial separability this has a profound effect on the design of the software and the algorithmic techniques used. Trust region methods typically make different demands from line search approaches. Small scale sequential quadratic programming techniques do not have much in common with the approach taken in LANCELOT.

Parallelism is starting to have its effect on optimization. Firstly, numerical linear algebra plays a fundamental role in optimization. Recently there has been much work implementing such methods on advanced architectures. An overview of many of the major issues is given in [74]. A lucid description, illustrated by considering the Cholesky factorization, is given in [130]. Secondly, parallelism can be exploited at the basic level of the optimization algorithm. Simple examples are the computation of 'extra' quantities in parallel (for example, speculative steps), or independent runs with different choices for some of the algorithm parameters. Often the most effective algorithms for very large problems are those which are very simple but not efficient if implemented in a sequential environment. However, with the possibility of intelligently running what amount to several instances in parallel, a much more effective algorithm results.

We think that although they are not scale invariant, truncated Newton methods are especially important in the context of large-scale nonlinear optimization. When incorporated with an iterative technique such as preconditioned conjugate directions, they enable us to handle the difficulty of deciding whether to solve inner iterations accurately or inaccurately, with the limited information available, whilst still being able to ensure a satisfactory asymptotic convergence rate and global convergence. For example, it is not easy to see how to achieve the same ends in the context of sequential quadratic programming using active set strategies.

Anyone who has tried inputting significant problems appreciates the potential of automatic differentiation. Recently, this has become a very active area of research (see, for example, [77] and [78]), and efforts are underway to provide automatic differentiation tools that promise to make this technology readily usable to the optimization community ([11] and [80]).

After the great success of quasi-Newton methods there was much hope that such techniques could be adapted to the manipulation of large Hessian approximations. Unfortunately, this turned out not to be the case and the results have been disappointing (see [124]). On the other hand, sparse finite difference schemes have been successful. Since the pioneering work of Curtis, Powell and Reid [49] there has been significant progress, some of which exploits parallelism (see [21], [27], [28], [29], [30], [71] [113], and [116]).

An alternative approach is to use limited memory quasi-Newton updating. This uses the information of only a few, most recent, steps to define a variable metric approximation to the Hessian (see, for example [16] and [110]). These methods have proved to be very useful for solving certain large unstructured problems and Nocedal, [109], claims it is competitive with the partitioned quasi-Newton method on partially separable problems in which the number of element variables exceeds five or six, at least when the cost of evaluating the objective function is relatively low. Recently Bartholomew-Biggs and Hernandez ([7]) have been able to solve large problems using limited memory approximations to the inverse of the Lagrangian in the sequential quadratic programming framework of [8].

38

We should also mention that the current familiarity of users with sophisticated computer environments has created a high expectation for a user friendly interface to software. Unfortunately, developers have been too busy, as yet, coping with the algorithmic complexities to have devoted much time to the important practicalities of a first rate interface.

4 In conclusion

We would like to emphasize that it is possible to solve large nonlinear constrained problems in thousands of variables in acceptable time on reasonable workstations. At least two software packages, LANCELOT and MINOS, are available. Input is important, and significant progress in both modelling languages and a standard input format, have been made.

This is a vibrant, challenging and useful research area. Our hope is that, in the not too distant future, practitioners will be solving nonlinear models rather than linear ones, when the former is the most appropriate one to consider. Prefering to solve linear models, only because we understand how to solve large linear programs, should no longer be the normal practice.

Acknowledgements

This research was supported in part by the Advanced Research Projects Agency of the Department of Defense and was monitored by the Air Force Office of Scientific Research under Contract No F49620-91-C-0079. The United States Government is authorized to reproduce and distribute reprints for governmental purposes notwithstanding any copyright notation hereon.

References

[1] J. Abadie and J. Carpentier. Généralisation de la méthode du gradient réduit de Wolfe au cas des contraintes non-linéaires. In : D.B. Hertz and J. Melese (Eds.), *Proceedings IFORS Conference*, John Wiley, Amsterdam, pp 1041–1053, 1966.

[2] K. M. Anstreicher, D. den Hertog, C. Roos and T. Terlaky. A long step barrier method for convex quadratic programming. Technical Report 90-53, Faculty of Mathematics and Computer Science, Delft University of Technology, Delft, Netherlands, 1990.

[3] K. J. Arrow and R. M. Solow. Gradient methods for constrained maxima with weakened assumptions. In L. Hurwicz Arrow and Hirofumi Uzawa (Eds.), *Studies in Linear and Nonlinear Programming*, Stanford University Press, Stanford, CA, 1958.

[4] B. M. Averick, R. G. Carter and J. J. Moré. The MINPACK-2 test problem collection. Technical Report ANL/MCS-TM-150, Applied Mathematics Division, Argonne National Labs, Argonne, IL, 1991.

[5] O. Bahn, J. L. Goffin, J. P. Vial and O. Du Merle. Implementation and behaviour of an interior point cutting plane algorithm for convex programming: an application to geometric programming. Technical report, GERARD, Faculty of Management, McGill University, McGill University, Montreal, 1991.

[6] R. H. Bartels and G. H. Golub. The simplex method of linear programming using the LU decomposition. *Communications of the ACM*, **12**, pp 266–268, 1969.

[7] M. C. Bartholomew-Biggs and M. de F. G. Hernandez. Some improvements to the subroutine OPALQP for dealing with large problems. Technical report, Numerical Optimization Center, Hatfield Polytechnic, Hatfield, UK, 1992.

[8] M.C. Bartholomew-Biggs. Recursive quadratic programming methods based on the augmented Lagrangian function. *Mathematical Programming Study*, **31**, pp 21–42, 1987.

[9] M. Ben Daya and C. M. Shetty. Polynomial barrier function algorithm for convex quadratic programming. Report J85-5, School of ISE, Georgia Institute of Technology, Atlanta, Georgia, 1988.

[10] D. P. Bertsekas. Projected Newton methods for optimization problems with simple constraints. *SIAM J. Control & Optimization*, **20**, pp 221–246, 1982.

[11] C. Bischof, A. Carle, G. Corliss, P. Hovland and A. O. Griewank. ADIFOR: Generating derivative codes from Fortran programs. Technical Report MCS-P263-0991, Argonne National Labs, Argonne, IL, 1991.

[12] P. T. Boggs, P. D. Domich, J. E. Rogers and C. Witzgall. An interior point method for linear and quadratic programming problems. *Mathematical Programming Society COAL Newsletter*, August 1991.

[13] P. T. Boggs and J. W. Tolle. A truncated SQP algorithm for large scale nonlinear programming problems, January 1992. Presented at the sixth IIMAS-UNAM workshop on numerical analysis and optimization.

[14] A. Brooke, D. Kendrick and A. Meeraus. *GAMS: a User's Guide*. The Scientific Press, Redwood City, USA, 1988.

[15] A. G. Buckley. Test functions for unconstrained minimization. Technical Report CS-3, Computing Science Division, Dalhousie University, Dalhousie, Canada, 1989.

[16] A. G. Buckley and A. LeNir. QN-like variable storage conjugate gradients. *Mathematical Programming*, **27**, pp 155–175, 1983.

[17] S. Busovača. Handling degeneracy in a nonlinear l_1 algorithm. Technical Report Tech. Rept. CS-85-34, Univ. of Waterloo, Dept. of Computer Science, Univ. of Waterloo, Waterloo, Ontario N2L 3G1, 1985.

[18] A. Cauchy. Méthode générale pour la résolution des systèmes d'équations simultannées. *Comptes Rendus de l'Académie des Sciences*, pp 536–538, 1847.

[19] T. F. Coleman. Large Sparse Numerical Optimization. In : *Lecture Notes in Computer Science #165*. Springer-Verlag, Berlin, 1984.

[20] T. F. Coleman. Large Scale Numerical Optimization: Introduction and overview. In : *Encyclopedia of Computer Science and Technology*. Marcel Dekker, Inc., New York, 1992 (to appear).

[21] T. F. Coleman and J-Y. Cai. The cyclic coloring problem and estimation of sparse Hessian matrices. *SIAM J. Appl. Math.*, **20**, pp 187–209, 1983.

[22] T. F. Coleman and L. Hulbert. A globally and superlinearly convergent algorithm for convex quadratic programs with simple bounds. Technical Report TR 90-1092, Department of Computer Science, Cornell University, Ithaca, NY, 1990. To appear in *SIAM J. Optimization*.

[23] T. F. Coleman and Y. Li. A global and quadratic affine scaling method for (augmented) linear l_1 problems. In : G.A. Watson (Ed.), *Proceedings of the 13th Biennial Numerical Analysis Conference Dundee 1989*. Longmans, 1989.

[24] T. F. Coleman and Y. Li (Eds.). *Large-Scale Numerical Optimization*. SIAM, Philadelphia, 1990.

[25] T. F. Coleman and Y. Li. A global and quadratic affine scaling method for linear l_1 problems. *Mathematical Programming*, 1992 (to appear).

[26] T. F. Coleman and Y. Li. A global and quadratically convergent method for linear l_∞ problems. *SIAM J. Optimization*, 1992 (to appear).

[27] T. F. Coleman and J. J. Moré. Estimation of sparse Jacobian matrices and graph coloring problems. *SIAM J. Numer. Anal.*, **20**, pp 187–209, 1983.

[28] T. F. Coleman and J. J. Moré. Estimation of sparse Hessian matrices and graph coloring problems. *Mathematical Programming*, **28**, pp 243–270, 1984.

[29] T. F. Coleman and J. J. Moré. Software for estimating sparse Jacobian matrices. *TOMS*, **10**, pp 329–345, 1984.

[30] T. F. Coleman and P. E. Plassman. Solution of nonlinear least-squares problems on a multiprocessor. *SIAM J. Sci. Stat. Comp.*, **13**, 1992.

[31] A. R. Conn, N. I. M. Gould and Ph. L. Toint. Global convergence of a class of trust region algorithms for optimization with simple bounds. *SIAM J. Numer. Anal.*, **25**, pp 433–460, 1988. See also *SIAM J. Numer. Anal.*, **26**, pp 764-767, 1989.

[32] A. R. Conn, N. I. M. Gould and Ph. L. Toint. Testing a class of methods for solving minimization problems with simple bounds on the variables. *Mathematics of Computation*, **50**, pp 399–430, 1988.

[33] A. R. Conn, N. I. M. Gould and Ph. L. Toint (Eds.). *Large-Scale Optimization*, Volume 45, No. 3 of *Mathematical Programming*. North-Holland, 1989.

[34] A. R. Conn, N. I. M. Gould and Ph. L. Toint. An introduction to the structure of large scale nonlinear optimization problems and the lancelot project. In : R. Glowinski and A. Lichnewsky (Eds.), *Computing Methods in Applied Sciences and Engineering*, SIAM, Philadelphia, pp 42–51, 1990.

[35] A. R. Conn, N. I. M. Gould and Ph. L. Toint (Eds.). *Large-Scale Optimization — Applications*, Volume 48, No. 1 of *Mathematical Programming*. North-Holland, 1990.

[36] A. R. Conn, N. I. M. Gould and Ph. L. Toint. A proposal for a standard data input format for large-scale nonlinear programming problems. Report CS-89-61, Department of Computer Science, University of Waterloo, Waterloo, Ontario, Canada N2L 3G1, 1990.

[37] A. R. Conn, N. I. M. Gould and Ph. L. Toint. Convergence of quasi-Newton matrices generated by the symmetric rank one update. *Mathematical Programming*, **50**, pp 177–195, 1991.

[38] A. R. Conn, N. I. M. Gould and Ph. L. Toint. A globally convergent augmented Lagrangian algorithm for optimization with general constraints and simple bounds. *SIAM J. Numer. Anal.*, **28**, pp 545–572, 1991.

[39] A. R. Conn, N. I. M. Gould and Ph. L. Toint. Convergent properties of minimization algorithms for convex constraints using a structured trust region. Technical report, Department of Mathematics, FUNDP, Namur, Belgium, 1992.

[40] A. R. Conn, N. I. M. Gould and Ph. L. Toint. CUTE: a collection of constrained and unconstrained test examples for nonlinear programming, 1992.

[41] A. R. Conn, N. I. M. Gould and Ph. L. Toint. A globally convergent Lagrangian barrier algorithm for optimization with general inequality constraints and simple bounds. Technical report, IBM Thomas J. Watson Research Center, P.O.Box 218, Yorktown Heights, NY 10598, U.S.A., 1992.

[42] A. R. Conn, N. I. M. Gould and Ph. L. Toint. LANCELOT: *a Fortran package for large-scale nonlinear optimization*. Springer-Verlag, Berlin, Heidelberg and New York, 1992.

[43] A. R. Conn, N. I. M. Gould and Ph. L. Toint. On the number of inner iterations per outer iteration of a globally convergent algorithm for optimization with general nonlinear constraints and simple bounds. In : D. F. Griffiths and G.A. Watson (Eds.), *Proceedings of the 14th Biennial Numerical Analysis Conference Dundee 1991*. Longmans, 1992 (to appear).

[44] A. R. Conn, N. I. M. Gould, M. Lescrenier and Ph. L. Toint. Performance of a multifrontal scheme for partially separable optimization. In : S. Gómez, J. P. Hennart and R. A. Tapia (Eds.), *Proceedings of the Sixth Mexico-United States Numerical Analysis Workshop*, SIAM, Philadelphia, 1992 (to appear).

42

[45] A. R. Conn, N. I. M. Gould, A. Sartenaer and Ph. L. Toint. Global convergence of a class of trust region algorithms for optimization using inexact projections on convex constraints. *SIAM J. Optimization*, 1992 (to appear).

[46] L. Contesse and J. Villavicencio. Resolución de un modelo económico de despacho de carga eléctrica mediante el método de penalización Lagrangeana con cotas. *Revista del Instituto Chileno de Investigación Operativa*, pp 80–112, 1982.

[47] International Business Machines Corporation. Mathematical programming system/360 version 2, linear and separable programming-user's manual. Technical Report H20-0476-2, IBM Corporation, 1969.

[48] R. Courant. Variational methods for the solution of problems of equilibrium and vibrations. *Bull. Amer. Math. Soc.*, **49**, pp 1–23, 1943.

[49] A. R. Curtis, M. J. D. Powell and J. K. Reid. On the estimation of sparse Jacobian matrices. *IMA J. Appl. Math.*, **13**, pp 117–120, 1974.

[50] A. Dax. A note on optimality conditions for the Euclidean multifacility location problem. *Mathematical Programming*, **36**, pp 72–80, 1986.

[51] R. S. Dembo and J. G. Klincewicz. A scaled reduced gradient algorithm for network flow problems with convex separable costs. *Mathematical Programming*, **15**, pp 125–147, 1981.

[52] R. S. Dembo and U. Tulowitski. On the minimization of quadratic functions subject to box constraints. Technical Report B71, Yale School of Management, Yale University, New Haven, USA, 1983.

[53] D. den Hertog, C. Roos and T. Terlaky. A potential reduction method for a class of smooth convex programming problems. Technical Report 90-01, Faculty of Mathematics and Computer Science, Delft University of Technology, Delft, Netherlands, 1990.

[54] J. E. Dennis Jr. and R. B. Schnabel. *Numerical Methods for Unconstrained Optimization and Nonlinear Equations*. Prentice-Hall, Englewood Cliffs, NJ, 1983. Russian edition, Mir Publishing Office, Moscow, 1988, O. Burdakov, translator.

[55] J. E. Dennis Jr. and H. Wolkowicz. Sizing and least-change secant methods. *SIAM J. Numer. Anal.*, 1992 (to appear).

[56] I. S. Duff, A. M. Erisman and J. K. Reid. *Direct methods for sparse matrices*. Clarendon Press, Oxford, UK, 1986.

[57] I. S. Duff, R. G. Grimes and J. G. Lewis. Sparse matrix test problems. *ACM Transactions on Mathematical Software*, **15**, pp 1–14, 1989.

[58] S. K. Eldersveld. *Large-scale Sequential Quadratic Programming Algorithms*. PhD thesis, Department of Operations Research, Stanford University, Stanford, CA, 1991. Also available as a technical report.

[59] J. E. Falk and G. P. McCormick. Computational aspects of the international coal trade model. In : P.T. Harker (Ed.), *Spacial price equilibrium: Advances in theory, computation and application, Lecture Notes in Economics and Mathematical Systems #249*, Springer-Verlag, Berlin, Heidelberg and New York, pp 73–117, 1986.

[60] A. V. Fiacco and G. P. McCormick. *Nonlinear Programming: Sequential Unconstrained Minimization Techniques*. John Wiley & Sons, New York, NY, 1968. Reprinted as *Classics in Applied Mathematics 4*, SIAM, 1990.

[61] R. Fletcher. *Practical Methods of Optimization*. J. Wiley and Sons, Chichester, second edition, 1987.

[62] R. Fletcher. Degeneracy in the presence of roundoff errors. *Linear Algebra and its Applications*, **106**, pp 149–183, 1988.

[63] R. Fletcher. Resolving degeneracy in quadratic programming. Technical Report NA/135, Department of Mathematical Sciences, University of Dundee, 1991.

[64] R. Fletcher and E. Sainz de la Maza. Nonlinear programming and non-smooth optimization by successive linear programming. *Mathematical Programming*, **43**, pp 235–256, 1989.

[65] R. Fourer, D. M. Gay and B. W. Kernighan. AMPL: A mathematical programming language. Computer science technical report, AT&T Bell Laboratories, Murray Hill, USA, 1987.

[66] D. M. Gay. Electronic mail distribution of linear programming test problems. Mathematical Programming Society COAL Newsletter, December 1985.

[67] P. E. Gill and W. Murray. Conjugate-gradient methods for large-scale nonlinear optimization. Technical Report SOL79-15, Operations Research Department, Stanford University, Stanford, USA, 1979.

[68] P. E. Gill, W. Murray, M. A. Saunders and M. H. Wright. Shifted barrier methods for linear programming. Technical Report SOL88-9, Operations Research Department, Stanford University, Stanford, USA, 1988.

[69] P. E. Gill, W. Murray, M. A. Saunders and M. H. Wright. A practical anti-cycling procedure for linearly constrained optimization. *Mathematical Programming*, **45**(3), pp 437–474, 1989.

[70] P. E. Gill, W. Murray and M. H. Wright. *Practical Optimization*. Academic Press, New York, London, Toronto, Sydney and San Francisco, 1981.

[71] D. Goldfarb and Ph. L. Toint. Optimal estimation of Jacobian and Hessian matrices that arise in finite difference calculations. *Math. Comp.*, **43**, pp 69–88, 1984.

[72] D. Goldfarb and S. Liu. An $O(n^3L)$ primal interior point algorithm for convex quadratic programming. *Mathematical Programming*, **49**, pp 325–343, 1991.

[73] D. Goldfarb and M. J. Todd. Linear programming. In : G.L. Nemhauser, A.H.G. Rinnooy Kan and M.J. Todd (Eds.), *Optimization*, Volume 1 of *Handbooks in Operations Research and Management Science*, chapter 2. North-Holland, Amsterdam, 1989.

[74] G. H. Golub and C. F. Van Loan. *Matrix Computations*. Johns Hopkins University Press, Baltimore, Maryland, second edition, 1989.

[75] N. I. M. Gould. An algorithm for large-scale quadratic programming. *IMA J. Numer. Anal.*, **11**, pp 299–324, 1991.

[76] N. I. M. Gould and J. K. Reid. New crash procedures for large systems of linear constraints. *Mathematical Programming*, **45**, pp 475–502, 1989.

[77] A. O. Griewank. Direct calculation of Newton steps without accumulating Jacobians. In : T. F. Coleman and Yuying Li (Eds.), *Large-Scale Numerical Optimization*, SIAM, Philadelphia, pp 115–137, 1990.

[78] A. O. Griewank and G. F. Corliss. *Automatic Differentiation of Algorithms: Theory, Implementation, and Application*. SIAM, Philadelphia, 1991.

[79] A. O. Griewank and Ph. L. Toint. On the unconstrained optimization of partially separable functions. In : M.J.D. Powell (Ed.), *Nonlinear Optimization*. Academic Press, London, 1982.

[80] A. O. Griewank, D. Juedes, J. Srinivasan and C. Tyner. ADOL-C, a package for the automatic differentiation of algorithms written in C/C++. *ACM Transactions on Mathematical Software*, to appear. Also appeared as Preprint MCS–P180–1190, Mathematics and Computer Science Division, Argonne National Laboratory, 9700 S. Cass Ave., Argonne, IL 60439, 1990.

[81] R. E. Griffith and R. A. Stewart. A nonlinear programming technique for the optimization of continuous processing systems. *Management Sci.*, **7**, pp 379–392, 1961.

[82] M. D. Grigoriadis. An efficient implementation of the network simplex method. *Mathematical Programming*, **26**, pp 83–111, 1986.

[83] C-G. Han, P. M. Pardalos and Y. Ye. Computational aspects of an interior point algorithm for quadratic programming problems with box constraints. In : T.F. Coleman and Y. Li (Eds.), *Large-Scale Numerical Optimization*, SIAM, Philadelphia, pp 92–102, 1990.

[84] C-G. Han, P. M. Pardalos and Y. Ye. Solving some engineering problems using an interior-point algorithm. Technical Report CS-91-04, Computer Science Department, The Pennsylvania State University, University Park, PA, 1991.

[85] M. R. Hestenes. Multiplier and gradient methods. *J. Optim. Theory Appl.*, **4**, pp 303–320, 1969.

[86] F. Jarre and M. A. Saunders. Practical aspects of an interior-point method for convex programming. Technical Report SOL91-9, Department of Operations Research, Stanford University, Stanford, USA, 1991.

[87] A. P. Jones. The chemical equilibrium problem: An application of SUMT. Technical Report RAC-TP-272, Reserach Analysis Corporation, McLean, Virginia, 1967.

[88] H. Khalfan, R. H. Byrd and R. B. Schnabel. A theoretical and experimental study of the symmetric rank one update. *SIAM J. Optimization*, 1992 (to appear).

[89] J. L. Lagrange. *Théorie des Fonctions Analytiques*. Impr. de la République, Paris, 1797.

[90] L. S. Lasdon, A. D. Waren, A. Jain and M. Ratner. Design and testing of a generalized reduced gradient code for nonlinear programming. *TOMS*, 4, pp 34–50, 1978.

[91] E.S. Levitin and B.T. Polyak. Constrained minimization problems. *USSR Comput. Math. and Math. Phys.*, 6, pp 1–50, 1966.

[92] D. Mahidhara and L. Lasdon. An SQP algorithm for large sparse nonlinear programs. Working paper, MSIS Department, School of Business, University of Texas, Austin, TX 78712, 1990.

[93] H. M. Markowitz. The elimination form of the inverse and its application to linear programming. *Management Sci.*, 3, pp 255–269, 1957.

[94] G. P. McCormick. Anti-zig-zagging by bending. *Management Sci.*, 15, pp 315–320, 1969.

[95] G. P. McCormick. Computational aspects of nonlinear programming solutions to large-scale inventory problems. Technical Report Technical Memorandum Serial T-63488, George Washington University, Institute of Management Science and Engineering, Washington,DC, 1972.

[96] G. P. McCormick and A. Sofer. Optimization with unary functions. *Mathematical Programming*, 52, pp 167–179, 1991.

[97] S. Mehrotra and J. Sun. An interior point algorithm for solving smooth convex programs based on Newton's method. In : J.C. Lagarias and M.J. Todd (Eds.), *Contemporary Mathematics*, AMS, Providence, Rhode Island, pp 265–284, 1990.

[98] R. C. Monteiro. The global convergence of a class of primal potential reduction algorithms for convex programming. Report, Systems and Industrial Engineering Department, University of Arizona, Tucson, Arizona, 1991.

[99] R. C. Monteiro and I. Adler. Interior path following primal-dual algorithms, part ii: convex quadratic programming. Report, Department of IEOR, University of California, Berkeley, California, 1987.

[100] R. C. Monteiro, I. Adler and M. G. Resende. A polynomial-time primal-dual affine scaling algorithm for linear and convex quadratic programming and its power series extension. Report ESRC 88-8, Department of IEOR, University of California, Berkeley, California, 1987.

[101] J. J. Moré. Recent developments in algorithms and software for trust region methods. In : A. Bachem, M. Grötschel and B. Korte (Eds.), *Mathematical Programming: The State of the Art*, Springer Verlag, Berlin, pp 258–287, 1983.

[102] J. J. Moré, B. S. Garbow and K. E. Hillstrom. Testing unconstrained optimization software. *ACM Transactions on Mathematical Software*, 7(1), pp 17–41, 1981.

[103] J. J. Moré and G. Toraldo. Algorithms for bound constrained quadratic programming problems. *Numerische Mathematik*, 14, pp 14–21, 1979.

[104] J. J. Moré and G. Toraldo. On the solution of large quadratic programming problems with bound constraints. *SIAM J. Optimization*, 1, pp 93–113, 1991.

[105] W. Murray and F. J. Prieto. A sequential quadratic programming algorithm using an incomplete solution of the subproblem. Technical Report SOL90-12, Operations Research Department, Stanford University, Stanford, USA, 1990.

[106] B. A. Murtagh and M. A. Saunders. MINOS 5.1 user's guide. Technical Report SOL83-20R, Department of Operations Research, Stanford University, Stanford, USA, 1987.

[107] S. G. Nash and A. Sofer. A barrier method for large-scale constrained optimization. Technical Report 91-10, Department of Operations Research and Applied Statistics, George Mason University, Fairfax, Virginia 22030, 1991.

[108] G. L. Nemhauser, A. H. G. Rinnooy Kan and M. J. Todd (Eds.). *Optimization*, Volume 1 of *Handbooks in Operations Research and Management Science*. North-Holland, Amsterdam, 1989.

[109] J. Nocedal. Theory of algorithms for uncontrained optimization. *Acta Numerica*, 1, 1992 (to appear).

[110] J. Nocedal and D. C. Liu. On the limited memory BFGS method for large scale optimization. *Mathematical Programming, Series B*, 45, pp 503–528, 1989.

[111] M. R. Osborne and L. P. Sun. A new approach to the symmetric rank-one updating algorithm. *Mathematical Programming*, 1992 (to appear).

[112] P. M. Pardalos, C-G. Han and Y. Ye. Interior point algorithms for solving nonlinear optimization problems. *Mathematical Programming Society COAL Newsletter*, August 1991.

[113] P.E.Plassman. Sparse Jacobian estimation and factorization on a multiprocessor. In : T.F. Coleman and Y. Li (Eds.), *Large-Scale Numerical Optimization*, SIAM, Philadelphia, pp 152–179, 1990.

[114] R. Polyak. Modified barrier functions (theory and methods). Research report RC 15886 (#70630), IBM T. J. Watson Research Center, P.O.Box 218, Yorktown Heights, NY 10598, USA, 1990.

[115] M. J. D. Powell. A method for nonlinear constraints in minimization problems. In : R. Fletcher (Ed.), *Optimization*, Academic Press, New York, NY, pp 283–298, 1969.

[116] M. J. D. Powell and Ph. L. Toint. On the estimation of sparse Hessian matrices. *SIAM J. Numer. Anal.*, **16**, pp 1060–1074, 1979.

[117] J. K. Reid. Fortran subroutines for handling sparse linear programming base. Technical Report AER-R-8269, AERE Harwell Laboratory, Harwell, UK, 1976.

[118] S. M. Robinson. A quadratically convergent algorithm for general nonlinear programming problems. *Mathematical Programming*, **3**, pp 145–156, 1972.

[119] J. B. Rosen and J. Kreuser. A gradient projection algorithm for nonlinear constraints. In : F. Lootsma (Ed.), *Numerical Methods for Nonlinear Optimization*, Academic Press, New York, NY, pp 297–300, 1972.

[120] A. S. Ruzinsky and E. T. Olson. l_1 and l_∞ minimization via a variant of Karmarkar's algorithm. *IEEE Trans. Acoustics, Speech and Signal Processing*, **37**, pp 245–253, 1989.

[121] D. M. Ryan and M. R. Osborne. On the solution of highly degenerate linear programs. *Mathematical Programming*, **41**, pp 385–392, 1988.

[122] A. Sartenaer. *On some strategies for handling constraints in nonlinear optimization.* PhD thesis, Department of Mathematics, FUNDP, Namur, Belgium, 1991.

[123] D. A. Schrady and U. C. Choe. Models for multi-item continuous review inventory policies subject to constraints. *Naval Research Logistics Quarterly*, **18**, pp 451–463, 1971.

[124] D. C. Sorensen. An example concerning quasi-Newton estimation of a sparse Hessian. *SIGNUM Newsletter*, **16**, pp 8–10, 1981.

[125] Ph. L. Toint. Call for test problems in large scale nonlinear optimization. *COAL Newsletter*, **16**, pp 5–10, 1987.

[126] Ph. L. Toint and D. Tuyttens. On large scale nonlinear network optimization. *Mathematical Programming, Series B*, **48**(1), pp 125–159, 1990.

[127] Ph. L. Toint and D. Tuyttens. LSNNO: a Fortran subroutine for solving large scale nonlinear network optimization problems. *ACM Transactions on Mathematical Software*, 1990 (to appear).

[128] J. A. Tomlin. The influences of algorithmic and hardware developments on computational mathematical programming. In : M. Iri and K. Tanabe (Eds.), *Mathematical Programming— Recent Developments and applications*, Tokyo, KTK Scientific Publishers, pp 159–175, 1988.

[129] J. A. Tomlin and J. S. Welch. Mathematical programming systems. Technical Report RJ 7400 (69202), IBM Thomas J. Watson Research Center, P.O.Box 218, Yorktown Heights, NY 10598, USA, 1990.

[130] C. F. Van Loan. A survey of matrix computations. Technical Report CTC90TR26, Cornell Theory Center, Cornell University, Ithaca, NY, 1990.

[131] P. Werbos. Backpropagation: past and future. *Proceedings of the 2nd International Conference on Neural Networks*, IEEE, New York, 1988.

[132] H. Wolkowicz. Measures for symmetric rank-one updates. *Mathematical Programming*, 1992 (submitted).

[133] M. H. Wright. Optimization and large scale computation. In : J.P. Mesirov (Ed.), *Very Large Scale Computation in the 21st Century*, SIAM, Philadelphia, 1991.

[134] M. H. Wright. Interior methods for constrained optimization. *Acta Numerica*, 1, 1992 (to appear).

[135] S. J. Wright. An interior-point algorithm for linearly constrained optimization. Technical Report MCS-P162-0790, Argonne National Labs, Argonne, IL, 1990.

[136] Y. Ye. *Interior point algorithms for linear, quadratic and linearly constrained convex programming*. PhD thesis, Engineering and Economic Systems Department, Stanford University, Stanford, CA, 1987.

[137] Y. Ye. Interior point algorithms for quadratic programming. In : S. Kumar (Ed.), *Recent Developments in Mathematical Programming*. Gordan and Breach Scientific Publishers, London, to appear 1992.

TRADING OFF PARALLELISM AND NUMERICAL STABILITY

J. W. DEMMEL
Computer Science Division and Mathematics Department
University of California
Berkeley, CA 94720
demmel@cs.berkeley.edu

ABSTRACT. The fastest parallel algorithm for a problem may be significantly less stable numerically than the fastest serial algorithm. We illustrate this phenomenon by a series of examples drawn from numerical linear algebra. We also show how some of these instabilities may be mitigated by better floating point arithmetic.

KEYWORDS. Parallel numerical linear algebra, numerical stability, floating point arithmetic.

Introduction

The most natural way to design a parallel numerical algorithm is to take an existing numerically stable algorithm and parallelize it. If the parallel version performs the same floating point operations as the serial version, and in the same order, one expects it to be equally stable numerically. In some cases, such as matrix operations, one expects that the parallel algorithm may reorder some operations (such as computing sums) without sacrificing numerical stability. In other cases, reordering sums could undermine stability, e.g. ODEs and PDEs.

Our purpose in this paper is to point out that designing satisfactorily fast and stable parallel numerical algorithms is not so easy as parallelizing stable serial algorithms. We identify two obstacles:

1. An algorithm which was adequate on small problems may fail once they are large enough. This becomes evident when the algorithm is used on a large parallel machine to solve larger problems than possible before. Reasons for this phenomenon include roundoff accumulation, systematically increasing condition numbers, and systematically higher probability of "random instability."

49

M. S. Moonen et al. (eds.), Linear Algebra for Large Scale and Real-Time Applications, 49–68.
© 1993 *Kluwer Academic Publishers.*

2. A fast parallel algorithm for a problem may be significantly less stable than a fast serial algorithm. In other words, there is a tradeoff between parallelism and stability.

We also discuss two techniques which sometimes remove or mitigate these obstacles. The first is *good floating point arithmetic*, which, depending on the situation, may mean carefully rounding, adequate exception handling, or the availability of extra precision without excessive slowdown. The second technique is as follows:

1. Solve the problem using a fast method, provided it is rarely unstable.

2. Quickly and reliably confirm or deny the accuracy of the computed solution. With high probability, the answer just (quickly) computed is accurate enough to keep.

3. Otherwise, recompute the desired result using a slower but more reliable algorithm.

This paradigm lets us combine a fast but occasionally unstable method with a slower, more reliable one to get guaranteed reliability and usually quick execution. One could also change the third step to just issue a warning, which would guarantee fast execution, guarantee not to return an unreliable answer, but occasionally fail to return an answer at all. Which paradigm is preferable is application dependent.

The body of the paper consists of a series of examples drawn both from the literature and from the experience in the LAPACK project [3]. As our understanding of problems improves, the status of these tradeoffs will change. For example, until recently it was possible to use a certain parallel algorithm for the symmetric tridiagonal eigenvalue problem only if the floating point arithmetic was accurate enough to simulate double the input precision [19, 35, 73, 10]. Just recently, a new formulation of the inner loop was found which made this unnecessary [48]. The fact remains that for a number of years, the only known way to use this algorithm stably was via extra precision. So one can say that the price of insufficiently accurate arithmetic was not an inability to solve this problem, but several years of lost productivity because a more straightforward algorithm could not be used.

Section 1 describes how algorithms which have been successful on small or medium sized problems can fail when they are scaled up to run on larger machines and problems. Section 2 describes parallel algorithms which are less stable than their serial counterparts. The benefit of better floating point arithmetic will be pointed out while discussing the relevant examples, and overall recommendations for arithmetic summarized in section 3

1 Barriers to scaling up old algorithms

1.1 SPARSE CHOLESKY ON THE CRAY Y-MP AND CRAY 2

We discuss the experience of Russell Carter in porting an existing code for sparse Cholesky factorization to a Cray Y-MP [15]. Cholesky is a very stable algorithm, and this code had been in use for some time. The Cray Y-MP was larger than machines previously available, and Carter ran it on a large linear system $Ax = b$ from a structural model. A

Computer	Bits	Nominal precision	Displacement
Cray 2	128	1.e-29	.447440341
Convex 220	64	1.e-16	.447440339
IRIS	64	1.e-16	.447440339
IBM 3090	64	1.e-17	.447440344
Cray 2	64	4.e-15	.447440303
Cray Y-MP	64	4.e-15	.447436106

Table 1: *Sparse Cholesky Results*

had dimension 16146. Results are shown in table 1. The first column is the computer with which the problem is solved, the second is the number of bits in the floating point format, the third column is the approximate relative accuracy with which the floating point arithmetic can represent numbers (which is not the accuracy of computation on the Cray [55]), and the last column records one of the solution components of interest. The top line, which is done to about twice the accuracy of the others, is accurate in all the digits shown. In the other results the incorrect digits are underlined.

It can be seen that the Cray Y-MP loses two more digits than the Cray 2, even though both are using 64 bit words, and their 48-fraction-bit arithmetics are quite similar. The reason for this discrepancy is that both the Cray 2 and Cray Y-MP subtract incorrectly, but the Cray 2 does so in an unbiased manner. In particular, the inner loop of Cholesky computes $a_{ii} - \sum_{j=1}^{i-1} a_{ij}^2$, where a_{ii} is positive and the final result is also positive. Whenever the Cray 2 subtracts an a_{ij}^2, the average error is 0; the computed difference is too large as often as it is too small. On the Cray Y-MP, on the other hand, the difference is always a little too big. So the error accumulates with each subtract, instead of averaging out as on the Cray 2. The accumulating error is very small, and makes little difference as long as there are not too many terms in the sum. But $n = 16146$ was finally large enough to cause a noticeable loss of 2 decimal places in the final answer. The fix used by Carter was to use the single precision iterative refinement routine SGERFS in LAPACK [3].

The lessons of this example are that instability may become visible only when a problem's dimension becomes large enough, and that accurate arithmetic would have mitigated the instability.

1.2 INCREASING CONDITION NUMBERS

The last section showed how instability can arise when errors accumulate in the course of solving larger problems than ever attempted before. Another way this can arise is when the condition number of the problem grows too rapidly with its size. This may happen, for example, when we increase the mesh density with which we discretize a particular PDE. Consider the biharmonic equation $u_{xxxx} + u_{yyyy} = f$ on an n by n mesh, with boundary conditions chosen so that it represents the displacement of a square sheet fixed at the edges. The linear system $Ax = b$ resulting from the discretization has a condition number which grows like n^4. Suppose that we want to compute the solution correct to 6 decimal digits (a relative accuracy of 10^{-6}).

Generally one can solve $Ax = b$ with a backward error of order ε, the machine precision. Write $\varepsilon = 2^{-p}$, where p is the number of bits in the floating point fraction. This means the relative accuracy of the answer will be about $\varepsilon n^4 = 2^{-p}n^4$. For this to be less than or equal to 10^{-6}, we need $2^{-p}n^4 \leq 10^{-6}$ or $p \geq 4\log_2 n + 6\log_2 10 \approx 4\log_2 n + 20$. In IEEE double precision, $p = 52$ so we must have $n \leq 259$, which is fairly small.

One might object that for the biharmonic equation, Laplace's equation, and others from mathematical physics, if they have sufficiently regularity, then one can use techniques like multigrid, domain decomposition and FFTs to get accurate solutions for larger n (for the biharmonic, use boundary integral methods or [12]). This is because these methods work best when the right hand side b and solution x are both reasonably smooth functions, so that the more extreme singular values of the differential operators are not excited, and the bad conditioning is not visible. One often exploits this in practice. So in the long run, clever algorithms may become available which mitigate the ill-conditioning. In the short run, more accurate arithmetic (a larger p) would have permitted conventional algorithms to scale up to larger problems without change and remain useful longer. We will see this phenomenon later as well.

1.3   INCREASING PROBABILITY OF RANDOM INSTABILITIES

Some numerical instabilities only occur when exact or near cancellation occurs in a numerical process. In particular, the result of the cancellation must suffer a significant loss of relative accuracy, and then propagate harmfully through the rest of the algorithm. The best known example is Gaussian elimination without pivoting, which is unstable precisely when a leading principal submatrix is singular or nearly so. The set of matrices where this occurs is defined by a set of polynomial equations: $\det(A_r) = 0$, $r = 1, ..., n$, where A_r is a leading r by r principal submatrix of the matrix A. More generally, the set of problems on or near which cancellation occurs is an *algebraic variety* in the space of the problem's parameters, i.e. defined by a set of polynomial equations in the problem's parameters. Geometrically, varieties are smooth surfaces except for possible self intersections and cusps. Other examples of such varieties include polynomials with multiple roots, matrices with multiple eigenvalues, matrices with given ranks, and so on [23, 24, 40, 41].

Since instability arises not just when our problem lies on a variety, but when it is near one, we want to know how many problems lie near a variety. One may conveniently reformulate this as a probabilistic question: given a "random" problem, what is the probability that it lies within distance η of a variety? We may choose η to correspond to an accuracy threshold, problems lying outside distance η being guaranteed to be solved accurately enough, and those within η being susceptible to significant inaccuracy. For example, we may choose $\eta = 10^d \varepsilon$ (where ε is the machine precision) if we wish to guarantee at least d significant decimal digits in the answer.

It turns out that for a given variety, we can write down a simple formula that estimates this probability as a function of several simple parameters [24, 41]: the probability per second P of being within η of an instability is [55]

$$P = C \cdot M^k \cdot S \cdot \eta$$

where C and k are problem-dependent constants, M is the memory size in words, and S is the machine speed in flops per second.

For example, consider an SIMD machine where we assign each processor the job of LU decomposition of an independent random real matrix of fixed size n, and repeat this. We choose LU without pivoting in order to best match the SIMD architecture of the machine. We assume that each processor has an equal amount of memory, so that M is proportional to the number of processors $M = p \cdot M_p$. ¿From [41], we use the fact that the probability that a random n by n real matrix has a condition number $\|A\|_F\|A^{-1}\|_2$ exceeding $1/\eta$ is asymptotic to $n^{3/2}\eta$. Finally, suppose that we want to compute the answer with d decimal digits of accuracy, so that we pick $\eta = 10^d\varepsilon$. Combining this information, we get that the probability per second that an instability occurs (because a matrix has condition number exceeding $1/\eta$) is at least about

$$P = p \times \frac{S}{\frac{2}{3}n^3} \times n^{3/2}10^d\varepsilon = \frac{3}{2n^{3/2}M_p} \cdot M \cdot S \cdot 10^d \cdot \varepsilon$$

The important features of this formula is that is grows with increasing memory size M, with increasing machine speed S, and desired accuracy d, all of which are guaranteed to grow. We can lower the probability, however, by shrinking ε, i.e. by using more accurate arithmetic.

One might object that a better solution is to use QR factorization with Givens rotations instead of LU, because this is guaranteed to be stable without pivoting, and so is amenable to SIMD implementation. However, it costs three times as many flops. So we see there is a tradeoff between speed and stability.

If we instead fill up the memory with a single matrix of size $M^{1/2}$ by $M^{1/2}$, then the probability changes to $P = 1.5 \cdot M^{-3/4} \cdot S \cdot 10^d \cdot \varepsilon$. Interestingly, the probability goes down with M. The reason is that the time to solve an $M^{1/2}$ by $M^{1/2}$ matrix grows like $M^{3/2}$, so that the bigger the memory, the fewer such problems we can solve per second.

Another consequence of this formula is that random testing intended to discover instabilities in a program is more effective when done at low precision.

2 Trading off numerical accuracy and parallelism in new algorithms

2.1 FAST BLAS

The BLAS, or Basic Linear Algebra Subroutines, are building blocks for many linear algebra codes, and so they should be as efficient as possible. We describe two ways of accelerating them that sacrifice some numerical stability to speed. The stability losses are not dramatic, and a reasonable BLAS implementation might consider using them.

Strassen's method is a fast way of doing matrix multiplication based on multiplying 2-by-2 matrices using 7 multiplies and 15 or 18 additions instead of 8 multiplies and 4 additions

[1]. Strassen reduces n by n matrix multiplication to $n/2$ by $n/2$ matrix multiplication and addition, and recursively to $n/2^k$ by $n/2^k$. Its overall complexity is therefore $O(n^{\log_2 7}) \approx O(n^{2.81})$ instead of $O(n^3)$. The constant in the $O(\cdot)$ is, however, much larger for Strassen's than for straightforward matrix multiplication, and so Strassen's is only faster for large matrices. In practice once k is large enough so the $n/2^k$ by $n/2^k$ submatrices fit in fast memory, conventional matrix multiply may be used. A drawback of Strassen's method is the need for extra storage for intermediate results. It has been implemented on the Cray 2 [9, 8] and IBM 3090 [50].

The conventional error bound for matrix multiplication is as follows:

$$|fl_{\text{Conv}}(A \cdot B) - A \cdot B| \leq n \cdot \varepsilon \cdot |A| \cdot |B|$$

where the absolute values of matrices and the inequality are meant componentwise. The bound for Strassen's [13, 14, 49] is

$$\|fl_{\text{Strassen}}(A \cdot B) - A \cdot B\|_M \leq f(n) \cdot \varepsilon \cdot \|A\|_M \cdot \|B\|_M + O(\varepsilon^2)$$

where $\| \cdot \|_M$ denotes the largest component in absolute value, and $f(n) = O(n^{\log_2 12}) \approx O(n^{3.6})$. This can be extended to all the other BLAS, such as triangular system solving with many right hand sides [49], as well as many methods besides Strassen's [11].

These bounds differ when there is significant difference in the scaling of A and B. For example, changing A to AD and B to $D^{-1}B$ where D is diagonal does not change the error bound for conventional multiplication, but can make Strassen's arbitrarily large. Also, if $A = |A|$ and $B = |B|$, then the conventional bound says each component of $A \cdot B$ is computed to high relative accuracy; Strassen's can not guarantee this.

On the other hand, most error analyses of Gaussian elimination and other matrix routines based on BLAS do not depend on this difference, and remain mostly the same when Strassen based BLAS are used [27]. Only when the matrix or matrices are strongly graded (the diagonal matrix D above is ill-conditioned) will the relative instability of Strassen's be noticed.

Strictly speaking, the tradeoff of speed and stability between conventional and Strassen's matrix multiplication does not depend on parallelism, but on the desire to exploit memory hierarchies in modern machines. The next algorithm, a parallel algorithm for solving triangular systems, could only be of interest in a parallel context because it uses significantly more flops than the conventional algorithm.

The algorithm may be described as follows. Let T be a unit lower triangular matrix (a nonunit diagonal can easily be scaled to be unit). For each i from 1 to $n - 1$, let T_i equal the identity matrix except for column i where it matches T. Then it is simple to verify $T = T_1 T_2 \cdots T_{n-1}$ and so $T^{-1} = T_{n-1}^{-1} \cdots T_2^{-1} T_1^{-1}$. One can also easily see that T_i^{-1} equals the identity except for the subdiagonal of column i, where it is the negative of T_i. Thus T_i^{-1} comes free, and the work to be done is to compute the product $T_{n-1}^{-1} \cdots T_1^{-1}$ in $log_2 n$ parallel steps using a tree. Each parallel step involves multiplying n by n matrices (which are initially quite sparse, but fill up), and so takes about $log_2 n$ parallel substeps, for a total of $log_2^2 n$. Error analysis of this algorithm [66] yields an error bound proportional to $\kappa(T)^3 \varepsilon$ where $\kappa(T) = \|T\| \cdot \|T^{-1}\|$ is the condition number and ε is machine precision; this is in

Figure 1: *Parallel Prefix on 16 Data Items*

0	1	2	3	4	5	6	7	8	9	a	b	c	d	e	f
	0:1		2:3		4:5		6:7		8:9		a:b		c:d		e:f
			0:3				4:7				8:b				c:f
							0:7								8:f
															0:f
											0:b				
					0:5				0:9				0:d		
		0:2		0:4		0:6		0:8		0:a		0:c		0:e	

contrast to the error bound $\kappa(T)\varepsilon$ for the usual algorithm. The error bound for the parallel algorithm may be pessimistic — the worst example we have found has an error growing like $\kappa(T)^{1.5}\varepsilon$ — but shows that there is a tradeoff between parallelism and stability.

2.2 PARALLEL PREFIX

This parallel operation, also called *scan*, may be described as follows. Let $x_0, ... x_n$ be data items, and $\cdot$ any associative operation. Then the scan of these n data items yields another n data items defined by $y_0 = x_0$, $y_1 = x_0 \cdot x_1$, ... , $y_i = x_0 \cdot x_1 \cdots x_i$; thus y_i is the reduction of x_0 through x_i. The attraction of this operation, other than its usefulness, is its ease of implementation using a simple tree of processors. We illustrate in figure 1 for $n = 15$, or f in hexadecimal notation; in the figure we abbreviate x_i by i and $x_i \cdots x_j$ by $i : j$. Each row indicates the values held by the 16 processors; after the first row only the data that changes is indicated. Each updated entry combines its current value with one a fixed distance to its left.

Parallel prefix may be used to solve linear recurrence relations. For example, to evaluate $z_{i+1} = a_i z_i + b_i$, $i \geq 0$, $z_0 = 0$, we do the following operations:

Compute $p_i = a_0 \cdots a_i$ using parallel prefix multiplication
Compute $\beta_i = b_i / p_i$ in parallel
Compute $s_i = \beta_0 + \cdots + \beta_{i-1}$ using parallel prefix addition
Compute $z_i = s_i \cdot p_{i-1}$ in parallel

This approach extends to n term linear recurrences $z_{i+1} = \sum_{j=0}^{n-2} a_{i,j} z_{i-j} + b_i$, but the associative operation becomes $n - 1$ by $n - 1$ matrix multiplication. Basic linear algebra operations which can be solved this way include tridiagonal Gaussian elimination (a three term recurrence), solving bidiagonal linear systems (two terms), Sturm sequence evaluation for the symmetric tridiagonal eigenproblem (three terms), and the bidiagonal dqds algorithm for singular values (three terms) [63].

The numerical stability of these algorithms is not completely understood. For some applications, it is easy to see the error bounds are rather worse than the those of the sequential implementation [20]. For others, such as Sturm sequence evaluation [76], empirical evidence suggests it is stable enough to use in practice.

Another source of instability besides roundoff is susceptibility to over/underflow, because of the need to compute extended products (such as $p_i = a_0 \cdots a_i$ above). These over/underflows are often unessential because the output will eventually be the solution scaled to have unit norm (inverse iteration for eigenvectors). But to use parallel prefix, one must either scale before multiplication, or deal with over/underflow after it occurs; the latter requires reasonable exception handling [25]. In the best case, a user-level trap handler would be called to deal with scaling after over/underflow, requiring no overhead if no exceptions occur. Next best is an exception flag that could be tested, provided this can also be done quickly. The worst situation occurs when all exceptions require a trap into operating system code, which is then hundreds or thousands of times slower than a single floating point operation; this is the case on the DEC α chip, for example. In this case it is probably better to code defensively by scaling every step to avoid all possibility of over/underflow. This is unfortunate because it makes portable code so hard to write: what is fastest on one machine may be very slow on another, even though both formally implement IEEE arithmetic.

2.3 LINEAR EQUATION SOLVING

In subsection 2.1, we discussed the impact of implementing LU decomposition using BLAS based on Strassen's method. In this section we discuss other variations on linear equation solving where parallelism (or just speed) and numerical stability trade off.

Parallelism in LU decomposition (and others) is often attained by blocking. For example, if A is symmetric and positive definite, its Cholesky factorization $A = R^T R$ may be divided into three blocks as follows:

$$A = R^T R = \begin{bmatrix} R_{11}^T & 0 & 0 \\ R_{12}^T & R_{22}^T & 0 \\ R_{13}^T & R_{23}^T & R_{33}^T \end{bmatrix} \cdot \begin{bmatrix} R_{11} & R_{12} & R_{13} \\ 0 & R_{22} & R_{23} \\ 0 & 0 & R_{33} \end{bmatrix}$$

LAPACK uses the Level 3 BLAS which perform matrix multiplication and triangular system solving in its implementation of this algorithm [3]. On some machines, solving triangular systems is rather less efficient than matrix multiplication, so that an alternative algorithm using only matrix multiplication is preferred. This can be done provided we compute the following block decomposition instead of standard Cholesky:

$$A = LU = \begin{bmatrix} I & 0 & 0 \\ L_{21} & I & 0 \\ L_{31} & L_{32} & I \end{bmatrix} \cdot \begin{bmatrix} U_{11} & U_{12} & U_{13} \\ 0 & U_{22} & U_{23} \\ 0 & 0 & U_{33} \end{bmatrix}$$

Pivoting Method	Pivot Search Cost (serial)	Worst Pivot Growth	Average Pivot Growth
Complete	n^2	$O(n^{1+x})$	$n^{1/2}$
Partial	n	2^{n-1}	$n^{2/3}$
Pairwise	1	4^{n-1}	$O(n)$
Parallel	1	2^{n-1}	$e^{n/4\log n}$

Table 2: *Stability of various pivoting schemes in LU decomposition*

In [28] it is shown that using this block LU to solve the symmetric positive definite system $Ax = b$ yields a solution $\hat{x}$ satisfying $(A + \delta A)\hat{x} = b$, with $\|\delta A\| = O(\varepsilon)(\kappa(A))^{1/2}\|A\|$, where $\kappa(A) = \|A\| \cdot \|A^{-1}\|$ is the condition number. This contrasts with the standard backward stability analysis of Cholesky which yields $\|\delta A\| = O(\varepsilon)\|A\|$. So the final error bound from block LU is $O(\varepsilon)(\kappa(A))^{3/2}$, much bigger than $O(\varepsilon)\kappa(A)$ for Cholesky. This is the price paid in stability for speed up.

Another tradeoff occurs in the choice of pivoting strategy [77]. The standard pivot strategies are complete pivoting (where we search for the largest entry in the remaining submatrix), partial pivoting (the usual choice, where we only search the current column for the largest entry), pairwise pivoting [72] (where only rows n and $n - 1$ engage in pivoting and elimination, then rows $n - 1$ and $n - 2$ and so on up to the top) and parallel pivoting (where the remaining rows are grouped in pairs, and engage in pivoting and elimination simultaneously). Neither pairwise nor parallel pivoting require pivot search outside of two rows, but pairwise pivoting is inherently sequential in its access to rows, whereas parallel pivoting (as its name indicates) parallelizes easily. Table 2 summarizes the analysis in [77] of the speed and stability of these methods[1]. The point is that in the worst case partial, pairwise and parallel pivoting are all unstable, but on average only parallel pivoting is unstable. This is why we can using partial pivoting in practice: its worst case is very rare, but parallel pivoting is so often unstable as to be unusable. We note that an alternate kind of parallel pivoting discussed in [42] appears more stable, apparently because it eliminates entries in different columns as well as rows simultaneously. A final analysis of this problem remains to be done. We also note that, on many machines, the cost of partial pivoting is asymptotically negligible compared to the overall computation; the benefit of faster pivoting is solving smaller linear systems more efficiently.

We close by describing the fastest known parallel algorithm for solving $Ax = b$ [18]. It is also so numerically unstable as to be useless in practice. There are four steps:

1) Compute the powers of A (A^2, A^3, ... , A^{n-1}) by repeated squaring ($\log_2 n$ matrix multiplies of $\log_2 n$ steps each).
2) Compute the traces $s_i = \text{tr}(A^i)$ of the powers in $\log_2 n$ steps.
3) Solve the Newton identities for the coefficients a_i of the characteristic poly-

---

[1]Some table entries have been proven, some are empirical with some theoretical justification, and some are purely empirical. Alan Edelman believes the $n^{2/3}$ average case pivot growth for partial pivoting should really be $n^{1/2}$.

nomial; this is a triangular system of linear equations whose matrix entries and right hand side are known integers and the s_i (we can do this in $\log_2^2 n$ steps as described above).

4) Compute the inverse using Cayley-Hamilton Theorem (in about $\log_2 n$ steps).

The algorithm is so unstable as to lose all precision in inverting $3I$ in double precision, where I is the identity matrix of size 60 or larger.

2.4 THE SYMMETRIC EIGENVALUE PROBLEM AND SINGULAR VALUE DECOMPOSITION

The basic parallel methods available for dense matrices are summarized as follows. We assume the reader is acquainted with methods discussed in [47].

1. Jacobi, which operates on the original (dense) matrix.

2. Reduction from dense to tridiagonal (or bidiagonal) form, followed by

   (a) Bisection (possibly accelerated), followed by inverse iteration for eigenvectors (if desired).

   (b) Cuppen's divide and conquer method.

   (c) QR iteration (and variations).

Jacobi has been shown to be more stable than the other methods on the list, provided it is properly implemented, and only on some classes of matrices (essentially, those whose symmetric positive definite polar factor H can be diagonally scaled as $D \cdot H \cdot D$ to be well-conditioned [30, 71]; for the SVD we use the square of the polar factor). In particular, Jacobi is capable of computing tiny eigenvalues or singular values with higher relative accuracy than methods relying on tridiagonalization. So far the error analyses of these proper implementations have depended on their use of 2-by-2 rotations, as used in conventional Jacobi. Therefore, the inner loop of these algorithms perform operations on pairs of rows or columns, i.e. Level 1 BLAS [56]. On many machines, it is more efficient to do matrix-matrix operations like level 3 BLAS [31], so one is motivated to use *block Jacobi* instead, where groups of Jacobi rotations are accumulated into a single larger orthogonal matrix, and applied to the matrix with a single matrix-matrix multiplication [67, 68, 70]. It is unknown whether this blocking destroys the subtler error analyses in [30, 71]; it is easy to show that the conventional norm-based backward stability analysis of Jacobi is not changed by blocking.

Reduction from dense to tridiagonal form is eminently parallelizable too. Having reduced to tridiagonal form, we have several parallel methods from which to choose. Bisection and QR iteration can both be reformulated as three-term linear recurrences, and so implemented using parallel prefix in $O(\log_2 n)$ time as described in section 2.2 The stability is unproven. Experiments with bisection [76] are encouraging, but the only published analysis [20] is very pessimistic. Initial results on the dqds algorithm for the bidiagonal SVD, on the other hand, indicate stability may be preserved in some cases [63]. On the other hand, bisection can easily be parallelized by having different processors refine disjoint intervals, evaluating the Sturm sequence in the standard serial way. This involves much less communication, and is preferable in most circumstances, unless there is special support for parallel prefix.

Having used bisection to compute eigenvalues, we must use inverse iteration to compute eigenvectors. Simple inverse iteration is also easy to parallelize, with each processor independently computing the eigenvectors of the eigenvalues it owns. However, there is no guarantee of orthogonality of the computed eigenvectors, in contrast to QR iteration or Cuppen's method [53]. In particular, to achieve reasonable orthogonality one must reorthogonalize eigenvectors against those of nearby eigenvalues. This requires communication to identify nearby eigenvalues, and to transfer the eigenvectors [51]. In the serial implementation in [53], each iterate during inverse iteration is orthogonalized against previously computed eigenvectors; this is not parallelizable. The parallel version in [51] completes all the inverse iterations in parallel, and then uses modified Gram-Schmidt in a pipeline to perform the orthogonalization. To load balance, vector j was stored on processor $j \bmod p$ (p is the number of processors), and as a result reorthogonalization took a very small fraction of the total time; however, this may only have been effective because of the relatively slow floating point on the machine used (iPSC-1). In any event, the price of guaranteed orthogonality among the eigenvectors is reduced parallelism.

Cuppen's method has been analyzed by many people [19, 35, 73, 51, 54, 10, 48]. At the center of the algorithm is the solution of the *secular equation* $f(\lambda) = 0$, where f is a rational function in λ whose zeros are eigenvalues. This algorithm, while simple and attractive, proved hard to implement stably. The trouble was that to guarantee the computed eigenvectors were orthogonal, it appeared that the roots of $f(\lambda) = 0$ had to be computed in double the input precision [10, 73]. When the input is already in double precision (or whatever is the largest precision supported by the machine), then quadruple would be needed, which may be simulated using double provided double is accurate enough [22, 64]. So the availability of Cuppen's algorithm hinged on having sufficiently accurate floating point arithmetic [73, 10]. Recently, however, Gu and Eisenstat [48] have found a new way to implement this algorithm which makes extra precision unnecessary. Thus, even though carefully rounded floating point turned out not to be necessary to use Cuppen's algorithm, it took several years of research to discover this, so the price paid for poorly rounded floating point was several years of delay.

2.5 THE NONSYMMETRIC EIGENPROBLEM

Five kinds of parallel methods for the nonsymmetric eigenproblem have been investigated:

1. Hessenberg QR iteration [6, 79, 78, 21, 45, 37, 82, 81, 75],

2. Reduction to nonsymmetric tridiagonal form [46, 32, 43, 44],

3. Jacobi's method [38, 39, 74, 61, 69, 65, 80],

4. Divide and conquer based on Newton's method or homotopy continuation [16, 17, 83, 57, 58, 34]

5. Divide and conquer based on the matrix sign-function [59, 7, 60]

In contrast to the symmetric problem or SVD, no guaranteed stable and highly parallel algorithm for the nonsymmetric problem exists. Reduction to Hessenberg form (the prerequisite to methods (1) and (4) above) can be done efficiently [33, 36], but Hessenberg QR is hard to parallelize, and the other approaches are not guaranteed to converge and/or produce stable results. We summarize the tradeoffs among these methods here; for a more detailed survey, see [26].

Hessenberg QR is the serial method of choice for dense matrices. There have been a number of attempts to parallelize it, all of which maintain numerical stability since they continue to apply only orthogonal transformations to the original matrix. They instead sacrifice convergence rate or perform more flops in order to introduce higher level BLAS or parallelism. So far the parallelism has been too modest or too fine-grained to be very advantageous. In the paradigm described in the introduction, where we fall back on a slower but more stable algorithm if the fast one fails, Hessenberg QR can play the role of the stable algorithm.

Reduction to nonsymmetric tridiagonal form (followed by the tridiagonal LR algorithm) requires nonorthogonal transformations. The algorithm can break down, requiring restarting with different initial conditions [62]. Even if it does not break down, the nonorthogonal transformations required can be arbitrarily ill-conditioned, so sacrificing stability. By monitoring the condition number and restarting if it exceeds a threshold, some stability can be maintained at the cost of random running time. The more stability is demanded, the longer the running time may be, and there is no upper bound.

Jacobi's method can be implemented with orthogonal transformations only, maintaining numerical stability at the cost of linear convergence, or use nonorthogonal transformations which retain asymptotic quadratic convergence but can be arbitrarily ill-conditioned, and so possibly sacrifice stability. Orthogonal Jacobi could play the role of a slow but stable algorithm, but linear convergence makes it quite slow. The condition number of the transformation in nonorthogonal Jacobi could be monitored, and another scheme used if it is too large.

Divide and conquer using Newton or homotopy methods is applied to a Hessenberg matrix, setting the middle subdiagonal entry to zero, solving the two independent subproblems in parallel, and merging the answers of the subproblems using either Newton or a homotopy. There is parallelism in solving the independent subproblems, and in solving for the separate eigenvalues; these are the same sources of parallelism as in Cuppen's method. These methods can fail to be stable for the following reasons. Newton's method can fail to converge. Both Newton and homotopy may appear to converge to several copies of the same root without any easy way to tell if a root has been missed, or if the root really is multiple. To try to avoid this with homotopy methods requires communication to identify homotopy curves that are close together, and smaller step sizes to follow them more accurately. The subproblems produced by divide and conquer may potentially be more ill-conditioned than the original problem. See [52] for further discussion.

Divide and conquer using the matrix sign function (or a similar function) computes an orthogonal matrix $Q = [Q_1, Q_2]$ where Q_1 spans a right invariant subspace of A, and then divides the spectrum by forming $QAQ^T = \begin{bmatrix} A_{11} & A_{12} \\ 0 & A_{21} \end{bmatrix}$. To attain reasonable efficiency, Q_1 should have close to $n/2$ columns, where n is the dimension, or if the user only wants some eigenvalues, it should span the corresponding, or slightly larger, invariant subspace. One way to form Q is via the QR decomposition of the identity matrix plus the *matrix sign function* $s(A)$ of A, a function which leaves the eigenvectors alone but maps left half plane eigenvalues to -1 and right half plane eigenvalues to $+1$. A globally and asymptotically quadratically convergent iteration to compute $s(A)$ is $A_{i+1} = .5(A_i + A_i^{-1})$. This divides the spectrum into the left and right half planes; by applying this function to $A - \sigma I$ or $(A - \sigma I)^2$ or $e^{i\theta}A - \sigma I$, the spectrum can be separated along other lines.

This method can fail if the iteration fails to converge to an accurate enough approximation of $s(A)$. This will happen if some eigenvalue of A is too close to the imaginary axis (along which the iteration behaves chaotically). A symptom of this may be an intermediate A_i which is very ill-conditioned, so that A_i^{-1} is very inaccurate. It may require user input to help select the correct spectral dividing line. It can monitor its own accuracy by keeping track of the norm of the (2,1) block of QAQ^T; since the method only applies orthogonal transformations to A, it will be stable if this (2,1) block is small.

We close with some comments on finding eigenvectors, given accurate approximate eigenvalues; this is done if only a few eigenvectors are desired. The standard method is *inverse iteration*, or solving $(A - \lambda)x_{i+1} = \alpha_i x_i$ until x_i converges to an eigenvector; α_i is chosen to keep $\|x_{i+1}\| = 1$. This involves triangular system solving with a very ill-conditioned matrix, the more so to the extent that λ is an accurate eigenvalue. This ill-conditioning makes overflow a reasonable possibility, even though we only want the scaled unit vector at the end. This means the code is to compute the answer despite possible overflow, since this overflow does not mean that the eigenvector is ill-posed or even ill-conditioned. To do this portably currently requires a "paranoid" coding style, with testing and scaling in the inner loop of the triangular solve [2], making it impossible to use machine optimized BLAS. If one could defer the handling of overflow exceptions, it would be possible to run the fast BLAS, and only redo the computation with relatively slow scaling when necessary. This is an example of the paradigm of the introduction. IEEE standard floating point arithmetic [5] provides this facility in principle. However, if exception handling is too expensive (on the DEC α chip, ∞ arithmetic requires a trap into the operating system, which is quite slow), overflow can cause a slowdown of several orders of magnitude.

For the generalized nonsymmetric eigenproblem $A - \lambda B$ we do not even know how to perform generalized Hessenberg reduction using more than the Level 1 BLAS. The sign-function and related techniques [60, 7] promise to be helpful here.

3 Recommendations for floating point arithmetic

We summarize the recommendations we have made in previous sections regarding floating point arithmetic support to mitigate the tradeoff between parallelism (or speed) and stability: accurate rounding, support for higher precision, and efficient exception handling. The IEEE floating point standard [5], *efficiently implemented*, is a good model. We emphasize the efficiency of implementation because if it is very expensive to exercise the features we need, it defeats the purpose of using them to accelerate computation.

Accurate rounding attenuates or eliminates roundoff accumulation in long sums as described in section 1.1 It also permits us to simulate higher precision cheaply, which often makes it easier to design stable algorithms quickly (even though a stable algorithm which does not rely on higher precision may exist, it may take a while to discover). This was the case for Cuppen's method (section 2.4), and also for many of the routines for 2-by-2 and 4-by-4 matrix problems in the inner loops of various LAPACK routines, such as slasv2, which computes the SVD of a 2-by-2 triangular matrix [3, 29]. Higher precision also makes it possible to extend the life of codes designed to work on smaller problems, as they are scaled to work on larger ones with larger condition numbers (section 1.2), or with more random instabilities (section 1.3). It is important that the extra precision be as accurate as the basic precision, because otherwise promoting a code to higher precision can introduce bugs where none were before. A simple example is that $\arccos(x/(x^2 + y^2)^{1/2})$ can fail because the argument of arccos can exceed 1 if rounding is inaccurate in division or square root [15]. Extra range and precision are very useful, since they permit us us to forego some testing and scaling to avoid over/underflow in common computations such as $\sqrt{\sum_i x_i^2}$.

Efficient exception handling permits us to run fast "risky" algorithms which usually work, without fear of having program execution terminated. Indeed, in some cases such as condition estimation, overflow permits us to finish early (in this case overflow implies that 0 is an excellent approximate reciprocal condition number). In particular, it lets us use optimized BLAS, thereby taking advantage of the manufacturer's effort in writing them (see section 2.5). In analogy to the argument for using RISC ("reduced instruction set computers"), we want algorithms where the most common case — no exceptions — runs as quickly as possible.

This is not useful if the price of exception handling is too high; we need to be able to run with ∞ and NaN (Not a Number) arithmetic at nearly full floating point speed. The reason is that once created, an ∞ or NaN propagates through the computation, creating many more ∞'s or NaN's. This means, for example, that the DEC α implementation of this arithmetic, which uses traps to the operating system, is too unacceptably slow to be useful. The LAPACK 2 project will produce codes assuming reasonably efficient exception handling, since this is the most common kind of implementation [4].

Acknowledgements

The author acknowledges the support of NSF grant ASC-9005933, NSF PYI grant CCR-9196022, DARPA grant DAAL03-91-C-0047 via subcontract ORA-4466.02 from the University of Tennessee, and DARPA grant DM28E04120 via subcontract W-31-109-ENG-38 from Argonne National Laboratory. He also thanks W. Kahan for numerous comments on an earlier draft of this paper.

References

[1] A. Aho, J. Hopcroft, and J Ullman. *The design and analysis of computer algorithms.* Addison-Wesley, 1974.

[2] E. Anderson. Robust triangular solves for use in condition estimation. Computer Science Dept. Technical Report CS-91-142, University of Tennessee, Knoxville, 1991. (LAPACK Working Note #36).

[3] E. Anderson, Z. Bai, C. Bischof, J. Demmel, J. Dongarra, J. Du Croz, A. Greenbaum, S. Hammarling, A. McKenney, S. Ostrouchov, and D. Sorensen. *LAPACK Users' Guide, Release 1.0.* SIAM, Philadelphia, 1992.

[4] E. Anderson, C. Bischof, J. Demmel, J. Dongarra, J. Du Croz, S. Hammarling, and W. Kahan. Prospectus for an extension to LAPACK: a portable linear algebra library for high-performance computers. Computer Science Dept. Technical Report CS-90-118, University of Tennessee, Knoxville, 1990. (LAPACK Working Note #26).

[5] ANSI/IEEE, New York. *IEEE Standard for Binary Floating Point Arithmetic*, Std 754-1985 edition, 1985.

[6] Z. Bai and J. Demmel. On a block implementation of Hessenberg multishift QR iteration. *International Journal of High Speed Computing*, 1(1), pp 97–112, 1989. (also LAPACK Working Note #8).

[7] Z. Bai and J. Demmel. Design of a parallel nonsymmetric eigenroutine toolbox. Computer Science Dept. preprint, University of California, Berkeley, CA, 1992.

[8] D. H. Bailey. Extra high speed matrix multiplication on the Cray-2. *SIAM J. Sci. Stat. Comput.*, 9, pp 603–607, 1988.

[9] D. H. Bailey, K. Lee, and H. D. Simon. Using Strassen's algorithm to accelerate the solution of linear systems. *J. Supercomputing*, 4, pp 97–371, 1991.

[10] J. Barlow. Error analysis of update methods for the symmetric eigenvalue problem. to appear in *SIAM J. Mat. Anal. Appl.* Tech Reprot CS-91-21, Computer Science Department, Penn State University, August 1991.

[11] D. Bini and D. Lotti. Stability of fast algorithms for matrix multiplication. *Num. Math.*, 36, pp 63–72, 1980.

[12] P. Bjorstad. *Numerical solution of the biharmonic equation.* PhD thesis, Stanford University, 1980.

[13] R. P. Brent. Algorithms for matrix multiplication. Computer Science Dept. Report CS 157, Stanford University, 1970.

[14] R. P. Brent. Error analysis of algorithms for matrix multiplication and triangular decomposition using Winograd's identity. *Num. Math*, 16, pp 145–156, 1970.

[15] R. Carter. Cray Y-MP floating point and Choleksy decomposition. to appear in *Int. J. High Speed Computing*, 1992.

[16] M. Chu. A note on the homotopy method for linear algebraic eigenvalue problems. *Lin. Alg. Appl*, **105**, pp 225–236, 1988.

[17] M. Chu, T.-Y. Li, and T. Sauer. Homotopy method for general λ-matrix problems. *SIAM J. Mat. Anal. Appl.*, **9**(4), pp 528–536, 1988.

[18] L. Csanky. Fast parallel matrix inversion algorithms. *SIAM J. Comput.*, **5**, pp 618–623, 1977.

[19] J.J.M. Cuppen. A divide and conquer method for the symmetric tridiagonal eigenproblem. *Numer. Math.*, **36**, pp 177–195, 1981.

[20] Kuck D. and A. Sameh. A parallel QR algorithm for symmetric tridiagonal matrices. *IEEE Trans. Computers*, **C-26**(2), 1977.

[21] G. Davis, R. Funderlic, and G. Geist. A hypercubde implementation of the implicit double shift QR algorithm. In *Hypercube Multiprocessors 1987*, pp 619–626, Philadelphia, PA, 1987. SIAM.

[22] T. Dekker. A floating point technique for extending the available precision. *Num. Math.*, **18**, pp 224–242, 1971.

[23] J. Demmel. On condition numbers and the distance to the nearest ill-posed problem. *Num. Math.*, **51**(3), pp 251–289, 1987.

[24] J. Demmel. The probability that a numerical analysis problem is difficult. *Math. Comput.*, **50**(182), pp 449–480, 1988.

[25] J. Demmel. The inherent inaccuracy of implicit tridiagonal QR. Technical report, IMA, University of Minnesota, 1992.

[26] J. Demmel, M. Heath, and H. van der Vorst. Parallel numerical linear algebra. In : A. Iserles (Ed.), *Acta Numerica, volume 2*. Cambridge University Press, 1993 (to appear).

[27] J. Demmel and N. J. Higham. Stability of block algorithms with fast Level 3 BLAS. to appear in *ACM Trans. Math. Soft.*

[28] J. Demmel, N. J. Higham, and R. Schreiber. Block LU factorization. in preparation.

[29] J. Demmel and W. Kahan. Accurate singular values of bidiagonal matrices. *SIAM J. Sci. Stat. Comput.*, **11**(5), pp 873–912, 1990.

[30] J. Demmel and K. Veselić. Jacobi's method is more accurate than QR. Computer Science Dept. Technical Report 468, Courant Institute, New York, NY, October 1989. (also LAPACK Working Note #15), to appear in *SIAM J. Mat. Anal. Appl.*

[31] J. Dongarra, J. Du Croz, I. Duff, and S. Hammarling. A set of Level 3 Basic Linear Algebra Subprograms. *ACM Trans. Math. Soft.*, **16**(1), pp 1–17, 1990.

[32] J. Dongarra, G. A. Geist, and C. Romine. Computing the eigenvalues and eigenvectors of a general matrix by reduction to tridiagonal form. Technical Report ONRL/TM-11669, Oak Ridge National Laboratory, 1990. to appear in *ACM TOMS*.

[33] J. Dongarra, S. Hammarling, and D. Sorensen. Block reduction of matrices to condensed forms for eigenvalue computations. *JCAM*, **27**, pp 215–227, 1989. (LAPACK Working Note #2).

[34] J. Dongarra and M. Sidani. A parallel algorithm for the non-symmetric eigenvalue problem. Computer Science Dept. Technical Report CS-91-137, University of Tennessee, Knoxville, TN, 1991.

[35] J. Dongarra and D. Sorensen. A fully parallel algorithm for the symmetric eigenproblem. *SIAM J. Sci. Stat. Comput.*, **8**(2), pp 139–154, 1987.

[36] J. Dongarra and R. van de Geign. Reduction to condensed form for the eigenvalue problem on distributed memory computers. Computer Science Dept. Technical Report CS-91-130, University of Tennessee, Knoxville, 1991. (LAPACK Working Note #30), to appear in *Parallel Computing*.

[37] A. Dubrulle. The multishift QR algorithm : is it worth the trouble? Palo Alto Scientific Center Report G320-3558x, IBM Corp., 1530 Page Mill Road, Palo Alto, CA 94304, 1991.

[38] P. Eberlein. A Jacobi method for the automatic computation of eigenvalues and eigenvectors of an arbitrary matrix. *J. SIAM*, **10**, pp 74–88, 1962.

[39] P. Eberlein. On the Schur decomposition of a matrix for parallel computation. *IEEE Trans. Comput.*, **36**, pp 167–174, 1987.

[40] A. Edelman. Eigenvalues and condition numbers of random matrices. *SIAM J. on Mat. Anal. Appl.*, **9**(4), pp 543–560, 1988.

[41] A. Edelman. On the distribution of a scaled condition number. *Math. Comp.*, **58**(197), pp 185–190, 1992.

[42] K. A. Gallivan, R. J. Plemmons, and A. H. Sameh. Parallel algorithms for dense linear algebra computations. *SIAM Review*, **32**, pp 54–135, 1990.

[43] G. A. Geist. Parallel tridiagonalization of a general matrix using distributed memory multiprocessors. *Proceedings of the Fourth SIAM Conference on Parallel Processing for Scientific Computing*, pp 29–35, SIAM, Philadelphia, PA, 1990.

[44] G. A. Geist. Reduction of a general matrix to tridiagonal form. *SIAM J. Mat. Anal. Appl.*, **12**(2), pp 362–373, 1991.

66

[45] G. A. Geist and G. J. Davis. Finding eigenvalues and eigenvectors of unsymmetric matrices using a distributed memory multiprocessor. *Parallel Computing*, **13**(2), pp 199–209, 1990.

[46] G. A. Geist, A. Lu, and E. Wachspress. Stabilized reduction of an arbitrary matrix to tridiagonal form. Technical Report ONRL/TM-11089, Oak Ridge National Laboratory, 1989.

[47] G. Golub and C. Van Loan. *Matrix Computations*. Johns Hopkins University Press, Baltimore, MD, 2nd edition, 1989.

[48] Ming Gu and S. Eisenstat. A stable and efficient algorithm for the rank-1 modification of the symmetric eigenproblem. Computer Science Dept. Report YALEU/DCS/RR-916, Yale University, August 1992.

[49] N. J. Higham. Exploiting fast matrix multiplication within the Level 3 BLAS. *ACM Trans. Math. Soft.*, **16**, pp 352–368, 1990.

[50] IBM. *Engineering and Scientific Subroutine Library, Guide and Reference, Release 3, Program 5668-863*, 4 edition, 1988.

[51] I. Ipsen and E. Jessup. Solving the symmetric tridiagonal eigenvalue problem on the hypercube. *SIAM J. Sci. Stat. Comput.*, **11**(2), pp 203–230, 1990.

[52] E. Jessup. A case against a divide and conquer approach to the nonsymmetric eigenproblem. Technical Report ONRL/TM-11903, Oak Ridge National Laboratory, 1991.

[53] E. Jessup and I. Ipsen. Improving the accuracy of inverse iteration. *SIAM J. Sci. Stat. Comput.*, **13**(2), pp 550–572, 1992.

[54] E. Jessup and D Sorensen. A divide and conquer algorithm for computing the singular value decomposition of a matrix. *Proceedings of the Third SIAM Conference on Parallel Processing for Scientific Computing*, pp 61–66, SIAM, Philadelphia, PA, 1989.

[55] W. Kahan. How Cray's arithmetic hurts scientific computing. Presented to Cray User Group Meeting, Toronto, April 10, 1991.

[56] C. Lawson, R. Hanson, D. Kincaid, and F. Krogh. Basic linear algebra subprograms for fortran usage. *ACM Trans. Math. Soft.*, **5**, pp 308–323, 1979.

[57] T.-Y. Li and Z. Zeng. Homotopy-determinant algorithm for solving nonsymmetric eigenvalue problems. to appear in *Math. Comp.*

[58] T.-Y. Li, Z. Zeng, and L. Cong. Solving eigenvalue problems of nonsymmetric matrices with real homotopies. *SIAM J. Num. Anal.*, **29**(1), pp 229–248, 1992.

[59] C-C. Lin and E. Zmijewski. A parallel algorithm for computing the eigenvalues of an unsymmetric matrix on an SIMD mesh of processors. Department of Computer Science TRCS 91-15, University of California, Santa Barbara, CA, July 1991.

[60] A. N. Malyshev. Parallel aspects of some spectral problems in linear algebra. Dept. of Numerical Mathematics Report NM-R9113, Centre for Mathematics and Computer Science, Amsterdam, July 1991.

[61] M.H.C. Pardekooper. A quadratically convergent parallel Jacobi process for diagonally dominant matrices with distinct eigenvalues. *J. Comput. Appl. Math.*, **27**, pp 3–16, 1989.

[62] B. Parlett. Reduction to tridiagonal form and minimal realizations. *SIAM J. Mat. Anal. Appl.*, **13**(2), pp 567–593, 1992.

[63] B. Parlett and V. Fernando. Accurate singular values and differential QD algorithms. Math Department PAM-554, University of California, Berkeley, CA, July 1992.

[64] D. Priest. Algorithms for arbitrary precision floating point arithmetic. In : P. Kornerup and D. Matula (Eds.), *Proceedings of the 10th Symposium on Computer Arithmetic*, pp 132–145, Grenoble, France, June 26-28 1991. IEEE Computer Society Press, 1991.

[65] A. Sameh. On Jacobi and Jacobi-like algorithms for a parallel computer. *Math. Comp.*, **25**, pp 579–590, 1971.

[66] A. Sameh and R. Brent. Solving trangular systems on a parallel computer. *SIAM J. Num. Anal.*, **14**, pp 1101–1113, 1977.

[67] R. Schreiber. Solving eigenvalue and singular value problems on an undersized systolic array. *SIAM J. Sci. Stat. Comput.*, **7**, pp 441–451, 1986. first block Jacobi reference?

[68] R. Schreiber. Block algorithms for parallel machines. In : M. Schultz (Ed.), *Numerical Algorithms for Modern Parallel Computer Architectures*, IMA Volumes in Mathmatics and its Applications, v. 13, Springer-Verlag, 1988.

[69] G. Shroff. A parallel algorithm for the eigenvalues and eigenvectors of a general complex matrix. *Num. Math.*, **58**, pp 779–805, 1991.

[70] G. Shroff and R. Schreiber. On the convergence of the cyclic Jacobi method for parallel block orderings. *SIAM J. Mat. Anal. Appl.*, **10**, pp 326–346, 1989.

[71] I. Slapničar. *Accurate symmetric eigenreduction by a Jacobi method*. PhD thesis, Fernuniversität - Hagen, Hagen, Germany, 1992.

[72] D. Sorensen. Analysis of pairwise pivoting in Gaussian elimination. *IEEE Trans. Comput.*, **34**, pp 274–278, 1984.

[73] D. Sorensen and P. Tang. On the orthogonality of eigenvectors computed by divide-and-conquer techniques. *SIAM J. Num. Anal.*, **28**(6), pp 1752–1775, 1991.

[74] G. W. Stewart. A Jacobi-like algorithm for computing the Schur decomposition of a non-Hermitian matrix. *SIAM J. Sci. Stat. Comput.*, **6**, pp 853–864, 1985.

[75] G. W. Stewart. A parallel implementation of the QR algorithm. *Parallel Computing*, **5**, pp 187–196, 1987.

[76] P. Swarztrauber. A parallel algorithm for computing the eigenvalues of a symmetric tridiagonal matrix. To appear in *Math. Comp.*, 1992.

[77] L. Trefethen and R. Schreiber. Average case analysis of Gaussian elimination. *SIAM J. Mat. Anal. Appl.*, 11(3), pp 335–360, 1990.

[78] R. van de Geijn and D. Hudson. Efficient parallel implementation of the nonsymmetric QR algorithm. In : J. Gustafson (Ed.), *Hypercube Concurrent Computers and Applications*, ACM, 1989.

[79] Robert van de Geijn. *Implementing the QR Algorithm on an Array of Processors*. PhD thesis, University of Maryland, College Park, August 1987. Computer Science Department Report TR-1897.

[80] K. Veselić. A quadratically convergent Jacobi-like method for real matrices with complex conjugate eigenvalues. *Num. Math.*, 33, pp 425–435, 1979.

[81] D. Watkins. Shifting strategies for the parallel QR algorithm. Dept. of pure and applied math. report, Washington State Univ., Pullman, WA, 1992.

[82] D. Watkins and L. Elsner. Convergence of algorithms of decomposition type for the eigenvalue problem. *Lin. Alg. Appl.*, 143, pp 19–47, 1991.

[83] Zhonggang Zeng. *Homotopy-determinant algorithm for solving matrix eigenvalue problems and its parallelizations*. PhD thesis, Michigan State University, 1991.

SUBBAND FILTERING: CORDIC MODULATION AND SYSTOLIC QUADRATURE MIRROR FILTER TREE

E.F. DEPRETTERE
Department of Electrical Engineering
Delft University of Technology
2628 CD Delft, The Netherlands
ed@dutentd.et.tudelft.nl

ABSTRACT. The decomposition (analysis) of a finite-energy signal into a relatively small number of mutually independent signals which allows reconstruction (synthesis) of the original signal is called subband filtering. Subbands can be processed in parallel or recursively. In the latter case, one obtains a so-called quadrature mirror filter bank tree. The former case leads to cosine-modulated filter banks. The paper presents a Cordic based cosine-modulated bank and a systolic algorithm for quadrature mirror tree filtering.

KEYWORDS. Subband filtering, parallel algorithms, real-time signal processing.

Introduction

A conceptual scheme for uniform subband signal analysis/synthesis is shown in figure 1, in which $u(n)$ is a finite-energy signal, bandwidth B, that is input to a set of M filters (figure 1.a) with rational transfer functions $H_i(z)$, $i = 0, \ldots, M-1$, with ideally non-overlapping bandwidths $B_i = B/M$, and

$$\sum_{i=0}^{M-1} |H_i(e^{j\theta})|^2 = 1$$

The boxes $\downarrow M$ in figure 1.b are decimators [1], taking every M-th sample from the signals $u_i(n)$, reducing the filter outputs $u_i(n)$ to their Nyquist sampled versions $x_i(n)$:

$$x_i(n) = u_i(Mn) \qquad i = 0, \ldots, M-1$$

---

[1]Basic concepts and definitions can be found in e.g., [1]

69

M. S. Moonen et al. (eds.), Linear Algebra for Large Scale and Real-Time Applications, 69–89.

or, taking Fourier transforms:

$$X_i(e^{j\theta}) = \frac{1}{M} \sum_{k=0}^{M-1} U_i(e^{j\theta/M} W_M^k) \qquad i = 0, \ldots, M - 1$$

The term $U_i(e^{j\theta/M})$ $(k = 0)$ is the M-scale stretched version of $U_i(e^{j\theta})$. The terms with $k > 0$ are uniformly shifted versions of $U_i(e^{j\theta/M})$. They do overlap, since rational functions can only approximate ideal rectangular windows. In this paper, the block in figure 1.c will

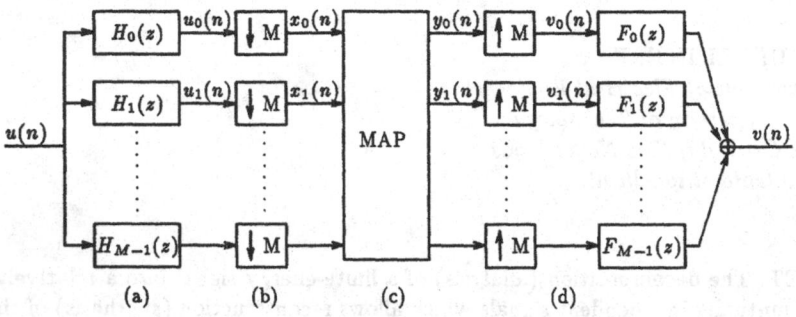

Figure 1: *Conceptual scheme of a subband filtering system*

be assumed throughout to be the identity map, that is, $y_i(n) \equiv x_i(n)$. The boxes $\uparrow M$ insert $M - 1$ zero valued samples in between every 2 samples of the received signals $y_i(n)$:

$$v_i(n) = \begin{cases} y_i(\frac{n}{M}) & \text{if } n \text{ is a multiple of } M \\ 0 & \text{otherwise} \end{cases} \qquad i = 0, \ldots, M - 1$$

or, taking Fourier transforms:

$$V_i(e^{j\theta}) = Y_i(e^{jM\theta}) \qquad i = 0, \ldots, M - 1$$

Thus, $V_i(e^{j\theta})$ is an M-scale compressed and M-fold copied version of $Y_i(e^{j\theta})$. The boxes $F_i(z)$ are copy suppressing interpolation filters, with rational transfer functions $F_i(z)$, which should be so chosen that the summed output signal $v(n)$ is a reconstruction of the input signal $u(n)$ in the sense that

$$|V(e^{j\theta})| = |U(e^{j\theta})|$$

In case when the subband filtering system is required to be built on real transfer functions, 2 approaches can be taken. The first one leads to the so-called *cosine-modulated filter banks* and the second one leads to the so-called *quadrature mirror filter tree banks*. The next 2 sections are devoted to cosine modulated filter banks and quadrature mirror filter tree banks, respectively.

1   Cosine-modulated filter banks

In cosine-modulated filter banks, the subband filters $H_i(z)$, $i = 0\dots M-1$, are modulated versions of a real-coefficient, low-pass prototype filter $H(z)$, impulse response $h(n)$, approximating the ideal window:

$$H(e^{j\theta}) = \begin{cases} 1 & \text{if} \quad |\theta| \leq \pi/2M \\ 0 & \text{if} \quad \pi/2M < |\theta| \leq \pi \end{cases}$$

Taking $H_i(z) = \alpha H(z\, e^{(2i+1)\pi/2M}) + \alpha^* H(z\, e^{-(2i+1)\pi/2M})$, where the superscript $*$ denotes complex conjugation and $|\alpha| = 1$, the condition that the impulse responses $h_i(n)$ should be real leads to the following general expression for cosine modulation [2]:

$$h_i(n) = 2h(n)\cos(f(i)(n+c) + \varphi_i)$$

where $c, \varphi_i \in \mathcal{R}$ and $f(i) = (2i+1)\pi/2M$. The impulse response $h(n)$ may be interpreted as the interlacing of $2M$ sequences which are up sampled versions of $2M$ sequences $e_n(m)$ which are obtained by down sampling $2M$ single step shifted versions of $h(n)$:

$$E_n(z) = \sum_{m=0}^{\infty} h(n+2mM)z^{-m} \qquad n = 0\dots 2M-1$$

and

$$H(z) = \sum_{n=0}^{2M-1} E_n(z^{2M})z^{-n}$$

The functions $E_n(z)$ are called the *polyphase components* of the function $H(z)$. The transfer functions $H_i(z)$ can now all be expressed in terms of the polyphase components of the prototype filter [2]:

$$H_i(z) = 2\sum_{n=0}^{2M-1} \cos((2i+1)(n+c)\pi/2M + \varphi_i)E_n(-z^{2M})z^{-n}$$

The resulting block diagram for the analysis part is shown in figure 2. A similar block diagram for the synthesis part is shown in figure 3. The block C_{M*2M} is an $M*2M$ matrix map with entries $[C_{i,n}]$ given by:

$$[C_{i,n}] = 2\cos((2i+1)(n+c)\pi/2M + \varphi_i)$$

and satisfying $C^t C = I_M$, where the superscript t denotes transpose and I_M is the identity matrix of order M. Notice that the down samplers and up samplers now appear at the input branch and output summer respectively. The consequence of this operator shift is that the power of z in the argument of the filters has droped from $2M$ to 2.

In the next theorem, it is assumed that $c = -rM + 0.5$, r integer, and $\varphi_i = (-1)^i\pi/4$.

---

[2]Notice that $\cos(f(i)(n+c+2Mm) + \varphi_i) = (-1)^m \cos(f(i)(n+c) + \varphi_i)$

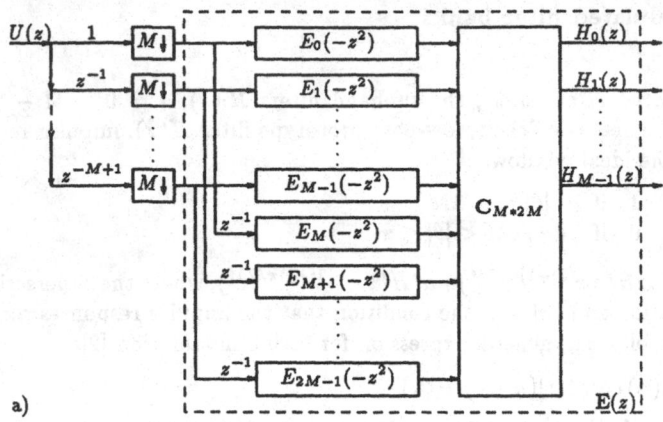

Figure 2: *Analysis polyphase filter bank*

Theorem.
Let $\mathbf{E}_n$ be the 2×2 matrices ($0 \leq n \leq M/2 - 1$):

$$\mathbf{E}_n = \sqrt{2M} \left[\begin{array}{cc} E_n(z) & -E_{M-1-n}(z) \\ E_{n+M}(z) & E_{2M-1-n}(z) \end{array} \right]$$

If $\mathbf{E}_n$ is 2×2 all-pass, then a) the polyphase matrix $\mathbf{E}(z)$ (figure 2) is paraunitary [3], and b) for equal analysis and synthesis filters, that is, $D_{M*2M} = C_{M*2M}$ and $F_n(-z^2) = E_n(-z^2)$, the analysis/synthesis bank (cascade of figures 2 and 3) is transparent in the sense $|V(e^{j\theta})| = |U(e^{j\theta})|$.
Proof: See proof in [3].
Both the all-pass functions $\mathbf{E}_n$ and the matrix C_{M*2M} can be implemented using Givens rotations as computational primitives.
Example ($M = 2$)
For cosine-modulation with $r = 0$ we have:

$$\left[\begin{array}{c} H_0 \\ H_1 \end{array} \right] = \left[\begin{array}{cc} \cos(\pi/8) & -\sin(\pi/8) \\ \cos(3\pi/8) & \sin(3\pi/8) \end{array} \right] \left[\begin{array}{c} e_0 \\ e_1 \end{array} \right] = \left[\begin{array}{cc} \cos(\pi/8) & -\sin(\pi/8) \\ \sin(\pi/8) & \cos(\pi/8) \end{array} \right] \left[\begin{array}{c} e_0 \\ e_1 \end{array} \right]$$

where the following properties have been used (for even M):

$$\cos((2(M - 1 - i) + 1)(n + c)\pi/2M + \varphi_i) = (-1)^n \sin((2i + 1)(n + c)\pi/2M + \varphi_i)$$

$$\sin((2(M - 1 - i) + 1)(n + c)\pi/2M + \varphi_i) = (-1)^n \cos((2i + 1)(n + c)\pi/2M + \varphi_i)$$

and e_0 and e_1 follow from:

$$E_n + (-1)^{r+1} E_{M-1-n} \quad \rightarrow \quad e_n \quad 0 \leq n \leq M/2 - 1$$

$$E_{2n} + (-1)^r E_{M-1-2n} \quad \rightarrow \quad e_n \quad M/2 \leq n \leq M - 1$$

---

[3] A square real-coefficient matrix $\mathbf{A}(z)$ is said to be paraunitary if $\mathbf{A}(z) \times \mathbf{A}^t(1/z) = I = \mathbf{A}^t(1/z) \times \mathbf{A}(z)$

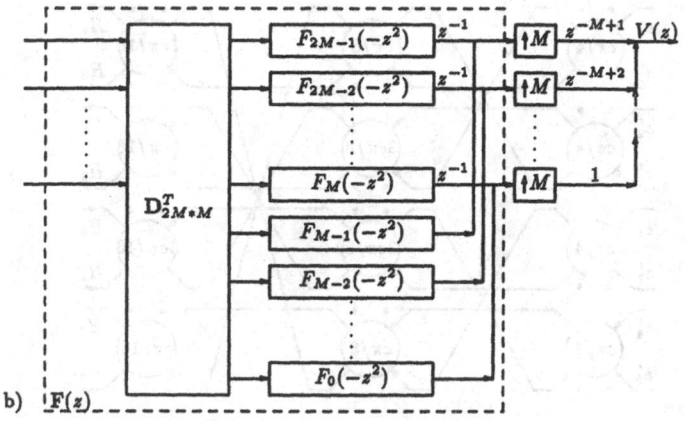

Figure 3: *Synthesis polyphase filter bank*

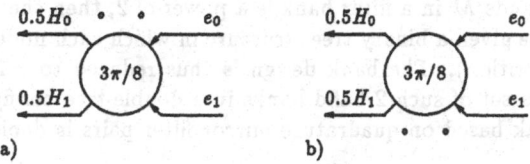

Figure 4: *Cordic implementation for $M = 2$*

The second 2×2 matrix in the above matrix equation is indeed an elementary (Givens) rotation, which can be efficiently implemented using Cordic arithmetic (figure 4a). For sine-modulation we will have the rotation matrix with row permutation, which is indicated by a double dot below the Cordic symbol (figure 4b).

End example ($M = 2$)

Here, we shall not derive the *Fast CM Transform* for general M, but leave cosine-modulated filter banks with the Transform for $M = 8$, figure 5, where:

$$e'_0 = e_0 + (-1)^r e_1 + e_2 + (-1)^r e_3 \qquad e'_4 = e_0 - (-1)^r e_1 + e_2 - (-1)^r e_3$$
$$e'_1 = e_4 + (-1)^r e_5 + e_6 + (-1)^r e_7 \qquad e'_5 = e_4 - (-1)^r e_5 + e_6 - (-1)^r e_7$$
$$e'_2 = e_0 + (-1)^r e_1 - e_2 - (-1)^r e_3 \qquad e'_6 = e_0 - (-1)^r e_1 - e_2 + (-1)^r e_3$$
$$e'_3 = e_4 + (-1)^r e_5 - e_6 - (-1)^r e_7 \qquad e'_7 = e_4 - (-1)^r e_5 - e_6 + (-1)^r e_7$$

See [3] for more details. The results on cosine-modulated analysis/synthesis banks summerized is this section include the special case when the reference filter - hence all filters - are polynomial filters. Part of the results for the polynomial filter case have been presented in [4].

74

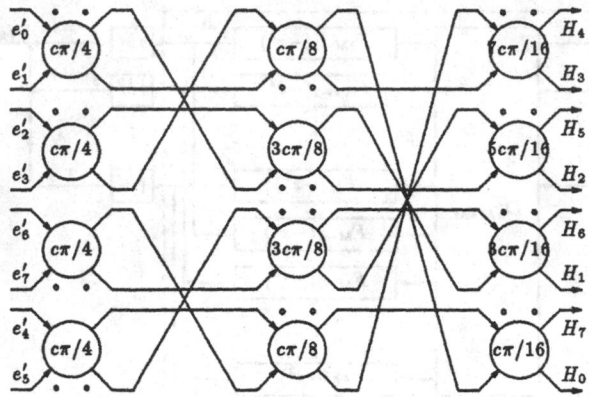

Figure 5: *Cordic implementation for M = 8*

2   QMF tree banks

If the number of subbands M in a filter bank is a power of 2, then the analysis (synthesis) part of the bank can be given a binary tree structure in which each node performs a 2-band decomposition (composition). The bank design is thus reduced to a 2-band bank design and the placement of a set of such 2-band banks in a double-tree configuration. A 2-band analysis/synthesis bank based on quadrature mirror filter pairs is depicted in figure 6. In

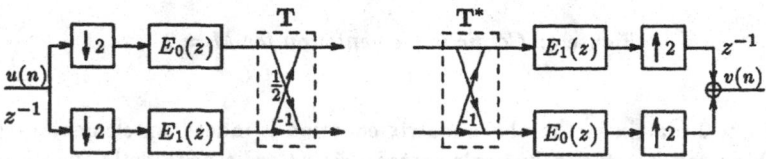

Figure 6: *Practical 2-band QMF bank.*

this figure, the boxes $E_0(z)$ and $E_1(z)$ are all-pass functions which originate from polyphase component functions $E_0(z^2)$ and $z^{-1}E_1(z^2)$ which emerge after shifting the 2 decimators to the right of the all-pass boxes. The derivation of these functions is as follows. There exists [5] a 2×2 all-pass function [4] [5]

$$\mathbf{E} = \frac{1}{g(z)} \left[\begin{array}{cc} h(z) & \sigma f(z) \\ z^p f_*(z) & z^p h_*(z) \end{array} \right]$$

where $|\sigma| = 1$, p is the degree of the polynomial $g(z)$, and the degrees of the polynomials $h(z)$ and $f(z)$ are at most p, and such that the function $H_0(z) = h(z)/g(z)$ has

---

[4] A square matrix $\mathbf{A}(z)$ is all-pass if it is unitary in $|z| = 1$ and contractive in $|z| > 1$.

[5] If $p(z)$ is a real-coefficient polynomial, then $p_*(z) = p(1/z)$.

all its zeros on the unit circle, has low-pass power spectral density function with cut-off frequency $\pi/2$, and the function $H_1(z) = \sigma f(z)/g(z)$ satisfies $H_1(z) = H_0(-z)$. These 2 functions are called quadrature mirror pairs. Notice that for quadrature mirror filters (QMF), $z^p f_*(z) = \sigma f(z)$ and $z^p h_*(z) = h(z)$. From the above properties, it follows that
$$|H_0(e^{j\theta})|^2 + |H_1(e^{j\theta})|^2 = 1 \quad \text{and} \quad H_0(e^{j\theta})H_1(e^{-j\theta}) + H_1(e^{j\theta})H_0(e^{-j\theta}) = 0$$

Adding and subtracting these 2 relations, we have :
$$[H_0(e^{j\theta}) + H_1(e^{j\theta})][H_0(e^{-j\theta}) + H_1(e^{-j\theta})] = 1 = [H_0(e^{j\theta}) - H_1(e^{j\theta})][H_0(e^{-j\theta}) - H_1(e^{-j\theta})]$$
whence the 2 functions $A_0(z) = H_0(z) + H_1(z)$ and $A_1(z) = H_0(z) - H_1(z)$ are stable all-pass functions which are even and odd in z respectively. We can thus write $A_0(z) = E_0(z^2)$ and $A_1(z) = z^{-1}E_1(z^2)$, where $E_0(z)$ and $E_1(z)$ are the transfer functions of the filter boxes in figure 6. The output signal $v(n)$ and the input signal $u(n)$ are related as $|V(e^{j\theta}| = |U(e^{j\theta})|$. To show this, let $u(n) = e^{jn\theta}$. We have to show that $v(n) = e^{j(n\theta + \varphi(\theta))}$, where $\varphi(\theta)$ is some phase function. Now we have $u_i(n) = H_i(e^{j\theta})e^{jn\theta}$, i=0,1. After down-sampling and up sampling, we have $v_i(n) = \frac{1}{2}(1 + (-1)^n)u_i(n)$, or equivalently $v_i(n) = \frac{1}{2}H_i(e^{j\theta})(e^{jn\theta} + e^{jn(\theta + \pi)})$, for i=0,1.
And finally,

$$v(n) = \quad \frac{1}{2}A_0(e^{j\theta})A_1(e^{j\theta})e^{jn\theta} + \frac{1}{2}A_0(e^{j\theta})A_1(e^{j(\theta+\pi)})e^{jn(\theta+\pi)} + $$
$$\frac{1}{2}A_0(e^{j\theta})A_1(e^{j\theta})e^{jn\theta} + \frac{1}{2}A_0(e^{j(\theta+\pi)})A_1(e^{j\theta})e^{jn(\theta+\pi)}$$

Now, $A_0(e^{j(\theta+\pi)}) = A_0(-e^{j\theta}) = A_0(e^{j\theta})$ and $A_1(e^{j(\theta+\pi)}) = A_1(-e^{j\theta}) = -A_1(e^{j\theta})$, since $A_0(z)$ and $A_1(z)$ are even and odd in z respectively. Therefore $v(n) = A_0(e^{j\theta})A_1(e^{j\theta})e^{jn\theta}$ and $|A_0(e^{j\theta})||A_1(e^{j\theta})| = 1$ since $A_0(z)$ and $A_1(z)$ are all-pass functions. It can be shown that the degrees of $A_0(z)$ and $A_1(z)$ add up to the common degree of $H_0(z)$ and $H_1(z)$ which is p, the degree of the polynomial $g(z)$.

In the next section we shall systematically develop a systolic algorithm for a QMF analysis tree for 2-dimension signals $u(n, m)$. The subband structure (in the frequency domain) could be as shown in figure 7. Moreover, we shall construct a separable bank tree, that is,

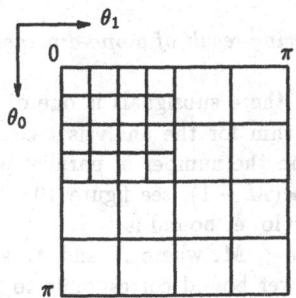

Figure 7: *Splitting scheme to obtain 28 subbands. (zero frequency is upper left)*

we shall interlace 2 1-dimensional QMF tree banks.

76

3   QMF tree bank algorithm

For simplicity, but without impairing generality, we shall assume that the all-pass functions $E_i(z)$ are of degree 1 and have a state-space realization

$$\left[\begin{array}{c} x_i(n+1) \\ y_i(n) \end{array}\right] = \Psi_i \left[\begin{array}{c} x_i(n) \\ u_i(n) \end{array}\right]$$

where Ψ_i is the usual (A, B, C, D) matrix map, relating next state variable $x(n+1)$ and current output variable $y(n)$ to current state variable $x(n)$ and current input variable $u(n)$. This is structurally represented as shown in figure 8. After 1 pass through a single 2-band

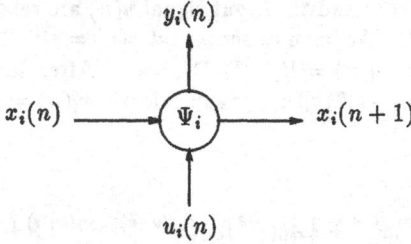

Figure 8: *Schematic representation of state-variable behaviour of the filters*

decomposition, the output signal is an interlacing of 4 subsignals which we shall label *ll*, *lh*, *hl* and *hh*. This is shown in figure 9 for a 1-dimensional signal. Parallel sorting of the

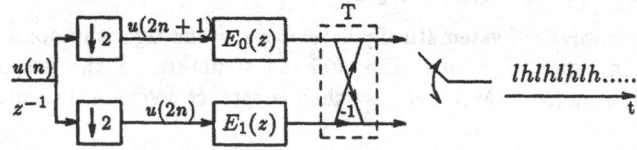

Figure 9: *Filtering result of a one-dimensional signal*

output signal samples to retrieve the 4 subsignals is one of the problems to be overcome in the design of a systolic algorithm for the analysis tree. This sorting is essentially a permutation. The lower bound on the number of parallel permutation steps is, for a 1D $(M+1)$-sample signal, $b = 0.5 * (M-1)$; see figure 10. It can be shown, [6], that for 2-dimensional signals $u(n,m)$, the lower bound is
$b = \sum_{l=0}^{L-1} 2^{-(l+1)}(M+1) - 1$, $N \le M$, where N and M are the numbers of time steps in n and m respectively. This lower bound corresponds to the case when filtering in the second dimension does not wait for sorting in the other dimension. From now on, we shall interpret the signal $u(n,m)$ as an $N \times M$ image ($N \le M$) with (i,j) pixel plane and pixel values $p(i,j) = u(n,m)$, $i = n$, $j = m$.
Now, to prepare for the designing of the systolic algorithm, we partition the pixel plane into nine regions R_p, $p = 1, \ldots, 9$, which are sets of pixels whose coordinates satisfy a finite

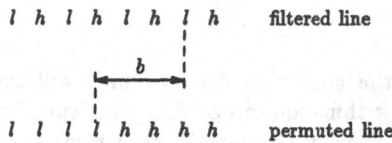

Figure 10: *Sorting of l and h components.*

number of inequalities as follows

$$R_p = \{(i,j) : A_p(i,j)^t \geq b_p\}$$

where $A_p \in Z^{n_p \times 2}, b_p \in Z^2$, and n_p the number of inequalities. This is shown in figure 11,

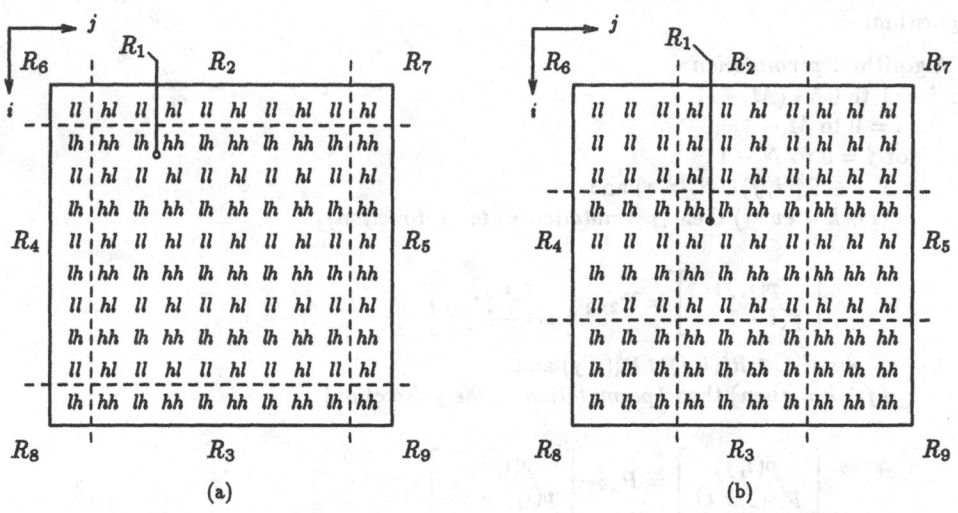

Figure 11: *Partitioning of the image into several regions*

in which region R_1 contains all pixels which have to be permuted crosswise, regions R_2 and R_3 contain all pixels which have to be permuted in the j-direction only, regions R_4 and R_5 contain all pixels which have to be permuted in the i-direction only, and regions $R_6, \ldots, R_9$ contain all correctly ordered pixels. Figure 11.a shows the regions after the image has been filtered for the first time (1st 2-band filtering). Figure 11.b shows the regions after 2 permutation steps. Hence, the vectors b_p, $p = 1, \ldots, 9$ change after each permutation step k, and so do the $|R_p|$, where $||$ denotes the cardinality of the set R_p. To make this k-dependency explicit, we write $R_p^k(i,j)$, $k \geq 1$ instead of $R_p(i,j)$. Finally, we introduce the region $R_0^0(i,j) = \{(i,j) : 0 \leq i \leq M-1, 0 \leq j \leq N-1\}$ which refers to the situation after filtering and before permutation.

78

4 Tree bank algorithm

We are now ready to start the algorithm design which will start out with preliminary filtering and permutation algorithms and proceed along a specific refinement methodology. The reference algorithms are given in the next 3 subsections.

4.1 PERMUTATION ALGORITHM

Let $\mathbf{P}_{2\times2} = \begin{bmatrix} 0 & 1 \\ 1 & 0 \end{bmatrix}$ and $\mathbf{P}_{4\times4} = \mathbf{P}_{2\times2} \otimes \mathbf{P}_{2\times2}$, where the symbol $\otimes$ denotes the kronecker product. Denoting by $A \cup B$ the union of the sets A and B, and calling the vector $I = (i, j, k)^t$ *the iteration vector*, we can describe the permutation with the aid of the following sequential algorithm:

Algorithm: *permutation*
for $k = 1$ to $0.5 * (M - 1)$
  for $i = 0$ to $M - 1$
    for $j = 0$ to $N - 1$
      if $I \in R_4^k(i,j) \cup R_5^k(i,j)$ and
      $(i + k = even)$ then {*permutation in the i-direction*}

$$\begin{bmatrix} p(i,j) \\ p(i+1,j) \end{bmatrix} = \mathbf{P}_{2\times2} \begin{bmatrix} p(i,j) \\ p(i+1,j) \end{bmatrix}$$

      else if $I \in R_2^k(i,j) \cup R_3^k(i,j)$ and
      $(j + k = even)$ then {*permutation in the j-direction*}

$$\begin{bmatrix} p(i,j) \\ p(i,j+1) \end{bmatrix} = \mathbf{P}_{2\times2} \begin{bmatrix} p(i,j) \\ p(i,j+1) \end{bmatrix}$$

      else if $I \in R_1^k(i,j)$ and
      $(i + k = even)$ and $(j + k = even)$ then {*cross-permutation*}

$$\begin{bmatrix} p(i,j) \\ p(i+1,j) \\ p(i,j+1) \\ p(i+1,j+1) \end{bmatrix} = \mathbf{P}_{4\times4} \begin{bmatrix} p(i,j) \\ p(i+1,j) \\ p(i,j+1) \\ p(i+1,j+1) \end{bmatrix}$$

      end(if)
     end(if)
     end(if)
    end
   end
  end

4.2 FILTERING ALGORITHM

Next, combining the 2 all-pass state-space realizations Ψ_i, $i = 0, 1$, given before, into a single state-space map

$$\mathbf{F} = \begin{bmatrix} A_0 & 0 & B_0 & 0 \\ 0 & A_1 & 0 & B_1 \\ C_0 & C_1 & D_0 & D_1 \\ C_0 & -C_1 & D_0 & -D_1 \end{bmatrix}$$

the filtering can also be described in terms of the following sequential algorithm.

Algorithm: *filtering*

for $i = 0$ to $M - 1$

   for $j = 0$ to $N - 1$

      if $(j = even)$ then {*filtering of the rows*}

$$\begin{bmatrix} x(i, j+3) \\ x(i, j+2) \\ p(i, j+1) \\ p(i, j) \end{bmatrix} = \mathbf{F} \begin{bmatrix} x(i, j+1) \\ x(i, j) \\ p(i, j+1) \\ p(i, j) \end{bmatrix}$$

      end(if)

   end

end

for $i = 0$ to $M - 1$

   for $j = 0$ to $N - 1$

      if $(i = even)$ then {*filtering of the columns*}

$$\begin{bmatrix} x(i+3, j) \\ x(i+2, j) \\ p(i+1, j) \\ p(i, j) \end{bmatrix} = \mathbf{F} \begin{bmatrix} x(i+1, j) \\ x(i, j) \\ p(i+1, j) \\ p(i, j) \end{bmatrix}$$

      end(if)

   end

end

4.3 COMPLETE ALGORITHM

Finally, the filtering and permutation algorithms must be repeated to obtain 4^L subbands. Let $l = 0, \ldots, L - 1$ be the depth in the tree bank and let m and n, $m, n = 0, \ldots, l$, be parameters to distinguish the different subbands at level l. The embedding of the algorithms *permutation* and *filtering* in the tree hierarchy is then like so:

   for $l = 0$ to $L - 1$

      for $m = 0$ to l

for $n = 0$ to l

{*filtering and permutation*}

   end
  end
end

In the complete algorithm, the different regions $R_p^k(i,j)$ thus not only depend on the permutation step k and the pixel coordinates (i,j), but also on the parameters m, n, and l. In the following complete algorithm we express these dependences by denoting the regions by $R_p^{k,l,m,n}(i,j)$ instead of $R_p^k(i,j)$.

Algorithm: *tree filtering*
for $k = 0$ to $M - 2^{-(l+1)}(M+1)$
 for $i = 0$ to $M - 1$
  for $j = 0$ to $N - 1$
   if $I \in R_0^{k,l,m,n}(i,j)$ and
   $(i + k = even)$ and $(j + k = even)$ then

    {*filtering of both rows and columns*}

   else if $I \in R_4^{k,l,m,n}(i,j) \cup R_5^{k,l,m,n}(i,j)$ an d
   $(i + k = even)$ then

    {*permutation in i-direction*}

   else if $I \in R_2^{k,l,m,n}(i,j) \cup R_3^{k,l,m,n}(i,j)$ an d
   $(j + k = even)$ then

    {*permutation in j-direction*}

   else if $I \in R_1^{k,l,m,n}(i,j)$ and
   $(i + k = even)$ and $(j + k = even)$ then

    {*cross-permutation*}

   end(if)
  end(if)
 end(if)
end(if)

```
    end
  end
end
```

5 Systolic QMF tree bank

The above algorithm is in the form of a *nested loop* program (NLP) which has the following generic structure.

For $i_1 = f_{l,1}$ to $f_{u,1}$ (step m_1)	else if *condition$_K$* then B_K
$\vdots$	end(if)
For $i_s = f_{l,s}$ to $f_{u,s}$ (step m_s)	$\vdots$
begin	end(if)
If *condition$_1$* then B_1	end
else if $\cdots$	end(for)
$\vdots$	end(for)
	$\vdots$
	end(for)
NLP Algorithm model	

where $i_l, l = 1,\ldots,s$ are loop indices, and $f_{l,j}, f_{u,j}$ are integer-valued boundary functions possibly involving loop indices $i_1,\ldots,i_{l-1}$. The constants $m_l \in \mathbf{Z} \setminus \{0\}$, $l = 1,\ldots,s$, where $\mathbf{Z} \setminus \{0\}$ denotes the set of nonzero integers, are increment or decrement steps of the loop indices $i_1,\ldots,i_s$. *Condition$_i$*, $i = 1,\ldots,K$ is a condition on the loop indices $i_1,\ldots,i_s$, which is in the form of a finite number of linear inequalities. B_k, $k = 1,\ldots,K$ are program bodies, which may be NLPs themselves. The collection of all if-then-else blocks is called the loopbody. The NLP is assumed to be parametrized with respect to its size. The vector $I = (i_1,\ldots,i_s)^t$ is called the iteration vector. In the next few subsections, we derive a systolic algorithm suited for execution on a 2-dimensional array of processors.

5.1 ALGORITHM FRAMEWORK

Parallelization of an NLP is based on the assumption that all variables in the loopbody are of the form $v[f(I)]$ [7], where v is the name of the variable, and $f(I)$ is the indexing function of the variable which is of the form $AI + b$, where $A \in \mathbf{Z}^{m \times s}$ (A is called the indexing matrix of the variable) and $b \in \mathbf{Z}^m$, $m \le s$.

Moreover, the assignment statements $S_{i,j}$ are of the form

$$v_l[f(I)] = F_{ij}\{\cdots, v_k[g(I)], \cdots, v_m[h(I)], \cdots\}$$

with $1 \le l, k, m \le L$, $1 \le i \le K$, $1 \le j \le J_i$. The maps $F_{i1},\ldots,F_{iJ_i}$ can all together be seen as a single map F_i which is evaluated in each and every iteration point $I \in \mathbf{I}_i$, provided *condition$_i$* is true. Here, the *iteration domain* $\mathbf{I}_i$ is a bounded polyhedron [8] of the form

$$\mathbf{I}_i = \{I : I \in L(\tilde{A}_i, \tilde{b}_i); A_i I \ge b_i\}$$

where $A_i \in \mathbf{Z}^{n_i \times s}, b_i \in \mathbf{Z}^{n_i}$, n_i is the number of linear inequalities and $L(\tilde{A}_i, \tilde{b}_i)$ is an integral lattice in $\mathbf{R}^s$, defined as

$$L(\tilde{A}, \tilde{b}) = \{I \in \mathbf{Z}^s : I = \tilde{A}\kappa + \tilde{b}; \ \kappa \in \mathbf{Z}^p\}$$

where $\tilde{A} \in \mathbf{Z}^{s \times p}$ is an integer matrix with non-parallel columns $a_1, \ldots, a_p$ and $\tilde{b} \in \mathbf{Z}^s$ is an offset vector. The maps F_i can also be grouped in a single map F by introducing selection variables, [9], which can identify particular iteration domains $\mathbf{I}_i$ and their maps F_i. This map is obtained as follows. If S_{ij} is an arbitrary statement of the loopbody, then denote by $L(S_{ij})$ the set of left-hand-side variables of S_{ij} and by $R(S_{ij})$ the set of right-hand-side variables of S_{ij}, and define

$$\mathcal{U} = \bigcup_{i=1}^{K} \bigcup_{j=1}^{J_i} L(S_{ij})$$

and

$$\mathcal{W} = \bigcup_{i=1}^{K} \bigcup_{j=1}^{J_i} [R(S_{ij}) \ominus \{\bigcup_{k=0}^{i-1} \bigcup_{l=1}^{J_k} L(S_{kl})\} \ominus \{\bigcup_{l=0}^{j-1} L(S_{il})\}]$$

where S_{ij}, $i.j = 0$, are dummy statements with $L(S_{ij}) = \emptyset$. The set $\mathcal{U}$ contains all the variables that are updated in the loopbody, and the set $\mathcal{W}$ contains all the variables whose first occurrence in the loopbody is at the right-hand-side of a statement. In this way, the loopbody can be viewed as a map $F : \mathcal{U}(I) = F(\mathcal{W}(I))$, which takes the variables $\mathcal{W}(I)$ as its arguments and assigns the result of each evaluation (iteration) to the variables in $\mathcal{U}(I)$. Notice that the program loop specification defines a (lexicographical) ordering: for any $I, J \in \mathbf{I}_i$, $i = 1, \ldots, K$, $I \prec J$ if iteration I is evaluated before iteration J. An NLP can thus be characterized by the tuple $< \{\mathbf{I}_i, \ i = 1, \ldots, K\}, \prec, F, \mathcal{U}, \mathcal{W} >$.
Denoting by $\Sigma = \{I_1, I_2, \ldots, I_M\}$ the set of all iteration points, $I_1 \prec I_2 \prec, \ldots, \prec I_M$, the execution of the NLP will proceed according to the following sequence Π of iterations:

$$
\Pi : \quad
\begin{aligned}
I_1 &: \quad \mathcal{W}(I_1) \mapsto \mathcal{U}(I_1) \\
I_2 &: \quad \mathcal{W}(I_2) \mapsto \mathcal{U}(I_2) \\
&\vdots \\
I_M &: \quad \mathcal{W}(I_M) \mapsto \mathcal{U}(I_M)
\end{aligned}
$$

Using this iteration sequence, an iteration J depends directly on an iteration I, via a variable, say $v[\xi(I)]$, if

- There is a variable $v[\xi(I)] \in \mathcal{U}(I) \cap \mathcal{W}(J)$

- $I \prec J$

- $v[\xi(I)] \notin \mathcal{U}(I')$ for all $I' \in \Sigma$, such that $I \prec I' \prec J$

The program inputs are defined as follows:

- if $v[\xi(J)] \in \mathcal{W}(J)$, and $v[\xi(I)] \notin \mathcal{U}(I)$ for all $I : I \prec J$, then $v[\xi(J)]$ in iteration J is an input of the program.

The program outputs are defined as

- if $v[\xi(I)] \in \mathcal{U}(I)$, and $v[\xi(J)] \notin \mathcal{U}(J)$ for all $I : I \prec J$, then the value of $v[\xi(I)]$ in the iteration I will not be changed, and is a (possible) output of the program.

Direct dependencies are assumed to be of the form $d(I) = (I. - D)I + d_0$, where $I.$ is the identity matrix of appropriate order, D is a dependency matrix and d_0 is a constant displacement vector.

With the above definitions, we can represent the NLP by means of a *dependence graph*, as follows:

- There are $|\Sigma|$ nodes in the dependence graph, each node represents one unique iteration. The node representing the iteration I is labeled I.

- The directed edges of the dependence graph correspond to the direct dependencies of the program.

5.2 SYSTOLIC TREE FILTERING ALGORITHM

In the NLP *tree filtering*, there are a lot of global communications. For example, iteration $J = (0, 0, \frac{1}{2}(M + 1))^t$ depends directly on iteration $I = (0, 0, 0)^t$. In fact, due to the pyramid structure of the region of the dependence graph in which cross-permutation is taking place, the lengths and directions of the dependence vectors depend on the pixel plane coordinates; there are dependence vectors $d_i = (\delta_{i1}, \delta_{i2}, \delta_{i3})^t$; $\delta_{i1}, \delta_{i2} \in \{-1, 0, 1\} \wedge \delta_{i3} \geq 2$. Such dependencies are not appealing and may be *localized* by introducing local *propagation vectors*, as follows ($\mathcal{D} = \{d_i : i = 1, \ldots, P\}$ is the set of all dependencies):

$$\forall d_i \in \mathcal{D} \qquad d_i = \sum_{j=1}^{\delta_{i3}-1} a_{ij} b_{ij}$$

where

$$a_{ij} = 1 \qquad 1 \leq j \leq \delta_{i3} - 1$$

$$b_{ij} = \begin{cases} (\delta_{i1}, \delta_{i2}, 1)^t & j = 1 \\ (0, 0, 1)^t & j \geq 2 \end{cases}$$

In doing so, additional intermediate nodes, hence iteration domains, will eventually have to be added to the dependence graph, because if a variable is to be sent from a node $I \in \mathbf{I}_v$ to a node $J \in \mathbf{I}_w$, using propagation vectors b_{ij} only, then nodes at distances b_{ij} may not exist, hence will have to be introduced. At this point, it is convenient to take a closer look at the dependence graph. It is shown in figure 12 for a pixel plane of size 8×8, and in the form of a series of snapshots which are intersections of the dependence graph perpendicular to the $k-$axis.

The dependences from iteration step $k - 1$ to iteration step k are drawn in solid lines, the dependencies within a single iteration step k are drawn in dashed lines. The different regions $R_0, \ldots, R_9$ are also indicated. The lattices defining the various iteration domains are:

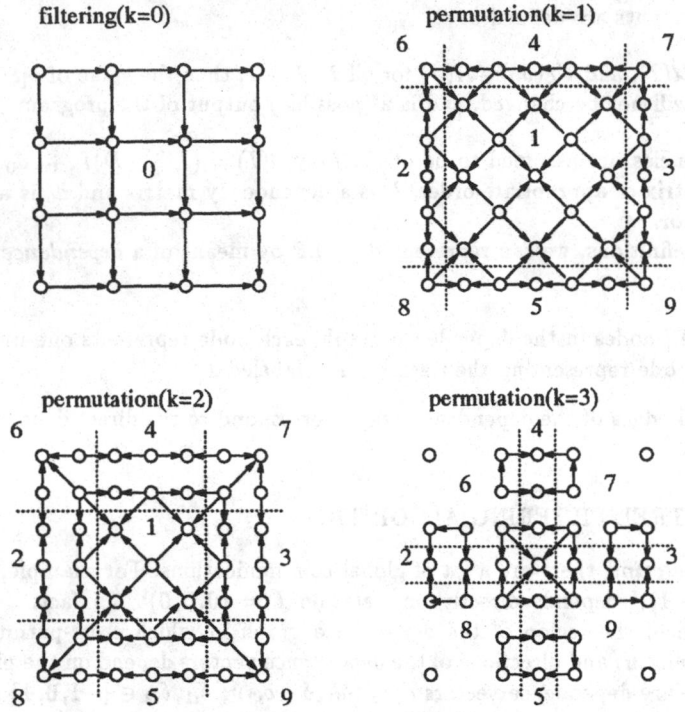

Figure 12: *A few snapshots of the dependence graph going with the algorithm* tree filtering

- *filtering and cross permutation*

$$L(\tilde{A}_0, \tilde{b}_0) = \{I \in \mathbf{Z}^3 : I = \tilde{A}_0 \kappa + \tilde{b}_0;\ \kappa \in \mathbf{Z}^3\}$$

with

$$\tilde{A}_0 = \begin{bmatrix} 2 & 0 & 1 \\ 0 & 2 & 1 \\ 0 & 0 & 1 \end{bmatrix}, \qquad \tilde{b}_0 = 0$$

Which is a direct consequence of the fact that both filtering nodes and cross-permutation nodes process groups of four pixels at each and every k-level in the dependence graph, see algorithm *tree filtering*.

- *permutation in i-direction*

$$\tilde{A}_1 = \begin{bmatrix} 2 & 0 & 1 \\ 0 & 1 & 0 \\ 0 & 0 & 1 \end{bmatrix}, \quad \tilde{b}_1 = 0$$

- *permutation in j-direction*

$$\tilde{A}_2 = \begin{bmatrix} 1 & 0 & 0 \\ 0 & 2 & 1 \\ 0 & 0 & 1 \end{bmatrix}, \quad \tilde{b}_2 = 0$$

- *localization*

$$\tilde{A}_3 = \begin{bmatrix} 2 & 0 & 0 \\ 0 & 2 & 0 \\ 0 & 0 & 1 \end{bmatrix}, \quad \tilde{b}_3 = 0$$

Now, it turns out that if we would like to let an iteration domain be functionally characterized by a single node function, as was proposed in [10], then we would arrive at as many as 75 different iteration domains. As a consequence, the algorithm can not be called a systolic algorithm in the traditional sense. To overcome this problem, the algorithm must be changed again (as was already done in relation to the localization of the dependences) to reduce the number of local propagation vectors and the number of iteration domains. This is called *regularization* of the algorithm, or equivalently of the dependence graph. One type of regularization was already alluded to before when it was noted that the maps $F_1, \ldots, F_K$ in the NLP could be merged into a single map F, provided new selection variables are introduced to distinguish between the actual validity of the conditions, hence the maps $F_{i1}, \ldots, F_{iJ_i}$. One way to do this in an optimal way is described in [9]. Another form of regularization which is convenient here is to minimize the actual number of local propagation vectors. In the algorithm *tree filtering*, 10 different regions $R_0, \ldots, R_9$ can be distinguished in the dependence graph. Therefore, the minimum number of iteration domains has a lower bound of 10. Moreover, there are essentially 5 basic node functions, namely filtering, cross-permutation, permutation in i-direction, permutation in j-direction, and propagation of pixel values which do not have to be permuted anymore. The lower bound of the number of lattices is of course 1. All these lower bounds can be achieved [6] with a set C of regularized propagation vectors as follows:

$$
\begin{aligned}
c_1 &= (2, 0, 0)^t \\
c_2 &= (0, 2, 0)^t \\
c_3 &= (1, 1, 1)^t \\
c_4 &= (1, -1, 1)^t \\
c_5 &= (-1, 1, 1)^t \\
c_6 &= (-1, -1, 1)^t
\end{aligned}
$$

The resulting dependence graph (in the same snapshot form as before) is shown in figure 14. All iteration domains have the same lattice $L(\tilde{A}_0, \tilde{b}_0)$, which is possible due to a slight -and in practice negligible- increase of node count along the boundaries. A formal specification of this dependence graph, including the iteration domains $I_0, \ldots, I_9$ and the selection mechanism of regularized node maps can be found in [6]. The resulting dependence graph thus corresponds to an orthodox systolic algorithm [11] from which systolic architectures can be obtained in the usual way. The interior of a 2D systolic processor array is schematically shown in figure 13.a. This graph is strictly regular, that is, it is characterized by

- an iteration space $\tilde{I}_p$.

- a single node map M (see figure 13.b).

- the set $\tilde{C}$ of projected dependences from C for the node.

- the set of time-weights $\tilde{W} = \lambda^t c_i$ on the outgoing arcs of the node.

which results from [12] chosing a systolic schedule vector $\lambda = (1,1,3)^t$ and a projection direction $u = (1,1,1)^t$.

The node map consist of the functions filtering, cross-permutation, permutation in i-direction, permutation in j-direction and passing of the incoming arguments. A specific function is selected by the variables q_{ij}. The size of the systolic array is $s_1 \times s_2 = [M+1-2^{-(L+2)}(M+1)] \times [\frac{1}{2}(M+N+2) - 2^{-(L+2)}(M+1)]$. The pixel values of the "image" to be processed are loaded in groups of 4 from a buffer (not shown) into nodes $\{(i',j')\}$, $i' \geq 0 \wedge j' \geq 0$. The other nodes are initialized with an arbitrary value, say zero. The selection variables are initialized in an initializing domain which is derived in [6].

For practical applications, an array of the above size will in general not be feasible. Further steps to reduce the node count (at the cost of extra memory and selection mechanisms) will have to be taken. For lack of space, we can not go into such steps in this paper. The interested reader may consult [13] for more information about them. However, it is worthwhile to notice that the number of nodes in the systolic array presented here can anyhow be reduced by a factor of 5 since the utilization of the array nodes is only $|\lambda^t u| = 1/5$. It is possible to find, in a systematic way, groups of 5 nodes which can be clustered together into a single node. Since this clustering does not introduce extra memory, it is called passive clustering. The theory behind passive clustering can be found in [13] and a slight generalization of it together with its application to the above array can be found in [6]. In case the then resulting array is still too large, another clustering technique, called active clustering, can be invoked to reduce the size of the array, now at the cost of extra memory and a reduced operation speed. See [13] for more details.

Acknowledgements

The author wishes to express his appreciation and thanks for the contributions of Richard Heusdens and Rik Theunis in preparing the paper and deriving some of the results. The

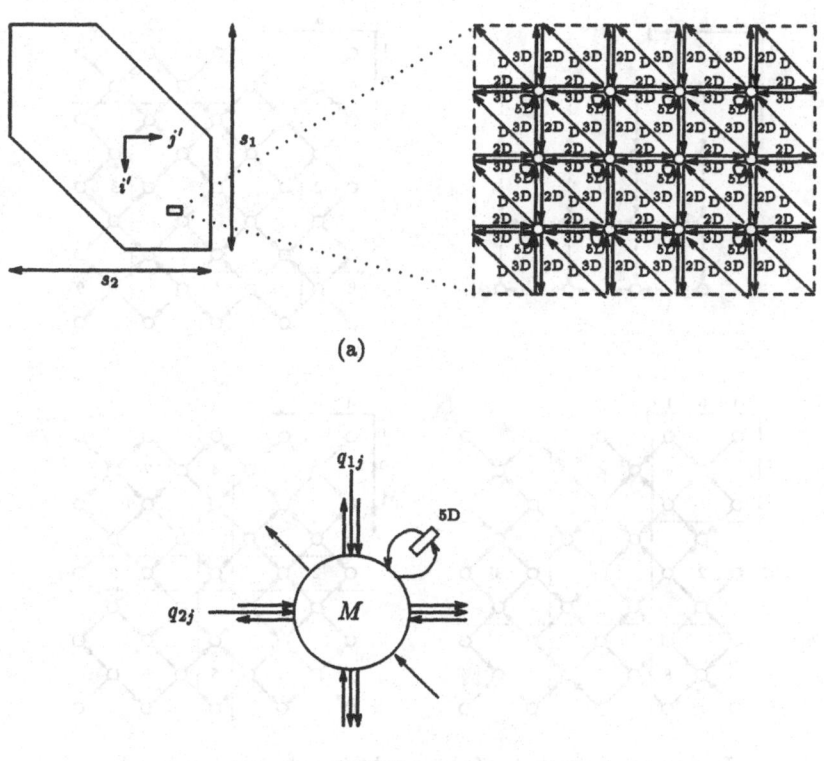

(a)

(b)

Figure 13: *The projected dependence graph of the complete filtering and permutation algorithm*

research was supported in part by the commission of the EC under the ESPRIT BRA program 6632 (NANA2), and in part by the Dutch National Technology Foundation under contract STW22.2614.

References

[1] P.P. Vaidyanathan. Multirate Digital Filters, Filter Banks, Polyphase Networks, and Applications: A Tutorial. *Proceedings of the IEEE*, **78**(1), pp 56–93, 1990.

[2] J.H. Rothweiler. Polyphase Quadrature Filters - A new Subband Coding Technique. *Proc. Int. Conf. Acoust., Speech and Signal Processing*, volume ICASSP'83, pp 1280–1283, 1983.

[3] R. Theunis. 2D Cosign Modulated Filter Banks. Technical report, Dept. Electrical Engineering, Delft University of Technology, 1992.

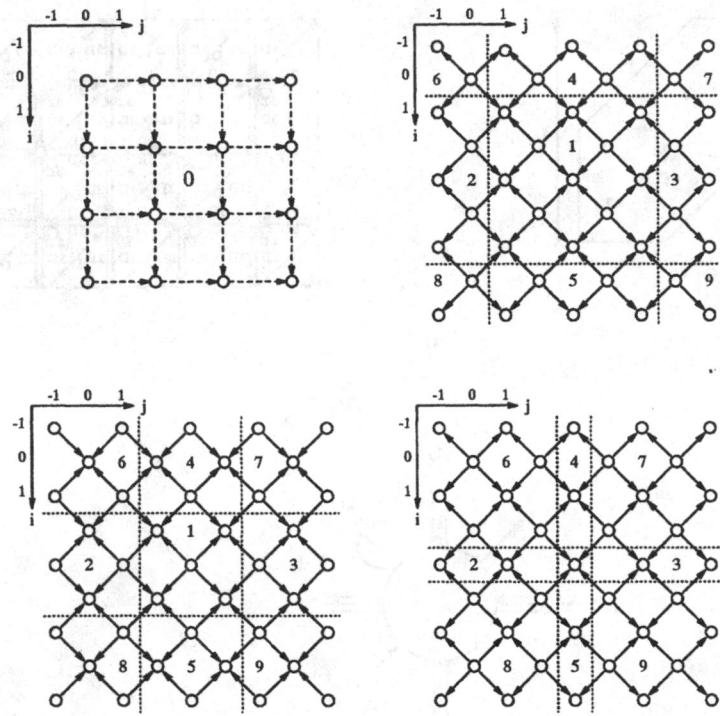

Figure 14: *Snapshots of the regularized dependence graph*

[4] R.D. Koilpillai and P.P. Vaidyanathan. New Results on Cosine-Modulated FIR Filter-banks Satisfying Perfect Reconstruction. *Proc. Int. Conf. Acoust., Speech and Signal Processing*, volume ICASSP'91, pp 1793–1796, 1991.

[5] E.F. Deprettere, P. Dewilde and P. Rao. Orthogonal Filter Design and VLSI Implementation. *Proc. Int. Conf. Comp., Systems, Signal Processing*, Bangalore, India, pp 779–790, 1984.

[6] R. Heusdens. A Parallel QMF Image Analysis Architecture. Technical report, Dept. Electrical Engineering, Delft University of Technology, 1992.

[7] J. Bu, E.F. Deprettere and L. Thiele. Systolic Array Implementation of Nested Loop Programs. *Proceedings Int. Conf. Application Specific Array Processing*, Volume 4, pp 31–42, 1990.

[8] G.L. Nemhauser and L.A. Wolsey. *Integer and Combinatorial Optimization*. John Wiley & Sons, 1988.

[9] J. Teich, M. Huber and L. Thiele. Design of Configurable Processor Arrays. *Proc. Int. Symp. Circuits and Systems*, Volume 1, pp 970–973, 1990.

[10] E.F. Deprettere. Example of combined algorithm development and architecture design. *Proc. SPIE Advanced Signal Processing Algorithms, Architect ures, and Implementations III*, 1992.

[11] H.T. Kung and C.E. Leiserson. Systolic Arrays for VLSI. *Sparse Matrix Proceedings*, SIAM, Philadelphia, pp 245–282, 1980.

[12] S.K. Rao and T. Kailath. Regular Iterative Algorithms and Their Implementation on Processor Arrays. *Proceedings of the IEEE*, 76(3), 1988.

[13] J. Bu and E.F. Deprettere. Processor Clustering for the Design of Optimal Fixed-Size Systolic Arrays. In : E.F. Deprettere and Alle-Jan van der Veen (Eds.), *Algorithms and Parallel VLSI Architectures*, Volume A, North Holland, Elsevier, Amsterdam, pp 341–362, 1991.

A PARALLEL IMAGE RENDERING ALGORITHM AND ARCHITECTURE BASED ON RAY TRACING AND RADIOSITY SHADING

E.F. DEPRETTERE and L.-S. SHEN
Department of Electrical Engineering
Delft University of Technology
2628 CD Delft, The Netherlands
ed@dutentd.et.tudelft.nl

ABSTRACT. Algorithms for rendering color scenes on a video screen potentially belong to the class of massively parallel algorithms. Developing effective and efficient data-driven architectures for algorithms from this class is in general a hard problem. However, in case of application specific algorithms, such as the rendering algorithm described in this paper, feasible solutions are conceivable. The parallel algorithm/architecture presented in this paper is a linear speed-up accelerator for the rendering of photo realistic scenes in interaction time.

KEYWORDS. Rendering, radiosity, ray-tracing ,parallel algorithms, data-driven architectures.

Introduction

An abstract scene is a collection of objects in a closed environment. Visualization of scenes, in particular realistic but non-existing scenes, such as designs of offices, show-rooms or factory halls on a high-resolution color monitor, is an emerging computer graphics technique which aims at photo-realism and instant rendering. In this technique, a display screen is located between the viewer's eye-point and the scene which are in front of and behind the screen, respectively. The objects in the scene are modeled by means of finite elements, called *patches*. Examples of commonly used patches are planar polygons and bi-cubic Bezier patches [1]. A Bezier curve is a (cubic) curve which is uniquely defined by four *control points* $P_i(x_i, y_i), i = 1, \cdots, 4$, two of which are curve extremal points, the other two lying on the curve tangents through the two curve extremal points. This is shown in figure 1. [1] A bi-cubic Bezier patch is similarly defined by means of 16 control points. The viewpoint

---

[1]The curve can be reconstructed from the control points by a technique called *subdivision*, see [1].

M. S. Moonen et al. (eds.), Linear Algebra for Large Scale and Real-Time Applications, 91–108.
© 1993 *Kluwer Academic Publishers.*

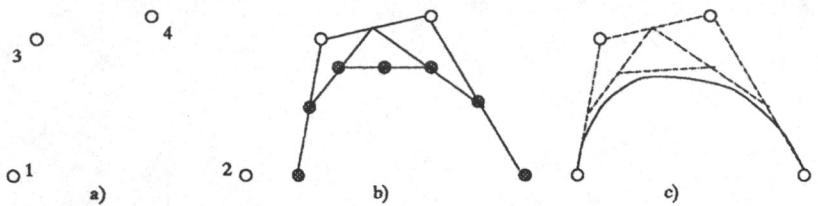

Figure 1: *Bezier curve. Control points (a), subdivision (b) and curve (c).*

in front of the screen defines projection directions- lines connecting the viewpoint and the screen pixels- along which the scene's object patches behind the screen are projected on the screen's image plane. Scene patches and patch property information- such as color, reflectivity and translucency- are stored in a data base which will be assumed in this paper to have a spatial enumerated data structure. This is shown in figure 2 for a 2-dimensional scene. Thus, given a scene and a viewpoint, the rendering problem is to find all visible

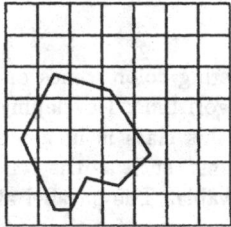

Figure 2: *Spatial enumerated data structure.*

patches and to project them on the image plane on the screen. A sampling technique which has proven to be appealing is called *ray tracing* and is schematically illustrated in figure 3. It shows the inverse path of a light ray propagating from a light emitting source via reflecting patches, through the image plane and to the viewpoint. The viewpoint-to-patch path segment is called the *primary ray*. Patch-to-patch and patch-to-source path segments are called *secondary rays*. A secondary ray which hits a light source, as shown in the figure, is called a *shadow ray*. Now, patches may have reflectivity characteristics ranging from perfectly specular to perfectly diffuse and the constellation of patches and sources in a scene may give rise to steep shade gradients, as a result of which the amount of secondary rays which have to be traced may be overwhelming. The idea, then, is to distinguish between specular patches, for which point sampling is necessary, and diffuse patches, for which patch sampling may be sufficient, because the light energy on a diffusely reflecting patch is likely to be constant over the patch. Patch sampling of diffusely reflecting objects is called *radiosity shading* and is schematically illustrated in figure 4. In this figure, D are diffuse patches and S are specular patches. For the time being, we shall neglect the presence of S-patches and assume that all patches are D-patches. We shall come back to

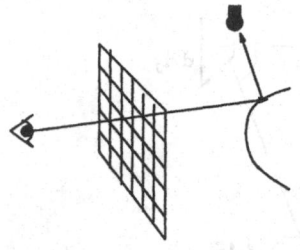

Figure 3: *The principle of ray tracing*

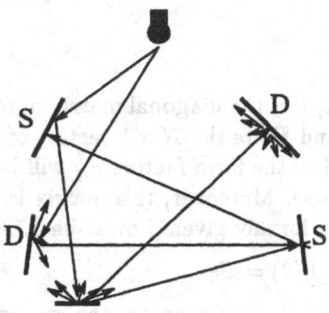

Figure 4: *Radiative transfer of light between patches*

the S-patches later on. Now diffuse patches P_i of area A_i reflect energies per unit area B_i, called radiosities, and emit energies per unit area E_i, if they are light sources. The so-called radiosity equations [2] then are: [2]

$$B_i A_i = E_i A_i + \rho_i \sum_{j=1}^{N} B_j A_j F_{ji} \quad i = 1.\ldots, N$$

where N is the total number of patches in a closed environment, ρ_i is the reflection coefficient of patch P_i and F_{ji} is the so-called *form factor* from patch P_j to Patch P_i. F_{ij} is a geometric visibility factor expressing the fraction of energy leaving patch P_i and arriving at patch P_j:

$$F_{ij} = \frac{1}{A_i} \int_{A_i} \int_{A_j} \frac{\cos \theta_i \cos \theta_j}{\pi r^2} \delta_{ij} dA_i dA_j$$

where θ_i, θ_j and r are as shown in figure 5 and δ_{ij} is 0 if the patches cannot see each other directly and is 1 otherwise. There is an obvious reciprocity relation between form factors

---

[2]The technique is borrowed from heat-transfer theory, see also [3].

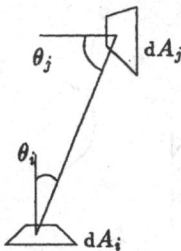

Figure 5: *Geometrical relation between 2 patches*

F_{ji} and F_{ij}, that is, $F_{ji}A_j = F_{ij}A_i$, whence the above radiosity equations can be expressed as follows:

$$(I - RF)B = E$$

where I is the $N \times N$ identity matrix, R is the diagonal matrix of reflection coefficients, F is the form factor matrix $[F_{ij}]$ and B and E are the $N \times 1$ vectors of radiosities and emissions respectively. In practice, almost 90% of the form factors F_{ij} will be zero because of $\delta_{ij} = 0$. The matrix $I - RF$ is thus very sparse. Moreover, this matrix is diagonal dominant since the sum of form factors F_{ij} over all i for any given j must be 1. As a result, the iteration

$$B(n+1) = RFB(n) + E \quad ; \quad B(0) = B_0$$

will converge very fast to $(I - RF)^{-1}E$. But even so, the storage of the matrix RF is a problem since in practice we can typically have $N = 10^5$ à 10^6, whence the amount of form factors to be stored is still tremendous. However, instead of computing row wise, that is

$$B_i(n+1) = \rho_i \sum_{j=1}^{N} F_{ij}B_j(n) + E_i \qquad i = 1\ldots,N$$

one can also compute incrementally column wise, that is:

$$B(n+1) = B(n) + Rf_i(\Delta B)_i(n) \quad ; \quad B(0) = E$$

$$\Delta B(n+1) = \Delta B(n) + Rf_i(\Delta B)_i(n) \quad ; \quad \Delta B(0) = E$$

$$\Delta B_i(n+1) = 0$$

where f_i is the i-th column of F,that is, the column of form factors $F_{ji}, j = 1,\ldots,N$, or equivalently $F_{ij}A_i/A_j$, $j = 1,\ldots,N$. The row-wise computation using form factors $F_{ij}, j = 1,\ldots,N$ corresponds to *light gathering*, see figure 6.a; the column-wise computation using form factors $F_{ji}, j = 1,\ldots,N$, corresponds to *light shooting* , see figure 6.b. The column approach is called *progressive refinement*, [4]. The idea is to select only those patches P_i for which $(\Delta B_i)(n)$ is significant and to compute the corresponding form factors f_i instead of selecting them from the stored matrix F. So the expensive part of the radiosity method is the computation of the form factors. An efficient way to compute them is by casting rays through a hemisphere placed on the viewing or source patch P_j which observes the environment in the direction of its positive normal vector. This is shown in figure 7 where the form factor F_{ij} is seen to be the sum of precomputed delta form factors ΔF :

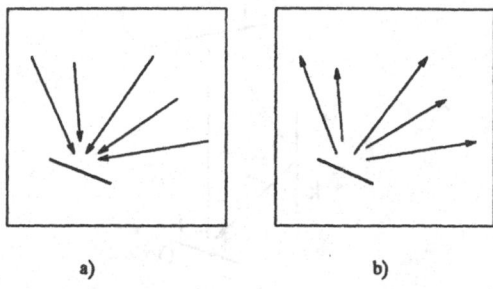

Figure 6: *Light gathering (a) and light shooting (b)*

each ray intersecting patch P_i contributes ΔF to the form factor F_{ij}, where ΔF is the form factor corresponding to a delta-area on the hemisphere. Notice that $F_{ij} = F_{kj}$, because patches P_i and P_k , when projected on the hemisphere, cover the same projected area. In a

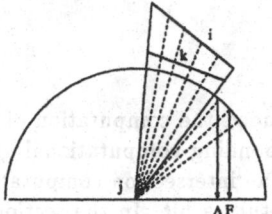

Figure 7: *Decomposition of form factors into a sum of delta form factors*

realistic scene, not all patches are diffuse patches so that patch sampling (radiosity method) and point sampling (ray-tracing) must be combined. This leads to a 2-pass approach in which the first pass is a viewpoint independent radiosity computation and the second pass adds point-sampled but viewpoint dependent specular effects and sharp shadow gradients. The two passes are not completely disjoint in the sense that mirrors play a role in both passes. Indeed, the first pass includes diffuse-to-diffuse and specular-to-diffuse (D-D and S-D, see figure 4) radiosity computations. Thus, a diffuse patch may see another diffuse patch directly and through a specular patch. This is shown in figure 8. Thus, if F_{ji} is the form factor from diffuse patch P_j to diffuse patch P_i and if F_{ki} is the form factor between the specular patch P_k and the diffuse patch P_i , then the actual form factor between the 2 diffuse patches is taken to be $F_{ji} + W \times F_{ki}$ where W is a weight factor which depends on the reflectivity property of the specular patch P_k and is 1 if it is a perfect mirror. For lack of space, we can not go into the details of the complete 2-pass radiosity algorithm, [5]. Instead, we shall concentrate on the computation of form factors needed to allow the progressive refinement radiosity algorithm to proceed. The *naive* algorithm then is as follows:

```
for patch P_i ;
  for all patches P_j ;
```

96

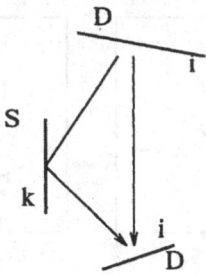

Figure 8: *The effect of the presense of specular patches on form factors between diffuse patches*

```
  for all rays Rₖ ;
    compute intersection ( Pⱼ , Rₖ);
    compute distance ( Pⱼ , Pᵢ);
  end for ;
 end for ;
end for
```

In this algorithm we have not included the computation of minimum distances - that is the actual visibility δ_{ij}. The corresponding computational graph is shown in figure 9 in which each and every node performs an intersection computation. Only a small fraction of the nodes - the shaded ones - will return a hit. In the sections to follow we shall address

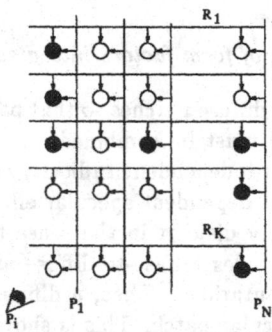

Figure 9: *Naive algorithm: all rays visit all patches*

the problem of how to order the patches and rays so that the execution time is minimized, given that computational resources and communication bandwidths are limited.

1 Space partitioning

In the computational graph of figure 9, there are no dependencies between the computations. This is due to the fact that the sets of input patches and input rays are not ordered with respect to the source patch P_i, whereas in object space, there is an ordering in the sense that patches have a distance relation and rays can only intersect a subset of the patches. Therefore, almost 90% of the nodes in the graph will return a *no hit* message which is actually no information since their hit probabilities are zero. These wasteful computations can be avoided by ordering the input sets of patches and rays in a way which is imposed by a certain partitioning of the object space. This is shown in figure 10 for a uniform partitioning of the space in conical *sections*. The result is that the computational graph is

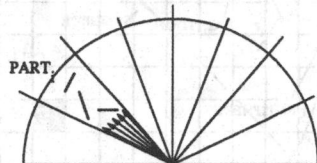

Figure 10: *Uniform partitioning of object space*

transformed from rectangular to block-diagonal form, at least in the ideal case when patches do not belong to more than 1 partition. See figure 11. Each block has again an all-ray

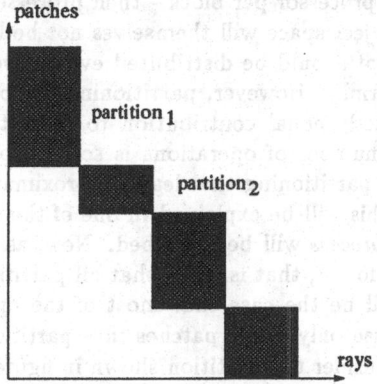

Figure 11: *Block diagonal patch-ray relationship due to uniform space partitioning*

to all-patch relationship, but the patches and rays that enter a block are now depending on the underlying space partitioning. The relationship between space partitioning and ray partitioning is trivial but the corresponding patch partitioning is not. Indeed, patches residing in a space partition must be found by searching, that is by traversing the spatial enumerated data structure. This is shown in figure 12 in which a ray r is traversing cells confined to the partition: when r hits a cell, this cell is opened and patches contained in it are input to the block, in the block-diagonal computational graph, that has r as an input ray. The price to be paid for this ordering is cell-traversal which is relatively cheap due to the uniform grid data structure. As mentioned before, each block has again an all-ray to all-patch relationship which is likely to be inefficient again. Moreover, the blocks have different sizes, that is, the amount of computations is not block invariant, because the amount of

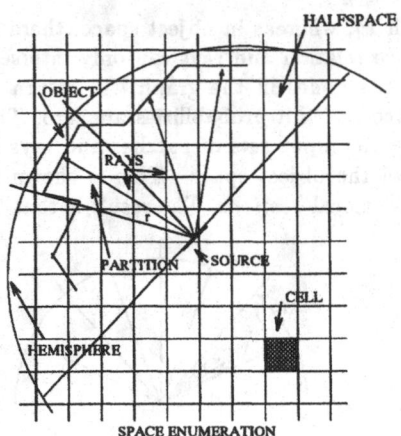

Figure 12: *Searching patches in a partition by cell traversal*

patches per space partition is scene dependent. In other words, if partitions are to be mapped on processors - one processor per block - then processors assigned to partitions of the uniformly partitioned object space will themselves not be loaded uniformly. It is thus the amount of operations that should be distributed evenly over the processors instead of partitioning the space uniformly. However, partitioning the object space such that each and every partition has exactly equal contribution to the total amount of operations is not feasible since the total number of operations is scene dependent. Nevertheless, it is possible to estimate a space partitioning that leads approximately to a uniform operation count over the partitions. This will be explained in one of the sections to come in which a *low-density ray-casting pre-process* will be described. Now, assuming that an appropriate space partitioning has been found, that is, such that all partitions have (almost) identical operation counts, it will still be the case that most of the operations in the blocks will appear to be wasteful because only a few patches in a partition are actually seen by the source patch. To see this, consider the partition shown in figure 13 in which P_1 and P_2 are 2 patches belonging to the partition. Also shown in figure 13 are *spherical bounding boxes*

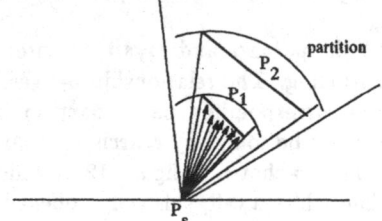

Figure 13: *Visibility ordering of patches in a partition*

of the patches P_1 and P_2. From this figure, it is easily seen that hemisphere rays that are confined to the spherical bounding box of patch P_1 will in general not intersect patch P_2

since these rays almost all die collectively on P_1. [3] However, even when patches P_1 and P_2 are known to reside in the partition, distance or visibility information will not, in general, be available. To avoid blind all-ray to all-patch intersection computations in the space partitions, further dependences have to be introduced: rays must not be instantaneously present in the entire partition, but should rather have to evaluate their relative distances at boundaries which are partitioning the depth of the object space relative to the source patch. This is shown in figure 14 where the concentric spheres define *shells* which force rays in a current shell to evaluate their distances to the source patch before entering the next shell. This leads to a *shelling technique* which we explain in the next section.

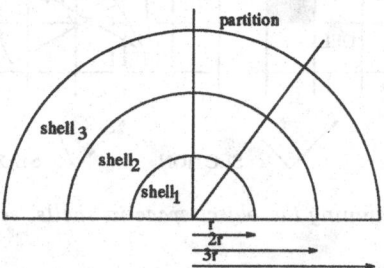

Figure 14: *Equidistance partitioning of the hemisphere*

2 The shelling technique

The shelling technique [7] is a space partitioning approach in which the object space seen by a source patch P_s is partitioned as shown in figure 15. In this figure, the shell structure relative to P_s is superimposed on the object-space uniform grid data structure. The shell structure consists of concentric spheres centered at the source patch's sampling point and having fixed relative radial distances. The areas between 2 successive spheres are called *shells*. The shells are radially partitioned into essentially independent *sections*, each section being further partitioned into a number of *sectors*. Sectors and sections are characterized by spanning solid angles. Sections are ideally so defined that the amount of ray-patch intersection computations is section invariant. Similarly, sectors are ideally so defined that the amount of ray-cell intersection computations is sector invariant. Again, these operation counts are not independent and are scene dependent: a ray which hits a visible patch dies on this patch and should not participate in neither ray-cell nor ray-patch intersection operations in the space behind this visible patch. However, visibility can only be determined by comparing distances, along a ray, between the source patch casting the ray and the patch hit by this ray, so that it seems unavoidable to compute all potential intersections in the entire section. It is here that the shells come into play. They define distance landmarks in the sense that the computation of potential intersections within a section can be restricted to parts of it, called *subshells*, which are the intersections of the section and the shells. This is shown in figure 16 where rays entering subshell SS_2 are known to have survived

---

[3] In actual 3-dimensional space, the spherical bounding box can be only computed approximately, because an exact computation is too expensive, see [6].

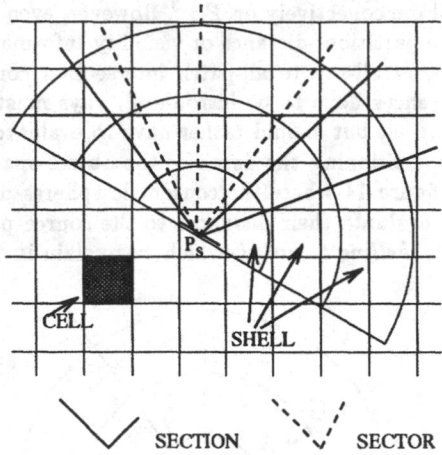

Figure 15: *Partitioning the object space in shells, sections and sectors*

all ray-patch intersection tests in subshell SS_1. The introduction of subshells corresponds

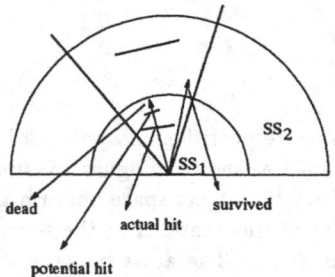

Figure 16: *Classifying survival rays at subshell boundaries*

to a further refinement in the ordering of patches, or equivalently a further structuring of
the blocks in the block-diagonal computational graph. The question that remains to be
answered is how to compute the spanning angles of the sectors and the sections. This is the
subject of the section following the next one in which a preliminary parallel architecture is
described.

3 Preliminary architecture

Recall that sections correspond to blocks in the computational graph and that blocks are
assigned to processors which are essentially intersection computation units (ICUs). It is
convenient to assume that the ICUs are pipeline processors which compute the intersections
of a bundle of rays with a patch. The patches that are delivered to an ICU are those that are
found by rays, confined to the section's sectors, which traverse the object space's uniform

grid data structure. Thus, each sector is assigned to a processor which is essentially a cell traversal unit (CTU). It is again convenient to assume that CTUs are pipeline processors. A section ICU and its sectors CTUs form a *cluster* as shown in figure 17, where the boxes labeled LCM and LPM are local memories, cell memory and patch memory respectively. The working of a cluster is roughly as follows. If for a given subshell, a sector ray hits a cell,

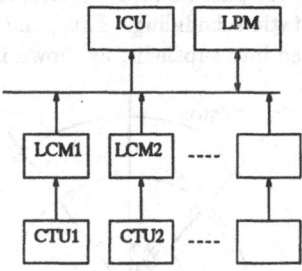

Figure 17: *A processor cluster*

then the patches residing in this cell are moved from LPM [4] to ICU. Also moved to ICU are, for each of these patches, all rays confined to the spherical bounding box of the patch that are alive in the current subshell. Sectors and sections are assumed to partition the visible hemisphere of the source patch P_s so that all CTUs and all ICUs are equally loaded and have a pipeline efficiencies of 1. Such a partitioning is certainly not static and, therefore, further pre-processing must be envisaged to estimate processor loads, or equivalently to determine the sector and section spanning angles. The next section is devoted to such a pre-processing procedure.

4 Low-density ray-casting pre-process

In parallel processing, the task of scheduling operations may be defined to consist of 1) assigning operations to operators, 2) ordering of the operations that have been assigned to an operator, and 3) determining the time instants at which the operators execute the operations. A schedule is said to be fully static if assignment, ordering and execution instants are known at compile time. It is said to be fully dynamic if none of these 3 subtasks can be decided upon at compile time. If the execution of a parallel algorithm needs a fully dynamic scheduler, then it is in general very difficult, if not impossible, to design a scheduler which does not obstruct a time-efficient execution of the algorithm. However, it is sometimes possible to avoid fully dynamic scheduling by introducing a pre-process which approximately determines process assignments and orderings thereby converting the fully dynamic scheduler into an estimated self-timed scheduler, that is a schedule for which only execution time instants are not known. Such a schedule may be called quasi-dynamic, and will not in general lead to an optimal work-load balancing since it is based on estimated assignments and orderings of operations. On the other hand, a dynamic, or run-time, load balance adjustment is always possible and the cost of pre-computing assignment and

---

[4] We assume, for the time being, that all patches are stored in local memory.

ordering of processes and run-time load balancing adjustment may be much lower than the cost of a plain dynamic scheduler. Of course, pre-computing a semi-static scheduling is not generally applicable, but it will certainly be a feasible option for many application specific algorithms, such as our image rendering algorithm. For this algorithm, the pre-processing algorithm is essentially a low-density ray-casting incomplete version of the actual algorithm. It is to be repeated for each source patch and, therefore, the quasi-dynamic scheduling can as well be called an iterative static scheduling. Thus, the visible hemisphere of the source patch P_s is uniformly partitioned into subshells, as shown in figure 18. Within each section,

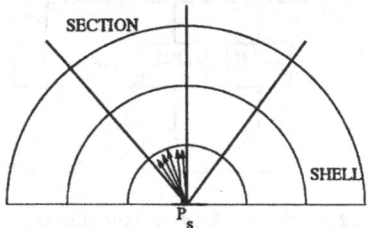

Figure 18: *Pre-process uniform space partitioning*

a limited number of rays traverse the data structure. Cells that are hit by rays release their patches and the number of high density rays, within the patch's spherical bounding box, that potentially hit the patch are estimated. Based on these estimates, sections are redefined so that the amount of estimated intersection computations is section invariant. Similarly, sections are partitioned into sectors so that the amount of estimated high-density ray cell traversal hits is sector invariant. Relevant patches that are released by this low-density ray-casting pre-process are classified according to the sector they belong to and are moved from global memory (GM) to local memory (LM), see figure 19. This procedure provides

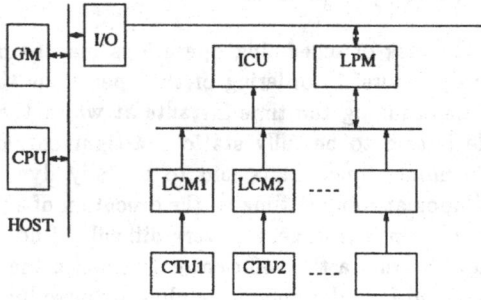

Figure 19: *Relationship between global and local memories*

estimated assignments and orderings of CTU and ICU operations. Global memory (GM) must contain all patch data in the scene to be rendered. [5] GM is shown in figure 20. It consists of a cell memory (GCM) and a patch memory (GPM). GCM in turn consists of a

---

[5] For a realistic storage capacity estimate, one may assume that a typical scene consists of 10^5 àlI 10^6 polygons, each of which requires 100 bytes of storages.

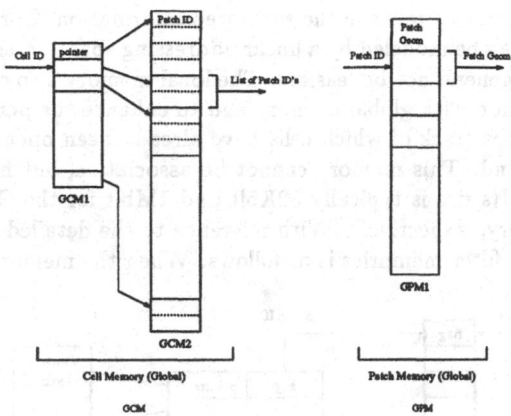

Figure 20: *Global memory structure*

primary address space (GCM1) and a secondary address space (GCM2). GCM1 is a linear storage space which is addressed by a Cell Identifier (Cell Id) and returns a pointer to GCM2 which in turn returns a list of patch Identifiers (Patch Id). GCM2 is approximately $3N_p$ big, where N_p is the total number of polygon patches and a patch is assumed to extend over 3 cells on the average. [6] A Patch Id requires only 20 bits of storage. GPM is addressed by a Patch Id and returns patch geometry data. The setup for local memory (LM) follows that of global memory, see figure 21. In this figure, the memory at the left is the local cell

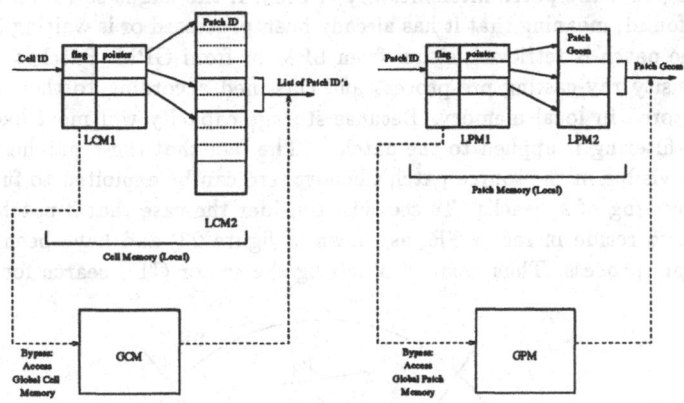

Figure 21: *Local memory structure*

memory (LCM). LCM1 is an addressing space of present flags and pointers which point to LCM2 which returns a list of patch Ids. A similar local patch memory (LPM) is shown at the right in the figure. If a cell or patch is not present in local memory, then a bypass

---

[6]Typical features are: maximum number of cells $(16\ldots32)^3$; maximum number of patches $0.1M - 1.0M$; total number of rays per hemisphere $1M$ and average number of cells per patch 3.

is made to global memory to retrieve the requested information. For the cell memory, the associative memory may be modeled by a linear addressing space. A similar modeling of the patch memory will in general not be feasible. The local memory also contains *filter* memory to reduce communication with global memory and to enhance the performance of the ICU. The filter memory keeps track of which cells have already been opened and which patches have already been found. This memory cannot be associative, but has to be fully, that is linearly, addressable. Its size is typically 32Kbit and 1Mbit for the Cell filter memory and the Patch filter memory, respectively. With reference to the detailed LPM shown in figure 22, the working of the filter memories is as follows. When the memory receives a Patch Id,

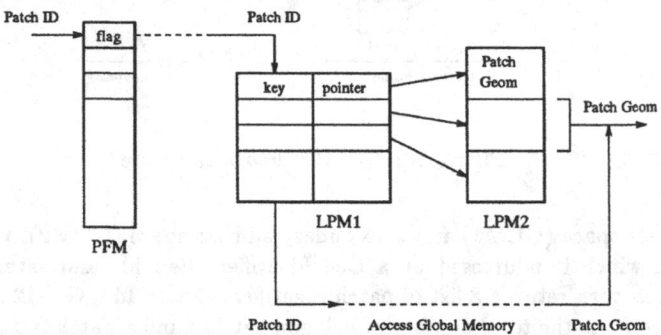

Figure 22: *A detail of the Local Patch Memory*

it is checked against the patch filter memory (PFM). If the flag is set, then this patch has already been found, meaning that it has already been processed or is waiting for processing. Otherwise, the patch is retrieved, either from LPM or from GPM. Patches that are found in the low-density ray-casting pre-process and classified according to their appearance in sections are stored in local memory. Because storage capacity will most likely be limited a priority pre-filtering is applied to the patches. The fact that these patches are known to be potentially visible in the source patch's hemisphere can be exploited to further refine or order the processing of subshells. To see this, consider the case that 3 patches P_1, P_2 and P_3 are known to reside in sector SR, as shown in figure 23, and have been found in this order by the pre-process. Thus, instead of letting the sector CTU search for all patches in

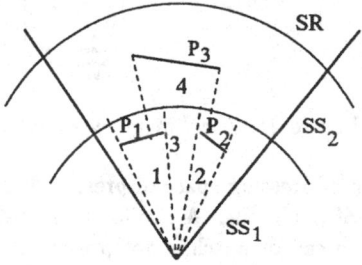

Figure 23: *Introducing secondary subshells*

subshell SS1, following some lexicographical scan order, we let the CTU search for patches within the area defined by the spherical bounding box of patch P_1, as shown in figure 23, while, at the same time, patch P_1 and the (survived) rays within this spherical bounding box are input to the section ICU for intersection testing. One could say that patch P_1 defines a *secondary subshell* within subshell $SS1$, and so do patches P_2 and P_3. Once the CTU has finished its searching for patches in $P_1's$ secondary subshell, it moves on to $P_2's$ secondary subshell. Also, once the ICU has returned all rays in $P_1's$ secondary subshell that have survived intersection tests, these rays are ready to traverse secondary subshells in subshell SS_2. Observe that individual rays are not allowed to proceed from one shell to the next one. All rays in a secondary subshell must have been processed by the ICU before they can move on to the next shell as a result of which the terminations of the processing of secondary subshells are defining *synchronization points*. Notice that patch P_3 defines secondary subshells in subshell SS2 as well as in subshell SS1. In the ideal case, the entire set of secondary subshells, defined by the patches in sector SR that were found by the low-density ray-casting pre-process, will completely cover the set of subshells in sector SR. In practice, however, this may not be the case (the pre-process being subsampling) so that some "holes" may remain unprocessed. Notice, however, that such holes will be uniquely determined by remaining rays that have not intersected any patches. Thus, once all known secondary subshells have been processed, the remaining rays will define final secondary subshells which will remove both the section holes and the unprocessed rays. These rays can be found by inspecting the *ray memory*, which is discussed in the next section.

5   Ray memory

Let P_s be a polygonal source patch in (x, y)-plane, with normal in the direction of the positive z-axes. Any point (x, y, z) on the "northern" hemisphere, with radius r, put on P_s satisfies $x^2 + y^2 + z^2 = r^2$, and has polar coordinates (r, θ, y) defined by

$$x = r \sin \phi \cos \theta$$
$$y = r \sin \phi \sin \theta$$
$$z = r \cos \phi$$

where θ is the angle in (x, y)-plane between the positive x-axes and the radius ρ : $x^2 + y^2 = \rho^2$, and ϕ is the angle in (z, ρ)-plane between the positive z-axes and the radius r : $x^2 + y^2 + z^2 = r^2$. Similarly, sectors, sections, subshells, secondary subshells and spherical bounding boxes of patches have *footprints* on (θ, ϕ)-plane defined by minimum and maximum values of θ and ϕ respectively. Now, to each section, there is assigned a ray memory which stores rays $R(\theta, \phi)$ residing within the footprint of the section. If a sector CTU has to traverse cells within a secondary subshell, then the rays which are involved in the process are read from ray memory by addressing the secondary subshell's footprint - defined by its spherical bounding box. The memory then returns all rays within this footprint that are *alive*, that is, that have not been labeled *dead* as a result of previous true hits, which are ray-patch intersections with minimum intersection distance or equivalently for which the visibility factor δ is 1. After having processed all known secondary subshells in the entire section, the ray memory must be inspected for rays that are still alive. These rays will find (through cell traversal) the remaining - unfound - patches which will eventually close the environment. For lack of space, we can not go into all the details of the complete parallel algorithm. Nevertheless, some important problems that have not been addressed so far will be briefly touched upon in the next closing section.

6 Concluding remarks

In this paper, we have given a rough description of a parallel algorithm/architecture for rendering photorealistic scenes based on ray casting (ray tracing and radiosity shading). We have concentrated on the first pass of the algorithm, which is mainly a viewpoint independent patch sampling procedure. However, the second pass of the algorithm, which is mainly a viewpoint dependent point sampling procedure, can be shown to have a similar effective and efficient parallel ray-casting solution within the realm of the proposed *shelling technique*. There are, nonetheless, a few crucial issues which have not been addressed so far. Among them are the placement of hemispheres on source patches, the actual number of CTUs and ICUs, and the design of the network in which the processors/memory clusters are embedded. We shall now briefly comment on these issues, taking others, such as the utmost important patch sampling accuracy (hemisphere ray distribution) to be beyond the scope of the paper.

hemisphere placement.   In order to keep the discussion tractable, we have made the assumption that a source patch P_s was looking into the environment through a single hemisphere. This is actually not a realistic assumption and many hemispheres have to be placed on each and every source patch, one at each sampling point. Conceptually, this would mean that for each and every hemisphere on a single source patch P_s, a low-density ray-casting pre-process would have to be run, including the loading of patches found in the pre-process to the local memories of the architecture. However, space partitions corresponding to neighboring hemispheres on a source patch will most likely be very similar in the sense that the resulting memory contents will be identical up to some shifts. Therefore, only 1 low-density ray-casting pre-process is executed for a group of neighboring hemispheres at the minor cost of having to fetch a small percentage of patches from a neighboring memory instead of a strictly local memory.

actual number of processor units.   Again, we assumed that each section is assigned to a single pipelined ICU. The objective, of course, is to avoid bubbles in the pipeline stages. However, even if pipeline efficiency can be guaranteed to be 100%, then it may still be the case that many operations are waiting in a queue to be served. Therefore, true efficiency is obtained if also queue-lengths are minimized. This can be achieved by tuning the number of CTUs and ICUs within a cluster.

cluster network.   The network in which the clusters, each of them consisting of an ICU, a few CTUs and local memory are embedded may in principle be of type *mesh-connected*, *hypercube*, *ring* or *torus*, [8]. For the application at hand, a hypercube and a mesh have both good traffic requirements. Simulations have shown that the mesh or torus connected network achieve the highest price/performance ratio. A mesh-connected network of cluster processors is shown in figure 24. Patches that are found in the low-density ray-casting pre-process may reside in more than one section. The question whether such patches should be stored in more than one local memory (cluster) depends on the memory management policy.

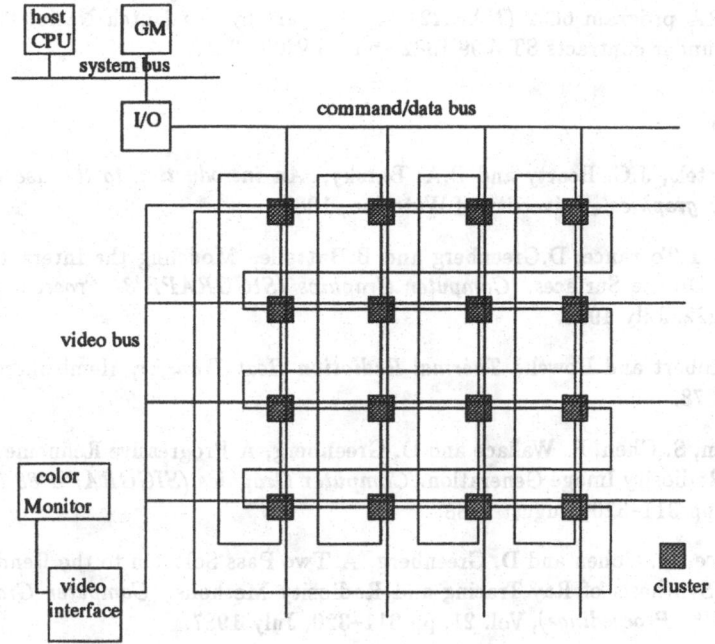

Figure 24: *System configuration*

If only local communication between clusters is allowed, then patch storage duplication will be necessary in case when the corresponding section memories are too far apart, provided memory capacity is large enough. If memories that have to be shared are not far apart, then such patches can be stored in only 1 section memory, and neighboring memories can then be informed about the actual storage location of such patches. In terms of the banded patch-ray computation structure, this means that off-band processing is kept near to band. **simulation results.** Simulations have shown that the parallel algorithm/architecture achieves nearly linear speedup with pipeline efficiency of 80% on the average. With a 20MHz pipeline clocking, a throughput of 16M rays/sec can be sustained. With a network of 32/64 clusters, this means a throughput of up to 512/1024 M rays/sec. In the simulations, a hierarchical memory was used, that is, global memory, local resident memory and (small) local cache memory, [6].

Acknowledgements

The authors wish to thank Prof. F.W. Jansen and A.J. Kok for their support in developing basic ray-tracing and radiosity algorithms. Thanks are also due to G.J. Hekstra for his contributions in defining and designing the critical (VLSI) functions in the architecture. Finally, M.T. Verelst's assistance in setting up the experimental environment is greatly appreciated. The research was supported in part by the commission of the EC under the

ESPRIT BRA program 6632 (NANA2) and in part by the Dutch National Technology Foundation under contracts STW99.1982 and STW00.2331.

References

[1] R.H. Bartels, J.C. Beatty and B.A. Barsky, *An introduction to the use of splines in computer graphics*. University of Waterloo, 1985.

[2] C.Goral, T.Torrance, D.Greenberg and B.Battaile, Modeling the Interaction of Light Between Diffuse Surfaces. *Computer Graphics (SIGGRAPH'84 Proceedings)*, Vol. 18, pp 213–222, July 1984.

[3] Siegel, Robert and Howell, *Thermal Radiation Heat Transfer*. Hemisphere Publishing Corp., 1978.

[4] M. Cohen, S. Chen, R. Wallace and D. Greenberg, A Progressive Refinement Approach to Fast Radiosity Image Generation. *Computer Graphics (SIGGRAPH'88 Proceedings)*, Vol. 22, pp 311–320, August 1988.

[5] J. Wallace, M. Cohen and D. Greenberg, A Two Pass Solution to the Rendering Equation: A Synthesis of Ray-Tracing and Radiosity Methods. *Computer Graphics (SIGGRAPH'87 Proceedings)*, Vol. 21, pp 311–320, July 1987.

[6] Li-Sheng Shen and Ed F. Deprettere, A Hierarchical Memory Structure for the 3D Shelling Technique. *Proceedings IEEE International Conference on Computer Systems and Software Engineering*, pp 244–249, May 1992.

[7] Li-Sheng Shen, F. Laarakker and Ed F. Deprettere, A New Space Partitioning for Mapping Computations of the Radiosity Method onto a Highly Pipelined Parallel Architecture: Part II. *Proceedings Sixth Eurographics Workshop on Graphics Hardware*, September 1991.

[8] K. Hwang and F.A. Briggs, *Computer Architecture and Parallel Processing*. McGraw-Hill, 1984.

REDUCTION AND APPROXIMATION OF LINEAR COMPUTATIONAL CIRCUITS

P. DEWILDE and A.-J. VAN DER VEEN
Delft University of Technology
Department of Electrical Engineering
2628 CD Delft
The Netherlands
dewilde@dutentb.et.tudelft.nl

ABSTRACT. In this tutorial, we present a survey of recent results on the approximation of matrices and operators with siblings of lower complexity. Our main method will be 'approximation via interpolation'. The approach has been successfully used in complex function theory to approximate analytic functions in a given domain by rational functions of low degree such that the approximation is analytic in the domain of interest and satisfies a norm constraint. The new theory is able to extend these results beyond the realm of complex function theory to general matrices and operators. The type of approximation obtained falls in two categories: Nevanlinna-Pick, Schur and Caratheodory-Fejer on the one hand whereby a space in which the approximant must lay is specified, and on the other hand the Adamjan-Arov-Krein type in which minimal state complexity is strived for. We shall show that both types can be brought into one unified framework, in which the properties of J-unitary (simplectic) operators play a major role.

KEYWORDS. Hankel-norm model reduction, time-varying linear systems, computational linear algebra.

1 Introduction

A matrix or operator represents typically the action of a linear map from one vector space into another. It occurs e.g. in the physical modeling of the distribution of a field quantity (electrical, magnetic, elastic field). The result of the modeling is either the computation of a matrix-vector product or the solution of a set of linear equations. We shall call the device (computer, dedicated piece of hardware) which performs the calculation, the *computational circuit*. We shall adopt a simplified model for it, for which a simplified complexity theory

109

M. S. Moonen et al. (eds.), Linear Algebra for Large Scale and Real-Time Applications, 109–135.
© 1993 *Kluwer Academic Publishers.*

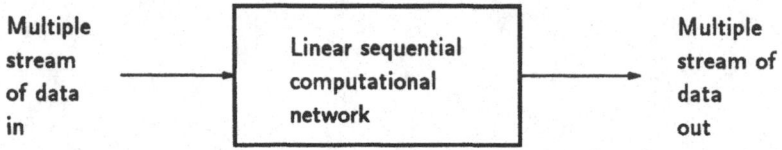

Figure 1: *Black box representation of a computational circuit.*

is valid. Our aim will be to replace the given computational circuit by one of much lower complexity, keeping the result within a given tolerance region.

Our *computational circuit* will be a calculator which performs computations on a possibly non-uniform stream of data, using a controlled amount of 'intermediate' internal data (its state), and which produces a possibly non-uniform stream of data at its output — see figure 1. Here, non-uniform means that the number of input tokens that are consumed in one unit of time, or the number of output tokens that are produced, need not be constant in time, and the same is true for the number of state variables. The complexity of the computational circuit is directly related to the number of state variables that are needed at each time instant.

The central problem that we shall consider in this paper is: *how can the computational circuit be replaced by one of lower complexity and still achieve an input-output map that is within a given tolerance of the original?* A special case of this problem would occur when the computation is 'time-invariant': the computational circuit executes the same computation at each point in its history (which would mean that the circuit exists for all times, but then the computation in a far past could just be trivially processing zero inputs and producing zero outputs). One may represent the operation of the circuit by a 'transfer function' in this special case. It will be a function which is analytic in some domain of the complex plane **C**. The low complexity approximation problem then becomes: find a rational approximant to the transfer function of as low degree as possible. The degree stands for the dimension of the state space needed to execute the approximate computation. One systematic method which produces approximants is due to I. Schur [23]. It holds for the case where the domain of analyticity is the open unit disc, the approximant has to be of the all-pole type, and has to be uniformly bounded by one. The Schur algorithm can be extended to cover more general type of functions, e.g. functions with specified zeros. This then leads to approximation problems of the Nevanlinna-Pick type or some generalization of it. A different approach was chosen by Adamjan, Arov and Krein [1] who explored the properties of the Hankel operator connected to a series expansion of the analytic function, and its relation to a generalized kind of interpolation theory, the Schur-Takagi interpolation problem. This work, and related efforts in functional analysis on the lifting theorem, has led to major breakthroughs in system and filtering theory, namely the solution of the robust control problem [27] for linear, stationary plants and an attractive solution to the broadband matching problem in linear filter theory [16].

In the light of the preceding theory the following question pops up: would it be possible to approximate a general computational circuit much as is possible with a time-invariant circuit? Proofs and properties in the classical case seem very dependent on the properties of spaces of analytic functions, although the questions make sense in the more general context. For example: the inverse of a band matrix (a matrix whose elements are zero outside a band of diagonals) is a full matrix. It makes sense to approximate a full matrix by the inverse of a band matrix of least possible band size. This problem has the same flavor as the Schur problem. Does 'interpolation' make sense in the more general context? Can one translate an approximation problem to an interpolation problem? We shall see that, indeed, these questions can be treated in a purely algebraic fashion.

Our presentation will be structured as follows: in section 2 we derive a mathematical model for linear, sequential computational circuits, and in section 3 we develop an appropriate 'evaluation' or transform theory for them. It will allow us to study the dynamics of a computational circuit that varies in time. The evaluation theory plays the role the z-transform theory plays for classical filters. It will allow us to define generalized interpolation properties. In section 4 we show how these interpolation problems can be solved using the computational model applied to simplectic or J-orthogonal matrices and operators. In section 5 we show how the generalized model reduction theory can be set up and solved in the framework. We wrap up the discussion in section 6 where we present some new results on parametrization and unicity.

2   The computational model

Our computational circuit is a processor which operates in a sequential mode. At the beginning of each stage in its cycle of operations it will have remembered some data from its past, and will take in some new data as input. We call the data remembered from the past, the *state* of the circuit. In the course of the execution of the stage at hand, the computational circuit will calculate (we assume linear computations) a new state for use in the next stage, and an output. This mode of operation can be represented by a signal flow diagram as depicted in figure 2. Assuming linear operations, we shall have, at stage i:

$$\begin{cases} x_{i+1} &= x_i A_i \;+\; u_i B_i \\ y_i &= x_i C_i \;+\; u_i D_i \end{cases} \tag{1}$$

(Note the operator convention: xA for the application of the matrix representation A of an operator to an argument or 'input signal' x — this for consistency with the literature). The dimension of any vector is arbitrary. Computations may start at some time (with an empty state) and end at some time (producing an empty state). This will be the case for computations on finite matrices, and a gentle introduction of the connection of the resulting computational linear algebra with realization theory is given in [25]. Time-invariant infinite Toeplitz operators are also part of the picture, and would correspond to a time-invariant circuit. Let's define sequences of vector spaces as follows:

$$\mathcal{B} = \{\mathcal{B}_i : j \in \mathbf{Z}\} \;=\; \{\cdots, \mathcal{B}_{-1}, [\mathcal{B}_0], \mathcal{B}_1, \cdots\} \tag{2}$$

where $\mathbf{Z}$ is the set of integers and each $\mathcal{B}_j$ is a vector space of infinite dimension (we distinguish the zeroth element by bracketing it). For computations with matrices, the $\mathcal{B}_j$

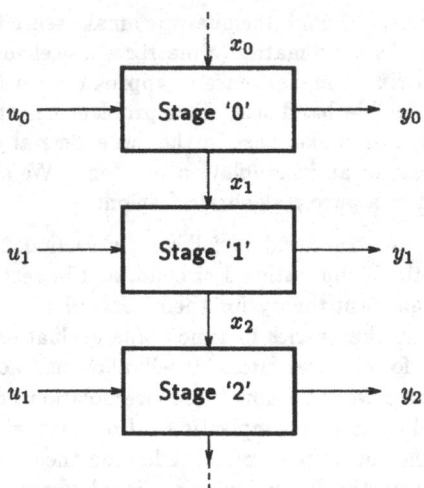

Figure 2: *A computational network.*

will be empty for $j < N_-$ and $j > N_+$ and some numbers $N_- < N_+$. However, there are interesting cases where this is not so: the time-invariant case as a borderline one, cases of periodic or semi-periodic entries .... In the sequel we shall denote the input sequence of spaces (the spaces to which the u_i of figure 2 belong) as $\mathcal{M} = \{\mathcal{M}_j\}$, the state sequence by $\mathcal{B} = \{\mathcal{B}_j\}$ and the output sequence by $\mathcal{N} = \{\mathcal{N}_j\}$.

A linear input-output operator $T : \mathcal{M} \to \mathcal{N}$ can be represented by a *tableau* (i.e. a generalized block-matrix of possibly infinite dimensions) of entries:

$$T = \begin{bmatrix} & \vdots & \vdots & \vdots & \\ \cdots & T_{-1,-1} & T_{-1,0} & T_{-1,1} & \cdots \\ & T_{-1,0} & [T_{0,0}] & T_{0,1} & \\ \cdots & T_{1,-1} & T_{1,0} & T_{1,1} & \cdots \\ & \vdots & \vdots & \vdots & \end{bmatrix} \tag{3}$$

Each T_{ij} represents the effect of the i-th input on the j-th output. It need not be square and may be empty. When the operator T is not a finite matrix, we shall assume that it represents a bounded operator on sequences of bounded quadratic norm. The corresponding input space is then denoted by $\ell_2^{\mathcal{N}}$. Such series and their operators carry with them an interpretation of *energy* which often allows for nice physical insights and provides a connection with passive network theory. This fact has often been exploited in the past to great benefit. We shall say that a transfer operator is *causal* if a future input has no effect on a past output. This will be the case if the tableau is upper triangular (although non-square) i.e. if all entries T_{ij} with $i > j$ are either empty or zero. For example, if $\mathcal{M} = \{\mathbf{C}^2, \emptyset, [\emptyset], \mathbf{C}\}$ and $\mathcal{N} = \{\mathbf{C}, [\mathbf{C}], \mathbf{C}\}$ then a causal operator has the form

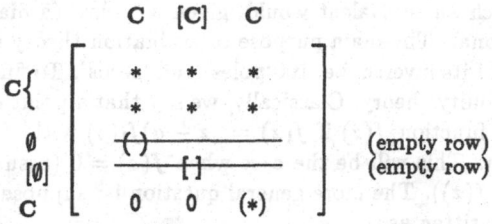

in which the diagonal is indicated, $T_{-1,-1}$ and $T_{0,0}$ vanish, and $T_{-2,-1}$ is a 2×1 matrix. Although this looks like a square matrix, its main diagonal has a rather peculiar form. The map corresponds to a causal computing network that takes in two values at $t = -2$, none at $t = -1, 0$, and one at $t = 1$ while producing one value at $t = -1, 0, 1$ each. The usefulness of computing networks with various sets of inputs and outputs will soon appear. The generic symbol $\mathcal{U}(\mathcal{M}, \mathcal{N})$ indicates bounded upper triangular operators $\mathcal{M} \to \mathcal{N}$ in a given context. Bounded lower triangular operators will form the set $\mathcal{L}(\mathcal{M}, \mathcal{N})$, $\mathcal{D}(\mathcal{M}, \mathcal{N})$ stands for diagonal operators, and $\mathcal{X}(\mathcal{M}, \mathcal{N})$ for the complete collection (warning: $\mathcal{X}(\mathcal{M}, \mathcal{N})$ is *not* the sum of $\mathcal{U}(\mathcal{M}, \mathcal{N})$ and $\mathcal{L}(\mathcal{M}, \mathcal{N})$, except in the finite dimensional case). Elementary shift operators on these spaces are defined as follows:

$B^{(k)}$ for the kth shift rightwards in a sequence, as in
$$B^{(1)} = \{ \cdots B_{-2}, [B_{-1}], B_0, \cdots \},$$

$Z : B \to B^{(1)}$ as the generic shift operator:
$$[\cdots, x_{-1}, [x_0], x_1, \cdots] Z = [\cdots, [x_{-1}], x_0, \cdots].$$

(4)

The product of k shifts will be indicated by $Z^{[k]}$. It is an operator $B \to B^{(k)}$. Notice that Z^k would be an incorrect notation since the subsequent Z's are not the same (the spaces on which they act are different). In a context of infinite time series we shall use the quadratic norm. Operators in $\mathcal{X}, \mathcal{U}, \mathcal{L}, \mathcal{D}$ will map $\ell_2^{\mathcal{M}} \to \ell_2^{\mathcal{N}}$ boundedly. However, our theory works equally well for finite dimensional spaces, and a reader who is not well versed in Hilbert space theory can ignore the functional calculus. $\mathcal{X}_2, \mathcal{U}_2, \mathcal{L}_2, \mathcal{D}_2$ will indicate elements of, respectively, $\mathcal{X}, \mathcal{U}, \mathcal{L}, \mathcal{D}$ whose entries are square summable with that sum as their norm squared. When the spaces are of finite dimensions then this norm is commonly called a *Frobenius norm*, otherwise a *Hilbert-Schmidt* norm. In contrast to $\mathcal{X}, \mathcal{U}, \mathcal{L}, \mathcal{D}$, such spaces form a Hilbert space or an Euclidean space with an inner product and orthogonal projections. E.g., **P** will indicate the orthogonal projection of $\mathcal{X}_2$ on $\mathcal{U}_2$.

Let $T \in \mathcal{U}$, then we can formally decompose T in a sum of shifted diagonal operators as in

$$T = \sum_{k=0}^{\infty} Z^{[k]} T_{[k]}, \qquad T_{[k]} \in \mathcal{D}(\mathcal{M}^{(k)}, \mathcal{N}).$$

(5)

When T is not a finite operator, then (5) may not converge and the sum must be considered 'formal'. In our time-varying theory, diagonals will play the role of the scalars of classical complex function theory. An important point is whether there is an equivalent to the classical complex function evaluation or, as electrical engineering would say, to the z-transform. Such an equivalent would, given a 'point' (a diagonal), map the transfer operator to a diagonal. The main purpose of evaluation theory is the discovery of the modes of the system and its inverse, i.e. its 'poles' and 'zeros'. To find what we need, we start out with some divisibility theory. Classically, we say that a point $a \in \mathbf{C}$ is a zero of a (bounded causal) transfer function $f(z)$ if $f(z) = (z - a)f_1(z)$ with $f_1(z)$ a new (bounded causal) transfer function. This will be the case when $f(a) = 0$ (assuming that a is in the domain of analyticity of $f(z)$). The more general question is: suppose that T is an upper operator, when can it be written as

$$T = (Z - D)T_1 \tag{6}$$

where D is diagonal, Z has appropriate dimensions and T_1 is again upper? The question finds an answer in terms of a new type of evaluation, the W-transform, which we now introduce.

Let $A : \mathcal{M} \to \mathcal{N}$ be an arbitrary operator, then we define the kth (South-East) shift of A as

$$A^{(k)} = (Z^{[k]})^* A Z^{[k]} : \quad \mathcal{M}^{(k)} \to \mathcal{N}^{(k)} : \tag{7}$$

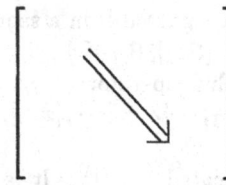

Let $W : \mathcal{M} \to \mathcal{M}^{(1)}$ be a diagonal operator, and let us define the subsequent products

$$W^{[0]} = I, \qquad W^{[1]} = W, \qquad W^{[n]} = WW^{(1)} \ldots W^{(n-1)}. \tag{8}$$

Definition 2.1. *Let $T = \sum_{k=0}^{\infty} Z^{[k]} T_{[k]}$ be a diagonal decomposition of $T : \mathcal{M} \to \mathcal{N}$ and let $W : \mathcal{M} \to \mathcal{M}^{(1)}$ be a diagonal operator. The W-transform of T is given by*

$$\hat{T}(W) = \sum_{k=0}^{\infty} W^{[k]} T_{[k]}, \tag{9}$$

whenever the sum converges [2].

The sum will always converge when T has finite dimensions. When T is infinite but bounded, then the sum will converge if $\ell_W := \lim_{n \to \infty} \| W^{[n]} \|^{1/n} < 1$. In terms of the W-transform, the divisibility theorem is as follows.

Theorem 2.2. *Let T be bounded upper, and let D be such that $\ell_D < 1$. Then $T = (Z - D)T_1$ with T_1 bounded upper, iff $\hat{T}(D) = 0$.*

A proof of the scalar property is in [2], a general proof in [3].

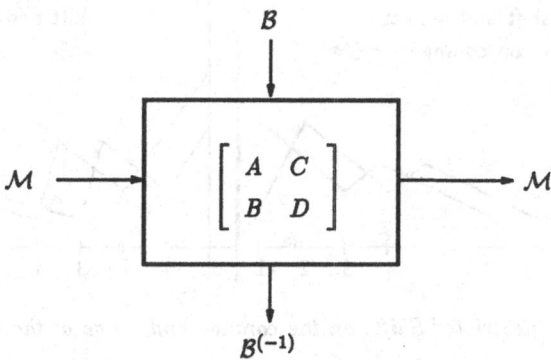

$\mathcal{B}$

$\mathcal{M} \longrightarrow$ $\begin{bmatrix} A & C \\ B & D \end{bmatrix}$ $\longrightarrow \mathcal{M}$

$\mathcal{B}^{(-1)}$

Figure 3: *Global state space representation.*

The W-transform does not inherit all the properties of the z-transform, but just enough to make things work. E.g. it is not true in general that $(TS)^{\wedge}(W) = \hat{T}(W)\hat{S}(W)$. The weaker but very useful property is $(TS)^{\wedge}(W) = (\hat{T}(W)S)^{\wedge}(W)$, see [3].

State descriptions

A computing circuit produces a causal, thus upper input-output map. But it does so in stages. The complexity of each stage can be measured by the size of the local state. In our matrix approximation theory we shall aim at reducing that size. Let's express the state representation in our diagonal algebra. The state space at the input of stage i we define as the vector space $\mathcal{B}_i$. Let $\mathcal{B} = \{i \in \mathbf{Z} \mapsto \mathcal{B}_i\}$ be the sequence of the state spaces (where $\mathcal{B}_i$ may be $\emptyset$). The *state transition map* then maps $\mathcal{B} \to \mathcal{B}^{(-1)}$ (the anticausal, backward shift). It is a diagonal operator:

$$A = \mathrm{diag}[\cdots, A_{-1}, [A_0], A_1, \cdots]. \tag{10}$$

Similarly, B will be a diagonal map $\mathcal{M} \to \mathcal{B}^{(-1)}, C : \mathcal{B} \to \mathcal{N}$ and $C : \mathcal{M} \to \mathcal{N}$ — see figure 3. If x is a state sequence, then the computational circuit will execute the transformation

$$\begin{cases} xZ^{-1} & = xA + uB \\ y & = xC + uD \end{cases} \tag{11}$$

in which $\{A, B, C, D\}$ are all diagonal operators and x is the state sequence. The transfer operator can be recovered from (11) when $Z^{-1} - A$ or equivalently $I - ZA$ is invertible, in which case

$$T = D + B(I - ZA)^{-1}ZC. \tag{12}$$

A sufficient condition for bounded invertibility is that $\ell_A < 1$, which is always the case when A is finite, and is otherwise known as the condition that the system is *strictly stable*.

116

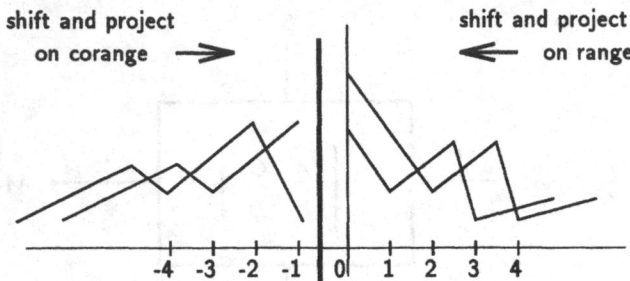

Figure 4: *Restricted Shifts on the corange and range of the Hankel operator.*

Realization theory

There is a large literature on realization theory, even for time-varying systems [26, 19, 20, 4, 14, 24]. We suffice here with some essential results. *Realizing* a causal operator means: deriving a computational circuit with its state description that reproduces T exactly. The classical, one-port case was solved by Kronecker [21] who stated that a transfer function $T(z) = t_0 + t_1 z + t_2 z^2 + \cdots$, with z a complex variable, is a rational of degree n, and hence realizable with a finite-dimensional state space system of dimension n if and only if the related Hankel operator:

$$
H_T = \begin{bmatrix} t_1 & t_2 & t_3 & \cdots \\ t_2 & t_3 & & \\ t_3 & & \ddots & \\ \vdots & & & \end{bmatrix}
\tag{13}
$$

has finite rank n. The one-port case was extended to the multiport case by Ho and Kalman [17]. As remarked by Kalman [18], the Hankel operator is nothing but a matrix representation of an operator that links the past of the system to its future. Take a context of ℓ_2 time series, let the 'current' point in time be $t = 0$, and let $\mathbf{P}$ be the orthogonal projection of $\ell_2(-\infty, \infty)$ to $\ell_2(0, \infty)$. Assume that T is a bounded operator and consider inputs with carrier on the semi-interval $t \leq -1$ (i.e. $u \doteq [\cdots, u_{-2}, u_{-1}, [0], 0, \cdots]$). Then the Hankel operator is defined as $H_T = \mathbf{P}Tu$. If H_T has finite range, then its corange is finite dimensional as well (the corange of an operator A between Hilbert spaces or Euclidean vector spaces is the range of its adjoint A^*) and lives in the space $\ell_2(-\infty, \infty)$. A minimal state space realization can be set up by choosing a basis either in the range (for an *output normal form*) or in the corange (for an *input normal form*). The A-matrix then becomes a representation of a type of restricted shift (i.e. shift followed by projection) in that space — see figure 4 for an illustration of the concept of *natural evolution* of the state. We could call the corange and the range of H_T the *natural state spaces* of the system, in which the unforced transition to a new state acts by 'shift and project'. If an orthonormal basis is chosen in the corange, then the restricted shift operator will become contractive. In fact, more is true for the input normal form: the resulting realization $\{A, B, C, D\}$ will have the pair A, B isometric: $A^*A + B^*B = I$. This is due to the fact that the matrix $\begin{bmatrix} A \\ B \end{bmatrix}$ stands for an orthonormal projection of past inputs to the corange. If the matrix is augmented to form a unitary matrix: $\begin{bmatrix} A & C_l \\ B & D_l \end{bmatrix}$, then the corresponding transfer function $U_l(z) = D_l + B(I - zA)^{-1}zC_l$ is allpass, i.e. analytic in the unit disc and unitary on the unit-circle (it is a Blaschke product), and forms a *left coprime factor* of $T(z)$ in the sense that $T(z) = U_l(z)\Delta_l^*(z)$ with $\Delta_l(z)$ analytic in the unit disc of smallest possible degree. (Alternatively: $U_l^*(z)$ pushes $T(z)$ to conjugate analyticity in a minimal fashion.) If n is the dimension of the state space (i.e. the dimension of the corange or range of H_T), then one can show that the corange is generated by $\mathbf{C}^n z^{-1}(I - A^*z^{-1})^{-1}B^* = \mathbf{C}^n(z - A^*)^{-1}B^*$. We shall call that space the *controllability space* of $T(z)$ with A, B its controllability pair. Related is the map which projects the input space to the state space and which is given by the controllability operator $\mathcal{C}$ for which $\mathcal{C}^* = [B^*\ A^*B^*\ (A^*)^2B^* \cdots]$ — assuming the natural basis for the input space.

Likewise and dually, an output normal representation can be based on an orthonormal basis in the range of H_T, which is also a restricted-shift invariant subspace. In this case one shall obtain an isometric $[A\ C]$ *observability pair* and an allpass $U_r(z) = D_r + B_r(I - zA)^{-1}zC$ with

$$\begin{bmatrix} A & C \\ B_r & D_r \end{bmatrix}$$

unitary and $T(z) = \Delta_r^*(z)U_r(z)$ a right coprime factorization, an observability space (the range of H_T) generated by $\mathbf{C}^n(I - zA)^{-1}C$ and an observability operator $\mathcal{O}$ given by $\mathcal{O} = [C\ AC\ A^2C \cdots]$.

In the case of a (time-varying) computational circuit, the situation is in principle much the same but mathematically more complex. A time-invariant system can be identified by looking at just one point in time: at all other points the system behaves just the same, at least as far as its past-inputs-to-future-outputs (i.e. Hankel) map is concerned. In a time-varying situation it will be necessary to look at each individual point separately. This is most elegantly achieved by 'stacking' inputs which are active up to the subsequent relevant points. Given a sequence of spaces $\mathcal{M} = \{\mathcal{M}_i\}$ we achieve this by considering as system input space the Hilbert-Schmidt spaces $\mathcal{X}_2^{\mathcal{M}} := \mathcal{X}_2(\mathbf{Z}, \mathcal{M}), \mathcal{L}_2^{\mathcal{M}} := \mathcal{L}_2(\mathbf{Z}, \mathcal{M})$, etc., i.e. for each time-point $i \in \mathbf{Z}$ in the computation a series in $\mathcal{M}$ which is bounded in energy, and such that the complete collection of input series considered is overall bounded in energy as well. Such a space is rich enough to test the behavior of the system, and small enough for tractability as a Hilbert space (or Euclidean space in Frobenius norm). The outputs are then collected in a similar Hilbert-Schmidt space of the type $\mathcal{X}_2^{\mathcal{N}}$. If the upper transfer operator T is ℓ_2-bounded it will map $\mathcal{X}_2^{\mathcal{M}}$ to $\mathcal{X}_2^{\mathcal{N}}$ with the same norm as before. The 'strict past' taken over all time points will now be represented by $\mathcal{L}_2^{\mathcal{M}}Z^{-1}$ while the future outputs, taken over all time points will lay in $\mathcal{U}_2^{\mathcal{N}}$. The Hankel operator is now defined as the system induced map $H_T : \mathcal{L}_2^{\mathcal{M}}Z^{-1} \to \mathcal{U}_2^{\mathcal{N}} : U \mapsto \mathbf{P}(UT)$ where $\mathbf{P}$

indicates the projection of $\mathcal{X}_2$ onto $\mathcal{U}_2$. H_T is a tensor: it maps matrices (or operators) to matrices (or operators). A *snapshot* of H_T is obtained when a time point k is fixed and the effect of strictly past inputs, zero from k on are considered on future outputs from time k on:

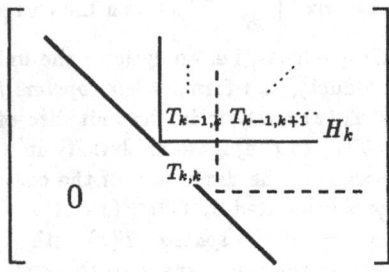

This results in an instantaneous Hankel snapshot H_k. In [24], the realization theory based on such snapshots was considered and it was shown there (and elsewhere...) that, modulo some technical conditions, a pointwise finite dimensional realization exists for the system if each H_k has finite rank. The minimal dimension of the state space at time point k will be precisely the rank of H_k. Notice that it cannot change very fast from one time point to the next since subsequent Hankel snapshots have a large set of data in common. With a scalar input, the change in rank can be at most 1. Assume now that indeed all Hankel snapshots have (uniformly bounded) finite rank. We shall then say that our system is *locally finite*. The columns of each H_k span a finite dimensional space which, when starred, will span the local corange, while the rows of each H_k span the local range. Choosing an orthonormal basis in these spaces and considering the natural evolution of the state will again lead to time-varying realizations in input and output normal form. In a canonical input normal form, each $\begin{bmatrix} A_k \\ B_k \end{bmatrix}$ is isometric at each k. Hence the collected block diagonal operator $\begin{bmatrix} A \\ B \end{bmatrix}$ will be isometric as well. Likewise, a realization in output normal form will have $[A \ C]$ isometric. Just as in the time-invariant case, we may consider global controllability and observability spaces which will be generated respectively by $\mathcal{D}_2^{\mathcal{B}}(Z - A^*)^{-1}B^*$ and $\mathcal{D}_2^{\mathcal{B}}(I - AZ)^{-1}C$ and represent the corange and range of H_T. Here also, coprime factorizations are possible and play an important role in approximation theory. For example, if the restricted shift realization is used in the output range, then the isometric $[A \ C]$ may be extended to form a unitary block diagonal operator $\begin{bmatrix} A & C \\ B_u & D_u \end{bmatrix}$. The corresponding transfer operator

$$U_r = D_u + B_u(I - ZA)^{-1}ZC \tag{14}$$

will be allpass (i.e. upper and unitary) and provide a coprime factorization

$$T = \Delta_r^* U_r \tag{15}$$

with Δ_r upper. U_r shares canonical output space with T.

State transformations

Suppose that you wish to transform the actual state sequence x to a new sequence $x = \hat{x}R$ where $R = \text{diag}[\cdots, R_{-1}, [R_0], R_{-1}, \cdots]$. We request the transformation to be bounded and boundedly invertible. Since x satisfies (11) we have for $\hat{x}$:

$$\begin{cases} \hat{x}RZ^{-1} & = \hat{x}RA + uB \\ y & = \hat{x}RC + uD \end{cases} \tag{16}$$

and the new transition operator becomes

$$\begin{bmatrix} RA(R^{(-1)})^{-1} & RC \\ B(R^{(-1)})^{-1} & D \end{bmatrix} \tag{17}$$

(notice the shift in the inverse!). When you start out from a non-normal controllability pair A, B and you wish to apply a state transformation R to bring the realization to input normal form, then R will have to satisfy

$$A^* R^* R A + B^* B = (R^{(-1)})^* R^{(-1)}$$

or, with $M = R^* R$,

$$A^* M A + B^* B = M^{(-1)}. \tag{18}$$

This equation is known as a *Lyapunov-Stein* equation. It has a unique solution when

$$\ell_A := \lim_{n \to \infty} \|A^{[n]}\|^{1/n} < 1, \tag{19}$$

which is given by

$$M = \left\{ \sum_{k=0}^{\infty} (A^*)^{[k]} (B^* B)^{(k)} ((A^*)^{[k]})^* \right\}^{(1)}. \tag{20}$$

R will exist when M is boundedly invertible, a condition of the system that we call *strictly controllable*.

3 J-Unitary operators

Symplectic, or equivalently, J-unitary transfer operators play an essential role in interpolation theory, and hence also in approximation and model-reduction theory as we shall see. We introduce them in this, and an important usage in the next section, where we shall explore the connection between between *lossless inverse scattering* and interpolation. A J-unitary transfer operator has split input and output terminals: $M = M_1 \oplus M_2$ and

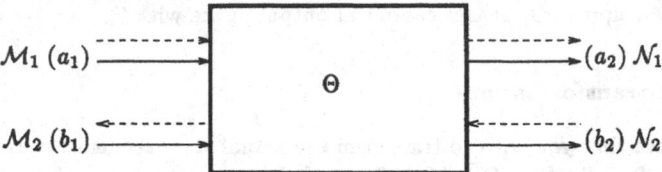

Figure 5: *Signal flow (solid arrow) and energy flow (dashed arrow) in a J-unitary operator.*

$\mathcal{N} = \mathcal{N}_1 \oplus \mathcal{N}_2$ where each of these symbols stand for a sequence of spaces. Input signals belonging to the $\mathcal{M}_2$ sequence are thought of as carrying positive energy to the device (we call them *incident waves*) while input signals belonging to the $\mathcal{M}_2$ sequence take energy away from the device (*reflected waves*). Similarly, a signal belonging to $\mathcal{N}_1$ carries energy to the next stage while one belonging to $\mathcal{N}_2$ brings in energy from the next stage. Connected to the input and the output there are signature matrices

$$J_1 = \begin{bmatrix} I_{\mathcal{M}_1} & \\ & -I_{\mathcal{M}_2} \end{bmatrix}, \qquad J_2 = \begin{bmatrix} I_{\mathcal{N}_1} & \\ & -I_{\mathcal{N}_2} \end{bmatrix}, \tag{21}$$

characterizing the energy flow at each port. We shall say that an operator Θ is J-unitary (with respect to the splitting (21)) if

$$\Theta^* J_1 \Theta = J_2, \qquad \Theta J_2 \Theta^* = J_1, \tag{22}$$

i.e. if the overall intake of energy of the device over all time is zero. If this is the case, the device is 'physically' *lossless*. Figure 5 displays the signal flow and energy flow in a Θ-operator. The intuitive reason why a J-unitary operator like Θ plays a role in interpolation theory has to do with the *scattering operator* Σ connected to Θ which links incident energy represented by the signals $[a_2 \ b_1]$ to reflected energy $[a_2 \ b_1]$. Σ is given in terms of Θ by the relation

$$\Sigma = \begin{bmatrix} I & -\Theta_{12} \\ 0 & I \end{bmatrix} \begin{bmatrix} \Theta_{11} & 0 \\ 0 & \Theta_{22}^{-1} \end{bmatrix} \begin{bmatrix} I & 0 \\ \Theta_{21} & I \end{bmatrix}. \tag{23}$$

Σ will exist as a bounded operator when Θ_{22}^{-1} exists and will be causal when both Θ and Θ_{22}^{-1} are causal. In that case, Σ is at the same time unitary and causal — it is called a *lossless scattering matrix*. It will happen that Θ_{22}^{-1} exists as a bounded operator but is not causal. In that case Σ will still exist as a unitary operator but will not be causal. It is also possible that Θ corresponds to a unitary and causal Σ, i.e. to a lossless network, but is itself non-causal (it is said not to be of *minimal phase* — in this paper we shall not use such Θ's). In the time-invariant case, the modes (poles) of Θ correspond to transmission zeros, i.e. zeros of the transfer function Σ_{11} which effectively shield the left port (in figure 5) from a passive (i.e. causal and contractive) load S_L at the right port. The resulting map $a_1 \mapsto b_1$ is called an *input scattering function*. The intrinsic input scattering function is obtained when $S_L = 0$ and is given by $\Sigma_{12} = -\Theta_{12}\Theta_{22}^{-1}$. An input scattering function corresponding to a

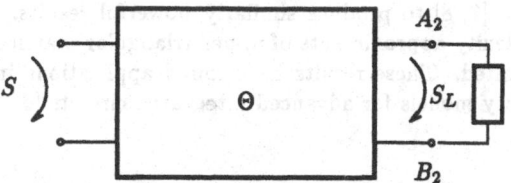

Figure 6: *A chain-scattering matrix connected to a load.*

different S_L will still exhibit the same scattering behavior when evaluated at the frequencies of the transmission zeros, hence will evaluate to the same value as Σ_{12} at those points. An additional property of the set-up is that contractivity is preserved: an input scattering function will be contractive as soon as the load S_L is — see figure 6. Θ is often called a *chain-scattering matrix*. Much of its time-invariant properties carry over to the time-variant case. In the next sections we shall make this statement precise, here we suffice with a brief description of the lossless inverse scattering philosophy.

Lossless inverse scattering

The lossless inverse scattering (LIS-)problem in the context of computational networks has been treated to some extent in the literature [3]: suppose that an input scattering function S is given which is causal and contractive, what could be a chain scattering matrix Θ and a causal and contractive load S_L such that S results from the combination as depicted in figure 6? Posed this way, the LIS problem has a very large collection of solutions. Consider

$$[I \quad S]\Theta = [A_2 \quad A_2 S_L] \tag{24}$$

in the time-invariant case (see [9, 10, 11] for details). If Θ is chosen to be causal, then A_2 and $A_2 S_L$ will be causal as well and inherit the zeros of Θ. It turns out that, because of J-unitarity, the zeros of Θ are precisely the reflections with respect to the unit circle of the complex plane of the poles of Θ and are hence located within the unit disc. It follows that $[A_2 \quad A_2 S_L]$ can be written as $U[A_2' \quad A_2' S_L]$ where U is a Blaschke product collecting the zeros of Θ, and $[A_2' \quad A_2' S_L]$ is still causal. Reversing the reasoning, give yourself any Blaschke product U and look for a Θ such that

$$U^*[I \quad S]\Theta \tag{25}$$

is causal. It turns out that the minimal Θ that achieves this feat is unique except for a right constant J-unitary factor, and that it defines a causal and contractive load S_L. In the next section we shall show how Θ is actually constructed even in the time-invariant case, and we shall use the state-space theory developed in the previous section and applied to a candidate Θ to achieve the result.

The simplest possible expression of the LIS principle is the celebrated Schur algorithm [23], in which $U = z^n$ is chosen. A time-varying version of this algorithm uses $U = Z^n$

122

and was developed in [7, 8] to produce similarly powerful results. In particular, it leads
to optimal low-complexity approximants of upper triangular matrices with matrices whose
inverses are band limited. These results have found applications in the determination of
optimal low-complexity models for advanced integrated circuits [6].

4 Interpolation

Keeping in line with the notation that 'diagonals' replace 'complex-points' in the time-
varying situation, and that divisibility is governed by the W-transform introduced earlier,
we can formulate an interpolation problem as follows:

> Given a set of diagonal 'points' $\{V_k : k = 1 \cdots n\}$ and a set of diagonal 'values'
> $\{s_k : k = 1 \cdots n\}$, find an $F \in \mathcal{U}$ such that $\| F \| < 1$ and $\hat{F}(V_k) = s_k$.

The remarkable property of Θ-matrices which makes them such a good vehicle for interpolation theory is the direct connection between their controllability space and the interpolation properties of its input scattering functions when it is connected to various (passive,
i.e. upper contractive) loads. When a Θ is loaded in a causally contractive load S_L, then
an input scattering operator which we shall call $S = T_\Theta(S_L)$ results, and is given by

$$S = T_\Theta(S_L) = (-AS_L + C)(BS_L - D)^{-1}. \tag{26}$$

(the inverse is guaranteed to exist when S_L is strictly contractive). The property needed is
stated in the following theorem.

Theorem 4.1. ([6]) *If* $S = T_\Theta(S_L)$, *where* $S_L \in \mathcal{U}$, *is strictly contractive and if the
causal, J-unitary matrix Θ is such that its controllability space is given by*

$$\text{span} \left\{ D_j(Z - V_j)^{-1}[I \quad -s_j] : D_j \in \mathcal{D}_2, j = 1 \cdots n \right\}, \tag{27}$$

then

$$\hat{S}(V_j) = s_j, \qquad j = 1 \cdots n. \tag{28}$$

Theorem 4.1 reduces the interpolation problem to finding the correct controllability space
for Θ. It also makes the intuitive notion precise that the modes of Θ determine the interpolation properties of an input scattering function. How can we now produce Θ with the
controllability space specified by (27)? Firstly, it is not hard to see that (27) is generated
by the controllability pair:

$$\begin{bmatrix} A \\ B_1 \\ B_2 \end{bmatrix} = \begin{bmatrix} \text{diag}(V_1^* & \cdots & V_n^*) \\ 1 & \cdots & 1 \\ -s_1^* & \cdots & -s_n^* \end{bmatrix}. \tag{29}$$

That such a controllability pair may qualify for a J-unitary Θ follows from J-orthogonal

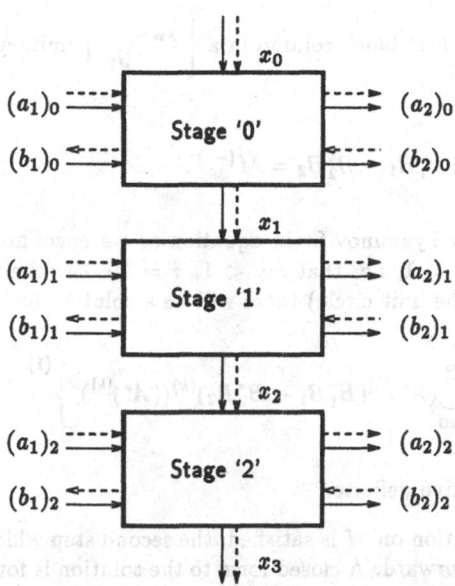

Figure 7: *Signal (solid arrow) and energy (dashed arrow) transfer in a realization of a time-varying Θ matrix.*

realization theory: use the sign splitting (21) on the input and output spaces and let I_n be a unit matrix corresponding to the dimensions of the interpolation problem. We have

Theorem 4.2. ([6]) *If the state space realization $\theta = \begin{bmatrix} A & C \\ B & D \end{bmatrix}$ for Θ satisfies*

$$\theta^* \begin{bmatrix} I_n & \\ & J_1 \end{bmatrix} \theta = \begin{bmatrix} I_n & \\ & J_2 \end{bmatrix}, \qquad \theta \begin{bmatrix} I_n & \\ & J_2 \end{bmatrix} \theta^* = \begin{bmatrix} I_n & \\ & J_1 \end{bmatrix} \tag{30}$$

then Θ is J-unitary and corresponds to a passive, lossless scattering operator Σ (i.e. an all-pass operator).

The J-unitarity of the realization θ means conservation of energy in the computations at each time step according to the energy flow displayed in figure 7. Notice that in this case the equations (30) force $J_1 = J_2$ because the two sides are congruent and hence have equal signatures. To determine θ which is J-unitary and has the correct controllability pair we proceed in two steps:

1. Find a state transformation operator X such that

$$\begin{bmatrix} \alpha \\ \beta_1 \\ \beta_2 \end{bmatrix} = \begin{bmatrix} XA \\ B_1 \\ B_2 \end{bmatrix} (X^{(-1)})^{-1} \tag{31}$$

qualifies as the first block column in a $\begin{bmatrix} I_n & \\ & J_1 \end{bmatrix}$-unitary matrix, i.e. such that, with $X^*X = M$,

$$A^*MA + B_1^*B_1 - B_2^*B_2 = M^{(-1)}. \tag{32}$$

This is again a Lyapunov-Stein equation as we encountered before. Under the condition that $\ell_A < 1$, i.e. that $\ell_{V_i} < 1$, $i = 1 \cdots n$ (the 'points' are in the diagonal equivalent of the unit circle) there will be a solution for R if and only if

$$M = \left\{ \sum_{k=0}^{\infty} (A^*)^{[k]} (B_1^*B_1 - B_2^*B_2)^{(k)} ((A^*)^{[k]})^* \right\}^{(1)} \tag{33}$$

is strictly positive definite.

2. Once the condition on M is satisfied, the second step which consists in completing θ is fairly straightforward. A closed form to the solution is found in [6]. Alternatively, one can perform the following computations: determine, for each k a matrix W_k whose columns span the orthogonal complement of $\begin{bmatrix} (\alpha)_k \\ (\beta_1)_k \\ (\beta_2)_k \end{bmatrix}$, and compute a sequence of congruence matrices R_k^{-1} such that

$$W_k^* \begin{bmatrix} (I_n)_k & \\ & (J_1)_k \end{bmatrix} W_k = R_k^{-*}(J_1)_k R_k^{-1} \tag{34}$$

and then remark that

$$\begin{bmatrix} (\alpha)_k \\ (\beta_1)_k \\ (\beta_2)_k \end{bmatrix} W_k R_k \tag{35}$$

satisfies the condition.

When we look more carefully at the crucial formula (33), we see that the necessary and sufficient condition for the existence of the solution becomes the positive definiteness of the operator with entries

$$M_{ij} = [(Z - V_i)^{-1}(I - s_i s_j^*)(Z^* - V_j^*)^{-1}]^\wedge(0). \tag{36}$$

Students of the classical Nevanlinna-Pick problem will recognize a generalization of the Pick matrix!

5 Hankel norm model reduction

In this section we address the problem of approximating a computational circuit of high (state) complexity with one of much lower complexity given a certain tolerance. The accuracy of the approximation will be measured in Hankel norm, which from experience is known to be a good strong norm (in the time-invariant case it is much stronger than the L_2 norm). For the time-invariant case, the solution of the problem is based on the work of Adamyan, Arov and Krein [1]. In the more general time-varying case, a solution was presented in [13]. Here we indicate the main ideas.

The numerical rank of a matrix or operator is defined as the number of singular values greater than a given precision ϵ. If A is an $m \times n$ matrix, then its singular value decomposition, which always exists, is given by [15]

$$A = U\Sigma V^*, \tag{37}$$

where U and V are, respectively, $m \times m$ and $n \times n$ unitary matrices, and Σ is an $m \times n$ matrix whose only non-zero elements are on its main diagonal and are called *singular values*:

$$\Sigma = \begin{bmatrix} \sigma_1 & & & 0 \\ & \sigma_2 & & \\ & & \ddots & \\ 0 & & & \sigma_{\min(m,n)} \end{bmatrix}, \tag{38}$$

with additionally $\sigma_1 \geq \sigma_2 \geq \cdots \geq \sigma_{\min(m,n)} \geq 0$. The numerical rank is $k = \max\{l : \sigma_l \geq \epsilon\}$ for some small $\epsilon > 0$. If A is replaced by $A_a = U\Sigma_a V$, with

$$\Sigma_a = \begin{bmatrix} \sigma_1 & & & 0 \\ & \sigma_2 & & \\ & & \ddots & \\ 0 & & & \sigma_k \end{bmatrix}, \tag{39}$$

then A_a has rank k and approximates A within ϵ both in operator norm:

$$\| A - A_a \| = \sup\{ \| (A - A_a)x \| : \|x\| \leq 1 \} \tag{40}$$

and in Hilbert-Schmidt or Frobenius norm:

$$\| A - A_a \|_{HS} = \{ \sum_{i=1,j=1}^{m,n} | [A - A_a]_{ij} |^2 \}^{1/2}. \tag{41}$$

For a time-invariant Hankel operator of finite rank H_T, a similar SVD decomposition exists with the number of non-zero singular values equal to the rank of the operator. Often, many of these singular values will be very small because the degree of T is largely overestimated.

If all singular values smaller than a given ϵ are neglected, an approximate operator H_a will result which, however, is not anymore of Hankel type and hence does not correspond to a reduced time-invariant circuit. Does it correspond to a reduced time-varying circuit? Not even that is likely, because the Hankel operator of a time-varying circuit must also satisfy shift-invariances. In [1] a construction is given of a Hankel operator H_{T_a} which approximates H_T in norm within ϵ and is of rank k. In the sequel we shall show how to obtain a similar construction in the time-varying case. Our construction does not follow the theory of AAK [1] which is based on the properties of analytic functions. It exploits the state space construction theory as explained in the previous sections and especially the properties of the controllability space of Θ-matrices. It is in the taste of the (time-invariant) method of Limebeer and Green [22], but exploits the interpolation technique initiated in [2], and further elaborated in [3, 12, 6], and the realization theory in [24].

We start out from a given, high order model for our computational network in output normal form:

$$T = B(I - ZA)^{-1}ZC, \tag{42}$$

with direct feed-through $D = 0$ and $AA^* + CC^* = I$. Our aim is to find a T_a which approximates T in Hankel norm to a given precision. Since we are working in the time-varying context, we allow ourselves a variable precision as well, given by a diagonal and hermitian operator Γ, say e.g. a sequence of ϵ's: $\Gamma = \text{diag}[\cdots, \epsilon_{-1}, [\epsilon_0], \epsilon_1, \cdots]$. Thus, we wish

$$\| \Gamma^{-1}(T - T_a) \|_H \leq 1. \tag{43}$$

There is an important connection between the Hankel norm and an operator norm which is known as a 'Nehari theorem'. Assume T_a bounded and let T' be an *extension* of T_a, i.e. T' coincides with T_a on the strictly upper part of indices, but is allowed to have a lower part as well (T_a is zero on the lower part). We have:

$$
\begin{aligned}
\| \Gamma^{-1}(T - T_a) \|_H &= \sup\{ \| \mathbf{P}[u(T - T_a)] \| : u \in \mathcal{L}_2 Z^{-1}, \|u\| = 1 \} \\
&= \sup\{ \| \mathbf{P}[u(T - T')] \| : u \in \mathcal{L}_2 Z^{-1}, \|u\| = 1 \} \\
&\leq \sup\{ \| [u(T - T')] \| : u \in \mathcal{L}_2 Z^{-1}, \|u\| = 1 \} \\
&\leq \| \Gamma^{-1}(T - T') \|.
\end{aligned}
\tag{44}
$$

This is the easy part of Nehari's theorem. The hard part, which we do not need here, is that the left hand side of (5.8) can be approximated as closely as desired by an appropriate T'.

Our procedure will consist in finding a T' such that

$$\| \Gamma^{-1}(T - T') \| \leq 1 \tag{45}$$

If T_a is taken to coincide with T' on its strictly upper part, zero elsewhere then it will satisfy $\| \Gamma^{-1}(T - T_a) \|_H \leq 1$ and a solution is found.

We start out by taking an output coprime factorization on $T = \Delta^* U$ with $\Delta \in \mathcal{U}$ and U allpass (the right coprime factorization of section 2). We know how to compute an orthogonal realization for $U : \{A, B_u, C, D_u\}$. Next, we look for a J-unitary and causal Θ such that

$$[U^* \quad -T^*\Gamma^{-1}]\Theta = [A' \quad -B'] \tag{46}$$

is upper again, i.e. such that $[U^* \quad -T^*\Gamma^{-1}]^*$ is in the controllability space of Θ. To see what this achieves, let us first assume that we have found $\Theta = \begin{bmatrix} \Theta_{11} & \Theta_{12} \\ \Theta_{21} & \Theta_{22} \end{bmatrix}$. Then, since by J-unitary we have that $\Theta_{22}^* \Theta_{22} = I + \Theta_{12}^* \Theta_{12}$, we will also have that Θ_{22}^{-1} exists as a (non-necessarily causal) contractive operator. It follows from (46) that

$$U^* \Theta_{12} - T^*\Gamma^{-1}\Theta_{22} = -B' \tag{47}$$

and putting $T' = \Gamma \Theta_{22}^{-*} B'^*$ we find

$$\Gamma^{-1}(T - T') = R^* U \tag{48}$$

and hence

$$\| \Gamma^{-1}(T - T') \| = \| R^* U \| \leq 1 \tag{49}$$

since R is contractive and U unitary. T' is a solution to our problem, provided it has the desired lowest complexity (we'll show what the complexity is in the next section).

The problem is thus reduced to finding Θ which satisfies (46). The technique that we shall use is a generalization of what we did in the previous section. Since Θ must share controllability space with $[U^* \quad -T^*\Gamma^{-1}]^*$, a minimal controllability pair is given by

$$\begin{bmatrix} A \\ B_u \\ -\Gamma^{-1}B \end{bmatrix}.$$

But Θ must be J-unitary as well! This could be achieved by a construction as in the previous section. However, the condition on J-unitary realizations stated there will be too restrictive for the present purposes, we must allow for more general constructions because this time, the interpolant is not upper anymore.

To prepare the grounds, assume a state space sequence $\mathcal{B}$ given and assume that each component in it has been decomposed in the direct sum of two subspaces. Let $\mathcal{B} = \mathcal{B}_+ \oplus \mathcal{B}_-$ the resulting decomposition in two sequences and let

$$J_{\mathcal{B}} = \begin{bmatrix} I_{\mathcal{B}_+} & \\ & -I_{\mathcal{B}_-} \end{bmatrix} \tag{50}$$

be the corresponding signature matrix. Then the following fact concerning Θ matrices generalizes theorem 4.2:

Theorem 5.1. ([13]) *If the state space realization* $\theta = \begin{bmatrix} A & C \\ B & D \end{bmatrix}$ *for* Θ *satisfies*

$$\theta^* \begin{bmatrix} J_B & \\ & J_1 \end{bmatrix} \theta = \begin{bmatrix} J_{B(-1)} & \\ & J_2 \end{bmatrix}, \qquad \theta \begin{bmatrix} J_{B(-1)} & \\ & J_2 \end{bmatrix} \theta^* = \begin{bmatrix} J_B & \\ & J_1 \end{bmatrix} \tag{51}$$

then Θ *is upper and J-unitary.*

(Remark: Θ will correspond to a causal, lossless scattering operator Σ only when B_- is empty.) The construction of a Θ with the announced controllability pair then proceeds as follows:

1. Solve the Lyapunov-Stein equation

$$A^* M A + B^* \Gamma^{-2} B = M^{(-1)}. \tag{52}$$

The solution is unique and given by (20). Let $\Lambda = I - M$ and remark that it satisfies

$$A^* \Lambda A + B_u^* B_u - B^* \Gamma^{-2} B = \Lambda^{(-1)} \tag{53}$$

since $A^* A + B_u^* B_u = I$. Suppose now that Λ is boundedly invertible and let J_B be its signature sequence, then we can write for some invertible diagonal (state transformation) operator $X : \Lambda = X^* J_B X$. It follows then that

$$\begin{bmatrix} \alpha \\ \beta_1 \\ \beta_2 \end{bmatrix} = \begin{bmatrix} XA \\ B_1 \\ B_2 \end{bmatrix} (X^{(-1)})^{-1} \tag{54}$$

is J_B-isometric as required by (51).

2. Make the state realization complete by finding output spaces $\mathcal{N}_1$ and $\mathcal{N}_2$ and a complete state realization

$$\theta = \begin{bmatrix} \alpha & \gamma_1 & \gamma_2 \\ \beta_1 & \delta_{11} & \delta_{12} \\ \beta_2 & \delta_{21} & \delta_{22} \end{bmatrix} \tag{55}$$

satisfying (51). In [13] it is shown that, given (54), this construction is always possible and amounts to the same numerical procedure as in section 4.

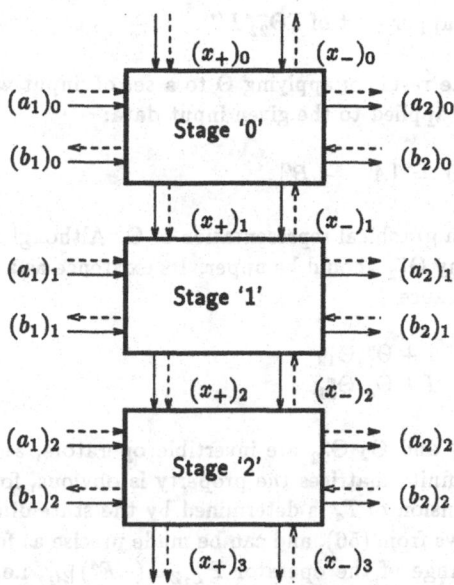

Figure 8: *The structure of the interpolating Θ for the time-varying model reduction. The solid arrows indicate data flow, the dashed arrows indicate the flow of energy.*

The solution of the model reduction problem thus hinges on the invertibility of $\Lambda = I - M$. This means that a solution exists when the sequence of local eigenvalues of M (i.e. the eigenvalues of each M_k) splits in two strictly disjoint sequences: one set of all the values that are uniformly greater than 1 (i.e. greater than $1 + \epsilon$ for some $\epsilon > 0$), and one set of all the values that are uniformly smaller than 1. This we call *dichotomy*. The sequence of numbers of eigenvalues larger than 1 determines the size of I_{B_-} while the sequence of numbers of eigenvalues smaller than 1 determines I_{B_+}. These sizes can be controlled by the magnitude of Γ^{-1}. When one or the other M_k has 1 as its eigenvalue, then the problem becomes singular and does not have the solution stated. This is very much like in the time-invariant case. The local eigenvalues of M are equal to the local singular values of the Hankel operator of $\Gamma^{-1}T$, see [5], so that they have a determining influence on the final size of B_-.

The resulting realization for Θ is shown in figure 8. To complete our work we must compute T_a from Θ and determine its local complexity. This is the subject of the next section.

6 Complexity and realization of the approximant

In the previous section we used the A and B matrices of a realization in output normal form for T to compute a J-unitary (symplectic) matrix or operator $\Theta = \begin{bmatrix} \Theta_{11} & \Theta_{12} \\ \Theta_{21} & \Theta_{22} \end{bmatrix}$ which contains the solution of our approximation problem in the form

$$T_a = \text{strictly upper part of } \Gamma\Theta_{22}^{-*}B'^{*}. \tag{56}$$

In (56), $B' \in \mathcal{U}$ is the result of applying Θ to a set of 'input waves' (i.e. an incident and a reflected wave), also applied to the given input data:

$$[U^* \quad -\Gamma T^*]\Theta = [A' \quad -B']. \tag{57}$$

See also figure 8 for a graphical representation of Θ. Although it is a block-upper operator, there is no reason why Θ_{22}^{-1} should be upper. Its existence as a bounded operator is assured from J-unitarity, because

$$\begin{cases} \Theta_{22}^*\Theta_{22} = I + \Theta_{12}^*\Theta_{12} \\ \Theta_{22}\Theta_{22}^* = I + \Theta_{21}\Theta_{21}^* \end{cases}. \tag{58}$$

Hence, both $\Theta_{22}^*\Theta_{22}$ and $\Theta_{22}\Theta_{22}^*$ are invertible operators, and this makes Θ_{22} invertible as well. (In case of finite matrices the property is obvious, for the more general case, see [3].) The state dimension of T_a is determined by the state dimension of the upper part of Θ_{22}^{-*}. This fact follows from (56), and can be made precise as follows: let us define by $\mathcal{H}(F)$ with $F \in \mathcal{X}$, the range of the operator $\mathbf{P}_{\mathcal{L}_2 Z^{-1}}(\cdot F^*)|_{\mathcal{U}_2}$, i.e. the corange of the Hankel operator H_F. F is now allowed to be a more general operator or matrix than in section 2: it need not be causal. From (56) it follows almost immediately that $\mathcal{H}(\Gamma^{-1}T_a) \subset \mathcal{H}(\Theta_{22}^{-*})$. A more detailed analysis shows that the two spaces are equal, see [13]. This space can be used to construct a state realization for $\Gamma^{-1}T_a$, and of course for T_a as well. The series of dimensions of $\mathcal{H}(\Theta_{22}^{-*})$ will determine the state complexity of T_a and an input normal form could be computed to obtain a state realization.

However, in the previous section we have already produced a state realization for Θ. An interesting question is then whether that state realization could be used, (1) to find the series of dimensions of $\mathcal{H}(\Theta_{22}^{-*})$, and (2) to derive the realization for T_a. An additional question is whether the minimal degree for the approximant is indeed achieved in this way. It turns out that we can answer these questions precisely, the losslessness of Θ will help us out. Detailed formulations and proofs can be found in [13], here we suffice with stating the most important results and giving our view of the philosophy behind them.

Look at figure 8 and focus on the time point $t = 0$ (using a 'diagonal formalism', all the other time points can be taken care of at the same token: we are after the division between strict past and future at every point in time). Let's concentrate on the 'flow of energy' — depicted in figure 9. Moreover, since we are only interested here in Θ_{22}^{-1}, we may as well take $a_1 \equiv 0$, it plays no role in the transfers sought. Let's now consider the operator $H' = \mathcal{H}(\Theta_{22}^{-*})$ in more detail. It will map the 'future' $b_{2f} = \mathbf{P}(b_2) \in \mathcal{U}_2$ to the 'strict past' $b_{1p} \in \mathcal{L}_2 Z^{-1}$ (we disregard other signals carrying outgoing energy). H' can be decomposed in two maps: $H' = \tau_1 \tau_2$ where

$$\begin{aligned} \tau_1 &: \quad b_{2f} \mapsto x_{-[0]} \\ \tau_2 &: \quad x_{-[0]} \mapsto b_{1p}. \end{aligned} \tag{59}$$

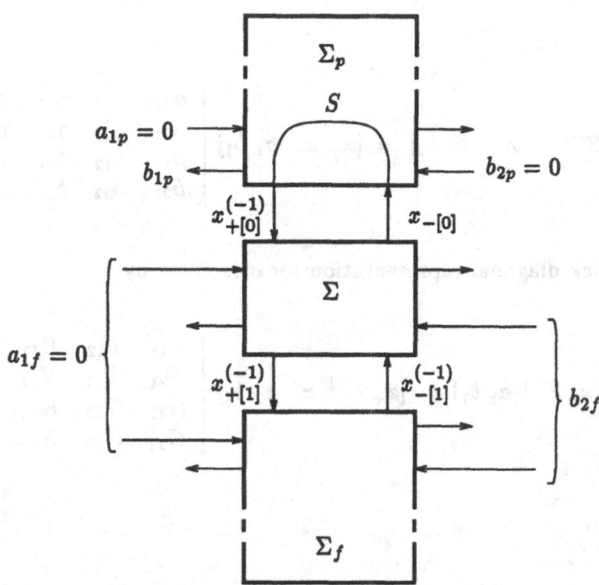

Figure 9: *Splitting* Θ *between strict past, present and strict future at each time point.*

It follows that the local dimension of H' cannot exceed the dimension of the space in which each $x_{-,k}$ lives. (Again, a more precise analysis shows that τ_1 is onto and τ_2 one-to-one). Figure 9 depicts a situation in which a set of signals for each time point is stacked. The state at time 0 for the i-th series is then given by the i-th entry of the diagonal $x_{[0]}$, assuming that also all the state series have been stacked. The range of H' will consist of a sequence of spaces of the dimension of $\mathcal{B}_{-,k}$ for the k-th space. Let $N = \{\cdots, N_{-1}, [N_0], N_1, \cdots\}$ be the sequence of these dimensions — we call it the s-dim of the range of H', it will become the s-dim of a realization of T_a.

A further analysis of (56) shows that there will be a very close relation between a realization of T_a and one for the strictly upper part of Θ_{22}^{-*}, which we will indicate by stup(Θ_{22}^{-*}). From the preceding discussion we can infer that $\Gamma^{-1}T_a$ and Θ_{22}^{-*} will share $\{A, B\}$ realization matrices, only C will be different. A key role in setting up a realization for stup(Θ_{22}^{-*}) is played by the diagonal operator S which maps $x_{-[0]}$ to $x_{+[0]}$ under the conditions $a_{1p} = 0$, $b_{2p} = 0$, and $R : x_{+[0]} \mapsto x_{-[0]}$ when $a_{1f} = 0$, $b_{2f} = 0$. These two operators (one working in the past, the other in the future) are contractive, because of energy loss through the neglected outputs. They are also diagonal by definition. They play a decisive role in the realization theory, because, as explained earlier, $x_{-[0]}$ can be taken as the state for stup(Θ_{22}^{-*}), and the definition of the state transition operator will require the propagation of $x_{-[0]}$ to its next value $x_{-[1]}^{(-1)}$. It takes some calculations to get there, and the scheme goes as follows.

1. From the realization θ for Θ, get a representation σ of the energy flow at each stage. This is easy and is computed locally, directly on θ. If a block diagonal representation

for θ is

$$[x_+ Z^{-1}\ x_- Z^{-1}\ a_2\ b_2] = [x_+\ x_-\ a_1\ b_1] \begin{bmatrix} \alpha_{11} & \alpha_{12} & \gamma_{11} & \gamma_{12} \\ \alpha_{21} & \alpha_{22} & \gamma_{21} & \gamma_{2\infty2} \\ \beta_{11} & \beta_{12} & \delta_{11} & \delta_{12} \\ \beta_{21} & \beta_{22} & \delta_{21} & \delta_{22} \end{bmatrix}, \tag{60}$$

then a block diagonal representation for σ is given by

$$[x_+\ x_- Z^{-1}\ a_1\ b_2] = [x_+ Z^{-1}\ x_-\ a_2\ b_1] \begin{bmatrix} F_{11} & F_{12} & H_{11} & H_{12} \\ F_{21} & F_{22} & H_{21} & H_{22} \\ G_{11} & G_{12} & K_{11} & K_{12} \\ G_{21} & G_{22} & K_{21} & K_{22} \end{bmatrix} \tag{61}$$

where e.g.,

$$\begin{bmatrix} F_{11} & H_{11} \\ G_{11} & K_{11} \end{bmatrix} = \begin{bmatrix} \alpha_{11} & \gamma_{11} \\ \beta_{11} & \delta_{11} \end{bmatrix} - \begin{bmatrix} \alpha_{12} & \gamma_{12} \\ \beta_{12} & \delta_{12} \end{bmatrix} \begin{bmatrix} \alpha_{22} & \gamma_{22} \\ \beta_{22} & \delta_{22} \end{bmatrix}^{-1} \begin{bmatrix} \alpha_{21} & \gamma_{21} \\ \beta_{21} & \delta_{21} \end{bmatrix} \tag{62}$$

etc. Note that σ is unitary and that all its block-diagonal entries are contractive.

2. The recursions for S and T become

$$\begin{cases} S = (F_{21} + F_{22}(I - SF_{12})^{-1} SF_{11})^{(+1)} \\ R = F_{12} + F_{11}(I - R^{(-1)}F_{21})R^{(-1)}F_{22} \end{cases} \tag{63}$$

The two recursions are well defined and converge (rather fast). The recursion for S runs forward in time, while the recursion for R runs backwards. (Notice that (61) does not represent a causal system: Σ corresponding to Θ is not causal.) Equation (63) stands for two continuous fractions in the F_{ij}'s, they are (lossy) input scattering matrices of strictly passive ladder networks. A realization for $\text{stup}(\Theta_{22}^{-*})$ is now readily obtained as $\{A_a,\ B_a,\ C_r\}$, with

$$\begin{cases} A_a = (F_{22}(I - SF_{12})^{-1})^* \\ B_a = (H_{22} + F_{22}(I - SF_{12})^{-1} SH_{12})^* \\ C_r = (I - SR)^{-*}(G_{22} + G_{21}(I - R^{(-1)}F_{21})^{-1} R^{(-1)}F_{22})^*. \end{cases} \tag{64}$$

The C_a in a realization for T_a is gleaned finally from a realization for B' — the precise formula can be found in [13].

Last but not least, we have to address the question of minimality of the representation and, if possible the parametrization of all possible solutions. This point is also treated in [13], and turns out to be a rather delicate matter, involving a generalization of a 'winding number theorem' used in [5] to the present algebraic setting. It turns out that the solution found is indeed minimal and that all T' satisfying (45) are given by loading Θ with an upper contractive S_L, as shown in figure 6.

7 Discussion and conclusions

How can the theory developed so far be used to find low complexity approximants of matrix calculations, at least in the state space sense? You can represent a vector-matrix multiplication $u.A = y$ with A an $n \times n$ matrix as a time-varying system in several ways, and then use the theory to reduce the calculations. If, e.g., the whole input vector u is thought to be available at $t = 0$, and the outputs can be processed in sequence, then you can take $\mathcal{M} = \{[C^n]\}$ and $\mathcal{N} = \{[\emptyset], C, \dots, C\}$. The approximation will involve the subsequent Hankel snapshots $H_1 = [A_{.1}, A_{.2}, \dots, A_{.n}]$, $H_2 = [A_{.2}, \dots, A_{.n}]$, $\dots$, $H_n = [A_{.n}]$. Alternatively, you could make an additive decomposition of $A = L + U$ into an upper and a lower part, and approximate each component separately, taking transposition for the lower component. You could also go for a multiplicative decomposition $A = LU$. The properties of these various approximations have not been carefully investigated yet!

There are still several interesting problems that merit investigation in the context of our theory:

- What happens in the *singular* case, i.e. when there are singular values of the Hankel operator equal to one?

- Given the desired s-dimension of the approximant, what is the corresponding minimal operator Γ, and how can it be computed?

- Is it possible to use partial continuous fractions to approximate the operators R and S, and can one find good but non-minimal approximants with less calculations?

- Can the sequential model of our paper be extended to parallel computations?

Also, many potential applications of the theory have not been explored yet. Computations in image processing and inverse imaging tend to be very complex. Can they be simplified in a systematic manner? We believe that the algebraic interpolation theory presented yields considerable promise as a new tool in numerical linear algebra and signal processing.

Acknowledgements

This research was supported in part by the commission of the EC under the ESPRIT BRA program 6632 (NANA2).

134

References

[1] V.M. Adamjan, D.Z. Arov and M.G. Krein. Analytic Properties of Schmidt Pairs for a Hankel Operator and the Generalized Schur-Takagi Problem. *Math. USSR Sbornik*, **15**(1), pp 31–73, 1971 (transl. of *Iz. Akad. Nauk Armjan. SSR Ser. Mat., 6, 1971*).

[2] D. Alpay and P. Dewilde. Time-varying Signal Approximation and Estimation. In : M.A. Kaashoek, J.H. van Schuppen and A.C.M. Ran (Eds.), *Signal Processing, Scattering and Operator Theory and Numerical Methods, Volume III, Proc. Int. Symp. MTNS-89*, Birkhäuser Verlag, pp 1–22, 1990.

[3] D. Alpay, P. Dewilde and H. Dym. Lossless Inverse Scattering and Reproducing Kernels for Upper Triangular Operators. In : I. Gohberg (Ed.), *Extension and Interpolation of Linear Operators and Matrix Functions*, Volume 47 of *Operator Theory, Advances and Applications*, Birkhäuser Verlag, pp 61–135, 1990.

[4] B.D.O. Anderson and J.B. Moore. Detectability and Stabilizability of Time-Varying Discrete-Time Linear Systems. *SIAM J. Control and Optimization*, **19**(1), pp 20–32, 1981.

[5] J.A. Ball, I. Gohberg and L. Rodman. *Interpolation of Rational Matrix Functions*. Volume 45 of *Operator Theory: Advances and Applications*, Birkhäuser Verlag, 1990.

[6] P.M. Dewilde. A Course on the Algebraic Schur and Nevanlinna-Pick Interpolation Problems. In : E.F. Deprettere and A.J. van der Veen (Eds.), *Algorithms and Parallel VLSI Architectures*, Elsevier, 1991.

[7] P. Dewilde and E. Deprettere. Approximative Inversion of Positive Matrices with Applications to Modeling. In : *NATO ASI Series, Vol. F34 on Modeling, Robustness and Sensitivity Reduction in Control Systems*, Springer Verlag, Berlin, 1987.

[8] P. Dewilde and E. Deprettere. The Generalized Schur Algorithm: Approximation and Hierarchy. In : *Operator Theory: Advances and Applications*, Volume 29, Birkhäuser Verlag, pp 97–115, 1988.

[9] P. Dewilde and H. Dym. Schur Recursions, Error Formulas and Convergence of Rational Estmators for Stationary Stochastic Processes. *IEEE Trans. Inf. Theory*, **27**, pp 446–461, 1981.

[10] P. Dewilde and H. Dym. Lossless Chain Scattering Matrices and Optimum Linear Prediction: The Vector Case. *Circuit Theory and Applications*, **9**, pp 135–175, 1984.

[11] P. Dewilde and H. Dym. Lossless Scattering, Digital Filters and Estimation Theory. *IEEE Trans. Inf. Theory*, **30**, pp 644–662, July 1984.

[12] P. Dewilde and H. Dym. Interpolation for Upper Triangular Operators. In : *Operator Theory: Advances and Applications*, Volume OT 56, Birkhäuser Verlag, pp 153–260, 1992.

[13] P.M. Dewilde and A.J. van der Veen. On the Hankel-Norm Approximation of Upper-Triangular Operators and Matrices. *to appear in Integral Equations and Operator Theory*, 1992.

[14] A. Feintuch and R. Saeks. *System Theory: A Hilbert Space Approach.* Academic Press, 1982.

[15] G. Golub and C.F. Van Loan. *Matrix Computations.* The Johns Hopkins University Press, 1984.

[16] J.W. Helton. Orbit Structure of the Möbius transformation Semigroup Acting on H_∞ (Broadband Matching). In : *Topics in Functional Analysis*, Volume 3 of *Adv. in Math. Suppl. Studies*, Academic Press, pp 129–133, 1978.

[17] B.L. Ho and R.E. Kalman. Effective Construction of Linear, State-Variable Models from Input/Output Functions. *Regelungstechnik*, 14, pp 545–548, 1966.

[18] R.E. Kalman, P.L. Falb and M.A. Arbib. *Topics in Mathematical System Theory.* Int. Series in Pure and Applied Math. McGraw-Hill, 1970.

[19] E.W. Kamen and K.M. Hafez. Algebraic Theory of Linear Time-Varying Systems. *SIAM J. Control and Optimization*, 24(6), pp 866–878, 1979.

[20] E.W. Kamen, P.P. Khargonekar and K.R. Poolla. A Transfer-Function Approach to Linear Time-Varying Discrete-Time Systems. *SIAM J. Control and Optimization*, 23(4), pp 550–565, 1985.

[21] L. Kronecker. Algebraische Reduction der Schaaren Bilinearer Formen. *S.B. Akad. Berlin*, pp 663–776, 1890.

[22] D.J.N. Limebeer and M. Green. Parametric Interpolation, H_∞-Control and Model Reduction. *Int. J. Control*, 52(2), pp 293–318, 1990.

[23] I. Schur. Uber Potenzreihen, die im Innern des Einheitskreises Beschränkt Sind, I. *J. Reine Angew. Math.*, 147, pp 205–232, 1917 Eng. Transl. *Operator Theory: Adv. Appl.*, Vol 18, Birkhäuser Verlag, pp 31-59, 1986.

[24] A.J. van der Veen and P.M. Dewilde. Time-Varying System Theory for Computational Networks. In : P. Quinton and Y. Robert (Eds.), *Algorithms and Parallel VLSI Architectures, II*, Elsevier, pp 103–127, 1991.

[25] A.J. van der Veen and P.M. Dewilde. Time-varying Computational Networks: Realization, Orthogonal Embedding and Structural Factorization. In : F.T. Luk (Ed.), *Proc. SPIE, Advanced Signal Processing Algorithms, Architectures and Implementations, III*, Volume 1770, San Diego, 1992.

[26] L.A. Zadeh. Time-Varying Networks, I. *Proc. IRE*, 49, pp 1488–1503, 1961.

[27] G. Zames. Feedback and Optimal Sensitivity: Model Reference Transformations, Multiplicative Semi-Norms and Approximate Inverses. *IEEE Trans. Aut. Control*, 23, pp 301–320, 1982.

THE LOOK-AHEAD LANCZOS PROCESS FOR LARGE NONSYMMETRIC MATRICES AND RELATED ALGORITHMS

R.W. FREUND

AT&T Bell Laboratories
600 Mountain Avenue, Room 2C-420
Murray Hill, New Jersey 07974-0636
U.S.A.
freund@research.att.com

ABSTRACT. The nonsymmetric Lanczos process can be used to compute approximate eigenvalues of large non-Hermitian matrices, or to obtain approximate solutions of large non-Hermitian linear systems. However, the Lanczos algorithm in its original form is susceptible to possible breakdowns and potential instabilities. We describe a look-ahead variant of the Lanczos process that remedies the problems of the original algorithm. We also discuss related algorithms for the construction of formally orthogonal and biorthogonal polynomials, and for computing inverse triangular factorizations of Hankel and Toeplitz matrices.

KEYWORDS. Lanczos process, non-Hermitian matrices, look-ahead techniques, approximate eigenvalues, systems of linear equations, orthogonal polynomials, inverse triangular factorizations, Hankel matrices, biorthogonal polynomials, Toeplitz matrices, Levinson algorithm.

1 Introduction

In 1950, Cornelius Lanczos [30] proposed an iterative procedure for the successive reduction of a given, in general non-Hermitian, square matrix A to a sequence of tridiagonal matrices. In contrast to direct methods that generate a tridiagonalization by eliminating elements of the matrix A, the Lanczos process does not modify A. In fact, the matrix is only used in the form of matrix-vector products with A and its transpose A^T. Because of this feature, the Lanczos process is well suited for large sparse matrices or large structured dense matrices for which matrix-vector products can be computed cheaply. The Lanczos process can be used to compute approximate eigenvalues of A, or to obtain approximate solutions of systems of linear equations, $Ax = b$, with A as coefficient matrix.

137

M. S. Moonen et al. (eds.), *Linear Algebra for Large Scale and Real-Time Applications*, 137–163.
© 1993 *Kluwer Academic Publishers.*

Unfortunately, in the classical nonsymmetric Lanczos method breakdowns—triggered by division by 0—may occur in the course of the process. In finite-precision arithmetic, such exact breakdowns are very unlikely; however, near-breakdowns—triggered by division by a number close to 0—may occur, which can then cause numerical instabilities in subsequent iterations. The possibility of breakdowns and near-breakdowns has brought the nonsymmetric Lanczos process into discredit, and it has certainly prevented many people from using the algorithm on non-Hermitian matrices. On the other hand, Cullum and Willoughby [8] demonstrated that the Lanczos process—even in its original form—can be a powerful tool for large sparse matrix computations. For the special case of Hermitian matrices A, the Lanczos process simplifies, and work per iteration and storage requirements are roughly halved. More importantly, for the resulting Hermitian Lanczos method, breakdowns and near-breakdowns can be excluded. We remark that, unlike the general Lanczos process, the Hermitian Lanczos method has found widespread use and has been studied extensively since the 1970s; we refer the reader to [33, 19] and the references given there.

On the other hand, it is possible to modify the Lanczos process so that it skips over those iterations in which an exact breakdown or a near-breakdown would occur in the standard method. Such a modified Lanczos method, called the *look-ahead Lanczos algorithm*, was first developed by Taylor [36] and Parlett, Taylor, and Liu [35] in the early 1980s. However, their procedure is fairly complicated, and in fact, details were only given for the simplest case of look-ahead steps of length 2. This may be the reason why the look-ahead idea did not receive much attention until 1989. Since then, there has been a true revival of the nonsymmetric Lanczos process and its look-ahead variants. It is due to these recent research efforts that the possible breakdowns and near-breakdowns in the classical algorithm are now well understood. Moreover, look-ahead Lanczos algorithms were developed that can handle look-ahead steps of any length, and that are not restricted to steps of length 2. The renewed interest in the Lanczos process is reflected in the number of papers that were written since 1989. For example, the papers [3, 4, 5, 10, 13, 14, 18, 23, 24, 26, 34] all deal with various aspects of the look-ahead Lanczos process.

The Lanczos process has intimate connections with many other areas of Mathematics, such as formally orthogonal polynomials (FOPs), Padé approximation, factorization of Hankel matrices, control theory, and coding theory. The problem of breakdowns has a corresponding formulation in all of these areas, and remedies for exact breakdowns in these different settings have been known for quite some time. For example, the breakdown in the Lanczos process is equivalent to a breakdown of the generic three-term recurrence relation for FOPs, and it is well known how to overcome such breakdowns by modifying the recursions for FOPs; see [20, 9, 23] and the references given there. In the context of the partial realization problem in control theory, remedies for breakdowns were given in [28, 21]. The Lanczos process is also closely related to fast algorithms for the factorization of Hankel matrices, and again, it was known how to overcome possible breakdowns of these algorithms; see [25, 18] and the references therein. Finally, a treatment of breakdowns in the context of coding theory is given in [2, 29, 7]. We stress that, in all these cases, only the problem of exact breakdowns was addressed. Nevertheless, in view of all these results, it is surprising that Numerical Analysts did not study the look-ahead Lanczos process earlier.

In this paper, we describe the look-ahead Lanczos process, sketch an actual implementation, and discuss some of its applications. Among the many possible approaches to the look-ahead Lanczos process, we have chosen the one based on the connection with FOPs. The reason for this is twofold. First, this approach appears to be the most elementary. Second, as a by-product, we immediately obtain a related look-ahead procedure for computing inverse block-factorizations of Hankel matrices. We also show that similar techniques, based on formally biorthogonal polynomials (FBOPs), can be used to derive a look-ahead variant of the classical Levinson algorithm for solving general Toeplitz systems.

The remainder of this paper is organized as follows. In Section 2, we briefly review the classical Lanczos algorithm. In Section 3, we introduce the concept of FBOPs and FOPs associated with general bilinear forms, and we show that these polynomials lead to inverse block-factorizations of the corresponding moment matrices. In Section 4, we present a look-ahead procedure for constructing FOPs associated with Hankel matrices. In Section 5, we describe the corresponding look-ahead Lanczos algorithm for general non-Hermitian matrices. In Section 6, we briefly discuss applications of this algorithm in matrix computations. In Section 7, we sketch a look-ahead procedure for computing inverse block-factorizations of Hankel matrices. We then turn to FBOPs associated with general Toeplitz matrices. In Section 8, we present recurrence relations for the construction of such FBOPs. Finally, in Section 9, we present a look-ahead Levinson algorithm for the solution of linear systems with Toeplitz matrices.

Throughout the paper, all vectors and matrices are allowed to have real or complex entries. As usual, $\overline{M} = [\overline{m_{jk}}]$, $M^T = [m_{kj}]$, and $M^H = \overline{M}^T = [\overline{m_{kj}}]$ denote the complex conjugate, the transpose, and the conjugate transpose, respectively, of the matrix $M = [m_{jk}]$. The vector norm $\|x\| := \sqrt{x^H x}$ is always the Euclidean norm. We use the notation

$$K_n(c, B) := \text{span}\{c, Bc, \ldots, B^{n-1}c\}$$

for the nth *Krylov subspace* of C^N generated by $c \in C^N$ and $B \in C^{N \times N}$. We denote by

$$\mathcal{P}_n := \{\varphi(\lambda) \equiv \sigma_0 + \sigma_1\lambda + \cdots + \sigma_n\lambda^n \mid \sigma_0, \sigma_1, \ldots, \sigma_n \in C\}$$

the set of all complex polynomials of degree at most n, and $\mathcal{P}$ is the set of all complex polynomials of arbitrary degree. Finally, the symbol 0 will be used for the number zero, the zero vector, and the zero matrix; its actual meaning and, in the case of the zero vector or the zero matrix, its dimension will always be apparent from the context.

2 The classical Lanczos process

In this section, we briefly review the classical Lanczos process for general square matrices.

2.1 THE ALGORITHM AND SOME OF ITS PROPERTIES

Let $A \in C^{N \times N}$, and let $v_0, w_0 \in C^N$, $v_0, w_0 \neq 0$, be two arbitrary nonzero vectors. Starting with v_0 and w_0, the classical Lanczos algorithm [30] generates two sequences of vectors

$$\{v_j\}_{j=0}^n \quad \text{and} \quad \{w_j\}_{j=0}^n, \quad n = 0, 1, \ldots, \tag{1}$$

140

such that, for all n,

$$\text{span}\{v_0, v_1, \ldots, v_n\} = K_{n+1}(v_0, A),$$
$$\text{span}\{w_0, w_1, \ldots, w_n\} = K_{n+1}(w_0, A^T). \tag{2}$$

and the two sequences are biorthogonal, $i.e.$,

$$w_j^T v_k = \begin{cases} 0, & \text{if } j \neq k, \\ \delta_k \neq 0, & \text{if } j = k, \end{cases} \quad \text{for all } j, k = 0, 1, \ldots, n. \tag{3}$$

It turns out that vectors (1) satisfying (2) and (3) can be constructed by means of simple three-term vector recurrences. This procedure is just the classical *Lanczos algorithm*.

Algorithm 2.1 (Classical Lanczos algorithm [30]).

0) Choose $v_0, w_0 \in \mathbb{C}^N$ with $v_0, w_0 \neq 0$.
Set $v_{-1} = w_{-1} = 0$ and $\delta_{-1} = 1$.

For $n = 0, 1, \ldots$, do:

1) Compute $\delta_n = w_n^T v_n$.
If $\delta_n = 0$, then stop.

2) Set

$$\alpha_n = \frac{w_n^T A v_n}{\delta_n}, \quad \beta_n = \frac{\delta_n}{\delta_{n-1}}. \tag{4}$$

3) Set

$$v_{n+1} = A v_n - v_n \alpha_n - v_{n-1} \beta_n,$$
$$w_{n+1} = A^T w_n - w_n \alpha_n - w_{n-1} \beta_n. \tag{5}$$

We remark that, in practice, one uses a scaled version of the recurrences (5), in order to avoid over- or underflow. A proven choice is to scale the Lanczos vectors to have unit length:

$$\|v_n\| = \|w_n\| = 1 \quad \text{for all} \quad n. \tag{6}$$

In this paper, we will always use the unscaled recurrences (5), since they simplify the exposition of the algorithms. However, for practical purposes, we recommend the scaling (6) of the Lanczos vectors.

Next, we list a few properties of Algorithm 2.1. The recursions (5) for all the Lanczos vectors $\{v_j\}_{j=0}^{n+1}$ and $\{w_j\}_{j=0}^{n+1}$ can be written compactly in matrix form as follows:

$$AV_n = V_n T_n + [0 \ \cdots \ 0 \ v_{n+1}],$$
$$A^T W_n = W_n T_n + [0 \ \cdots \ 0 \ w_{n+1}]. \tag{7}$$

Here
$$V_n := [\, v_0 \quad v_1 \quad \cdots \quad v_n \,] \quad \text{and} \quad W_n := [\, w_0 \quad w_1 \quad \cdots \quad w_n \,]$$
are matrices whose columns are just the Lanczos vectors (1), and

$$T_n := \begin{bmatrix} \alpha_0 & \beta_1 & 0 & \cdots & 0 \\ 1 & \alpha_1 & \ddots & \ddots & \vdots \\ 0 & \ddots & \ddots & \ddots & 0 \\ \vdots & \ddots & \ddots & \ddots & \beta_n \\ 0 & \cdots & 0 & 1 & \alpha_n \end{bmatrix} \tag{8}$$

is an $(n+1) \times (n+1)$ tridiagonal matrix whose entries are just the recurrence coefficients in (5).

The Lanczos process is mostly applied to large matrices A, and usually, the iteration index n is much smaller than the order N of A. From (7), it is apparent how the Lanczos algorithm can be used to obtain approximate eigenvalues of large matrices A or approximate solutions of large linear systems $Ax = b$. Indeed, if $v_{n+1} = 0$ or $w_{n+1} = 0$ in (7), then each eigenvalue of T_n would also be an eigenvalue of A. This suggests to use the eigenvalues of T_n as approximate eigenvalues of A. Similarly, an approximate solution of $Ax = b$ can be obtained by solving a suitable linear system with T_n as coefficient matrix, or a weighted least-squares problem with an extended version of T_n as coefficient matrix. This will be discussed in more detail in Section 6.

We stress that the termination check $w_n^T v_n = 0$ in step 1) of Algorithm 2.1 is indispensable to avoid division by 0 in (4). In exact arithmetic, this check will be satisfied after at most N iterations. Normally, the algorithm stops because $w_n^T v_n = 0$ is satisfied with
$$v_n = 0 \quad \text{or} \quad w_n = 0. \tag{9}$$
It is easy to verify that, in (9), the case $v_n = 0$ occurs if, and only if, $K_n(v_0, A)$ is an A-invariant subspace of C^N. In view of (2), the Lanczos vectors $\{v_j\}_{j=0}^{n-1}$ then build a basis for the A-invariant subspace $K_n(v_0, A)$. Similarly, if $w_n = 0$ in (9), then the vectors $\{w_j\}_{j=0}^{n-1}$ span the A^T-invariant subspace $K_n(w_0, A^T)$ of C^N. The event (9) is called *regular termination*.

Unfortunately, it can also happen that Algorithm 2.1 stops because
$$w_n^T v_n = 0, \quad \text{but} \quad v_n \neq 0 \quad \text{and} \quad w_n \neq 0. \tag{10}$$
In this case, the Lanczos process terminates prematurely, before it has constructed a basis for an A-invariant or an A^T-invariant subspace. Therefore, termination due to (10) constitutes a *breakdown* of the Lanczos algorithm. Of course, in finite-precision arithmetic, exact breakdowns (10) are unlikely; however, *near-breakdowns*, i.e.,
$$w_n^T v_n \approx 0, \quad \text{but} \quad v_n \not\approx 0 \quad \text{and} \quad w_n \not\approx 0, \tag{11}$$
may occur, which can then cause numerical instabilities in subsequent iterations. Look-ahead Lanczos methods are modifications of the classical Algorithm 2.1 that allow to continue the Lanczos process even in the case of an exact breakdown (10) or a near-breakdown (11).

We remark that, for the special case of Hermitian matrices A, the possibility of breakdowns can be excluded. In fact, if $A = A^H$, then one sets $w_0 = \overline{v_0}$, and from (4) and (5), it follows that $w_n = \overline{v_n}$ for all n. Hence, we have

$$w_n^T v_n = v_n^H v_n = \|v_n\|^2 > 0 \quad \text{if} \quad v_n \neq 0,$$

and this shows that breakdowns cannot occur. Furthermore, in the Hermitian Lanczos process, only the vectors v_n are generated, using the first relation in (5). In particular, work and storage requirements are roughly halved for Hermitian matrices.

There is another special case for which the two recurrences (5) reduce to one, namely complex symmetric matrices $A = A^T$. Here, one sets $w_0 = v_0$, and this implies that $w_n = v_n$ for all n. However, we stress that, in contrast to the Hermitian Lanczos process, exact breakdowns and near-breakdowns in the resulting complex symmetric Lanczos algorithm cannot be excluded. For a detailed discussion of the complex symmetric Lanczos process, we refer the reader to [10].

2.2 POLYNOMIAL FORMULATION

There is an equivalent formulation of Algorithm 2.1 in terms of polynomials, instead of vectors. In fact, from (5), it follows that each pair of Lanczos vectors v_n and w_n can be represented in the form

$$v_n = \varphi_n(A)v_0 \quad \text{and} \quad w_n = \varphi_n(A^T)w_0, \tag{12}$$

where φ_n is a *monic* polynomial of degree n, i.e.,

$$\varphi_n(\lambda) \equiv \lambda^n + \cdots \in \mathcal{P}_n. \tag{13}$$

We note that φ_n is uniquely determined by (12). The polynomials φ_n defined by (12) are called *Lanczos polynomials*. Using (12), we now rewrite Algorithm 2.1 and some of its properties in terms of the Lanczos polynomials.

To this end, we first define a bilinear form $\langle \cdot, \cdot \rangle$ on $\mathcal{P} \times \mathcal{P}$ by setting

$$\langle \psi, \varphi \rangle := w_0^T \psi(A)\varphi(A)v_0 \quad \text{for all} \quad \varphi, \psi \in \mathcal{P}. \tag{14}$$

From (12) and (14), we have

$$w_j^T v_k = (\varphi_j(A^T)w_0)^T \varphi_k(A)v_0$$
$$= w_0^T \varphi_j(A)\varphi_k(A)v_0 = \langle \varphi_j, \varphi_k \rangle \quad \text{for all} \quad j, k,$$

and thus the biorthogonality condition (3) of the Lanczos vectors can be rewritten as follows:

$$\langle \varphi_j, \varphi_k \rangle = \begin{cases} 0, & \text{if } j \neq k, \\ \langle \varphi_k, \varphi_k \rangle \neq 0, & \text{if } j = k, \end{cases} \quad \text{for all} \quad j, k = 0, 1, \ldots, n. \tag{15}$$

In view of (15), the Lanczos polynomials are orthogonal with respect to the bilinear form (14). Furthermore, the recursions (5) for the Lanczos vectors are just a reformulation of the usual three-term polynomial recurrence

$$\varphi_{n+1} = \lambda \varphi_n - \alpha_n \varphi_n - \beta_n \varphi_{n-1} \tag{16}$$

for orthogonal polynomials. Note that, by (4), (12), and (14), the coefficients in (16) are given by

$$\alpha_n = \frac{\langle \varphi_n, \lambda\varphi_n \rangle}{\langle \varphi_n, \varphi_n \rangle}, \quad \beta_n = \frac{\langle \varphi_n, \varphi_n \rangle}{\langle \varphi_{n-1}, \varphi_{n-1} \rangle}. \tag{17}$$

Finally, the conditions (10) and (11) for an exact breakdown or a near-breakdown of the classical Lanczos algorithm translate into the conditions

$$\langle \varphi_n, \varphi_n \rangle = 0 \quad \text{or} \quad \langle \varphi_n, \varphi_n \rangle \approx 0, \tag{18}$$

which signal an exact breakdown or a near-breakdown in the recursions (16) and (17). In Section 4, we present modified recurrences that allow to continue the construction of the Lanczos polynomials even if (18) occurs. These recurrences are the basis for the look-ahead Lanczos process.

3 Bilinear forms and formally biorthogonal polynomials

In Section 2.2, we saw that the Lanczos polynomials are orthogonal with respect to the particular bilinear form (14). In this section, we consider arbitrary bilinear forms, and we introduce the more general concepts of formally biorthogonal polynomials and formally orthogonal polynomials.

3.1 BILINEAR FORMS

A complex-valued functional

$$\langle \cdot, \cdot \rangle : \mathcal{P} \times \mathcal{P} \longmapsto \mathbb{C} \tag{19}$$

is called a *bilinear form* if

$$\begin{aligned}
\langle \psi, \sigma_1\varphi_1 + \sigma_2\varphi_2 \rangle = \sigma_1\langle \psi, \varphi_1 \rangle + \sigma_2\langle \psi, \varphi_2 \rangle \quad \text{for all} \quad \psi, \varphi_1, \varphi_2 \in \mathcal{P}, \ \sigma_1, \sigma_2 \in \mathbb{C}, \\
\langle \tau_1\psi_1 + \tau_2\psi_2, \varphi \rangle = \tau_1\langle \psi_1, \varphi \rangle + \tau_2\langle \psi_2, \varphi \rangle \quad \text{for all} \quad \psi_1, \psi_2, \varphi \in \mathcal{P}, \ \tau_1, \tau_2 \in \mathbb{C}.
\end{aligned} \tag{20}$$

The numbers

$$m_{jk} := \langle \lambda^j, \lambda^k \rangle, \quad j, k = 0, 1, \ldots, \tag{21}$$

are called the *moments* of $\langle \cdot, \cdot \rangle$. For $n = 0, 1, \ldots$, we set

$$M_n := [m_{jk}]_{j,k=0,\ldots,n} = \begin{bmatrix} m_{00} & m_{01} & \cdots & m_{0n} \\ m_{10} & m_{11} & & \vdots \\ \vdots & & \ddots & \vdots \\ m_{n0} & \cdots & \cdots & m_{nn} \end{bmatrix}. \tag{22}$$

The matrix (22) is called the nth *moment matrix* associated with $\langle \cdot, \cdot \rangle$.

With (20)–(22), it follows that for any two polynomials

$$\varphi(\lambda) \equiv \sum_{j=0}^{n} \sigma_j\lambda^j, \quad \psi(\lambda) \equiv \sum_{j=0}^{n} \tau_j\lambda^j$$

of degree at most n, $n = 0, 1, \ldots$, we have

$$\langle \psi, \varphi \rangle = v^T M_n u, \quad \text{where} \quad u := \begin{bmatrix} \sigma_0 \\ \sigma_1 \\ \vdots \\ \sigma_n \end{bmatrix}, \quad v := \begin{bmatrix} \tau_0 \\ \tau_1 \\ \vdots \\ \tau_n \end{bmatrix}. \tag{23}$$

This shows that a bilinear form is uniquely determined by its moments.

We note that the special form (14) underlying the Lanczos process clearly satisfies the bilinearity conditions (20). In view of (21) and (14), the moments of this bilinear form are given by

$$m_{jk} = w_0^T A^{j+k} v_0, \quad j, k = 0, 1, \ldots . \tag{24}$$

Finally, we remark that for the Lanczos bilinear form (14), we have

$$\langle \psi, \varphi \rangle = \langle \varphi, \psi \rangle \quad \text{for all} \quad \varphi, \psi \in \mathcal{P}. \tag{25}$$

In general, a bilinear form (19) with the additional property (25) is said to be *symmetric*. In view of (23), a bilinear is symmetric if, and only if,

$$M_n = M_n^T \quad \text{for all} \quad n = 0, 1 \ldots . \tag{26}$$

3.2 FORMALLY BIORTHOGONAL AND ORTHOGONAL POLYNOMIALS

We stress that, in general, a bilinear form is not an inner product. Indeed, it is possible that a nonzero polynomial φ has "norm" $\langle \varphi, \varphi \rangle = 0$ or $\langle \varphi, \varphi \rangle < 0$. Nevertheless, it turns out to be useful to study polynomials that are orthogonal with respect to a given bilinear form (19). We now give a definition of these so-called *formally biorthogonal polynomials* (FBOPs). Recall that a polynomial $\varphi_n \in \mathcal{P}_n$ is said to be *monic* if it is of the form (13).

Definition 3.1. A monic polynomial $\varphi_n \in \mathcal{P}_n$ is called a *right* FBOP (with respect to the bilinear form (19)) of degree n if

$$\langle \psi, \varphi_n \rangle = 0 \quad \text{for all} \quad \psi \in \mathcal{P}_{n-1}. \tag{27}$$

A monic polynomial $\psi_n \in \mathcal{P}_n$ is called a *left* FBOP (with respect to the bilinear form (19)) of degree n if

$$\langle \psi_n, \varphi \rangle = 0 \quad \text{for all} \quad \varphi \in \mathcal{P}_{n-1}. \tag{28}$$

A right or left FBOP φ_n or ψ_n is said to be *regular* if it is uniquely determined by (27) or (28), respectively.

In contrast to orthogonal polynomials associated with a true inner product, right and left FBOPs φ_n and ψ_n need not exist or need not be unique for every degree n. Next, we derive a necessary and sufficient condition for the existence of regular FBOPs.

In view of (27) and (23), a monic polynomial

$$\varphi_n(\lambda) \equiv \lambda^n + \sum_{i=0}^{n-1} u_{in} \lambda^i$$

is a right FBOP of degree n if, and only if, its coefficients u_{in} satisfy

$$M_{n-1} \begin{bmatrix} u_{0n} \\ u_{1n} \\ \vdots \\ u_{n-1,n} \end{bmatrix} = - \begin{bmatrix} m_{0n} \\ m_{1n} \\ \vdots \\ m_{n-1,n} \end{bmatrix}. \tag{29}$$

Analogously, by (28) and (23), a monic polynomial

$$\psi_n(\lambda) \equiv \lambda^n + \sum_{i=0}^{n-1} v_{in} \lambda^i$$

is a left FBOP of degree n if, and only if, its coefficients v_{in} fulfill

$$[\, v_{0n} \quad v_{1n} \quad \cdots \quad v_{n-1,n} \,] M_{n-1} = - [\, m_{n0} \quad m_{n1} \quad \cdots \quad m_{n-1,n} \,]. \tag{30}$$

As an immediate consequence of (29) and (30), we have the following result.

Lemma 3.2. *The following conditions are equivalent:*
(i) A regular right FBOP φ_n of degree n exists.
(ii) A regular left FBOP ψ_n of degree n exists.
(iii) The matrix M_{n-1} is nonsingular.

In view of Lemma 3.2, no regular FBOPs of degree n exist if M_{n-1} is singular. Furthermore, the proof of Lemma 3.2 suggests that, in finite-precision arithmetic, it is not advisable to compute regular FBOPs if M_{n-1} is nonsingular, but ill-conditioned. For the case of singular or ill-conditioned moment matrices M_{n-1}, we will need the concept of m-quasi-FBOPs that only satisfy relaxed versions of the biorthogonality conditions (27) and (28). The precise definition of m-quasi-FBOPs is as follows.

Definition 3.3. A monic polynomial $\varphi_n \in \mathcal{P}_n$ is called a *right m-quasi-FBOP* (with respect to the bilinear form (19)) of degree n if $m < n$ and

$$\langle \psi, \varphi_n \rangle = 0 \quad \text{for all} \quad \psi \in \mathcal{P}_{m-1}. \tag{31}$$

A monic polynomial $\psi_n \in \mathcal{P}_n$ is called a *left m-quasi-FBOP* (with respect to the bilinear form (19)) of degree n if $m < n$ and

$$\langle \psi_n, \varphi \rangle = 0 \quad \text{for all} \quad \varphi \in \mathcal{P}_{m-1}. \tag{32}$$

The next lemma gives a sufficient condition for the existence of m-quasi-FBOPs.

Lemma 3.4. *Let $1 \le m < n$. If M_{m-1} is nonsingular, then right and left m-quasi-FBOPs φ_n and ψ_n of degree n exist.*

Proof. Let φ_n be a monic polynomial of the form

$$\varphi_n(\lambda) \equiv \lambda^n + \sum_{i=0}^{m-1} u_{in} \lambda^i. \tag{33}$$

By (31) and (23), the polynomial (33) is a right m-quasi-FBOP if its coefficients

$u_{0n}, u_{1n}, \ldots, u_{m-1,n}$ satisfy

$$M_{m-1} \begin{bmatrix} u_{0n} \\ u_{1n} \\ \vdots \\ u_{m-1,n} \end{bmatrix} = - \begin{bmatrix} m_{0n} \\ m_{1n} \\ \vdots \\ m_{m-1,n} \end{bmatrix}. \tag{34}$$

If M_{m-1} is nonsingular, the linear system (34) is guaranteed to have a solution, and the corresponding polynomial (33) is a right m-quasi-FBOP.

Similarly, one shows that a left m-quasi-FBOP exists if M_{m-1} is nonsingular. $\Box$

We remark that, for symmetric bilinear forms $\langle \cdot, \cdot \rangle$, the conditions (27) and (28) are equivalent, and thus right and left regular FBOPs coincide, i.e.,

$$\varphi_n = \psi_n. \tag{35}$$

Furthermore, the conditions (31) and (32) for right and left m-quasi-FBOPs are also equivalent. Consequently, in the case of symmetric bilinear forms $\langle \cdot, \cdot \rangle$, we do not have to distinguish between right and left FBOPs, respectively right and left m-quasi FBOPs, and we can assume (35). Finally, for symmetric bilinear forms, the conditions (27) and (28), respectively (31) and (32), represent orthogonality relations, rather than biorthogonality relations. Hence, we refer to φ_n as a *formally biorthogonal polynomial* (FOP), respectively an m-quasi-FOP.

3.3 INVERSE BLOCK-FACTORIZATIONS OF MOMENT MATRICES

We now show how FBOPs and quasi-FBOPs can be used to construct inverse block-factorizations of moment matrices. In the sequel, we assume that $\langle \cdot, \cdot \rangle$ is an arbitrary, but fixed, bilinear form, and FBOPs and quasi-FBOPs refer to polynomials associated with this bilinear form.

First, we need to introduce some further notation, which will also be used later on. We denote by $\mathcal{I}$ the set of all integers n for which regular FBOPs of degree n exist. In view of Lemma 3.2, $\mathcal{I}$ just consists of all integers for which M_{n-1} is nonsingular. Moreover, we always have $0 \in \mathcal{I}$. Note that, for $n = 0$, the conditions (27) and (28) are void, and thus $\varphi_0(\lambda) \equiv 1$ and $\psi_0(\lambda) \equiv 1$ are regular FBOPs of degree 0. Now let $\{n_k\}_{k=0}^K \subseteq \mathcal{I}$ be an arbitrary subsequence of $\mathcal{I}$; here either $K = \infty$ or K is an integer. We always assume that $\{n_k\}_{k=0}^K$ includes 0, and that the n_k's are numbered in increasing order. Therefore, we have

$$0 =: n_0 < n_1 < \cdots < n_k < \cdots. \tag{36}$$

We refer to the numbers n_k in (36) as *regular indices*. Finally, if $K < \infty$, then we set $n_{K+1} := \infty$.

Based on the given regular indices (36), we now define two sequences of monic polynomials

$$\{\varphi_n\}_{n=0}^\infty \quad \text{and} \quad \{\psi_n\}_{n=0}^\infty \tag{37}$$

as follows. For $n = n_k$, $0 \le k \le K$, let φ_n and ψ_n be the regular right and left FBOPs of degree n. Since $n_k \in \mathcal{I}$, these regular FBOPs are guaranteed to exist. For $n_k < n < n_{k+1}$, $0 \le k \le K$, we choose φ_n and ψ_n as any pair of right and left n_k-quasi-FBOPs of degree

n. Note that $n_k \in \mathcal{I}$ implies that M_{n_k-1} is nonsingular, and thus, by Lemma 3.4, n_k-quasi-FBOPs of degree n are guaranteed to exist. However, we stress that these n_k-quasi-FBOPs are not unique.

We now show that, for each n, the subsequences

$$\{\varphi_j\}_{j=0}^n \quad \text{and} \quad \{\psi_j\}_{j=0}^n \tag{38}$$

of (37) induce an *inverse block-factorization* of the nth moment matrix M_n. In the sequel, let n be arbitrary, but fixed. Using (36), the polynomials (38) can be partitioned into blocks

$$\Phi^{(k)} := \begin{cases} [\,\varphi_{n_k} \quad \varphi_{n_k+1} \quad \cdots \quad \varphi_{n_{k+1}-1}\,], & \text{if } 0 \le k < l, \\ [\,\varphi_{n_l} \quad \varphi_{n_l+1} \quad \cdots \quad \varphi_n\,], & \text{if } k = l, \end{cases} \tag{39}$$

and

$$\Psi^{(k)} := \begin{cases} [\,\psi_{n_k} \quad \psi_{n_k+1} \quad \cdots \quad \psi_{n_{k+1}-1}\,], & \text{if } 0 \le k < l, \\ [\,\psi_{n_l} \quad \psi_{n_l+1} \quad \cdots \quad \psi_n\,], & \text{if } k = l. \end{cases} \tag{40}$$

Here, $l := l(n)$ is the unique integer defined by

$$n_l \le n < n_{l+1}, \quad 0 \le l \le K. \tag{41}$$

Note that $\Phi^{(k)}$ and $\Psi^{(k)}$ are row vectors of polynomials. We remark that the first polynomial in each block $\Phi^{(k)}$ and $\Psi^{(k)}$ is a regular FBOP, while the remaining polynomials are n_k-quasi-FBOPs.

Using the biorthogonality conditions (27) and (28) for FBOPs and the relations (31) and (32) for n_k-quasi FBOPs, one readily verifies that the blocks (39) and (40) satisfy the block-biorthogonality conditions

$$\langle \Psi^{(j)}, \Phi^{(k)} \rangle = \begin{cases} 0, & \text{if } j \ne k, \\ \langle \Psi^{(k)}, \Phi^{(k)} \rangle, & \text{if } j = k, \end{cases} \quad \text{for all } \ j, k = 0, 1, \ldots, l. \tag{42}$$

Here and in the sequel, we have used the notation

$$\langle \Psi, \Phi \rangle := \begin{bmatrix} \langle \psi_0, \phi_0 \rangle & \cdots & \langle \psi_0, \phi_m \rangle \\ \vdots & & \vdots \\ \langle \psi_i, \phi_0 \rangle & \cdots & \langle \psi_i, \phi_m \rangle \end{bmatrix} \in \mathbb{C}^{(i+1) \times (m+1)},$$

where

$$\Phi = [\,\phi_0 \quad \phi_1 \quad \cdots \quad \phi_m\,] \quad \text{and} \quad \Psi = [\,\psi_0 \quad \psi_1 \quad \cdots \quad \psi_i\,]$$

are row vectors of polynomials in $\mathcal{P}$.

After these preliminaries, we can now state the main result of this section.

Theorem 3.5. *Let $\{n_k\}_{k=0}^K \subseteq \mathcal{I}$ be a given sequence of regular indices (36). Let $n \ge 0$, and let $l = l(n)$ be defined by (41). Let*

$$\varphi_j(\lambda) \equiv \lambda^j + \sum_{i=0}^{j-1} u_{ij} \lambda^i \quad \text{and} \quad \psi_j(\lambda) \equiv \lambda^j + \sum_{i=0}^{j-1} v_{ij} \lambda^i, \quad j = 0, 1, \ldots, n, \tag{43}$$

be the subsequence (38) of the polynomials (37) associated with $\{n_k\}_{k=0}^K$. Then the nth

148

moment matrix M_n associated with the bilinear form $\langle \cdot, \cdot \rangle$ has the following inverse block-factorization:

$$V_n^T M_n U_n = D_n, \tag{44}$$

where

$$U_n := [\, u_{ij}\,]_{i,j=0,\ldots,n} = \begin{bmatrix} 1 & u_{01} & \cdots & u_{0n} \\ 0 & 1 & \ddots & \vdots \\ \vdots & \ddots & \ddots & u_{n-1,n} \\ 0 & \cdots & 0 & 1 \end{bmatrix} \tag{45}$$

and

$$V_n := [\, v_{ij}\,]_{i,j=0,\ldots,n} = \begin{bmatrix} 1 & v_{01} & \cdots & v_{0n} \\ 0 & 1 & \ddots & \vdots \\ \vdots & \ddots & \ddots & v_{n-1,n} \\ 0 & \cdots & 0 & 1 \end{bmatrix} \tag{46}$$

are unit upper triangular matrices whose columns are just the coefficients of the polynomials (43), *and*

$$D_n := \mathrm{diag}(\delta^{(0)}, \delta^{(1)}, \ldots, \delta^{(l)}), \quad \text{where} \quad \delta^{(k)} := \langle \Psi^{(k)}, \Phi^{(k)} \rangle, \tag{47}$$

is a block-diagonal matrix. Moreover, in (47), *the blocks*

$$\delta^{(k)} \quad \text{are nonsingular for all} \quad k = 0, 1, \ldots, l-1, \tag{48}$$

and

$$\delta^{(l)} \quad \text{is nonsingular if} \quad n = n_{l+1} - 1. \tag{49}$$

Proof. By means of (23), we can rewrite the block-biorthogonality conditions (42) in terms of M_n and the coefficients u_{ij} and v_{ij} of the polynomials (43). The result is just the inverse block-factorization (44) with matrices U_n, V_n, and D_n given by (45)–(47).

Using (44)–(47), we deduce that

$$\det M_{n_l-1} = \prod_{k=0}^{l-1} \det \delta^{(k)} \quad \text{and} \quad \det M_n = (\det M_{n_l-1})(\det \delta^{(l)}). \tag{50}$$

In view of Lemma 3.2, the matrix M_{n_l-1} is nonsingular, and, if $n = n_{l+1}-1$, the matrix M_n is nonsingular. Hence (48) and (49) follow directly from (50). $\square$

Finally, consider the case that $\langle \cdot, \cdot \rangle$ is a symmetric bilinear form. Then, in view of (35), we have $u_{ij} = v_{ij}$ for all i, j in (43), and (44) reduces to the symmetric inverse block-factorization

$$U_n^T M_n U_n = D_n$$

of the symmetric moment matrix M_n. Here, U_n is given by (45) and

$$D_n := \mathrm{diag}(\delta^{(0)}, \delta^{(1)}, \ldots, \delta^{(l)}), \quad \text{where} \quad \delta^{(k)} := \langle \Phi^{(k)}, \Phi^{(k)} \rangle, \tag{51}$$

is a symmetric block-diagonal matrix.

4 Constructing FOPs associated with Hankel matrices

Polynomials that are orthogonal with respect to a true inner product satisfy simple three-term recurrences, see, *e.g.*, [6]. This is no longer for FBOPs and FOPs associated with arbitrary bilinear forms. In general, the construction of these polynomials requires all previous polynomials. However, there are two important special cases, namely bilinear forms corresponding to Hankel and Toeplitz moment matrices, for which FOPs and FBOPs can be constructed by means of short recurrences involving only few of the previous polynomials. In this section, we discuss the Hankel case. The Toeplitz case is treated in Section 8.

In the following, we assume that $\langle \cdot, \cdot \rangle$ is a bilinear form whose moments m_{jk} depend only on the *sum* $j + k$ of their indices, *i.e.*,

$$m_{jk} = h_{j+k} \quad \text{for all} \quad j, k = 0, 1, \ldots . \tag{52}$$

Here $\{h_j\}_{j=0}^{\infty}$ is any sequence of real or complex numbers. The corresponding moment matrices (22) are then of the form

$$M_n = [\, h_{j+k} \,]_{j,k=0,\ldots,n} = \begin{bmatrix} h_0 & h_1 & h_2 & \cdots & h_n \\ h_1 & \cdot^{\cdot^{\cdot}} & & \cdot^{\cdot^{\cdot}} & \vdots \\ h_2 & & \cdot^{\cdot^{\cdot}} & & \vdots \\ \vdots & \cdot^{\cdot^{\cdot}} & & & h_{2n-1} \\ h_n & \cdots & \cdots & h_{2n-1} & h_{2n} \end{bmatrix}, \quad n = 0, 1, \ldots . \tag{53}$$

Matrices of the form (53) are called *Hankel matrices*. Note that the moment matrices (53) always satisfy the symmetry condition (26), and thus $\langle \cdot, \cdot \rangle$ is a symmetric bilinear form. Therefore, the corresponding polynomials are FOPs and quasi-FOPs. Finally, we remark that, by (24), the special bilinear form (14) underlying the Lanczos algorithm satisfies the condition (52) with

$$h_j = w_0^T A^j v_0. \tag{54}$$

In [18], Freund and Zha derived an algorithm for constructing a sequence

$$\{\varphi_n\}_{n=0}^{\infty} \tag{55}$$

of regular FOPs and quasi-FOPs associated with general moment matrices of the form (53). Next, we present a sketch of this algorithm. We use the notation introduced in Section 3.3. In particular, indices (36) are used to mark the regular FOPs, and the first $n+1$ polynomials $\{\varphi_j\}_{j=0}^{n}$ are partitioned into blocks $\Phi^{(k)}$, $k = 0, 1, \ldots, l$, given by (39). Here, $l = l(n)$ is defined in (41); note that n_l is just the index of the last regular FOP φ_{n_l} with index $\leq n$. Finally, $\delta^{(k)}$ are the blocks defined in (51).

With these preparations, the algorithm for constructing (55) can now be sketched as follows.

Algorithm 4.1 (Look-ahead process for constructing FOPs [18]).

0) Set $\varphi_0 = 1$, $\Phi^{(0)} = \varphi_0$, $\delta^{(0)} = \langle \Phi^{(0)}, \Phi^{(0)} \rangle$.
   Set $n_0 = 0$, $l = 0$, $\Phi^{(-1)} = 0$, $\delta^{(-1)} = 1$.

For $n = 0, 1, \ldots,$ do:

1) Decide whether to construct φ_{n+1} as a regular FOP or as an n_l-quasi-FOP and go to 2) or 3), respectively.

2) (*Constructing a regular* FOP) Compute

$$\alpha_n = (\delta^{(l)})^{-1}\langle\Phi^{(l)}, \lambda\varphi_n\rangle, \quad \beta_n = (\delta^{(l-1)})^{-1}\langle\Phi^{(l-1)}, \lambda\varphi_n\rangle, \tag{56}$$

$$\varphi_{n+1} = \lambda\varphi_n - \Phi^{(l)}\alpha_n - \Phi^{(l-1)}\beta_n, \tag{57}$$

set $n_{l+1} = n+1$, $l = l+1$, $\Phi^{(l)} = \emptyset$, and go to 4).

3) (*Constructing a quasi-*FOP) Compute

$$\beta_n = (\delta^{(l-1)})^{-1}\langle\Phi^{(l-1)}, \lambda\varphi_n\rangle, \tag{58}$$

$$\varphi_{n+1} = \lambda\varphi_n - \Phi^{(l-1)}\beta_n - \sum_{i=n_l}^{n} \xi_i^{(n)}\varphi_i. \tag{59}$$

4) Set $\Phi^{(l)} = [\, \Phi^{(l)} \quad \varphi_{n+1}\,]$, $\delta^{(l)} = \langle\Phi^{(l)}, \Phi^{(l)}\rangle$.

We remark that, in (59), $\xi_i^{(n)} \in C$ are arbitrary recurrence coefficients.

The strategy for deciding in step 1) of Algorithm 4.1 whether to build the next polynomial as a regular FOP or as a quasi-FOP is called the *look-ahead strategy*. In Section 5, we will briefly discuss the look-ahead strategy for one particular application of Algorithm 4.1. Here, we only remark that, in view of (49), for constructing φ_{n+1} as a regular FOP it is necessary that

$$\delta^{(l)} \quad \text{is nonsingular.} \tag{60}$$

Moreover, if the look-ahead strategy is only based on (60), then Algorithm 4.1 constructs all existing regular FOPs. Note that the condition (60) guarantees that the inverse matrices in (56), respectively (58), always exist.

Algorithm 4.1 shows that regular FOPs associated with general Hankel moment matrices can always be constructed by means of short recurrences (57), respectively (59), which involve only polynomials from the last two blocks $\Phi^{(l)}$ and $\Phi^{(l-1)}$. Finally, if all polynomials are constructed as regular FOPs, then (57) reduces to a simple three-term of the form (16).

5 A look-ahead Lanczos algorithm for nonsymmetric matrices

We now return to the Lanczos process for general nonsymmetric matrices, and as a first application of Algorithm 4.1, we obtain a look-ahead Lanczos algorithm.

5.1 THE ALGORITHM

As in Section 2.1, let $A \in C^{N \times N}$, and let v_0, $w_0 \in C^N$, v_0, $w_0 \neq 0$, be two arbitrary nonzero vectors. Let $\langle \cdot, \cdot \rangle$ be the Lanczos bilinear form defined by (54) and (52), and let $\{\varphi_n\}_{n=0}^{\infty}$ be the sequence (55) of FOPS and quasi-FOPs generated by Algorithm 4.1. Recall from (36) that the indices n_k mark the regular FOPs.

In analogy to (12), we use the polynomials φ_n to define Lanczos vectors by setting

$$v_n := \varphi_n(A)v_0 \quad \text{and} \quad w_n := \varphi_n(A^T)w_0. \tag{61}$$

If $n = n_k$ and thus φ_{n_k} is a regular FOP, then the vectors v_{n_k} and w_{n_k} are called *regular vectors*. If φ_n is a quasi-FOP, then we refer to v_n and w_n as *inner vectors*.

For each $n \geq 0$, let $l = l(n)$ be the integer defined in (41). Note that n_l is the index of the last pair of regular vectors v_{n_l} and w_{n_l} with index $\leq n$. In view of the transition rules (61), the blocks $\Phi^{(k)}$, $k = 0, 1, \ldots, l$, in (39) now correspond to blocks of Lanczos vectors defined by

$$V^{(k)} := \begin{cases} [\, v_{n_k} \quad v_{n_k+1} \quad \cdots \quad v_{n_{k+1}-1} \,], & \text{if } 0 \leq k < l, \\ [\, v_{n_l} \quad v_{n_l+1} \quad \cdots \quad v_n \,], & \text{if } k = l, \end{cases} \tag{62}$$

and

$$W^{(k)} := \begin{cases} [\, w_{n_k} \quad w_{n_k+1} \quad \cdots \quad w_{n_{k+1}-1} \,], & \text{if } 0 \leq k < l, \\ [\, w_{n_l} \quad w_{n_l+1} \quad \cdots \quad w_n \,], & \text{if } k = l. \end{cases} \tag{63}$$

Furthermore, for the blocks $\delta^{(k)}$ defined in (51), we obtain the following representations:

$$\delta^{(k)} = (W^{(k)})^T V^{(k)} \quad \text{for all} \quad k = 0, 1, \ldots, l. \tag{64}$$

Note that $V^{(k}$ and $W^{(k)}$ are $N \times \rho_k$ matrices, where

$$\rho_k := \begin{cases} n_{k+1} - n_k, & \text{if } 0 \leq k < l, \\ n + 1 - n_l, & \text{if } k = l. \end{cases} \tag{65}$$

The first vectors v_{n_k} and w_{n_k} in each of the blocks (62) and (63), respectively, are always regular vectors, while the remaining ones are inner vectors. The number ρ_k is called the *length* of the kth look-ahead step. If $\rho_k = 1$, then $V^{(k)} = v_{n_k}$ and $W^{(k)} = w_{n_k}$ just consist of a single regular vector. We refer to steps of length $\rho_k > 1$ as *true* look-ahead steps. We remark that the matrices (62) and (63) define a partitioning of the first $n + 1$ Lanczos vectors

$$\{v_j\}_{j=0}^n \quad \text{and} \quad \{w_j\}_{j=0}^n \tag{66}$$

into blocks, according to the look-ahead steps taken:

$$\begin{aligned} V_n &:= [\, v_0 \quad v_1 \quad \cdots \quad v_n \,] = [\, V^{(0)} \quad V^{(1)} \quad \cdots \quad V^{(l)} \,], \\ W_n &:= [\, w_0 \quad w_1 \quad \cdots \quad w_n \,] = [\, W^{(0)} \quad W^{(1)} \quad \cdots \quad W^{(l)} \,]. \end{aligned} \tag{67}$$

By means of (61)–(64), we can now rewrite Algorithm 4.1 in terms of the Lanczos vectors (61), instead of the polynomials (55). The resulting procedure is just the look-ahead Lanczos algorithm proposed by Freund, Gutknecht, and Nachtigal [14]. An outline of this algorithm is as follows.

Algorithm 5.1 (Sketch of the look-ahead Lanczos algorithm [14]).

    0) Choose $v_0, w_0 \in C^N$ with $v_0, w_0 \neq 0$.

Set $V^{(0)} = v_0$, $W^{(0)} = w_0$, $\delta^{(0)} = (W^{(0)})^T V^{(0)}$.

Set $n_0 = 0$, $l = 0$, $v_{-1} = w_{-1} = 0$, $V^{(-1)} = v_{-1}$, $W^{(-1)} = w_{-1}$, $\delta^{(-1)} = 1$.

For $n = 0, 1, \ldots$, do:

1) Decide whether to construct v_{n+1} and w_{n+1} as regular or inner vectors and go to 2) or 3), respectively.

2) (*Regular step*) Compute

$$\alpha_n = (\delta^{(l)})^{-1}(W^{(l)})^T A v_n, \quad \beta_n = (\delta^{(l-1)})^{-1}(W^{(l-1)})^T A v_n, \tag{68}$$

$$v_{n+1} = A v_n - V^{(l)}\alpha_n - V^{(l-1)}\beta_n, \tag{69}$$

$$w_{n+1} = A^T w_n - W^{(l)}\alpha_n - W^{(l-1)}\beta_n, \tag{70}$$

set $n_{l+1} = n + 1$, $l = l + 1$, $V^{(l)} = W^{(l)} = \emptyset$, and go to 4).

3) (*Inner step*) Compute

$$\beta_n = (\delta^{(l-1)})^{-1}(W^{(l-1)})^T A v_n,$$

$$v_{n+1} = A v_n - v_n \zeta_n - v_{n-1}\eta_n - V^{(l-1)}\beta_n, \tag{71}$$

$$w_{n+1} = A^T w_n - w_n \zeta_n - w_{n-1}\eta_n - W^{(l-1)}\beta_n. \tag{72}$$

4) If $v_{n+1} = 0$ or $w_{n+1} = 0$, then stop.
   Otherwise, set

$$V^{(l)} = [V^{(l)} \quad v_{n+1}], \quad W^{(l)} = [W^{(l)} \quad w_{n+1}], \quad \delta^{(l)} = (W^{(l)})^T V^{(l)}.$$

It remains to specify the look-ahead strategy that is used for the decision in step 1) of Algorithm 5.1. In [14], we developed a look-ahead strategy based on three different checks. If all these checks are satisfied, then the vectors v_{n+1} and w_{n+1} are constructed as regular vectors; otherwise, they are built as inner vectors. First, it is checked whether $\delta^{(l)}$ is nonsingular; recall from (60) that this is a necessary condition for performing a regular step. The other two checks compare the norms of α_n and β_n in (68) to some suitable estimate $n(A)$ of the norm of A. The vectors v_{n+1} and w_{n+1} are built as regular vectors only if the norms of α_n and β_n are not substantially bigger than $n(A)$. Roughly speaking, the motivation for these two checks comes from the recurrences (69) and (70), which show that the components $A v_n \in K_{n+1}(v_0, A)$ and $A^T w_n \in K_{n+1}(w_0, A^T)$ of the new Lanczos vectors v_{n+1} and w_{n+1} would be dominated by components in the old Krylov subspaces $K_n(v_0, A)$ and $K_n(w_0, A^T)$ if the norms of the coefficient vectors α_n and β_n are large compared to the norm of A. As a result, v_{n+1} and w_{n+1} would be poor basis vectors for $K_{n+1}(v_0, A)$ and $K_{n+1}(w_0, A^T)$ if they were built as regular vectors. For a more detailed description of this look-ahead strategy, we refer the reader to [14, 16]. Here, we only note that the outlined look-ahead strategy builds regular vectors in preference to inner vectors, and it takes as few true look-ahead steps as possible.

We remark that, in (71) and (72), $\zeta_n \in C$ and $\eta_n \in C$, $n = 0, 1, \ldots$, are arbitrary recurrence coefficients with $\eta_{n_k} = 0$, $k = 1, 2, \ldots$. The choice of these coefficients is not overly important, since Algorithm 5.1 usually performs only few inner steps anyway. We mostly used the algorithm with $\zeta_n := 1$ and, if $n \neq n_k$, $\eta_n := 1$. We stress that, as in the case of the classical Lanczos algorithm, in practice, the Lanczos vectors in Algorithm 5.1 should be scaled to have unit length.

Finally, we note that a FORTRAN 77 implementation of Algorithm 5.1 is available from netlib by sending an email message consisting of the single line "send lalqmr from misc" to netlib@ornl.gov or netlib@research.att.com.

5.2 PROPERTIES

Next, we list a few properties of the look-ahead Lanczos Algorithm 5.1.

Recall that the vectors generated by the classical Lanczos Algorithm 2.1 are character-ized by the properties (2) and (3). The Lanczos vectors (66) constructed by means of Algorithm 5.1 still satisfy (2), and thus, for $n = 0, 1, \ldots$, we have

$$\text{span}\{v_0, v_1, \ldots, v_n\} = K_{n+1}(v_0, A), \tag{73}$$

$$\text{span}\{w_0, w_1, \ldots, w_n\} = K_{n+1}(w_0, A^T).$$

However, instead of (3), the Lanczos vectors (66) are now only block-biorthogonal, i.e.,

$$(W^{(j)})^T V^{(k)} = \begin{cases} 0, & \text{if } j \neq k, \\ \delta^{(k)}, & \text{if } j = k, \end{cases} \quad \text{for all} \quad j \neq k, \quad j, k = 0, 1, \ldots, l. \tag{74}$$

We note that (74) is just a reformulation of the block-biorthogonality condition (42) for polynomials.

In analogy to (7), the recurrences (69)–(72) for the Lanczos vectors $\{v_j\}_{j=0}^{n+1}$ and $\{w_j\}_{j=0}^{n+1}$ can be summarized in matrix form as follows:

$$AV_n = V_n T_n + [0 \cdots 0 \ v_{n+1}], \tag{75}$$

$$A^T W_n = W_n T_n + [0 \cdots 0 \ w_{n+1}]. \tag{76}$$

Here, V_n and W_n are the matrices defined in (67), and T_n is now an $(n+1) \times (n+1)$ block-tridiagonal matrix of the form

$$T_n := \begin{bmatrix} \mu_0 & \nu_1 & 0 & \cdots & 0 \\ 1 & \mu_1 & \ddots & \ddots & \vdots \\ 0 & \ddots & \ddots & \ddots & 0 \\ \vdots & \ddots & \ddots & \ddots & \nu_l \\ 0 & \cdots & 0 & 1 & \mu_l \end{bmatrix}. \tag{77}$$

The diagonal blocks μ_k, $k = 0, 1, \ldots, l$, in (77) are square matrices of order ρ_k. Recall from (65) that ρ_k is the length of the kth look-ahead step. If no true look-ahead steps are performed, then Algorithm 5.1 reduces to the classical Lanczos Algorithm 2.1, and the matrix (77) is just the scalar tridiagonal matrix (8) associated with Algorithm 2.1.

We remark that the diagonal block μ_k in (77) are upper Hessenberg matrices with all 1's on their subdiagonals. Thus, in addition to being block tridiagonal, the matrix T_n is also upper Hessenberg with all 1's on its subdiagonal. This guarantees that the extended $(n+2) \times (n+1)$ matrix

$$T_n^{(e)} := \begin{bmatrix} T_n \\ e_n^T \end{bmatrix}, \quad \text{where} \quad e_n := [0 \ \cdots \ 0 \ 1]^T \in \mathbf{R}^n, \tag{78}$$

has always full column rank:

$$\text{rank}\, T_n^{(e)} = n+1 \tag{79}$$

Finally, we stress that the look-ahead Lanczos Algorithm 5.1 is different from the original look-ahead Lanczos algorithm proposed by Taylor [36] and Parlett, Taylor, and Liu [35].

6 Applications in matrix computations

In this section, we briefly discuss the application of Algorithm 5.1 to eigenvalue computations and to the iterative solution of linear systems.

6.1 APPROXIMATE EIGENVALUES

As we discussed in Section 2.1, relations of the type (75) and (76) suggest to use the eigenvalues of the matrix T_n as approximations to the eigenvalues of a large matrix $A \in \mathbf{C}^{N \times N}$. The resulting procedure thus consists of two steps. First, we run the look-ahead Lanczos Algorithm 5.1 for n iterations to obtain the block-tridiagonal Lanczos matrix (77), T_n. Second, we compute some or all of the eigenvalues of T_n. Note that, unless we are also interested in computing approximate right or left eigenvectors of T_n, it is not necessary to store the Lanczos vectors. For an actual implementation of this procedure, a number of important issues need to be addressed, such as the solution of the eigenvalue problem for T_n, convergence tests, and the detection of spurious eigenvalues caused by loss of biorthogonality. For a discussion of these practical aspects, we refer the reader to the paper by Cullum and Willoughby [8], where the classical Lanczos algorithm without look-ahead is considered. Numerical results of eigenvalue computations based on the look-ahead Lanczos Algorithm 5.1 can be found in [14].

6.2 ITERATIVE SOLUTION OF LINEAR SYSTEMS

Next, we turn to systems of linear equations,

$$Ax = b, \tag{80}$$

with $A \in \mathbf{C}^{N \times N}$ as coefficient matrix. We are interested in computing approximate solutions of (80) of the form

$$x_{n+1} \in x_0 + K_{n+1}(r_0, A), \quad n = 0, 1, \ldots. \tag{81}$$

Here, $x_0 \in \mathbf{C}^N$ is an arbitrary initial guess for (80), and $r_0 := b - Ax_0$ is the associated residual vector. In the sequel,

$$r_{n+1} := b - Ax_{n+1} \tag{82}$$

always denotes the residual vector corresponding to the iterate x_{n+1}.

We now choose

$$v_0 = r_0 \quad \text{and any} \quad w_0 \in \mathbb{C}^N, \quad w_0 \neq 0, \tag{83}$$

as the starting vectors for Algorithm 5.1. Then, by (73), the Lanczos vectors $\{v_j\}_{j=0}^n$ span the Krylov subspace $K_{n+1}(r_0, A)$ in (81). Therefore, using the matrix V_n defined in (67), any iterate (81) can be represented in the form

$$x_{n+1} = x_0 + V_n z_n, \quad \text{where} \quad z_n \in \mathbb{C}^{n+1}. \tag{84}$$

It remains to select the free parameter vector z_n in (84). Based on the block-tridiagonal Lanczos matrix (77), T_n, and its extended version (78), $T_n^{(e)}$, there are two natural choices for z_n. The first approach is to select z_n as the solution of the linear system

$$T_n z_n = e_1, \quad \text{where} \quad e_1 := [\,1 \quad 0 \quad \cdots \quad 0\,]^T \in \mathbb{R}^{n+1}. \tag{85}$$

To motivate (85), we remark that, by (82), (83), and (75), the residual vector corresponding to any iterate (84) satisfies

$$r_{n+1} = V_n(e_1 - T_n z_n) - v_{n+1}(z_n)_n. \tag{86}$$

Here, e_1 is the first unit vector defined in (85), and $(z_n)_n$ denotes the last component of the vector z_n. Hence the choice (85) of z_n just eliminates all components along V_n in the representation (86) of r_{n+1}. It can be shown that the resulting iterative procedure is mathematically equivalent to the classical *biconjugate gradient algorithm* (BCG) [31], provided that no look-ahead steps are performed in the underlying Lanczos Algorithm 5.1. The problem with this first approach is that the coefficient matrix T_n in (85) may be singular or ill-conditioned. In fact, this is one reason for the typical erratic convergence behavior of BCG with wild oscillations in the residual norm.

In order to obtain a method with better numerical and theoretical properties than BCG, Freund and Nachtigal [15] suggested to choose z_n as the solution of the least-squares problem

$$\|f_n - \Omega_n T_n^{(e)} z_n\| = \min_{z \in \mathbb{C}^{n+1}} \|f_n - \Omega_n T_n^{(e)} z\|. \tag{87}$$

Here, $f_n := [\,\|r_0\| \quad 0 \quad \cdots \quad 0\,]^T \in \mathbb{R}^{n+2}$, and Ω_n is a diagonal scaling matrix given by

$$\Omega_n := \text{diag}(\|v_0\|, \|v_1\|, \ldots, \|v_{n+1}\|).$$

Note that, in view of (79), the coefficient matrix $\Omega_n T_n^{(e)}$ of (87) has full column rank $n+1$, and thus (87) always has a unique solution z_n. The motivation for this choice of z_n is as follows. In view of (82), (83), (75), and (78), the residual vector corresponding to any iterate (84) satisfies

$$r_{n+1} = V_{n+1}(\Omega_n)^{-1}(f_n - \Omega_n T_n^{(e)} z_n). \tag{88}$$

Hence, in view of (87), the iterate x_{n+1} is characterized by a minimization of the second factor in the representation (88) of its residual vector r_{n+1}. This is called a *quasi-minimal residual* (QMR) property, and the resulting iterative scheme for solving linear systems (80) is the QMR method. We refer the reader to [15, 16] for a detailed description of two different implementations of the QMR method. Theoretical properties of the QMR method, such as error bounds, are derived in [15, 11]. Finally, we remark that the QMR approach can also be used to devise a *transpose-free* QMR algorithm (TFQMR) [12] that, in contrast to the standard QMR method based on Algorithm 5.1, only requires matrix-vector products with A, but no multiplications with the transpose A^T.

7 Inverse block-factorization of Hankel matrices

In this section, as a second application of Algorithm 4.1, we sketch a look-ahead procedure for inverse Hankel factorization.

Let $\{M_n\}_{n=0}^{\infty}$ be a sequence of Hankel matrices (53), and let $\{\varphi_n\}_{n=0}^{\infty}$ be the corresponding sequence (55) of FOPs and quasi-FOPs generated by Algorithm 4.1. In view of Theorem 3.5, these polynomials induce, for each n, an inverse block-factorization of M_n of the form

$$U_n^T M_n U_n = D_n. \tag{89}$$

Here, $U_n = [u_{ij}]_{i,j=0,\ldots,n}$ is the unit upper triangular matrix (45) whose entries are just the coefficients of the polynomials $\{\varphi_j\}_{j=0}^{n}$, and D_n is a symmetric block-diagonal matrix defined in (51). More precisely, denoting the last column of U_n by

$$u_n := \begin{bmatrix} u_{0n} \\ \vdots \\ u_{n-1,n} \\ 1 \end{bmatrix},$$

we have

$$\varphi_n(\lambda) \equiv [\,1 \quad \lambda \quad \cdots \quad \lambda^n\,]\, u_n. \tag{90}$$

Similarly, setting

$$U^{(k)} := \begin{cases} [\,u_{ij}\,]_{i=0,\cdots,n_{k+1}-1;j=n_k,\cdots,n_{k+1}-1}\,, & \text{if } 0 \le k < l, \\ [\,u_{ij}\,]_{i=0,\cdots,n;j=n_l,\cdots,n}\,, & \text{if } k = l, \end{cases}$$

the blocks $\Phi^{(k)}$ in (39) can be represented as follows:

$$\Phi^{(k)}(\lambda) \equiv [\,1 \quad \lambda \quad \cdots \quad \lambda^n\,]\, U^{(k)}, \quad k = 0,1,\ldots,l. \tag{91}$$

Adopting the notation introduced in Section 5, we call the vector u_n in (90) *regular* if it corresponds to a regular FOP φ_n, and u_n is said to be an *inner* vector, if φ_n is a quasi-FOP.

Finally, using (90) and (91) and setting $\xi_i^{(n)} = 0$ for all i, n in (59), we can rewrite Algorithm 4.1 in terms of the matrices U_n and D_n defining the inverse block-factorization (89) of M_n. This leads to the following algorithm, which was first proposed by Freund and Zha [18]. An outline of this algorithm is as follows.

Algorithm 7.1 (Sketch of a look-ahead algorithm for inverse Hankel factorization [18]).

   0) Set $u_0 = 1$, $U^{(0)} = u_0$, $\delta^{(0)} = (U^{(0)})^T M_0 U^{(0)}$.
      Set $n_0 = 0$, $l = 0$, $U^{(-1)} = 0$, $\delta^{(-1)} - 1$.

For $n = 0, 1, \ldots,$ do:

1) Decide whether to construct u_{n+1} as a regular or as an inner vector and go to 2) or 3), respectively.

2) (*Regular step*) Compute α_n and β_n by solving

$$\delta^{(l)}\alpha_n = \begin{bmatrix} U^{(l)} \\ 0 \end{bmatrix}^T M_{n+1} \begin{bmatrix} 0 \\ u_n \end{bmatrix} \quad \text{and} \quad \delta^{(l-1)}\beta_n = \begin{bmatrix} U^{(l-1)} \\ 0 \end{bmatrix}^T M_{n+1} \begin{bmatrix} 0 \\ u_n \end{bmatrix},$$

respectively.
Compute

$$u_{n+1} = \begin{bmatrix} 0 \\ u_n \end{bmatrix} - \begin{bmatrix} U^{(l)} \\ 0 \end{bmatrix} \alpha_n - \begin{bmatrix} U^{(l-1)} \\ 0 \end{bmatrix} \beta_n,$$

set $n_{l+1} = n + 1$, $l = l + 1$, $U^{(l)} = \emptyset$, and go to 4).

3) (*Inner step*) Compute β_n by solving

$$\delta^{(l-1)}\beta_n = \begin{bmatrix} U^{(l-1)} \\ 0 \end{bmatrix}^T M_{n+1} \begin{bmatrix} 0 \\ u_n \end{bmatrix}$$

and set

$$u_{n+1} = \begin{bmatrix} 0 \\ u_n \end{bmatrix} - \begin{bmatrix} U^{(l-1)} \\ 0 \end{bmatrix} \beta_n.$$

4) Set

$$U^{(l)} = \begin{bmatrix} U^{(l)} \\ 0 \end{bmatrix} \; u_{n+1} \end{bmatrix}, \quad \delta^{(l)} = (U^{(l)})^T M_{n+1} U^{(l)}.$$

For a discussion of the look-ahead strategy used in step 1) and other implementation details of Algorithm 7.1, we refer the reader to [18]. Furthermore, in [18], we also show how solutions of Hankel systems $M_n x_n = b_n$ can be computed, based on the factorization (89).

We remark that Algorithm 7.1 can be viewed as an extension of a classical procedure due to Trench [38] for computing inverse factorizations of a sequence of nonsingular Hankel matrices.

8 Constructing FBOPs associated with Toeplitz matrices

In this section, we consider the construction of FBOPs associated with Toeplitz matrices.

In the sequel, it is always assumed that $\langle \cdot, \cdot \rangle$ is a bilinear form whose moments m_{jk} depend only on the *difference* $j - k$ of their indices, *i.e.*,

$$m_{jk} = t_{j-k} \quad \text{for all} \quad j, k = 0, 1, \dots .$$

Here $\{t_j\}_{j=-\infty}^{\infty}$ is any biinfinite sequence of real or complex numbers. The corresponding moment matrices (22) are then of the form

$$M_n = [\,t_{i-j}\,]_{i,j=0,\ldots,n} = \begin{bmatrix} t_0 & t_{-1} & t_{-2} & \cdots & t_{-n} \\ t_1 & t_0 & \ddots & & \vdots \\ t_2 & \ddots & \ddots & \ddots & \vdots \\ \vdots & & \ddots & & t_{-1} \\ t_n & \cdots & \cdots & t_1 & t_0 \end{bmatrix}, \quad n = 0, 1, \ldots . \tag{92}$$

Matrices of the form (92) are called *Toeplitz matrices*. Note that, unless $t_{-j} = t_j$ for all j, the moment matrices (92) do not satisfy the symmetry condition (26). Therefore, in general, the bilinear form $\langle \cdot, \cdot \rangle$ is not symmetric, and we have to construct two sequences

$$\{\varphi_n\}_{n=0}^{\infty} \quad \text{and} \{\psi_n\}_{n=0}^{\infty}$$

of right, respectively left, FBOPs and quasi-FBOPs.

First, we consider the case that the matrices (92) are all nonsingular. By Lemma 3.2, this implies that right and left regular FBOPs of any degree exist. It turns out that these polynomials can be generated easily, by means of the celebrated Szegö recursions [22, 1, 27].

Algorithm 8.1 (The classical Szegö recursions).

0) Set $\varphi_0 = \psi_0 = 1$, $\delta_0 = t_0$.

For $n = 0, 1, \ldots$, do:

1) Compute $\rho_n = \langle 1, \lambda \varphi_n \rangle$, $\tau_n = \langle \lambda \psi_n, 1 \rangle$.

2) Set

$$\begin{aligned} \alpha_n &= \rho_n/\delta_n, & \varphi_{n+1} &= \lambda \varphi_n - \widehat{\psi}_n \alpha_n, \\ \beta_n &= \tau_n/\delta_n, & \psi_{n+1} &= \lambda \psi_n - \widehat{\varphi}_n \beta_n. \end{aligned} \tag{93}$$

3) Set $\delta_{n+1} = \delta_n(1 - \alpha_n \beta_n)$.

Here, in (93), the polynomial $\widehat{\psi}_n$ and $\widehat{\varphi}_n$ is the reverse of ψ_n and φ_n, respectively. Recall that, for any polynomial $\varphi \in \mathcal{P}$, its *reverse* $\widehat{\varphi}$ is given by

$$\widehat{\varphi}(\lambda) \equiv \lambda^{\partial(\varphi)} \varphi(1/\lambda), \tag{94}$$

where $\partial(\varphi)$ is the exact degree of $\varphi \in \mathcal{P}$, i.e., $\partial(\varphi)$ is the smallest integer $n \geq 0$ such that $\varphi \in \mathcal{P}_n$. Note that $\widehat{\varphi}$ is a polynomial of degree at most $\partial(\varphi)$.

In the following, we will also use the following extension of (94). For any row vector

$$\Phi = [\,\varphi_0 \quad \varphi_1 \quad \cdots \quad \varphi_j\,]$$

of polynomials in $\mathcal{P}$, we define its *block-reverse* by

$$\widehat{\Phi}(\lambda) \equiv \lambda^{n_\Phi} \Phi(1/\lambda), \quad \text{where} \quad n_\Phi := \max_{i=0,1,\ldots,j} \partial(\varphi_i).$$

Note that the entries of $\widehat{\Phi}$ are again polynomials.

In general, it cannot be guaranteed that the moment matrices (92) are all nonsingular. If some of the matrices M_n are singular or nonsingular, but ill-conditioned, then breakdowns or near-breakdowns will occur in Algorithm 8.1. Recently, Freund and Zha [17], derived extensions of the Szegö recursions for the stable construction of FBOPs and quasi-FBOPs associated with general Toeplitz moment matrices.

Next, we present a sketch of this algorithm. Again, we use the notation introduced in Section 3.3. In addition to the blocks (39) and (40), we define blocks of monomials

$$\Lambda^{(k)} := \begin{cases} [\lambda^{n_k} \quad \lambda^{n_k+1} \quad \cdots \quad \lambda^{n_{k+1}-1}], & \text{if } 0 \le k < l, \\ [\lambda^{n_k} \quad \lambda^{n_k+1} \quad \cdots \quad \lambda^n], & \text{if } k = l. \end{cases}$$

Furthermore, we set

$$F^{(k)} := \langle \Lambda^{(k)}, \Phi^{(k)} \rangle \quad \text{and} \quad G^{(k)} := \langle \Psi^{(k)}, \Lambda^{(k)} \rangle, \quad k = 0, 1, \dots, l.$$

Finally, for $k < l$, we define vectors f_k and g_k as follows. If $n_{k+1} - n_k > 1$, then f_k and g_k are given by

$$F^{(k)} = \begin{bmatrix} * & f_k \\ * & * \end{bmatrix} \quad \text{and} \quad (G^{(k)})^T = \begin{bmatrix} * & g_k \\ * & * \end{bmatrix}, \tag{95}$$

where the elements $*$ in the lower right corners in (95) are 1×1. If $n_{k+1} - n_k = 1$, then we set $f_k = g_k = \emptyset$.

After these preparations, our algorithm based on generalized Szegö recursions can be formulated as follows.

Algorithm 8.2 (Look-ahead process for constructing FBOPs [17]).

0) Set $\varphi_0 = \psi_0 = 1$, $\Phi^{(0)} = \Psi^{(0)} = 1$, $F^{(0)} = G^{(0)} = \langle 1, 1 \rangle$, $n_0 = 0$, $l = 0$.

For $n = 0, 1, \dots,$ do:

1) Compute

$$\rho_n = \langle 1, \lambda \varphi_n \rangle \quad \text{and} \quad \tau_n = \langle \lambda \psi_n, 1 \rangle.$$

2) Decide whether to construct φ_{n+1} and ψ_{n+1} as regular FBOPs or as inner quasi-FBOPs and go to 3) or 4), respectively.

3) (*Constructing a regular FBOP*) Set

$$\alpha_n = (G^{(l)})^{-T} \begin{bmatrix} 0 \\ \rho_n \end{bmatrix}, \quad \mu_n = (F^{(l)})^{-1} \begin{bmatrix} 0 \\ f_l \end{bmatrix}, \quad \varphi_{n+1} = \lambda \varphi_n - \widehat{\Psi}^{(l)} \alpha_n - \Phi^{(l)} \mu_n,$$

$$\beta_n = (F^{(l)})^{-1} \begin{bmatrix} 0 \\ \tau_n \end{bmatrix}, \quad \nu_n = (G^{(l)})^{-T} \begin{bmatrix} 0 \\ g_l \end{bmatrix}, \quad \psi_{n+1} = \lambda \psi_n - \widehat{\Phi}^{(l)} \beta_n - \Psi^{(l)} \nu_n.$$

Set $n_{l+1} = n + 1$, $l = l + 1$, $\Phi^{(l)} = \Psi^{(l)} = \emptyset$, and go to 5).

4) (*Constructing a quasi-FBOP*) Set

$$\alpha_n = (G^{(l-1)})^{-T} \begin{bmatrix} 0 \\ \rho_n \end{bmatrix}, \quad \varphi_{n+1} = \lambda\varphi_n - \widehat{\Psi}^{(l-1)}\alpha_n,$$

$$\beta_n = (F^{(l-1)})^{-1} \begin{bmatrix} 0 \\ \tau_n \end{bmatrix}, \quad \psi_{n+1} = \lambda\psi_n - \widehat{\Phi}^{(l-1)}\beta_n.$$

5) Set $\Phi^{(l)} = [\,\Phi^{(l)} \quad \varphi_{n+1}\,]$, $\Psi^{(l)} = [\,\Psi^{(l)} \quad \psi_{n+1}\,]$, $F^{(l)} = \langle\Lambda^{(l)}, \Phi^{(l)}\rangle$, $G^{(l)} = \langle\Psi^{(l)}, \Lambda^{(l)}\rangle$.

9 Inverse block factorization of Toeplitz matrices

In view of Theorem 3.5, the polynomials $\{\varphi_n\}_{n=0}^{\infty}$ and $\{\psi_n\}_{n=0}^{\infty}$ generated by Algorithm 8.2 induce inverse block-factorizations of the Toeplitz matrices (92) of the form

$$V_n^T M_n U_n = D_n. \tag{96}$$

Here, $U_n = [\,u_{ij}\,]_{i,j=0,\dots,n}$ and $V_n = [\,v_{ij}\,]_{i,j=0,\dots,n}$ is a unit upper triangular matrix (45) and (46) whose entries are just the coefficients of the polynomials $\{\varphi_j\}_{j=0}^{n}$ and $\{\psi_j\}_{j=0}^{n}$, respectively, and D_n is a block-diagonal matrix defined in (47). The translation of Algorithm 8.2 into a look-ahead procedure for constructing the matrices U_n, V_n, and D_n in (96) is analogous to the derivation of Algorithm 7.1 from Algorithm 4.1. Therefore, we omit the details, and we directly state the resulting algorithm.

Algorithm 9.1 (Sketch of a look-ahead algorithm for inverse Toeplitz factorization [17]).

0) Set $u_0 = v_0 = 1$, $U^{(0)} = V^{(0)} = 1$, $F^{(0)} = G^{(0)} = t_0$, $n_0 = 0$, $l = 0$.

For $n = 0, 1, \dots$, do:

1) Compute

$$\rho_n = s_n^T u_n, \quad \tau_n = v_n^T r_n. \tag{97}$$

2) Decide whether to construct u_{n+1} and v_{n+1} as regular vectors or as inner vectors and go to 3) or 4), respectively.

3) (*Regular step*) Compute $\alpha_l^{(1)}$, μ_n, $\beta_l^{(1)}$, ν_n by solving

$$(G^{(l)})^T \alpha_l^{(1)} = \begin{bmatrix} 0 \\ 1 \end{bmatrix}, \quad F^{(l)}\mu_n = \begin{bmatrix} 0 \\ f_l \end{bmatrix}, \quad F^{(l)}\beta_l^{(1)} = \begin{bmatrix} 0 \\ 1 \end{bmatrix}, \quad (G^{(l)})^T \nu_n = \begin{bmatrix} 0 \\ g_l \end{bmatrix},$$

respectively, and set

$$u_{n+1} = \begin{bmatrix} 0 \\ u_n \end{bmatrix} - \rho_n \begin{bmatrix} JV^{(l)} \\ 0 \end{bmatrix} \alpha_l^{(1)} - \begin{bmatrix} U^{(l)} \\ 0 \end{bmatrix} \mu_n,$$

$$v_{n+1} = \begin{bmatrix} 0 \\ v_n \end{bmatrix} - \tau_n \begin{bmatrix} JU^{(l)} \\ 0 \end{bmatrix} \beta_l^{(1)} - \begin{bmatrix} V^{(l)} \\ 0 \end{bmatrix} \nu_n.$$

Set $n_{l+1} = n + 1$, $l = l + 1$, $U^{(l)} = V^{(l)} = F^{(l)} = G^{(l)} = \emptyset$, and go to 5).

4) (*Inner step*) Set

$$u_{n+1} = \begin{bmatrix} 0 \\ u_n \end{bmatrix} - \rho_n \begin{bmatrix} JV^{(l-1)} \\ 0 \end{bmatrix} \alpha_{l-1}^{(1)}, \quad v_{n+1} = \begin{bmatrix} 0 \\ v_n \end{bmatrix} - \tau_n \begin{bmatrix} JU^{(l-1)} \\ 0 \end{bmatrix} \beta_{l-1}^{(1)}.$$

5) Set

$$U^{(l)} = \left[\begin{array}{c|c} U^{(l)} \\ 0 \end{array} \; \middle| \; u_{n+1} \right], \quad V^{(l)} = \left[\begin{array}{c|c} V^{(l)} \\ 0 \end{array} \; \middle| \; v_{n+1} \right],$$

and update $F^{(l)}$, $G^{(l)}$.

In (97), the vectors s_n, $r_n \in \mathbb{C}^{n+1}$ are defined by the partitioning

$$M_{n+1} = \begin{bmatrix} t_0 & s_n^T \\ r_n & M_n \end{bmatrix}$$

of the $(n+1)$st moment matrix M_{n+1}. Furthermore,

$$J = \begin{bmatrix} 0 & \cdots & 0 & 1 \\ \vdots & \iddots & 1 & 0 \\ 0 & \iddots & \iddots & \vdots \\ 1 & 0 & \cdots & 0 \end{bmatrix}$$

denotes the antidiagonal identity matrix.

For a discussion of the look-ahead strategy used in step 1) of Algorithm 9.1 and other implementation details, we refer the reader to [17]. Also, in [17], it is shown how solutions of Toeplitz systems $M_n x_n = b_n$ can be computed, based on the factorization (96).

Finally, we remark that Algorithm 9.1 is an extension of the classical Levinson-Trench algorithm [32, 37] for computing inverse factorizations of a sequence of nonsingular Toeplitz matrices.

References

[1] G. Baxter. Polynomials defined by a difference system. *J. Math. Anal. Appl.*, **2**, pp 223–263, 1961.

[2] E.R. Berlekamp. *Algebraic Coding Theory*. McGraw-Hill, New York, 1968.

[3] D.L. Boley, S. Elhay, G.H. Golub and M.H. Gutknecht. Nonsymmetric Lanczos and finding orthogonal polynomials associated with indefinite weights. *Numer. Algorithms*, **1**, pp 21–43, 1991.

[4] D.L. Boley and G.H. Golub. The nonsymmetric Lanczos algorithm and controllability. *Systems Control Lett.*, **16**, pp 97–105, 1991.

[5] C. Brezinski, M. Redivo Zaglia and H. Sadok. Avoiding breakdown and near-breakdown in Lanczos type algorithms. *Numer. Algorithms*, **1**, pp 261–284, 1991.

162

[6] T.S. Chihara. *An Introduction to Orthogonal Polynomials.* Gordon and Breach, New York, 1978.

[7] T.K. Citron. *Algorithms and architectures for error correcting codes.* Ph.D. thesis, Stanford University, Stanford, California, 1986.

[8] J. Cullum and R.A. Willoughby. A practical procedure for computing eigenvalues of large sparse nonsymmetric matrices. In: J. Cullum and R.A. Willoughby (Eds.), *Large Scale Eigenvalue Problems*, North-Holland, Amsterdam, pp 193–240, 1986.

[9] A. Draux. *Polynômes Orthogonaux Formels – Applications.* Lecture Notes in Mathematics, Vol. 974, Springer-Verlag, Berlin, 1983.

[10] R.W. Freund. Conjugate gradient-type methods for linear systems with complex symmetric coefficient matrices. *SIAM J. Sci. Statist. Comput.*, **13**, pp 425–448, 1992.

[11] R.W. Freund. Quasi-kernel polynomials and convergence results for quasi-minimal residual iterations. In: D. Braess and L.L. Schumaker (Eds.), *Numerical Methods of Approximation Theory, Vol. 9*, Birkäuser, Basel, pp 77–95, 1992.

[12] R.W. Freund. A transpose-free quasi-minimal residual algorithm for non-Hermitian linear systems. *SIAM J. Sci. Comput.*, **14**, 1993, to appear.

[13] R.W. Freund, G.H. Golub and N.M. Nachtigal. Iterative solution of linear systems. *Acta Numerica*, **1**, pp 57–100, 1992.

[14] R.W. Freund, M.H. Gutknecht and N.M. Nachtigal. An implementation of the look-ahead Lanczos algorithm for non-Hermitian matrices. *SIAM J. Sci. Comput.*, **14**, 1993, to appear.

[15] R.W. Freund and N.M. Nachtigal. QMR: a quasi-minimal residual method for non-Hermitian linear systems. *Numer. Math.*, **60**, pp 315–339, 1991.

[16] R.W. Freund and N.M. Nachtigal. An implementation of the QMR method based on coupled two-term recurrences. Technical Report 92.15, RIACS, NASA Ames Research Center, Moffett Field, California, June 1992.

[17] R.W. Freund and H. Zha. Formally biorthogonal polynomials and a look-ahead Levinson algorithm for general Toeplitz systems. Technical Report 91.27, RIACS, NASA Ames Research Center, Moffett Field, California, December 1991.

[18] R.W. Freund and H. Zha. A look-ahead algorithm for the solution of general Hankel systems. *Numer. Math.*, to appear.

[19] G.H. Golub and C.F. Van Loan. *Matrix Computations*, Second Edition. The Johns Hopkins University Press, Baltimore, 1989.

[20] W.B. Gragg. Matrix interpretations and applications of the continued fraction algorithm. *Rocky Mountain J. Math.*, **4**, pp 213–225, 1974.

[21] W.B. Gragg and A. Lindquist. On the partial realization problem. *Linear Algebra Appl.*, **50**, pp 277–319, 1983.

[22] U. Grenander and G. Szegö. *Toeplitz Forms and their Applications*, Second Edition. Chelsea, New York, 1984.

[23] M.H. Gutknecht. A completed theory of the unsymmetric Lanczos process and related algorithms, Part I. *SIAM J. Matrix Anal. Appl.*, **13**, pp 594–639, 1992.

[24] M.H. Gutknecht. A completed theory of the unsymmetric Lanczos process and related algorithms, Part II. *SIAM J. Matrix Anal. Appl.*, to appear.

[25] G. Heinig and K. Rost. *Algebraic Methods for Toeplitz-like Matrices and Operators*. Birkhäuser, Basel, 1984.

[26] W. Joubert. Lanczos methods for the solution of nonsymmetric systems of linear equations. *SIAM J. Matrix Anal. Appl.*, **13**, pp 926–943, 1992.

[27] T. Kailath, A. Vieira and M. Morf. Inverses of Toeplitz operators, innovations and orthogonal polynomials. *SIAM Rev.*, **20**, pp 106–119, 1978.

[28] S.-Y. Kung. *Multivariable and multidimensional systems: analysis and design*. Ph.D. thesis, Stanford University, Stanford, California, 1977.

[29] J.L. Massey. Shift-register synthesis and BCH decoding. *IEEE Trans. Inform. Theory*, **IT-15**, pp 122–127, 1969.

[30] C. Lanczos. An iteration method for the solution of the eigenvalue problem of linear differential and integral operators. *J. Res. Nat. Bur. Standards*, **45**, pp 255–282, 1950.

[31] C. Lanczos. Solution of systems of linear equations by minimized. *J. Res. Nat. Bur. Standards*, **49**, pp 33–53, 1952.

[32] N. Levinson. The Wiener RMS (root mean square) error criterion in filter design and prediction. *J. Math. Phys.*, **25**, pp 261–278, 1946.

[33] B.N. Parlett. *The Symmetric Eigenvalue Problem*. Prentice-Hall, Englewood Cliffs, N.J., 1980.

[34] B.N. Parlett. Reduction to tridiagonal form and minimal realizations. *SIAM J. Matrix Anal. Appl.*, **13**, pp 567–593, 1992.

[35] B.N. Parlett, D.R. Taylor and Z.A. Liu. A look-ahead Lanczos algorithm for unsymmetric matrices. *Math. Comp.*, **44**, pp 105–124, 1985.

[36] D.R. Taylor. *Analysis of the look ahead Lanczos algorithm*. Ph.D. thesis, University of California, Berkeley, California, 1982.

[37] W.F. Trench. An algorithm for the inversion of finite Toeplitz matrices. *J. Soc. Indust. Appl. Math.*, **12**, pp 515–522, 1964.

[38] W.F. Trench. An algorithm for the inversion of finite Hankel matrices. *J. Soc. Indust. Appl. Math.*, **13**, pp 1102–1107, 1965.

CASE STUDIES OF REAL-TIME PROCESSING IN ROBOTICS[*]

W. MORVEN GENTLEMAN
Institute for Information Technology
National Research Council of Canada
Ottawa, Canada K1A 0R8
gentleman@iit.nrc.ca

ABSTRACT. Examples are used to illustrate how robotics is computationally demanding, particularly with respect to time constraints within which the computation must complete. The examples are picked both from control algorithms for robots and from sensor data interpretation for robotics. The linear algebra problems involved are indicated.

KEYWORDS. Robotics, sensors, real-time, control, data interpretation

1. Introduction

Robotics is computationally demanding. The physical motions of the robot and objects in its environment often impose time constraints on the computation. Table 1 illustrates typical magnitudes of time intervals within which all relevant computation, including any numerical procedure, must complete. Although these times are independent of computer technology, and hence will become

TABLE 1. Typical Time Constraints

Source of constraint	Time
Preferred cycle time from mechanical bandwidth of manipulator	3.3 milliseconds
Existing servo system cycle time of manipulator	28 milliseconds
Video frame rate	33 milliseconds
Laser range-finder frames	74 milliseconds
Vehicle startup to full speed (13 cm)	400 milliseconds
Vehicle motion 1 cm along line	12.5 milliseconds
Vehicle motion 1° along arc	14 milliseconds

less of an issue in the future, they are challenging to meet today.

---

[*] NRC 35027

M. S. Moonen et al. (eds.), Linear Algebra for Large Scale and Real-Time Applications, 165–182.

Constraints on the kinds of computers that are appropriate further complicate the situation. The computers must be inexpensive enough that the whole robotic application is cost effective.

TABLE 2. Typical Costs of System Components

Component	Cost in $Can
Robot manipulator arm	50,000
Commercial Autonomous Guided Vehicle Platform	50,000
Camera	2,000
Laser Range-finder	60,000

Clearly, although there are exceptions, the computers in most robotic applications will have to be cheap. In many cases the computer must be on board the robot, which imposes size and weight constraints, as well as restrictions on power and cooling. A faster computer may not be a viable option. Supercomputer and massively parallel technology is irrelevant, although multiprocessors with a few or even a few tens of processors are commonplace.

Linear algebra plays a role, both conceptually and computationally, in robotics. Unlike the linear algebra problems arising in many other areas of science, the matrices involved in robotics are typically small, but there may be a great many to process in each time period.

We will consider two types of computations from robotics:

- control algorithms for robots, and
- sensor data interpretation for robotics

2. Control Algorithms for Robots

For robots such as a manipulator or an automatic guided vehicle (AGV), robot control algorithms are not usually expressed in traditional control theory terms. In control theory, the behaviour of systems is described by state vectors, which are defined by differential equations and typically cannot be observed or controlled directly. In robotics, the behaviour over time is thought of as being made up of a number of distinct segments, each segment consisting of interactions between the robot and objects in its environment. The required task may intrinsically be made up of several segments; limitations of the robot may make it impossible to perform an intended action in a single segment; or collision avoidance, where the finite sized robot must move through a cluttered workspace, may necessitate different segments to avoid different obstacles. The parts of the robot and the objects in the environment each have local properties, including a local coordinate system. Linear algebra occurs at three levels in robot control:

a) The relationships and transformations of the various coordinate systems, and the expression of physics in these coordinates.

b) The transitions between segments.

c) Path planning, that is, choosing a path made up of many segments.

2.1 MULTIPLE COORDINATE SYSTEMS

2.1.1 *Homogeneous Coordinates*. In general, we are interested not only in the location of an object in the workspace, but also in its orientation. Moreover, we are interested in it not in absolute terms, but rather in its position and orientation relative to some other object, and we may be interested in the position and orientation of other objects relative to it. This is best addressed by associating frames of reference (coordinate systems) with an object, and thinking of the transformations implied by moving from one of these frames of reference to another. Because for physical bodies we are primarily thinking in terms of rigid body motions, we usually think in terms of Cartesian coordinates, and need to allow for both translations and rotations. Denavit and Hartenberg[3],[12] provided an elegant solution to this problem long ago: use homogeneous coordinates. An arbitrary point [a b c] in 3 space is represented as a 4 vector:

$$
\mathbf{v} = \begin{bmatrix} x \\ y \\ z \\ w \end{bmatrix} \quad \text{where} \quad \begin{cases} a = x/w \\ b = y/w \\ c = z/w \end{cases}
\tag{EQ 1}
$$

A translation transformation by a vector [a b c] is then represented as the matrix

$$
\mathbf{H} = \begin{bmatrix} 1 & 0 & 0 & a \\ 0 & 1 & 0 & b \\ 0 & 0 & 1 & c \\ 0 & 0 & 0 & 1 \end{bmatrix}
\tag{EQ 2}
$$

A rotation of θ_x about the x axis is represented as

$$
\mathbf{R}_x = \begin{bmatrix} 1 & 0 & 0 & 0 \\ 0 & \cos\theta_x & -\sin\theta_x & 0 \\ 0 & \sin\theta_x & \cos\theta_x & 0 \\ 0 & 0 & 0 & 1 \end{bmatrix}
\tag{EQ 3}
$$

A rotation of θ_y about the y axis is represented as

$$
\mathbf{R}_y = \begin{bmatrix} \cos\theta_y & 0 & \sin\theta_y & 0 \\ 0 & 1 & 0 & 0 \\ -\sin\theta_y & 0 & \cos\theta_y & 0 \\ 0 & 0 & 0 & 1 \end{bmatrix}
\tag{EQ 4}
$$

A rotation of θ_z about the z axis is represented as

$$R_z = \begin{bmatrix} \cos\theta_z & -\sin\theta_z & 0 & 0 \\ \sin\theta_z & \cos\theta_z & 0 & 0 \\ 0 & 0 & 1 & 0 \\ 0 & 0 & 0 & 1 \end{bmatrix}$$ (EQ 5)

An arbitrary rigid body motion can thus be composed by multiplying together the matrices representing a translation and rotations about the three axis, yielding

$$T = \begin{bmatrix} n_x & o_x & a_x & p_x \\ n_y & o_y & a_y & p_y \\ n_z & o_z & a_z & p_z \\ 0 & 0 & 0 & 1 \end{bmatrix}$$ (EQ 6)

2.1.2 *Transformation Chains.* A particularly important use of such transformations is to form transformation chains, where the position and orientation of an object is specified as the product of a number of transformations, each of which represents the position and orientation of an intermediate object with respect to another object. For example, Figure 1 shows the Unimation

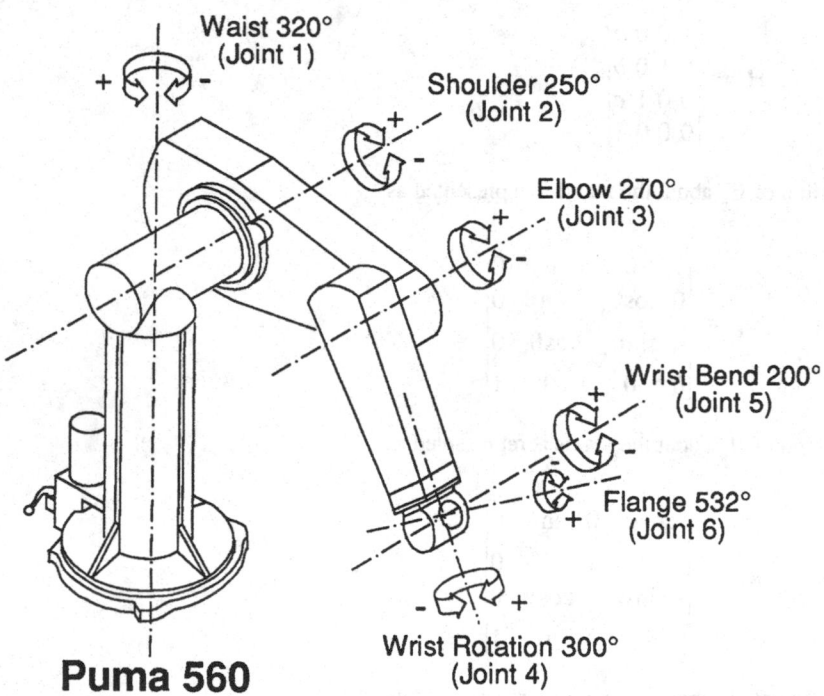

FIGURE 1. Typical industrial manipulator

PUMA 560, a robot arm with six joints that has been popular in robotics research laboratories for the last decade. The position and orientation of the end effector or load can be described by what is conventionally called the tool transformation E, a translation and rotation with respect to the end of the final link of the robot. The position and orientation of the end of the final link of the robot can be described by A_6, a translation and rotation with respect to joint 6, and each joint k+1 can similarly be described with respect to the previous joint k by A_k, a translation and rotation, right back to joint 1. Joint 1 itself is located with respect to the base of the robot by a translation and rotation described by A_0. The position and orientation of the base of the robot is described by a transformation P which is in the world coordinate system. For this design of manipulator, the translations are constant, so the A matrices only vary with the joint angles. The transformation X defining the position of the end effector or load in world coordinates can be written as

$$X = PT_6E \qquad \text{(EQ 7)}$$

where

$$T_6 = A_0A_1A_2A_3A_4A_5A_6 \qquad \text{(EQ 8)}$$

These are the kinematic equations. The velocity of the end effector can be determined by differentiating this matrix with respect to time. Motion systems such as this can involve many more links and so A matrices, for instance when a robot hand is mounted on the end of the arm or when multiple arms are coordinating in an activity.

As another example of transformation chains, consider an application of a robot in a chemistry laboratory. A test tube filled with liquid is to be lifted from a rack with slots for several test tubes, and the contents of the test tube poured into a beaker, as illustrated in Figure 2. The location and orientation of the rack can be represented by the transformation R, in world coordinates, and the location and orientation of the beaker can be represented by the transformation B, also in world coordinates. The location and orientation of the particular slot in the rack is represented by the transformation S, relative to the rack, and is a function of which slot it is. The initial location and orientation of the centroid of the test tube relative to the slot is represented by the transformation C, and the location of the grasp point of the test tube relative to its centroid is represented by the transformation G. The pouring point on the lip of the test tube relative to its centroid is represented by the transformation L. To grasp the test tube the end effector of the robot must be moved so that

$$T_6 = P^{-1}RSCGE^{-1} \qquad \text{(EQ 9)}$$

and the gripper closed. The tool transformation should then be thought of as being used to describe some point in the test tube, rather than some point in the gripper. That is, a new tool transformation should be used. One possibility would be to use the new tool transformation to describe the bottom of the test tube, and then use this to describe what T_6 must be changed to so that the bottom of the test tube will be lifted clear of the rack. Typically, instead, the current tool transformation is unchanged but an intermediate "approach" point is defined by V, a transformation describing an arbitrary "big enough" change in the Z coordinate of the departure point, and T_6 is simply changed so that

170

$$T_6 = P^{-1}RSCGVE^{-1}$$
(EQ 10)

The load is then moved (by changing T_6) to a similar approach point above B, that is T_6 is changed so that

$$T_6 = P^{-1}BVE^{-1}$$
(EQ 11)

However, here the tool transformation does need to change, because as the test tube is tipped to pour out its contents, we need to focus on the coordinates of the pouring point on the lip of the test tube, to ensure that the liquid pours into the beaker below. Tipping is perhaps best done as a rotation about this point, using the new tool transformation

$$E_{new} = E_{old}L$$
(EQ 12)

This example is typical of positional task definition as used in many robotic applications. A great many transformations must be defined, relating positions and locations of coordinate frames relative to other frames. The transformations often change with time. Some of these transformations are known from CAD data, some are measured off line, some must be dynamically deter-

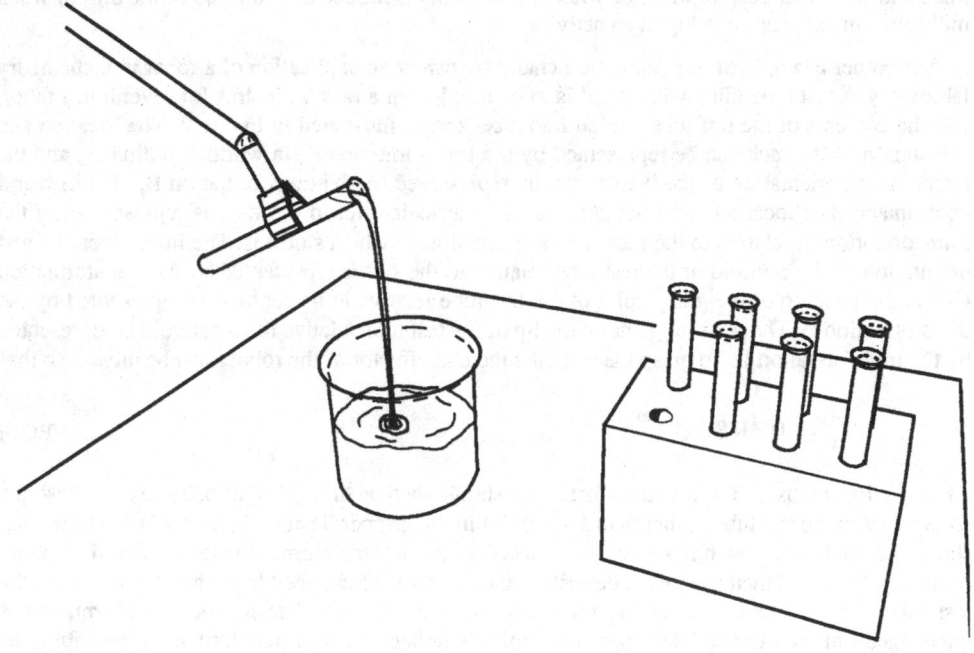

FIGURE 2. Chemistry Laboratory Application

mined on line by sensors, and some are under our control. The transformation chains used to define other transformations are different in different segments of the task.

2.1.3 *Limits, Stops, Singularities, and Redundancy.* Solving the kinematic equations to find positions, orientations, and velocities given joint angles is important, but even more important is solving the inverse kinematic equations to find the joint angles and joint velocities necessary to achieve desired objectives in the other coordinates.

It is essential to recognize that things can go wrong with this, things which may as an artifact exhibit numerical difficulties, but are actually intrinsic to the geometry.

First, because the relationship is highly nonlinear, instead of having a unique solution the inverse kinematics may have no solutions, may have multiple distinct solutions, or may have a continuum of solutions, i.e. there may be solution spaces. A desired objective may simply be unachievable. Choosing a particular solution when more than one is possible is likely to involve more than just numerical accuracy considerations. Redundancy, and consequently the implication of multiple solutions, is sometimes deliberately introduced in order to provide more degrees of freedom, e.g. to facilitate collision avoidance.

Second, joints typically cannot move over an arbitrary range, but instead have limits (see Figure 1). In some cases the joint angle is simply periodic, but it can matter which period the answer is taken from. In other cases, there is a physical limit on how far the joint can be turned, and going further may cause something to break. This often becomes an issue when motion between two achievable points in cartesian space would involve intermediate points that pass limits or singularities in joint space. Singularities may simply be connections between regions with multiple solutions, and reparameterizing may remove the difficulty, but there are circumstances where passing through a singularity can cause something to break.

2.1.4 *Cartesian/Joint Coordinates.* Returning to Equation 8, the position and orientation of the end effector in Cartesian coordinates can be considered to be a function of the controlled joint angles

$$p = f(\theta_1, \theta_2, \theta_3, \theta_4, \theta_5, \theta_6) \tag{EQ 13}$$

The differential change in position and orientation corresponding to a differential change in the joint coordinates depends on the Jacobian of this function.

$$\begin{bmatrix} d_x \\ d_y \\ d_z \\ \delta_x \\ \delta_y \\ \delta_z \end{bmatrix} = J(\theta_1, \theta_2, \theta_3, \theta_4, \theta_5, \theta_6) \begin{bmatrix} d\theta_1 \\ d\theta_2 \\ d\theta_3 \\ d\theta_4 \\ d\theta_5 \\ d\theta_6 \end{bmatrix} \tag{EQ 14}$$

This shows that inverting the Jacobian can be used solve inverse kinematics.

172

2.1.5 *Alternate Approaches*. The approach presented so far is convenient and systematic, but it is not the only one possible. Other coordinate systems can be chosen, each with advantages in some situations. For instance, instead of representing orientation by rotations about the x, y, and z axis, we could use Euler angles, cylindrical coordinates, or spherical coordinates. Instead of using homogeneous transformations we could have used quaternions, screw systems (Plücker coordinates)[10],[11] or others. Instead of using Gaussian elimination to invert the Jacobian, we could solve the inverse kinematics by ad hoc methods based on symmetries or other structure in the matrix.

An important example of an ad hoc method is the work by Featherstone, Hollerbach, and others[4],[5],[6],[7] exploiting coplanar links and a spherical wrist.

Operation counts differ significantly depending on representation and algorithm. Inverse kinematic velocities of the Stanford arm, deriving joint velocities to obtain Cartesian velocities, are shown in Table 3.[7],

TABLE 3. Operation counts for inverse kinematic velocities for Stanford arm

Method	Multiplications	Additions
6 by 6 matrix inversion	287	193
Gaussian elimination	141	98
Featherstone	36	20

It should be noted that achieving a 28 millisecond servo cycle time even with the Featherstone method is hard when the computation is done on a 25 MHz Motorola 68020/68881 combination. This has meant that in the past, and still even now, reduction in operation count has been viewed as much more important than issues such as mathematical simplicity or numerical accuracy. Fortunately, the methods have worked well.

2.1.6 *Physics*. So far we have presented only kinematics. When forces and accelerations are considered, we need to derive the equations of motion. This can be done in several ways, such as using the Hamiltonian formulation, the Lagrangian formulation, or the Newton-Euler formulation. Alternate coordinate systems can be used, and different methods can be applied for solving the systems. Again the cost differences are striking, and only those with the smallest operation counts would be computationally feasible today. For a 6 link arm, Table 4[8] shows the costs of different approaches

TABLE 4.

Method	Multiplications	Additions
Uicker/Kahn	66,271	51,548
Waters	7,051	5,652
Hollerbach (4 by 4)	4,388	3,586
Hollerbach (3 by 3)	2,195	1,719
Newton-Euler	852	738
Horn/Raibert	468	264

At one time it was believed that the formulation made a difference, and that the Newton-Euler formulation was most efficient. Of course the answer is unique, so it is not surprising that the formulations have been shown equivalent: what actually matters is the formulas by which things are computed, and how the answers are expressed. Recursive formulation is especially important.

2.1.7 *Effects of Hard to Measure Parameters.* Unfortunately, although the equations of motion are essential to understanding forces and accelerations, and to analyzing the motion of the robot at high speed, they turn out to be ineffective in practice, because of the effects of friction, backlash, and load are so great. In current industrial robots, for instance, friction is comparable to other forces.

One consequence of this has been the development of direct drive robots for robotics research[1]. Direct drive robots eliminate the above problems, so that valid physical models can be derived. On the other hand, so far direct drive robots are impractical for industrial use.

2.2 TRANSITIONS BETWEEN SEGMENTS

2.2.1 *The Chemistry Laboratory Example.* Motion within a segment may be straightforward to analyze, but the transition from one segment to another is often complicated or can only be approximated. The chemistry laboratory example above indicates some of the problems. Lifting the test tube from the rack is a segment, a precise vertical motion. Tipping the test tube once it is over beaker is another segment, a precise rotation about a horizontal axis. The detail of getting between these two segments should be unimportant, and the literature on robotic applications often suggests interpolation, either in joint or Cartesian coordinates. Unfortunately, there is a constraint here: the test tube is filled with liquid, and must not be tipped. Because of the nonlinearity of the kinematics, there is no simple guarantee with interpolation in joint coordinates that the test tube will not tip. Interpolation in Cartesian coordinates at first seems better, but this may not work either, because if we interpolate points in Cartesian coordinates, and then solve the inverse kinematics for these points, there is no guarantee that redundancy will be resolved for adjacent points the same way. Indeed, what typically happens as a sequence of points approach a joint limit is that when some critical value is passed, a different solution is chosen, and the effect on the robot trajectory is that that joint suddenly may reverse by 360°, spilling the liquid. If the velocities or accelerations are significant, care must be taken not just to avoid spilling, but also sloshing. Problems like these are avoided today by experimentally trying the setups, or by ad hoc examination of the configuration space, and this does not work well with sensor defined locations. Interpolation is a linear algebra process, and solving the kinematic equations is an algebraic process, so it seems there ought to be a better way.

2.2.2 *A Radio Controlled Model Truck.* As another example, we will look at causing a radio controlled model truck to change from travelling at constant speed in one straight line to travelling at the same speed in another straight line.

We use a small radio controlled model truck to demonstrate the abilities of our robot control and to investigate problems in navigation. The dimensions of the vehicle are 19 cm wide and 33.5 cm long, of which 26 cm is in front of the rear axle and 7.5 cm is behind it. An important characteristic of the truck is that it is available in models with two different radio frequencies, so that two different trucks can be controlled at the same time.

The radio controlled model truck has "bang-bang" control. That is, it can move forward at a single speed, or it can move backward at a single speed, or it can be stopped. If moving forward or backward, it can be steered straight ahead, or to the right, or to the left. The steering is not proportional: when turning, the steered front wheels are pointed at 15 degrees to the axis of the vehicle. Given that the vehicle wheelbase is 17.5 cm., this results in a fixed turning radius (measured to the centre of the rear axle) of 63.5 cm.

The truck accelerates at 200 cm per second per second to its maximum speed of 80 cm per second. This can also be viewed as a startup transient of about .25 second or of about 13 cm. Deceleration is the same. At constant speed, it takes 12.5 milliseconds to move 1 cm. When a turn is applied, it is effectively instantaneous.

More significantly, the trajectory of the truck is simple only when viewed from the midpoint of the rear axle. From this point, the trajectory is smoothly joined segments of straight lines and fixed radius circular arcs. For more interesting points on the vehicle, such as the centre of gravity or an outside front corner, the transitions from straight line motion to circular motion and back are not smooth: the sudden turning of the steered wheels produces a distinct kink in the trajectory. Homogeneous coordinates, this time restricted to 2D, are a convenient way to express the motions.

Since the turning radius is fixed, the only way to cause the vehicle to change direction by a given amount is to assert the turn for a specified time, about 14 milliseconds per degree. Performing a turn can take significant space.

Curve fitting to join two given segments, each say with the truck at a given location traveling in a given direction, is a significant nonstandard interpolation problem. Even such a simple problem as passing through a given point after starting at a given location when pointing in a given direction requires at least two straight line and one circular arc in general. Although approximation like this could be viewed as a linear algebra problem, it is perhaps best approached as a path planning problem as described below.

2.2.3 *A Commercial AGV.* The Cybermotion K2A is a three wheeled synchronous driven and synchronous steered platform. That is, all three wheels are connected via drive shafts to a drive motor minimizing the differences in speed between the wheels when they are driving. The same is true of the connections between the wheels and the steering motor. A constant linear platform speed is achieved by sending the drive motor controller constant drive values. A constant platform angular speed is attained by sending constant values to the steering motor controller. Both drive motor and steering motor can only change acceleration linearly. What this means is that the trajectory (as viewed from the center of the vehicle)[9] is more complicated than that of the radio controlled truck: straight line segments are smoothly connected to clothoids (cornu spirals) which may be smoothly connected to circular arcs or to other clothoids. Smooth joining in this case implies that at the join the curvature of the curves match. Again the curve fitting problem is a significant nonstandard approximation or interpolation problem, but here the equations for the fitting depend on Fresnel integrals which themselves necessitate approximation.

2.3 PATH PLANNING

2.3.1 *The Three-point Turn and Parallel Parking.* Navigating the radio controlled truck, in general, requires not one but a sequence of straight line and circular arc moves. This is especially true in a constrained environment. The classic such manoeuvre is the three-point turn, beloved of

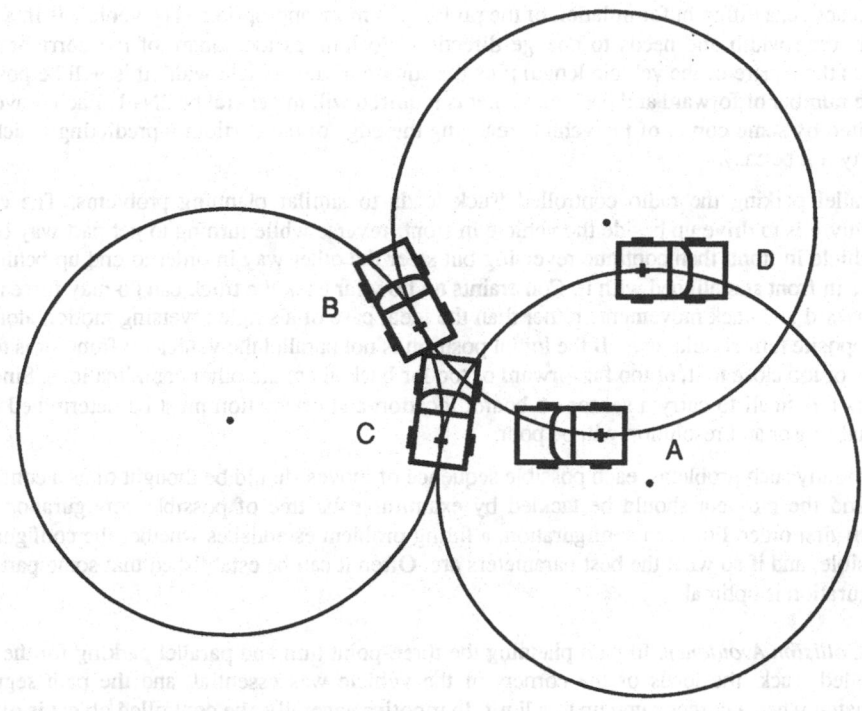

FIGURE 3. Optimal Three-point Turn

driving examinations. A vehicle can be made to travel forward in the opposite direction by the manoeuvre called a U-turn, simply making a circular arc of 180°. This requires a corridor the width of the vehicle plus twice the turning radius, which for the model truck is 146 cm. A narrower corridor than this can be used by making a circular arc of 90°, then reversing direction and steering the opposite way to make another circular arc of 90°, then reversing direction again and going straight to exit the turn.This requires a corridor only half the width of the vehicle plus the turning radius plus the length of the vehicle forward of the rear axle. For the model truck this takes a 99 cm corridor. The three-point turn does even better than this. As illustrated in Figure 3, the turn begins at position A by making a circular arc of angle $\pi/2 - \theta$ to position B. Reversing direction and steering the opposite way, a circular arc of angle $\phi - \theta$ is made to position C. Finally, reversing direction once more and steering again the first way, a circular arc of angle $\pi/2 - \phi$ is made to position D. Finding the optimum angles for the circular arcs to minimize the corridor width is a nice algebra problem depending on the turning radius R, the vehicle width $2w$, the length of the vehicle forward of the rear axle f, and the length of the vehicle behind the rear axle b. Equations 15 and 16 must hold simultaneously. For the model truck, the minimum corridor required is only 71 cm.

$$2R\sin\theta - R = (R - w)\sin\phi - b\cos\phi \qquad \text{(EQ 15)}$$

$$2R\sin\phi + R = (R + w)\sin\theta + f\cos\theta \qquad \text{(EQ 16)}$$

In practice, a different formulation of the problem is more appropriate. The vehicle is in a corridor of some width and needs to change direction. So long as the square of the corridor width exceeds the square of the vehicle length plus the square of the vehicle width this will be possible, but the number of forward and back movements required will in general be 2N+1. Each movement is limited by some corner of the vehicle reaching the edge of the corridor—predicting which corner may not be easy.

Parallel parking the radio controlled truck leads to similar planning problems. The classic manoeuvre is to drive up beside the vehicle in front, reverse while turning to get part way behind the vehicle in front, then continue reversing but steer the other way in order to end up behind the vehicle in front and aligned with it. Constraints on how far back the truck can go may force multiple forward and back movements rather than the ideal case of a single reversing motion along the two opposite turn circular arcs. If the initial position is not parallel the vehicle in front, or is too far from it or too close to it, or too far forward or too far back, there are other complications. Since the truck is too small to carry a sensor on board, location and orientation must be determined by an external sensor and resolution will be poor.

For many such problems, each possible sequence of moves should be thought of as a configuration, and the problem should be tackled by examining the tree of possible configurations in a breadth-first order. For each configuration, a fitting problem establishes whether the configuration is feasible, and if so what the best parameters are. Often it can be established that some particular configuration is optimal.

2.3.2 *Collision Avoidance*. In path planning the three-point turn and parallel parking for the radio controlled truck, the locus of the corners of the vehicle was essential, and the path segments terminated when a corner came up to a limit. In robotics generally, the controlled object is of finite size, and moves in a cluttered space with other objects of finite size. While the objective often is that the robot interact with some of these objects, it is mandatory that parts of the robot not accidentally collide with other objects. It is convenient to think of the tool transformation of a manipulator as a point with orientation, for instance, and to follow its motion, but we must actually check the locus of the upper links of the manipulator to ensure that they do not hit anything. Working with a map of instantaneous positions of all objects in the workspace, and looking for intersections of polygonal models of these objects, is an impractical solution for collision avoidance, even if optimizations are employed as with the clipping computations of computer graphics.

One conservative solution that has been proposed is to "expand" each potential obstacle by the size of the controlled object. The controlled object can then be treated as a point. This simplifies finding paths along which the point can move collision free. Unfortunately, objects often have an irregular shape, and indeed may change shape or orientation over time (consider a robot arm), thus "size" is poorly defined and "expand" is prohibitively expensive unless the bounds used are quite crude. While such bounds may suffice in some situations, they prohibit close interaction or optimal exploitation of space.

2.3.3 *Grasping a Swinging Object*. Causing a robot arm to grasp an object which is swinging appears to require establishing the trajectory of the swinging object as a function of time, as well as taking into account the time skews due to the sensor data acquisition and analysis, due to issuing a command to the robot, and due to the mechanical action of the robot. Doing this completely is not practical with today's robots and control computers. An alternative approach has, however, proved practical and effective. This approach uses a state space controller[2],[15].

The positions and velocities of the robot relative to the target form a state space, in which the desired goal is a single point. For each point in this state space, there is a command which could be issued to the robot that would be optimal in achieving this goal. A table can be computed, indexed by the state variables, containing these optimal commands. The table can be computed off-line by dynamic programing. Linear algebra problems arise in computing the table. The table is used by repeatedly observing the values of the state variables and issuing the corresponding command to the robot.

3. Sensor Data Interpretation for Robotics

The foregoing discussed how to tell the robot what to do if you know what it should do. However, only in rare situations when the environment is rigidly structured can this be done immediately. More typically, at least some aspects of the environment are unknown and must be empirically determined. This section discusses how to find out enough about the robot's environment to decide what to do. There are two ways of using sensors to direct motions. Look-and-move is where the sensor is used to assess the environment, and then the move is made open loop, i.e. without further reference to the sensor. This usually allows for more intensive processing of sensor data. Servo-on-sensor-data uses a closed loop where the move is incrementally modified in response to feedback from the sensor. This usually allows for better responsiveness to changing conditions. Combinations of the two ways of using sensors can be used, one for initial coarse movements, and the other for final fine movements. Of course sensors also must often be used to check that the actuator accomplished the intended movement: inadequate physical models as well as actuator malfunction can result in an issued command not producing the expected effect.

The simplest sensors used in robotics are contact sensors, or equivalents such as detecting when a light beam is cut. Most other sensors involve significant processing of the raw sensor data before it can be used. Sonar and force sensors tend to produce unreliable data because of confounding artifacts. One of the most effective sensors used in robotics is the TV camera These can be used singly, but multiple cameras are often used to increase the field of view, to provide stereoscopic vision, or for other reasons. The cameras may be mounted in a fixed position relative to the workplace, may be mounted on the moving robot, or may be independently mobile. The scene itself may be structured in some way, for instance by the way it is lighted, or may be completely general. Other kinds of sensors are often emulated by processing pixel data from cameras. Artificial skin, for simulating the sense of touch, produces pixel arrays that require analysis similar to camera data.

Another effective sensor is the 3D locator. There are several variations on this theme. Laser range-finders estimate distance by scanning the scene with a laser beam and measuring deflections of the reflected beam. Another type of device, the Selspot, is based on active infrared light emitting diodes attached to the target and cameras with lateral-effect photodiode detectors. Still another type of device, the Polymus, detects electromagnetic fields produced by emitters attached to the target.

There is a number of issues associated with interpreting sensor data. Many sensors suffer from occlusion, that is, objects in the scene can obstruct the sensor's detection of other objects, and so needed data may not be available. This is one of several reasons why another issue arises: sometimes data fusion is required, that is, data from multiple sensors (possibly even different types of

178

sensors) is needed to adequately assess the environment, and interpreting the combination of such data is complicated. Another issue that must be considered is that the objects in the scene often are in motion, and the sensor itself may be moving, so that observed sensor data is necessarily from the past, an effect exaggerated if significant processing is required for interpretation. Taking delays into account is essential.

Although camera data is often treated as a sequence of images, interpretation of sensor data for robotics differs from traditional image processing in several ways. One way is that objects in the scene are in motion, and typically it is the moving objects that are of interest. Another way is that there are time constraints on how much analysis can be done. A third way is that the scene will be reanalyzed with each new scan, and that significant savings in computation time can be achieved by assuming coherence between scans and treating the analysis as an update problem.

Processing of raw sensor data before it can be used often includes such operations as change of representation, calibration corrections, and filtering. A particularly common operation is extraction of signal from noise by some sort of fitting, either directly as a linear algebra problem or as a linearized approximation to some nonlinear fit[14].

3.1 MODEL FITTING OF MOVING OBJECTS

In simpler situations, the environment is controlled so that the objects in the scene have a known functional form, and sensor data is used to determine the unknown parameters in this form.

3.1.1 *Mapping the Model to What is Seen.* Many sensors do not view a three dimensional object as a whole, but merely a projection of it onto a surface. Even sensors that record three dimensional data can suffer from occlusion, where other objects or even the object of interest itself can hide aspects of interest.

A common situation is where an object is known to be in the scene, but its location and orientation are unknown. A simple example of this is where a camera is used as a sensor the scene contains a ball with a stripe around it. The location of the ball in the camera's field of view can be determined from the centroid of the pixels representing the image of the ball, and the distance to the ball can be determined by the diameter of the image. To find the orientation, however, one must fit sections of ellipses to the edges of the stripe, and compute from the parameters of the ellipse to find the axis about which the ball has been rotated, and how much it has been rotated. Obviously this problem is underdetermined, for rotation of the ball about its axis of symmetry produce no effect on the image.

Pose determination is the problem of categorizing all possible projections of an object into classes that are distinct in terms of aspects of interest. An interpretation tree of poses then gives projection models whose parameters can be fitted, to determine the best fitting pose. As a more complicated example, consider a teacup with handle: how many distinct poses are there in the full three dimensions of orientation when one takes into account the occlusion that can occur with respect to the handle, the inside of the cup, and the base of the cup? Linear algebra is used off-line to derive the poses, and on-line to fit the parameters of each pose.

3.1.2 *Slow Scan Sensors.* Traditional projections consider static objects. However if the object is in motion, or equivalently if the sensor is in motion, and if the sensor scans at a rate comparable with the relative object motion, then the projection seen by the sensor will be distorted, and that distortion must be modeled in the form to fit.

3.2 FEATURE EXTRACTION

3.2.1 *Objectives.* In more general situations, the environment is unknown, so the sensor data must be examined to extract features that might indicate characteristics of importance about objects possibly in the workspace. Many sensors, such as TV cameras, generate a flood of data from which the parameters of interest must be extracted. If the model is unknown, or if pose determination is thought to be easier in terms of recognized features, the sensor data is initially treated to extract recognizable features, such as line segments, circular arcs, segments of ellipses, etc. Although in some ways this resembles data fitting, there is a significant difference. This difference is that it is only assumed that a subset of the data fits the primitive being tested – the other data should not affect how well the primitive fits. This has some commonality with ideas of robust regression, but the difference is that there is no assumption that there are only a few outliers[13]. Once one feature is found, the data associated with it are removed from the image, and a search for the next feature is begun.

3.2.2 *Data Assumptions.* One common assumption, certainly true for laser range-finders and single cameras, is that the measurement error is very small. Consequently, a primitive can sensibly be fitted using the minimal subset of the data that will determine the parameters of the primitive. Unfortunately, for a large data set this leads to an unworkable number of possible primitives. Random sampling of the data can thus be used as a practical way to find plausible primitives, and the problem becomes one of comparing possible primitives. (The assumption of continuity across time can also provide plausible primitives.) Extending primitives, i.e. finding all points that are on that primitive and finding the best representation of it, can be done by template scoring, chaining, or genetic algorithms. Because genetic algorithms are uncommon in linear algebra, a description is given here. They are apparently quite effective on real image data.

3.2.3 *Genetic Algorithms.* The form of a genetic algorithm for feature extraction is as follows:

a) Initial candidate features are chosen and entered into a pool.

b) Each candidate feature is defined by "chromosomes". In the case of minimal subset fitting, the image points defining the basis can be the chromosomes.

c) Each candidate feature is given a score. In the case of minimal subset fitting, the score could be the number of image points close enough in the image to the candidate feature. (For nonlinear features, computing the distance from the image point to the feature can be hard.)

d) Pairs of candidates are drawn with probabilities proportional to their scores.

e) New candidates are generated by randomly interchanging chromosomes between the drawn pair, these new candidates are scored and added to the pool.

f) If some candidate becomes dominant, it and its equivalents are recognized and removed from the pool.

3.3 STEREO VISION

For many sensor data interpretation problems, the major cost is not the numerical computation, but the correspondence problem. An example of this occurs in the use of stereoscopic vision to measure distance to points. The underlying geometry is a simple model derived from considering a ray passing through the centre of each lens, as in Figure 4. Here Z is the unknown distance to the object, and X is its unknown displacement from the axis through the midpoint between the lenses. The separation $2D$ of the two lenses is known, the distance z from the lenses to the focal plane is known, and the displacements of each image from the central axis of its lens, $x + d$ and $x - d$ respectively, can be measured.

From similar triangles, we have

$$Z/z = (X + D) / (x + d) \qquad\qquad \text{(EQ 17)}$$

$$Z/z = (X - D) / (x - d) \qquad\qquad \text{(EQ 18)}$$

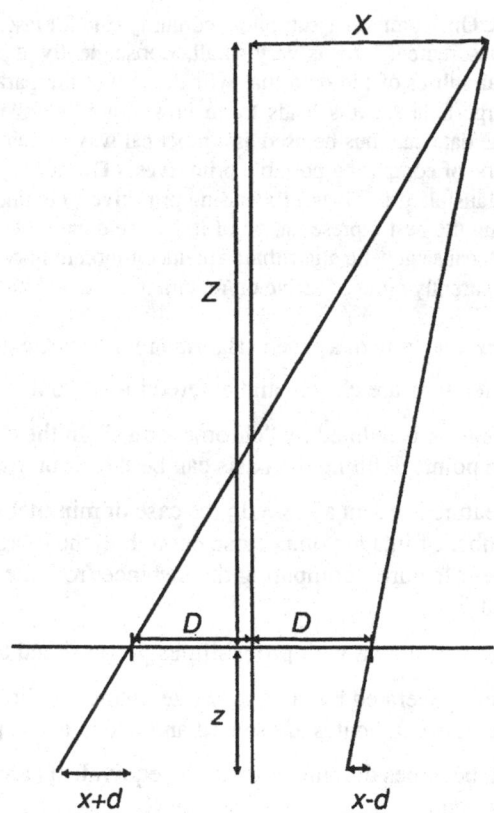

FIGURE 4. Rays through lens centres for stereo pair

which can be rewritten in matrix form as

$$\begin{bmatrix} \dfrac{1}{z} & -\dfrac{1}{(x+d)} \\ \dfrac{1}{z} & -\dfrac{1}{(x-d)} \end{bmatrix} \begin{bmatrix} Z \\ X \end{bmatrix} = \begin{bmatrix} D/(x+d) \\ -D/(x-d) \end{bmatrix} \qquad \text{(EQ 19)}$$

Even a 2 by 2 matrix can have numerical difficulties: it is obvious that this system can be ill conditioned if the object is far away. The other obvious problem is that quantization of the image into pixels can have serious effects.

However, the real computational problem is the correspondence problem: matching which points in the two images correspond to the same object in the scene so that the formulae apply. Coherence between successive images can be used to reduce the computational load of this.

4. Acknowledgments

The author gratefully acknowledges the assistance in the preparation of this paper by Colin Archibald, Martin Brooks, Shadia Elgazzar, David Green, Gerhard Roth, and Romiro Liscano, all of the Autonomous Systems Laboratory, Institute for Information Technology, National Research Council.

5. References

[1] An, C. H., Atkeson, C. G., & Hollerbach, J. M. *Model-Based Control of a Robot Manipulator*, The MIT Press, Cambridge, MA, 1988.

[2] Archibald, C. & van de Panne, M. "Tracking and Grasping Moving Objects Using Reflex Behaviour", Proceedings of the Fifth International Conference on Advanced Robotics (ICAR), Pisa, Italy. June 19-22, 1991. pp643-648.

[3] Denavit, J. & Hartenberg, R.S. "A Kinematic Notation for Lower-Pair Mechanisms Based on Matrices," ASME Journal of Applied Mechanics (June 1955), pp215-221.

[4] Elgazzar, S. "Efficient Solution for the Kinematic Positions for the PUMA Robot", ERB-973, National Research Council of Canada, December 1984. 36 pages.

[5] Elgazzar, S. "Efficient Solution for the Kinematic Velocities and Accelerations for the PUMA 560 Robot", ERB-974, National Research Council of Canada, December 1985. 45 pages.

[6] Featherstone, R. "Position and Velocity Transformations Between Robot End-effector coordinates and Joint Angles", International Journal of Robotics Research, Vol. 2, No. 2, (Spring 1983), pp35-45.

[7] Hollerbach, J. M. & Sahar, G. "Wrist-Partitioned, Inverse Kinematic Accelerations and Manipulator Dynamics", International Journal of Robotics Research, Vol. 2, No.

4, (Winter 1983), pp61-76.

[8] Hollerbach, J. M. "A Recursive Lagrangian Formulation of Manipulator Dynamics and a Comparative Study of Dynamics Formulation Complexity", IEEE Transactions on Systems, Man, and Cybernetics, Vol. SMC-10, No. 11, November 1980, pp730-736.

[9] Liscano, R. & Green, D. "A Method of Computing Smooth Transitions Between A Sequence of Straight Line Paths for an Autonomous Vehicle", ERB-1031, National Research Council of Canada, June 1990. 10 pages.

[10] Mason, M.T. & Salisbury, J. K. Jr. *Robot Hands and the Mechanics of Manipulation*, The MIT Press, Cambridge, MA, 1985.

[11] Ohwovoriole, M.S. "An Extension of Screw Theory and its Application t the Automation of Industrial Assemblies", Stanford Dept. of Computer Science Report No. STAN-CS-80-809, April 1980, 172 pages.

[12] Paul, R. P., *Robot Manipulators: Mathematics, Programming, and Control*, The MIT Press, Cambridge, MA, 1981.

[13] Roth, G. & Levine, M.D. "Geometric Primitive Extraction Using a Genetic Algorithm", Proceedings of Computer Vision and Pattern Recognition '92, June 15-18, 1992, Champaign, Illinois.

[14] Späth, H. *Mathematical Algorithms for Linear Regression*, Academic Press, Inc., San Diego, CA, 1987 (English Translation 1992).

[15] van de Panne, M., Fiume, E., and Vranesic, Z. "Reusable Motion Synthesis Using State-Space Controllers", Computer Graphics, Vol. 24, No. 4, (August 1990), pp225-234.

[16] Venkatesan, S. & Archibald, C. "Use of Wrist Mounted Range Profile Scanners for Real-time Tracking", Machine Vision and Applications, Vol. 5, (1992), pp1-16.

ADAPTIVE SIGNAL PROCESSING WITH EMPHASIS ON QRD-LEAST SQUARES LATTICE

S. HAYKIN
Communications Research Laboratory
McMaster University
1280 Main Street West
Hamilton, Ontario
Canada L8S 4K1
haykin@mcbuff.eng.mcmaster.ca

1 Introduction

An adaptive filter is a signal processing device that is *self-designing* in that the adaptive filter relies for its operation on an iterative algorithm. The algorithm makes it possible for the filter to perform satisfactorily in an environment when complete knowledge of the relevant signal characteristics is *not* available [1]. Adaptive filters find applications in such diverse fields as communications, radar, sonar, control, and biomedical engineering.

In this paper we present a general review of adaptive signal processing, and then focus on the recently developed adaptive filtering algorithm known as the *QR decomposition-least squares lattice* (QRD-LSL) *algorithm*. The algorithm is novel in several respects :

- It is modular in structure, with each module being in the form of a lattice.

- Its derivation relies on the method of QR-decomposition and recursive least squares estimation.

Accordingly, the algorithm has many of the distinctive features of these two mathematical procedures.

The QRD-LSL algorithm owes its origin to independent work by Proudler, McWhirter, and Shepherd [2], Regalia and Bellanger [3], and Ling [4]. A detailed derivation of the algorithm is presented in [1].

M. S. Moonen et al. (eds.), Linear Algebra for Large Scale and Real-Time Applications, 183–194.
© 1993 Kluwer Academic Publishers.

184

2.   Issues of Concern

The operation of an adaptive filter revolves around the error signal (estimation error), defined as the difference between some desired response and the actual filter output. The objective of the filter is to minimize the error signal in some statistical sense. The issues of concern in the analysis and design of an adaptive filter may be summarized as follows [1]:

- Rate of convergence, which refers to the number of iterations needed for the mean-squared error of the filter to approach its stead-state value.
- Misadjustment, which refers to the deviation of the steady-state value of the average squared error of the filter from the minimum mean-squared error produced by the corresponding Wiener filter.
- Robustness, with respect to changes in second-order statistics of the environment.
- Tracking of statistical variations of the input signals, when operating in a nonstationary environment.
- Computational requirements, with emphasis on the number of adds and multipliers needed to implement the algorithm.
- Structure, with emphasis on suitability for very-large-scale-integrated (VLSI) implementation.
- Numerical properties, with respect to the precision used in the computation.

Ideally, an adaptive filter would have a fast rate of convergence, a small misadjustment, good tracking behavior, and would be robust with respect to variations in second-order statistics of the environment; its computation requirements would be minimal, and the filter would have a modular structure suitable for VLSI implementation, and would offer good numerical properties and relative insensitivity to numerical precision. The QRD-LSL algorithm satisfies all the properties described here, with the exception of that referring to tracking. Being derived from the RLS algorithm that is based on a deterministic model, there is no guarantee that its fast convergence behavior would translate to a good tracking performance in a nonstationary environment. Essentially, the QRD-LSL algorithm combines many of the desirable features of the ubiquitous least-mean-square (LMS) algorithm and the recursive least-squares (RLS) algorithm, except for the fact that when the requirement is for operation in a nonstationary environment it is possible for the LMS algorithm to outperform it.

3.   Basic Configurations

Figure 1, taken from [1] presents four basic configurations of adaptive filters. These configurations relate to the following application areas:

- Identification (e.g., channel modeling)
- Inverse modeling (e.g., adaptive equalization)
- Prediction (e.g., linear predictive coding, and adaptive differential pulse-code modulation)
- Interference cancelling (e.g., adaptive noise cancelling, echo cancellation, and

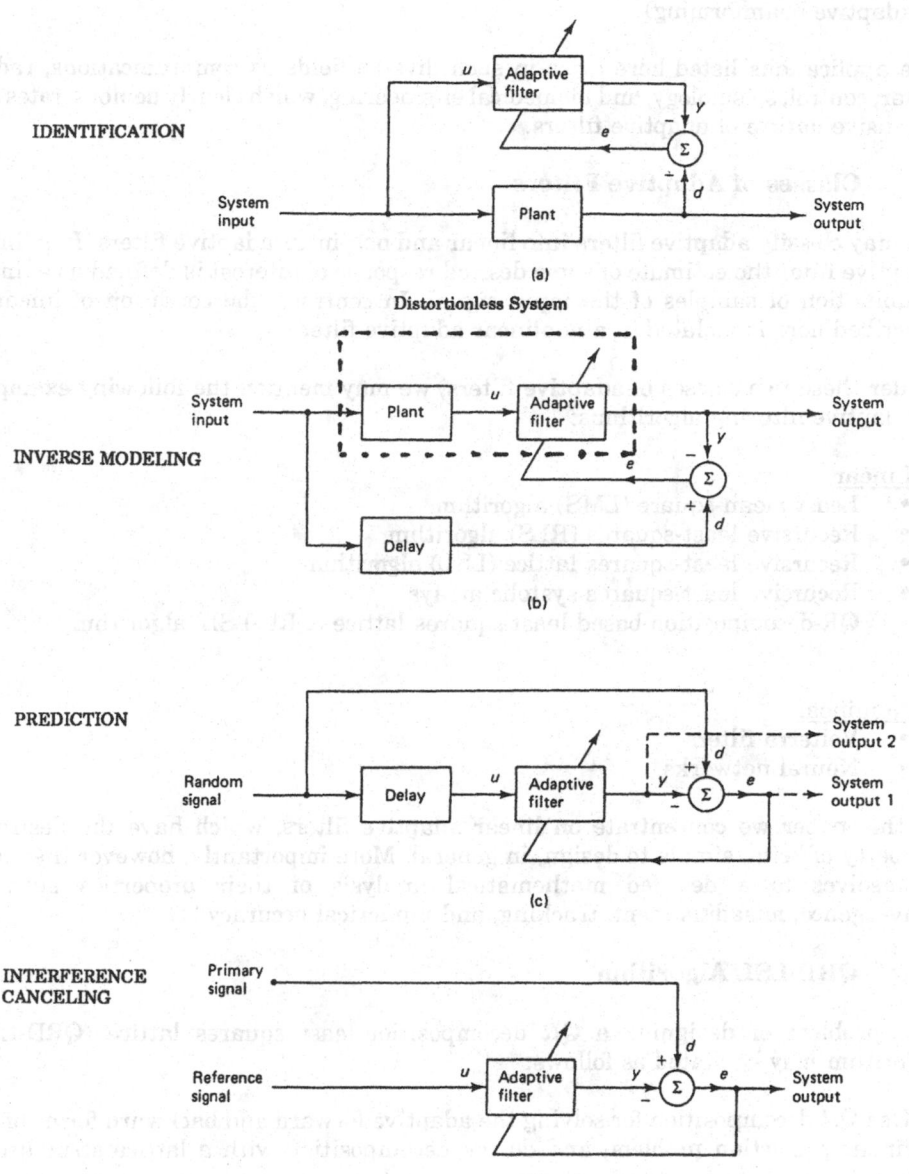

Figure 1 Four basic classes of adaptive filtering applications: (a) class I: identifcation; (b) class II: inverse modeling; (c) class III: prediction; (d) class IV: interference cancelling. (Taken from S. Haykin, Adaptive Filter Theory, 1991).

adaptive beamforming)

The applications listed here arise in such diverse fields as communications, radar, sonar, control, seismology, and biomedical engineering, which clearly demonstrates the pervasive nature of adaptive filters.

4.   Classes of Adaptive Filters

We may classify adaptive filters into linear and nonlinear adaptive filters. In a linear adaptive filter the estimate of some desired response of interest is defined as a linear combination of samples of the input signal. In contrast, the condition of linearity described here is violated in a nonlinear adaptive filter.

Under these two classes of adaptive filters, we may mention the following examples of adaptive filtering algorithms:

* Linear
    * Least mean-square (LMS) algorithm
    * Recursive least-squares (RLS) algorithm
    * Recursive least-squares lattice (LSL) algorithm
    * Recursive least-squares systolic arrays
    * QR-decomposition-based least-squares lattice (QRD-LSL) algorithm

* Nonlinear
    * Volterra filters
    * Neural networks

In this paper we concentrate on linear adaptive filters, which have the desirable property of being simple to design, in general. More importantly, however they lend themselves to a detailed mathematical analysis of their properties such as convergence, misadjustment, tracking, and numerical accuracy [1].

5.   QRD-LSL Algorithm

The problem of designing a QR decomposition-least squares lattice (QRD-LSL) algorithm may be stated as follows:

* Use QR-decomposition for solving the adaptive forward and backward forms of the linear prediction problem, and do the decomposition with a lattice structure in mind.

* Use this structure as the basis for computing the error signal (i.e., difference between the desired response and the actual filter output); desirably, the estimation should be based on backward prediction errors.

The motivation for point 1 is rooted in numerical analysis, where the QR-decomposition is well known for its good numerical properties, particularly when the filtering (estimation) problem is known to be ill-conditioned, that is, when the condition number of the covariance matrix of the incoming data matrix is high [5]. The motivation for point 2 follows from what we already know about the nice properties of lattice prediction-error filters and their use for joint-process estimation [1]. In effect, the QRD-LSL algorithm combines the desirable numerical and filtering properties that are inherent in these two points. It is noteworthy that the QR-decomposition is performed by using a sequence of Givens rotations to solve the forward and backward forms of linear prediction formulated as a recursive least-squares estimation problem.

Figure 2, taken from [1], presents a detailed layout of the QRD-LSL algorithm. This algorithm provides a fast implementation of recursive least squares filtering; the term *"fast"* is used in the sense that the computational complexity of the algorithm is *linear* with respect to the filter order, as it is in the LMS algorithm.

A Fortran code for the implementation of this algorithm is included at the end of the paper. For a detailed derivation of the algorithm, the reader is referred to [1].

6.   Important Features

The QRD-LSL algorithm offers the following list of useful features:

- Fast rate of convergence, which is inherent in recursive least-squares estimation.
- Numerical stability, which is inherent in QR-decomposition.
- High level of computational efficiency.
- Modular structure; hence, suitability for VLSI implementation.
- Integral set of auxiliary variables, represented by angle-normalized forward and backward prediction errors.

Moreover, backwards stability of the algorithm has been demonstrated by Regalia [6]. In backward error analysis, we use error bounds to show that the computed solution of a given problem is the exact solution of a slightly perturbed version of the problem. This is sufficient to ensure that the algorithm that performed the computation is numerically stable. In the context of linear adaptive filters, backward stability requires that the range of numerical reachability of each internal state variable of the algorithm remains *consistent* with least-squares interpretation. The QRD-LSL algorithm is backwards stable, and it is minimal in that the number of elements constituting the state vector of the filter (algorithm) is the strict minimum.

7.   Concluding Remarks

In this paper we have highlighted the general principles of linear adaptive filtering algorithms, and their area of applications. In specific terms, we focused on the QRD-LSL algorithm as a filtering algorithm that is rooted in (1) QR-decomposition used in numerical analysis, and (2) recursive least-squares (RLS) filtering rooted in linear estimation theory. The QRD-LSL algorithm has many desirable properties that befit its use for solving adaptive signal processing problems encountered in prediction,

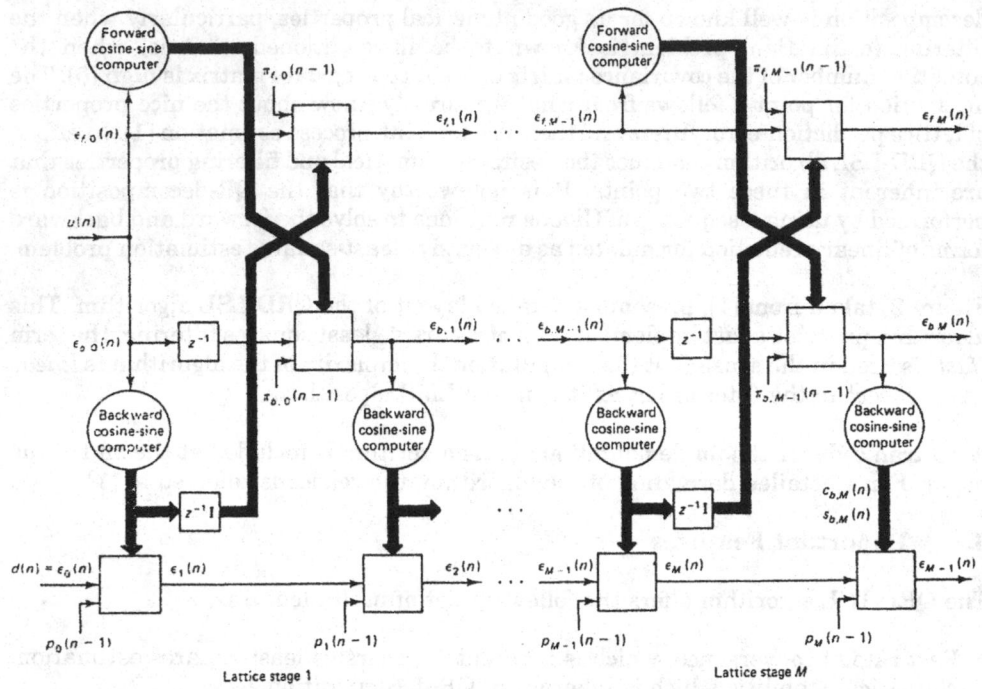

Figure 2 Signal-flow graph of QRD-RLS algorithm
(Taken from S. Haykin, Adaptive Filter Theory, 1991).

equalization, and echo cancellation.

References

[1]   S. Haykin, *Adaptive Filter Theory*, Second Edition, Prentice Hall, 1991.

[2]   I.K. Proudler, J.G. McWhirter, and T.J. Shepherd, QRD-Based Lattice Filter Algorithms, SPIE, San Diego, Calif., 1989.

[3]   P.A. Regalia and G. Bellanger, On the duality between fast QR methods and lattice methods in least squares adaptive filtering, IEEE Trans. Acoust. Speech Signal Process., vol. ASSP-39, 1991.

[4]   F. Ling, Efficient least-squares lattice algorithms based on Givens rotation with systolic array implementations, Proc. ICASSP, Glasgow, Scotland, pp. 1290-1293, 1989.

[5]   G.H. Golub and C.F. Van Loan, *Matrix Computations*, Second Edition, The Johns Hopkins University Press, Baltimore, Md., 1989.

[6]   P.A. Regalia, System Theoretic Properties in the Stability Analysis of QR-based Fast Least-squares Algorithms, Report DEC-0390-003, Département Electronique et Communications, Institut National des Télécommunications, 91011 Evry cedex, France, 1990.

PROGRAM FOR QRD-LSL ALGORITHM

```
      integer n,i,j,numits,kn,blah,z,wei,m,bigm
      real p(11,11),w(11),x(11),k(11),a(400)
      double precision zb(-1:12,-1:400)
      double precision zeb(-1:12,-1:400)
      double precision zcb(-1:12,-1:400)
      double precision zsb(-1:12,-1:400)
      double precision zef(-1:12,-1:400)
      double precision ze(-1:12,-1:400)
      double precision zpif(-1:12,-1:400)
      double precision zf(-1:12,-1:400)
      double precision zcf(-1:12,-1:400)
      double precision zsf(-1:12,-1:400)
      double precision zgamma(-1:12,-1:400)
      double precision zpi(-1:12,-1:400)
      double precision zpib(-1:12,-1:400)
      double precision zphi(-1:400)
      double precision zp(-1:12,-1:400)
      double precision zerr(400)
      real v(400),u(400), error(400)
      real pgam(400),pze(400)
      real h1,h2,h3,we
      real varnce,pi,output,sum
      real temp,desired,alpha
      double precision dseed

      pi=3.1415926
      z=0

      dseed = 15393537.0d0

C     to average over:
      n=100
C     number of iterations:
      numits=200
c     bigm
      bigm = 10

      do 70, wei=29,35,2

        we = (wei/10.0)
        dseed = 123457.0d0
        write(*,*) 'Eigenspread=',we

        z=z+1

        h1=0.5*(1+cos(((2.0*pi)/we)*-1.0))
        h2=1.0
        h3=0.5*(1+cos(((2.0*pi)/we)))

        do 32,kn=1,400
          error(kn)=0.00
          pgam(kn)=0.00
          pze(kn)=0.00
32      continue
```

```
      do 43, blah=1,n

c     write(*,*) 'eigen=',we,'ave no=',blah

      varnce=0.001

C     write(*,*) 'setting up data'

      call ggnml(dseed,400,v)

      do 80,i=1,400
        if (v(i) .ge. 0.00) then
          a(i)=1
        else
          a(i) = -1
        end if
80      continue

      call ggnml(dseed,400,v)

      do 567,i=1,numits
        v(i)=v(i)*(varnce**0.5)
c        v(i)=0.00
567     continue

      u(1)=v(1)
      u(2)=h1*a(1)+v(2)
      u(3)=h1*a(2)+h2*a(1)+v(3)
      u(4)=h1*a(3)+h2*a(2)+h3*a(1)+v(4)
      do 90,i=5,numits
        u(i)=h1*a(i-1)+h2*a(i-2)+h3*a(i-3)+v(i)
c     if (i .lt. 20) then
c     write(*,*) 'i,v(i),a(i),u(i)',i,v(i),a(i),u(i)
c     end if
90      continue

      do i=-1,12
        do j=-1,400
          zb(i,j) = 0.04
          zeb(i,j) = 0.0
          zcb(i,j) = 0.0
          zsb(i,j) = 0.0
          zef(i,j) = 0.0
          zpif(i,j) = 0.0
          zf(i,j) = 0.04
          zcf(i,j) = 0.0
          zsf(i,j) = 0.0
          zpib(i,j) = 0.0
          ze(i,j) = 0.0
          zpi(i,j) = 0.0
          zgamma(i,j) = 0.0
          zphi(j) = 0.0
          zp(i,j) = 0.0
        enddo
      enddo

      zeb(0,1) = u(1)
      zef(0,1) = u(1)
```

```fortran
      zgamma(0,1) = 1.0
      zgamma(0,0) = 1.0
      zgamma(0,-1) = 1.0
   do i=2,numits
      zef(0,i) = u(i)
      zeb(0,i) = u(i)
      zgamma(0,i) = 1.0
c     write(*,*) i,u(i),zef(0,i),zeb(0,i),zgamma(0,i)
   enddo
c     do 789,i=1,15
c     write(*,*) u(i)
C789  continue
c     write(*,*) 'done setting up data'

   do 67,kn=1,numits
      do 823,m=1,bigm
c     write(*,*) 'to use zeb(m-1,kn-1)',m-1,kn-1,zeb(m-1,kn-1)
          zb(m-1,kn-1) =
   +      zb(m-1,kn-2)+(zeb(m-1,kn-1)*zeb(m-1,kn-1))
c     write(*,*) 'zb(m-1,kn-1)',m-1,kn-1,zb(m-1,kn-1)
c         if (zb(m-1,kn-1) .eq. 0) then
c             zcb(m-1,kn-1) = 1.0
c             zsb(m-1,kn-1) = 0.0
c         else
              zcb(m-1,kn-1) =
   +      (zb(m-1,kn-2)**0.5)/(zb(m-1,kn-1)**0.5)
              zsb(m-1,kn-1) =
   +      zeb(m-1,kn-1)/(zb(m-1,kn-1)**0.5)
c         end if
c     write(*,*) 'after if',kn,m,zb(m-1,kn-1),
c     +       zcb(m-1,kn-1),zsb(m-1,kn-1)
c     write(*,*)'zef(m-1,kn)==>0',zef(m-1,kn)
              zef(m,kn) =
   +      zcb(m-1,kn-1)*zef(m-1,kn)-
   +      zsb(m-1,kn-1)*zpif(m-1,kn-1)
              zpif(m-1,kn) =
   +      zcb(m-1,kn-1)*zpif(m-1,kn-1)+
   +      zsb(m-1,kn-1)*zef(m-1,kn)
c     write(*,*) 'zef,zpif',kn,m,zef(m,kn),zpif(m-1,kn)
              zf(m-1,kn) = zf(m-1,kn-1) +
   +      zef(m-1,kn)*zef(m-1,kn)
c         if (zf(m-1,kn) .eq. 0) then
c             zcf(m-1,kn) = 1.0
c             zsf(m-1,kn) = 0.0
c         else
              zcf(m-1,kn) =
   +      (zf(m-1,kn-1)**0.5)/(zf(m-1,kn)**0.5)
          zsf(m-1,kn) =
   +      zef(m-1,kn)/(zf(m-1,kn)**0.5)
c         end if
c     write(*,*) 'after if',kn,m,zf(m-1,kn),zcf(m-1,kn),
c     +       zsf(m-1,kn)
              zeb(m,kn) =
   +      zcf(m-1,kn)*zeb(m-1,kn-1)-
   +      zsf(m-1,kn)*zpib(m-1,kn-1)
c     write(*,*) 'i have changed zeb(m,kn) to',m,kn,zeb(m,kn)
              zpib(m-1,kn) =
   +      zcf(m-1,kn)*zpib(m-1,kn-1)+
```

```
   +      zsf(m-1,kn)*zeb(m-1,kn-1)
c         write(*,*) 'zpib',kn,m,zpib(m-1,kn-1)
823    continue


c      filter
       if (kn-7 .ge. 1) then
          desired=a(kn-7)
       else
          desired = 0.00
       endif
          ze(0,kn) = desired

       do m=0,bigm
          zb(m,kn) =
   +      zb(m,kn-1)+(zeb(m,kn)*zeb(m,kn))
c          if (zb(m,kn) .eq. 0) then
c             zcb(m,kn) = 1.0
c             zsb(m,kn) = 0.0
c          else
             zcb(m,kn) =
   +      (zb(m,kn-1)**0.5)/(zb(m,kn)**0.5)
             zsb(m,kn) =
   +      zeb(m,kn)/(zb(m,kn)**0.5)
c          end if

       ze(m+1,kn) = zcb(m,kn)*ze(m,kn)-
   +   zsb(m,kn)*zp(m,kn-1)
       zp(m,kn) = zcb(m,kn)*zp(m,kn-1) +
   +   zsb(m,kn)*ze(m,kn)
       zgamma(m+1,kn) = zcb(m,kn)*zgamma(m,kn)
c      write(*,*) 'zgamma(11,kn)',11,kn,zgamma(bigm+1,kn)
       enddo

       if (zgamma(bigm+1,kn) .ne. 0) then
          zerr(kn) = ze(bigm+1,kn)/(zgamma(bigm+1,kn))
c         zerr(kn) = ze(bigm+1,kn)
       else
          zerr(kn) = 0.000000001
       end if
c      if (kn .lt. 20) then
c      write(*,*) kn,ze(bigm+1,kn),
c   +  zgamma(bigm+1,kn),zerr(kn),zerr(kn)*zerr(kn)
c      end if
c      write(*,*) ze(bigm+1,kn),zphi(kn),zgamma(bigm+1,kn)
c      write(*,*) 'error',kn,zerr(kn),zerr(kn)*zerr(kn)
       error(kn) = error(kn) + zerr(kn)*zerr(kn)
       pze(kn) = pze(kn) + ze(bigm+1,kn)*ze(bigm+1,kn)
       pgam(kn) = pgam(kn) + zgamma(bigm+1,kn)
67     continue

43     continue

       do kn=1,400
         if (error(kn) .ne. 0) then
          error(kn)=error(kn)/n
         else
          error(kn) = 1e-8
```

```
         end if
      enddo
      do kn=1,400
         pgam(kn)=pgam(kn)/n
      enddo
      do kn=1,400
         pze(kn)=pze(kn)/n
      enddo

      sum =0.0
      do 749, kn=180,190
      sum = sum + error(kn)
749   continue
      write (*,*) 'mean err=',sum/11.0

      open(unit=1,file='blast3a.dat',status='new')
c     do kn=8,numits
c        write(1,*) kn-7,error(kn)
      do kn=1,numits
         write(1,*) kn,error(kn)
      enddo
      close(unit=1)
      open(unit=1,file='blast3b.dat',status='new')
c     do kn=8,numits
c        write(1,*) kn-7,pgam(kn)
      do kn=1,numits
         write(1,*) kn,pgam(kn)
      enddo
      close(unit=1)
      open(unit=1,file='blast3c.dat',status='new')
c     do kn=8,numits
c        write(1,*) kn-7,pze(kn)
      do kn=1,numits
         write(1,*) kn,pze(kn)
      enddo
      close(unit=1)

70    continue
      stop
      end
```

A DIRECT METHOD FOR REORDERING EIGENVALUES IN THE GENERALIZED REAL SCHUR FORM OF A REGULAR MATRIX PAIR (A, B)

B. KÅGSTRÖM
Institute of Information Processing
University of Umeå
S-901 87 Umeå, Sweden
bokg@cs.umu.se

ABSTRACT. A direct orthogonal equivalence transformation method for reordering the eigenvalues along the diagonal in the generalized real Schur form of a regular matrix pair (A, B) is presented. Each swap of two adjacent eigenvalues (real, or complex conjugate pairs) involves solving a generalized Sylvester equation and the construction of two orthogonal transformation matrices from certain eigenspaces associated with the corresponding diagonal blocks. An error analysis of the direct reordering method is presented. Results from numerical experiments on well-conditioned as well as ill-conditioned problems illustrate the stability and the accuracy of the method. Finally, a direct reordering algorithm with controlled backward error is described.

KEYWORDS: Reordering of eigenvalues, generalized real Schur form, regular matrix pair, matrix pencil, generalized Sylvester equation, direct reordering method.

1 Introduction

In this paper we present a *direct* method for reordering eigenvalues in the generalized real Schur form of a regular matrix pair (A, B). The method performs an orthogonal equivalence transformation of the real matrix pair (A, B), where A is upper quasi-triangular and B upper triangular. (This form can be computed by an orthogonal equivalence transformation using the QZ algorithm [22].) A quasi-triangular matrix is triangular except, possibly, for 2×2 blocks along the diagonal. In the generalized Schur form the 2×2 blocks correspond to pairs of complex conjugate eigenvalues of the real *regular* pencil $A - \lambda B$ ($\det(A - \lambda B) = 0$ if and only if λ is an eigenvalue). The real eigenvalues are given by the ratios of the diagonal entries of A and B corresponding to 1×1 diagonal blocks in the generalized Schur form. So the problem of reordering eigenvalues is equivalent to swapping 1×1 and 2×2 diagonal blocks along the diagonal of (A, B). Typically a reordering comprises several swaps.

M. S. Moonen et al. (eds.), *Linear Algebra for Large Scale and Real-Time Applications*, 195–218.
© 1993 *Kluwer Academic Publishers*.

One source of motivation comes from different applications. For example, computing a pair of eigenspaces (deflating subspaces [26]) corresponding to a given set of eigenvalues of $A - \lambda B$ inside a region Γ of the complex plane [22], [32], [7, 8]. Estimating condition numbers and computing error bounds for a cluster of eigenvalues and/or their associated eigenspaces in the generalized eigenvalue problem $Ax = \lambda Bx, A - \lambda B$ regular [26], [19], [7, 8]. Several problems in control theory can be solved in terms of a generalized subspace problem with specified spectrum. Examples are different types of Riccati equations [20], [31] and the additive decomposition of a transfer matrix given by its generalized state-space realization [17].

The solution to these problems involves two initial steps:

- Compute a generalized real Schur form of the matrix pair (A, B) using the QZ algorithm [22].

- Reorder the specified k eigenvalues to appear in the $(1,1)$-block of the generalized real Schur form.

Now, the first k columns of the composite transformation matrices span a pair of eigenspaces corresponding to the spectrum of the $(1,1)$-block.

Several papers relating to the problem of reordering eigenvalues have been published. So far, most of the work concerns the standard $A - \lambda I$ problem. In [25], a direct method for reordering two 1×1 blocks (complex eigenvalues) in terms of a rotation is presented. The method was later implemented in the JNF software [16] as a part of the procedure for identifying clusters of close eigenvalues. An iterative method and software (EXCHNG), based on the implicit double-shift QR algorithm, that reorder 1×1 as well as 2×2 blocks, is presented in [27]. A hybrid direct/iterative method was used in [33] where a real version of JNF [16] is described. A direct method for swapping 2×2 blocks is presented in [23], [10]. The method in [23] has recently been improved in [2] giving a backward stable algorithm (by rejecting unstable swaps) for reordering eigenvalues in the real Schur form of A. For the regular $A - \lambda B$ problem, an iterative method [31] and software (EXCHQZ) [32] are presented that extend the work in [27] by using the implicit double-shift QZ algorithm. A reordering method based on the periodic Schur decompositon has been proposed recently [4].

The iterative methods (EXCHNG and EXCHQZ) are backward stable and finite in theory, but can fail to converge in practice even for only moderately ill-conditioned problems [2]. The purpose of this paper is to develop a direct reordering algorithm, with guaranteed backward stability, for the $A - \lambda B$ problem. We do this by extending and generalizing the direct algorithm in [2] to regular matrix pairs (A, B). The rest of the paper is outlined as follows. Section 2 presents the direct reordering method and its relation to the generalized Sylvester equation. In section 3, an error analysis of the direct reordering method is presented. Results from numerical experiments that illustrate the stability and the accuracy of the method on well-conditioned as well as ill-conditioned problems are shown and discussed in section 4. The analysis and results in sections 3 and 4 suggest that a practical implementation should reject a swap if it would result in too large backward errors (i.e. instability). In section 5 a direct algorithm with controlled backward error for swapping two diagonal blocks of a regular matrix pair (A, B) in generalized real Schur form is described. Finally, some conclusions are summarized in section 6.

The following notation is used in the paper. $\lambda(A, B)$ denotes the spectrum of a regular matrix pair (A, B) or pencil $A - \lambda B$. If $B = I$ we only use $\lambda(A)$. $\|A\|_2$ denotes the spectral norm (2-norm) of a matrix A induced by the Euclidean vector norm. $\|A\|_F$ denotes the Frobenius (or Euclidean) matrix norm. $\kappa_o(A) = \|A\|_o \cdot \|A^+\|_o$, where $o = 2$ or F, and A^+ is the pseudo inverse of A, denote the condition numbers of a matrix A with respect to the 2-norm and the Frobenius norm, respectively. $\sigma(A)$ denotes the set of singular values of a matrix A. Especially, $\sigma_{\max}(A)$ and $\sigma_{\min}(A)$ denote the largest and smallest singular values of A, respectively. For a square matrix A we have that $\|A\|_2 = \sigma_{\max}(A)$ and $\|A^{-1}\|_2 = \sigma_{\min}^{-1}(A)$. $A \otimes B$ denotes the Kronecker product of two matrices A and B whose (i, j)-th block element is $a_{ij} B$. The column vector $\mathrm{col}(A)$ denotes an ordered stack of the columns of A from left to right starting with the first column. A^T denotes the transpose of A.

2 Direct reordering of eigenvalues and the generalized Sylvester equation

Without loss of generality we consider the problem of reordering the diagonal blocks of a matrix pair (A, B) in the block form,

$$A = \begin{bmatrix} A_{11} & A_{12} \\ 0 & A_{22} \end{bmatrix}, \quad B = \begin{bmatrix} B_{11} & B_{12} \\ 0 & B_{22} \end{bmatrix}, \tag{1}$$

where (A_{11}, B_{11}) and (A_{22}, B_{22}) are of size $m \times m$ and $n \times n$, respectively, and $m, n = 1$ or 2. Troughout the paper we assume that (A_{11}, B_{11}) and (A_{22}, B_{22}) have no eigenvalues in common, otherwise, the diagonal blocks need not be swapped.

By solving for (L, R) in the generalized Sylvester equation [19]

$$\begin{aligned} A_{11} R - L A_{22} &= -A_{12}, \\ B_{11} R - L B_{22} &= -B_{12}, \end{aligned} \tag{2}$$

the matrix pair (A, B) can be block-diagonalized as

$$\left(\begin{bmatrix} A_{11} & A_{12} \\ 0 & A_{22} \end{bmatrix}, \begin{bmatrix} B_{11} & B_{12} \\ 0 & B_{22} \end{bmatrix} \right) =$$

$$\begin{bmatrix} I_m & L \\ 0 & I_n \end{bmatrix} \left(\begin{bmatrix} A_{11} & 0 \\ 0 & A_{22} \end{bmatrix}, \begin{bmatrix} B_{11} & 0 \\ 0 & B_{22} \end{bmatrix} \right) \begin{bmatrix} I_m & -R \\ 0 & I_n \end{bmatrix}. \tag{3}$$

Since (A_{11}, B_{11}) and (A_{22}, B_{22}) have non-intersecting spectra the solution (L, R) of (2) is unique [26]. Further, the diagonal blocks can be swapped by an orthogonal equivalence transformation:

$$P_l \left(\begin{bmatrix} A_{11} & 0 \\ 0 & A_{22} \end{bmatrix}, \begin{bmatrix} B_{11} & 0 \\ 0 & B_{22} \end{bmatrix} \right) P_r = \left(\begin{bmatrix} A_{22} & 0 \\ 0 & A_{11} \end{bmatrix}, \begin{bmatrix} B_{22} & 0 \\ 0 & B_{11} \end{bmatrix} \right), \tag{4}$$

where

$$P_l = \begin{bmatrix} 0 & I_m \\ I_n & 0 \end{bmatrix}, \quad P_r = \begin{bmatrix} 0 & I_n \\ I_m & 0 \end{bmatrix}, \quad P_l P_r = I_{m+n}.$$

Combining these two equivalence transformations we obtain

$$\left(\begin{bmatrix} A_{11} & A_{12} \\ 0 & A_{22} \end{bmatrix}, \begin{bmatrix} B_{11} & B_{12} \\ 0 & B_{22} \end{bmatrix} \right) = X \left(\begin{bmatrix} A_{22} & 0 \\ 0 & A_{11} \end{bmatrix}, \begin{bmatrix} B_{22} & 0 \\ 0 & B_{11} \end{bmatrix} \right) Y, \tag{5}$$

where

$$X = \begin{bmatrix} L & I_m \\ I_n & 0 \end{bmatrix}, \quad Y = \begin{bmatrix} 0 & I_n \\ I_m & -R \end{bmatrix}, \tag{6}$$

which is a non-orthogonal equivalence transformation that performs the required swapping. The final step is the construction of orthogonal transformation matrices. Since the first block column of X has full column rank and the last block row of Y has full row rank we can choose orthogonal matrices Q and Z of size $(m+n) \times (m+n)$ such that

$$Q^T \begin{bmatrix} L \\ I_n \end{bmatrix} = \begin{bmatrix} T_L \\ 0 \end{bmatrix}, \begin{bmatrix} I_m & -R \end{bmatrix} Z = \begin{bmatrix} 0 & T_R \end{bmatrix}, \tag{7}$$

where T_L is $n \times n$, T_R is $m \times m$ and both T_L, T_R are non-singular. If we partition Q and Z conformally with X and Y

$$Q = \begin{bmatrix} Q_{11} & Q_{12} \\ Q_{21} & Q_{22} \end{bmatrix}, \quad Z = \begin{bmatrix} Z_{11} & Z_{12} \\ Z_{21} & Z_{22} \end{bmatrix},$$

then

$$Q^T X = \begin{bmatrix} T_L & Q_{11}^T \\ 0 & Q_{12}^T \end{bmatrix}, \quad YZ = \begin{bmatrix} Z_{12} & Z_{22} \\ 0 & T_R \end{bmatrix}.$$

By applying Q and Z in an equivalence transformation of the original (A, B) (1), we obtain

$$Q^T \left(\begin{bmatrix} A_{11} & A_{12} \\ 0 & A_{22} \end{bmatrix}, \begin{bmatrix} B_{11} & B_{12} \\ 0 & B_{22} \end{bmatrix} \right) Z =$$

$$\left(\begin{bmatrix} T_L A_{22} Z_{21} & T_L A_{22} Z_{22} + Q_{11}^T A_{11} T_R \\ 0 & Q_{12}^T A_{11} T_R \end{bmatrix}, \begin{bmatrix} T_L B_{22} Z_{21} & T_L B_{22} Z_{22} + Q_{11}^T B_{11} T_R \\ 0 & Q_{12}^T B_{11} T_R \end{bmatrix} \right) \tag{8}$$

$$= \left(\begin{bmatrix} \hat{A}_{22} & \hat{A}_{12} \\ 0 & \hat{A}_{11} \end{bmatrix}, \begin{bmatrix} \hat{B}_{22} & \hat{B}_{12} \\ 0 & \hat{B}_{11} \end{bmatrix} \right) \equiv (\hat{A}, \hat{B}),$$

where (A_{ii}, B_{ii}) and $(\hat{A}_{ii}, \hat{B}_{ii})$ for $i = 1, 2$ are equivalent matrix pairs with the same eigenvalues but their positions are exchanged (swapped) along the block diagonal of (A, B).

In summary, we have the following direct method for swapping two diagonal blocks in the generalized real Schur form of (A, B):

Step 1 Solve for (L, R) in the generalized Sylvester equation (2):

$$A_{11}R - LA_{22} = -A_{12},$$
$$B_{11}R - LB_{22} = -B_{12}.$$

Step 2 Compute an orthogonal matrix Q (7):

$$Q^T \begin{bmatrix} L \\ I_n \end{bmatrix} = \begin{bmatrix} T_L \\ 0 \end{bmatrix}.$$

Step 3 Compute an orthogonal matrix Z (7):

$$\begin{bmatrix} I_m & -R \end{bmatrix} Z = \begin{bmatrix} 0 & T_R \end{bmatrix}.$$

Step 4 Perform an orthogonal equivalence transformation of (A, B) (8):

$$Q^T \left(\begin{bmatrix} A_{11} & A_{12} \\ 0 & A_{22} \end{bmatrix}, \begin{bmatrix} B_{11} & B_{12} \\ 0 & B_{22} \end{bmatrix} \right) Z = \left(\begin{bmatrix} \hat{A}_{22} & \hat{A}_{12} \\ 0 & \hat{A}_{11} \end{bmatrix}, \begin{bmatrix} \hat{B}_{22} & \hat{B}_{12} \\ 0 & \hat{B}_{11} \end{bmatrix} \right).$$

For step 1 we can use the generalized Schur methods in [18, 19], which are generalizations of the Schur method [3] and the Hessenberg-Schur method [12] for the standard Sylvester equation. In our case, $(A_{ii}, B_{ii}), i = 1, 2$ are already in generalized Schur form and we end up solving a $2mn \times 2mn$ linear system $Zx = b$, where

$$Z = \begin{bmatrix} I_n \otimes A_{11} & -A_{22}^T \otimes I_m \\ I_n \otimes B_{11} & -B_{22}^T \otimes I_m \end{bmatrix}, \quad x = \begin{bmatrix} \text{col}(R) \\ \text{col}(L) \end{bmatrix}, \quad b = \begin{bmatrix} -\text{col}(A_{12}), \\ -\text{col}(B_{12}) \end{bmatrix}. \tag{9}$$

Since $m, n = 1$ or 2 the linear system will be of size 2×2, 4×4 or 8×8. Steps 2 and 3 can be performed by using Householder or Givens transformations to compute a QR factorization and an RQ factorization, respectively. Finally, step 4 is just two matrix-matrix multiplication operations on A and B.

Notice that the method described above can also be used to reorder larger blocks or eigenvalue clusters when $m, n > 2$. The difference is that in order to restore the diagonal blocks $(\hat{A}_{ii}, \hat{B}_{ii}), i = 1, 2$ to generalized real Schur form we use the QZ algorithm [22], which typically involve an iterative process. If $m, n = 1, 2$ this can be done in a finite number of operations (see section 5).

3 Error analysis of the direct reordering method

This section is devoted to an error analysis of the direct swapping method described in the previous section. In the presence of rounding errors the conditioning and the solution of the generalized Sylvester equation will have the greatest impact on the stability of the direct swapping method. In the following analysis we consider the most interesting case, i.e., swapping 2×2 blocks.

Let $(\bar{L}, \bar{R})$ denote the computed solution of the generalized Sylvester equation (2), where $\bar{L} = L + \Delta L$, $\bar{R} = R + \Delta R$ and (L, R) is the exact solution. The residuals of the computed solution are

$$R_1 \equiv A_{11}\bar{R} - \bar{L}A_{22} + A_{12},$$
$$R_2 \equiv B_{11}\bar{R} - \bar{L}B_{22} + B_{12}. \tag{10}$$

In [15] the following normwise forward error bound for an approximate solution of the generalized Sylvester equation is derived:

$$\frac{\|(\Delta L, \Delta R)\|_F}{\|(L, R)\|_F} \leq \|Z^{-1}\|_2 \frac{\|(R_1, R_2)\|_F}{\|(L, R)\|_F}. \tag{11}$$

Here, Z is the coefficient matrix in (9). It shows that the error $(\Delta L, \Delta R)$ is bounded by the norm of the residuals (R_1, R_2) divided by the separation between the matrix pairs (A_{11}, B_{11}) and (A_{22}, B_{22}) [26, 9]:

$$\mathrm{Dif}[(A_{11}, B_{11}), (A_{22}, B_{22})] = \sigma_{\min}(Z) \tag{12}$$

Typically, $\sigma_{\min}(Z)$ is small if only a small perturbation is needed to make an eigenvalue in $\lambda(A_{11}, B_{11})$ coalesce with one in $\lambda(A_{22}, B_{22})$ [9]. Notice that if $A_{11} - \lambda B_{11}$ and $A_{22} - \lambda B_{22}$ have well-separated spectra and $\sigma_{\min}(Z)$ is small then this signals that the original regular pencil $A - \lambda E$ is close to a singular pencil $(\det(A - \lambda B) \equiv 0$ for all $\lambda)$ [7, 8].

Let $\bar{Q}, \bar{T}_L$ denote the computed QR factors in step 2:

$$\bar{G}_L \equiv \begin{bmatrix} \bar{L} \\ I_n \end{bmatrix} = \bar{Q} \begin{bmatrix} \bar{T}_L \\ 0 \end{bmatrix}, \tag{13}$$

where $\bar{Q} = Q + \Delta Q$, $\bar{T}_L = T_L + \Delta T_L$ and Q, T_L are the exact QR factors. Similarly, let $\bar{T}_R, \bar{Z}$ denote the computed RQ factors in step 3:

$$\bar{G}_R \equiv \begin{bmatrix} I_m & -\bar{R} \end{bmatrix} = \begin{bmatrix} 0 & \bar{T}_R \end{bmatrix} \bar{Z}^T, \tag{14}$$

where $\bar{Z} = Z + \Delta Z$, $\bar{T}_R = T_R + \Delta T_R$ and T_R, Z are the exact RQ factors.

If we use Householder transformations to compute the factorizations above, $\bar{Q}$ and $\bar{Z}$ are both orthogonal to machine precision [34]. The perturbation analyses of the QR factorization in [28, 30] show that the error matrices ΔT_L and ΔQ are essentially bounded by $\kappa_F(G_L)\frac{\|\Delta G_L\|_F}{\|G_L\|_F}$, i.e., the condition number of the matrix to factorize times the relative error (change) of that matrix. By applying the perturbation results to $\bar{G}_R^T$ we obtain corresponding bounds for the error matrices ΔT_R and ΔZ of the RQ factorization (14).

Given $(\bar{L}, \bar{R}), \bar{Q}$ and $\bar{Z}$, the following theorem shows how the errors in these quantities propagate to the results of the direct reordering method for swapping two 2×2 diagonal matrix pairs.

Theorem 3.1 *By applying the computed transformation matrices $\bar{Q}$ and $\bar{Z}$ in an equivalence transformation of (A, B) we get*

$$\bar{Q}^T A \bar{Z} = \begin{bmatrix} \hat{A}_{22} & \hat{A}_{12} \\ 0 & \hat{A}_{11} \end{bmatrix} + \begin{bmatrix} \Delta A_{22} & \Delta A_{12} \\ \Delta A_{21} & \Delta A_{11} \end{bmatrix} \equiv \hat{A} + \Delta A \tag{15}$$

and

$$\bar{Q}^T B \bar{Z} = \begin{bmatrix} \hat{B}_{22} & \hat{B}_{12} \\ 0 & \hat{B}_{11} \end{bmatrix} + \begin{bmatrix} \Delta B_{22} & \Delta B_{12} \\ \Delta B_{21} & \Delta B_{11} \end{bmatrix} \equiv \hat{B} + \Delta B, \tag{16}$$

where (A_{ii}, B_{ii}) and $(\hat{A}_{ii}, \hat{B}_{ii})$ for $i = 1,2$ are equivalent matrix pairs as in (8) and up to first order perturbations $O(\|(\Delta A, \Delta B)\|_2)$:

$$\|\Delta A_{11}\|_2 \leq \frac{1}{(1 + \sigma_{\min}^2(L))^{\frac{1}{2}}} \cdot \frac{\sigma_{\max}(R)}{(1 + \sigma_{\max}^2(R))^{\frac{1}{2}}} \cdot \|R_1\|_F, \tag{17}$$

$$\|\Delta A_{21}\|_2 \leq \frac{1}{(1 + \sigma_{\min}^2(L))^{\frac{1}{2}}} \cdot \frac{1}{(1 + \sigma_{\min}^2(R))^{\frac{1}{2}}} \cdot \|R_1\|_F, \tag{18}$$

$$\|\Delta A_{22}\|_2 \leq \frac{\sigma_{\max}(L)}{(1 + \sigma_{\max}^2(L))^{\frac{1}{2}}} \cdot \frac{1}{(1 + \sigma_{\min}^2(R))^{\frac{1}{2}}} \cdot \|R_1\|_F. \tag{19}$$

Similar bounds hold for $\|\Delta B_{11}\|_2, \|\Delta B_{21}\|_2, \|\Delta B_{22}\|_2$ with R_1 replaced by R_2.

Proof: Transform (A, B) with the equivalence transformation $(\bar{Q}, \bar{Z})$:

$$\bar{Q}^T A \bar{Z} = (Q + \Delta Q)^T A (Z + \Delta Z) = Q^T A Z + \Delta Q^T A Z + Q^T A \Delta Z + \Delta Q^T A \Delta Z.$$

By dropping the second order term and using that $\Delta Q^T Q = -Q \Delta Q^T$ up to first order, and $\hat{A} = Q^T A Z$ from (8) we get

$$\bar{Q}^T A \bar{Z} = \hat{A} + \hat{A}(Z^T \Delta Z) + (-Q \Delta Q^T) \hat{A} = \hat{A} + \hat{A} U + W \hat{A}.$$

Similarly, we have

$$\bar{Q}^T B \bar{Z} = \hat{B} + \hat{B} U + W \hat{B}.$$

Partition U, W conformally with $(\hat{A}, \hat{B})$ (15–16), i.e.,

$$U = \begin{bmatrix} U_{11} & U_{12} \\ U_{21} & U_{22} \end{bmatrix}, \quad W = \begin{bmatrix} W_{11} & W_{12} \\ W_{21} & W_{22} \end{bmatrix},$$

and we get the following expressions for ΔA_{ij}:

$$\Delta A_{11} = \hat{A}_{11} U_{22} + W_{22} \hat{A}_{11} + W_{21} \hat{A}_{12},$$
$$\Delta A_{12} = \hat{A}_{22} U_{12} + \hat{A}_{12} U_{22} + W_{11} \hat{A}_{12} + W_{12} \hat{A}_{11},$$
$$\Delta A_{21} = \hat{A}_{11} U_{21} + W_{21} \hat{A}_{22},$$
$$\Delta A_{22} = \hat{A}_{22} U_{11} + W_{11} \hat{A}_{22} + \hat{A}_{12} U_{21}.$$

Similar expressions hold for $\Delta B_{ij}, i, j = 1, 2$ with $\hat{A}_{ij}$ replaced by $\hat{B}_{ij}$. The perturbations $\Delta A_{21}, \Delta B_{21}$ are most crucial since their norms immediately reflect the stability of the swapping (they should ideally be of the order of the relative machine precision). The diagonal perturbations $\Delta A_{ii}, \Delta B_{ii}, i = 1, 2$ affect the eigenvalues directly. The matrices $\Delta A_{12}, \Delta B_{12}$ that perturb the $(1,2)$ blocks of A and B, respectively, are of less interest since they do not immediately affect the eigenvalues. Our task is now to derive norm bounds for the interesting perturbations. We start by expressing blocks of U and W in terms of $T_R, Z, \Delta R$ and $T_L, Q, \Delta L$, respectively.

Let us start to consider U. After post-multiplying (14) with Z we get

$$\begin{bmatrix} 0 & T_R + \Delta T_R \end{bmatrix} (Z^T + \Delta Z^T) Z = \begin{bmatrix} I_m & -R \end{bmatrix} Z + \begin{bmatrix} 0 & -\Delta R \end{bmatrix} Z,$$

or equivalently

$$\begin{bmatrix} 0 & T_R + \Delta T_R \end{bmatrix} (I_{m+n} + \Delta Z^T Z) = \begin{bmatrix} 0 & T_R \end{bmatrix} + \begin{bmatrix} -\Delta R Z_{21} & -\Delta R Z_{22} \end{bmatrix}. \qquad (20)$$

From the orthogonality of Z and $\bar{Z}$ we have up to first order terms that $\Delta Z^T Z = -Z^T \Delta Z$. Further, by pre-multiplying both sides of (20) with $(T_R + \Delta T_R)^{-1}$ we get

$$\begin{bmatrix} -U_{21} & I_m - U_{22} \end{bmatrix} = (T_R + \Delta T_R)^{-1} \begin{bmatrix} -\Delta R Z_{21} & T_R - \Delta R Z_{22} \end{bmatrix}.$$

Blockwise identification in the above equation give us, up to first order terms,

$$U_{21} = T_R^{-1} \Delta R Z_{21},$$
$$U_{22} = T_R^{-1} \Delta R Z_{22}.$$

In order to derive a similar first order expression for U_{11} we post-multiply both sides of (14) with Z^T, i.e.,

$$\begin{bmatrix} I_m & -R - \Delta R \end{bmatrix} (Z + \Delta Z) Z^T = \begin{bmatrix} 0 & T_R + \Delta T_R \end{bmatrix} Z^T$$

which can be expressed as

$$\begin{bmatrix} I_m & -R - \Delta R \end{bmatrix} (I_{m+n} + \Delta Z Z^T) = \begin{bmatrix} I_m & -R \end{bmatrix} + \begin{bmatrix} 0 & \Delta T_R \end{bmatrix} Z^T$$

Now, we first subtract $\begin{bmatrix} I_m & -R \end{bmatrix}$ from both sides of the equation above and then post-multiply with Z resulting in

$$\begin{bmatrix} 0 & -\Delta R \end{bmatrix} Z + \begin{bmatrix} I_m & -R - \Delta R \end{bmatrix} \Delta Z = \begin{bmatrix} 0 & \Delta T_R \end{bmatrix}.$$

Insert $I = Z Z^T$ before ΔZ in the equation above and it can be rewritten as

$$\begin{bmatrix} 0 & -\Delta R \end{bmatrix} Z + (\begin{bmatrix} 0 & T_R \end{bmatrix} + \begin{bmatrix} 0 & -\Delta R \end{bmatrix} Z) U = \begin{bmatrix} 0 & \Delta T_R \end{bmatrix}. \qquad (21)$$

By identifying the $(1,1)$ block element of (21) and assuming that the perturbation ΔR is non-singular we get up to first order terms

$$U_{11} = -Z_{21}^{-1} Z_{22} T_R^{-1} \Delta R Z_{21}.$$

First order results can be derived in a similar way for the corresponding W_{ij} blocks giving

$$W_{11} = -Q_{11}^T \Delta L T_L^{-1},$$
$$W_{21} = -Q_{12}^T \Delta L T_L^{-1},$$
$$W_{22} = Q_{12}^T \Delta L T_L^{-1} Q_{11}^T Q_{12}^{-T}.$$

By substituting the expressions for U_{ij} and W_{ij} in $\Delta A_{ij}, \Delta B_{ij}$ we obtain

$$\Delta A_{11} = Q_{12}^T R_1 Z_{22},$$
$$\Delta A_{21} = Q_{12}^T R_1 Z_{21},$$
$$\Delta A_{22} = Q_{11}^T R_1 Z_{21},$$

and

$$\Delta B_{11} = Q_{12}^T R_2 Z_{22},$$
$$\Delta B_{21} = Q_{12}^T R_2 Z_{21},$$
$$\Delta B_{22} = Q_{11}^T R_2 Z_{21}.$$

From the QR and RQ factorizations (7) we have that

$$Q_{21} = T_L^{-1}, \quad T_L^T T_L = I_n + L^T L \tag{22}$$

and

$$Z_{12}^T = T_R^{-1}, \quad T_R T_R^T = I_m + RR^T. \tag{23}$$

From (22), (23), we obtain the following relations between the singular values of T_L, T_R and L, R:

$$\sigma^2(T_L) = 1 + \sigma^2(L), \quad \sigma^2(T_R) = 1 + \sigma^2(R). \tag{24}$$

Further, from the CS (Cosine-Sine) decomposition (e.g., see [29], [13]) of Q and Z, respectively, we obtain the following norm relations between blocks of the exact transformation matrices:

$$\|Q_{12}^T\|_2 = \|Q_{21}\|_2, \quad \|Q_{22}\|_2 = \|Q_{11}\|_2$$
$$\|Z_{12}^T\|_2 = \|Z_{21}\|_2, \quad \|Z_{22}\|_2 = \|Z_{11}\|_2.$$

Combining these results we get

$$\|Q_{12}^T\|_2 = \|T_L^{-1}\|_2 = \frac{1}{\sigma_{\min}(T_L)} = \frac{1}{(1 + \sigma_{\min}^2(L))^{\frac{1}{2}}},$$

$$\|Q_{11}\|_2 = \frac{\sigma_{\max}(L)}{(1 + \sigma_{\max}^2(L))^{\frac{1}{2}}},$$

and similarly for the required Z_{ij} blocks

$$\|Z_{21}\|_2 = \|T_R^{-1}\|_2 = \frac{1}{\sigma_{\min}(T_R)} = \frac{1}{(1 + \sigma_{\min}^2(R))^{\frac{1}{2}}},$$

$$\|Z_{22}\|_2 = \frac{\sigma_{\max}(R)}{(1 + \sigma_{\max}^2(R))^{\frac{1}{2}}}.$$

The bounds in the theorem follow by applying the triangle inequality to the norms of the expressions for ΔA_{ij} and ΔB_{ij}. $\square$

What can we say about the size of the errors $(\Delta A, \Delta B)$ in Theorem 3.1? First of all, $\|\Delta A_{ij}\|_2$ and $\|\Delta B_{ij}\|_2$ depend on $\|R_1\|_F, \|R_2\|_F$, the norms of the residuals of the computed solution $(\bar{L}, \bar{R})$ to the generalized Sylvester equation, and on the conditioning of the exact solution (L, R). If $\sigma_{\min}(L)$ and $\sigma_{\min}(R)$ are small, the error can be as large as the norms of the residuals. We have experimental evidence that $\|R_1\|_F, \|R_2\|_F$ are large for large-normed (ill-conditioned) solutions $(\bar{L}, \bar{R})$. Notice that the separation between the two diagonal blocks (A_{11}, B_{11}) and (A_{22}, B_{22}) is not explicitly involved in the bounds for $(\Delta A_{ij}, \Delta B_{ij})$, as it is in the bound (11) for $(\Delta L, \Delta R)$. If $B = I$ (i.e., a standard reordering problem) then $L = R$ and the bounds for ΔA_{ij} are similar to the bounds derived in [2].

In [15] a perturbation analysis of the generalized Sylvester equation is presented that takes full account to the structure of the matrix equation, derives expressions for the backward error of an approximate solution $(\hat{L}, \hat{R})$, and derives condition numbers that measure the sensitivity of a solution to perturbations in A_{11}, A_{12}, A_{22} and B_{11}, B_{12}, B_{22}, respectively. The results in [15] extend the analysis in [14] of the standard Sylvester equation to the generalized Sylvester equation, and improve some results in [19] for the generalized Sylvester equation.

Due to the special structure of the (generalized) Sylvester equation the relation for linear systems "relative backward error = relative residual" [24] does not hold in general [14], [15]. Small relative backward errors will always result in small relative residuals. However, the analysis shows that for very ill-conditioned cases the norm of the relative backward errors can greatly exceed the norm of the relative residuals (in fact, by an arbitrary factor [14]). This situation appears when $(\hat{L}, \hat{R})$ is an ill-conditioned (i.e., $\sigma_{\min}(\hat{L})$ and $\sigma_{\min}(\hat{R})$ small) and large-normed (i.e., $\|(\hat{L}, \hat{R})\|_F$ large) solution to the generalized Sylvester equation. Consequently, these cases may give rise to large errors in the direct swapping method which also the error analysis above confirms. Such ill-conditioned examples will be shown in the coming section.

4 Numerical experiments

The direct reordering method has been implemented in the Matlab environment [21]. We use the "backslash operator", which implements Gaussian elimination with partial pivoting, for solving the linear system $Zx = b$ (9) and the QR built-in function for computing the QR and RQ factorizations (7).

Here we illustrate the stability and the accuracy of the direct method on problems with

- finite, real eigenvalues (simple as well as defective)

- pairs of complex conjugate eigenvalues

- infinite eigenvalues (simple as well as defective)

- $m, n = 1, 2$, i.e., only one swap is performed.

We assume that all problems are in generalized real Schur form, i.e.,

$$A = \begin{bmatrix} A_{11} & A_{12} \\ 0 & A_{22} \end{bmatrix}, \quad B = \begin{bmatrix} B_{11} & B_{12} \\ 0 & B_{22} \end{bmatrix}.$$

4.1 EXAMPLES

Example 1. Three random examples with finite, real eigenvalues. The elements of A_{ij} and B_{ij} are chosen as uniformly distributed random numbers on $[0, 1]$.

Example 2. Three examples with complex conjugate eigenvalues [2]:

$$A = \begin{bmatrix} 2 & -87 & -20000 & 10000 \\ 5 & 2 & -20000 & -10000 \\ 0 & 0 & 1 & -11 \\ 0 & 0 & 37 & 1 \end{bmatrix}, \quad B = \begin{bmatrix} 1 & 0 & 0 & 0 \\ 0 & 1 & 0 & 0 \\ 0 & 0 & 1 & 0 \\ 0 & 0 & 0 & 1 \end{bmatrix} \equiv I_4.$$

$$A = \begin{bmatrix} 1 & -3 & 3576 & 4888 \\ 1 & 1 & -88 & -1440 \\ 0 & 0 & 1.001 & -3 \\ 0 & 0 & 1.001 & 1.001 \end{bmatrix}, \quad B = I_4.$$

$$A = \begin{bmatrix} 1 & -100 & 400 & -1000 \\ 0.01 & 1 & 1200 & -10 \\ 0 & 0 & 1.001 & -3 \\ 0 & 0 & 100 & 1.001 \end{bmatrix}, \quad B = I_4.$$

Here we treat the swapping problem of the standard eigenvalue problem as a generalized problem which, e.g., means that we solve the Sylvester equations as generalized Sylvester equations. In this way we presumably make the reordering problem more ill-conditioned. As a consequence we should expect less favourable results compared to the standard case [2].

Example 3. Five examples with finite and infinite eigenvalues. The examples are constructed as $(A, B) = P(D_A, D_B)Q$ where P, Q are random orthogonal matrices. (D_A, D_B) are block diagonal matrices, expressed as direct sums of diagonal blocks, as follows

$$(D_A, D_B) = J_1(1) \oplus J_1(2) \oplus N_1,$$
$$(D_A, D_B) = J_1(1) \oplus J_1(1) \oplus N_1 \oplus N_1,$$
$$(D_A, D_B) = J_2(1) \oplus N_1,$$
$$(D_A, D_B) = J_1(1) \oplus J_1(2) \oplus N_2,$$
$$(D_A, D_B) = J_2(1) \oplus N_2,$$

where $J_k(\lambda)$ denotes a Jordan block of size k corresponding to the eigenvalue λ, and N_k denotes a Jordan block of size k corresponding to an infinite eigenvalue (i.e., $\lambda(N_k) = \{0\}$). If $k \geq 2$ the corresponding eigenvalue is said to be defective. In all cases we swap the finite and infinite eigenvalues.

Example 4. An (ill-conditioned) example from [14]:

$$A_{11} = \begin{bmatrix} 1 & -1 \\ 1 & -1 \end{bmatrix}, A_{22} = A_{11} - \alpha \begin{bmatrix} 1+\alpha & 0 \\ 0 & -1 \end{bmatrix}, B_{11} = B_{22} = \begin{bmatrix} 1 & 0 \\ 0 & 1 \end{bmatrix}.$$

In the first case, $\mathrm{col}(A_{12}, B_{12})$ is chosen as the right singular vector corresponding to the smallest singular value of Z (9). In the second case, $\mathrm{col}(A_{12})$ is chosen as the left singular vector of the smallest singular value of $I_n \otimes A_{11} - A_{22}^T \otimes I_m$, the coefficient matrix of the corresponding $Zx = b$ representation of the standard Sylvester equation [14].

In terms of the generalized Sylvester equation these examples represent a spectrum from well-conditioned to ill-conditioned problems.

4.2 BACKWARD STABILITY TEST CRITERIONS

In order to evaluate the stability of the direct swapping method we extend the backward stability tests in [2] to regular matrix pairs. Let $(\bar{A}, \bar{B})$ denote the computed matrix pair after the swapping of two diagonal blocks, i.e., $\bar{A} = \hat{A} + \Delta A$ (15) and $\bar{B} = \hat{B} + \Delta B$ (16) and we consider the following questions:

- How close is $(\bar{Q}\bar{A}\bar{Z}^T, \bar{Q}\bar{B}\bar{Z}^T)$ to the original matrix pair (A, B)?

- How nearly orthogonal are the computed transformation matrices $\bar{Q}$ and $\bar{Z}$?

m	n	$\|\bar{A}_{21}\|_F$	$\|\bar{B}_{21}\|_F$	$E_{A,B}$	E_Q	E_Z
2	1	5.55112e-17	2.77556e-17	1.5568	6.1237e-01	1.2247e+00
2	2	2.67395e-16	3.39935e-16	1.2038	1.5104e+00	1.6394e+00
1	2	2.38519e-16	9.6513e-16	6.9443	1.0420e+00	2.4398e+00

Table 1: *Example 1: Backward errors of the direct swapping*

- How close are the eigenvalues of (A_{ii}, B_{ii}) (before the swapping) and $(\bar{A}_{ii}, \bar{B}_{ii})$ (after the swapping)?

To answer the first two questions we measure the quantities

$$E_{A,B} = \frac{\|(A - \bar{Q}\bar{A}\bar{Z}^T, B - \bar{Q}\bar{B}\bar{Z}^T)\|_F}{\epsilon\|(A,B)\|_F}, \qquad (25)$$

$$\|\bar{A}_{21}\|_F, \|\bar{B}_{21}\|_F, \qquad (26)$$

$$E_Q = \frac{\|I_{m+n} - \bar{Q}^T\bar{Q}\|_F}{\epsilon}, E_Z = \frac{\|I_{m+n} - \bar{Z}^T\bar{Z}\|_F}{\epsilon}, \qquad (27)$$

where ϵ is the relative machine precision. Ideally, $E_{A,B}$ and E_Q, E_Z should be of size $O(1)$ and the norms of the $(2,1)$-blocks (26) should be of size $O(\epsilon)$.

It is well-known that the general unsymmetric and generalized eigenvalue problems are potentially ill-conditioned (e.g., see [6], [9], [29]). This means that there exist cases where the eigenvalues may change drastically even under small perturbations of the data. The best we can ask for is that the swapping of two diagonal blocks only results in $O(\epsilon\|(A,B)\|_F)$ changes in the original matrix pair. In certain (ill-conditioned) cases this perturbation is enough to change individual eigenvalues a lot (e.g., a real multiple eigenvalue λ of multiplicity k might spread around in a circle in the complex plane with center λ and radius $O(\epsilon^{1/k})$). However, for well-conditioned or only moderately ill-conditioned cases the change of the eigenvalues is an adequate measure on the reliability and accuracy of a reordering method. Therefore it is of interest to display the eigenvalues before and after the swapping.

4.3 TEST RESULTS

Example 1. The quantities (25), (26), (27) that together test the backward stability of the reordering method are shown in Table 4.3. All quantities show "ideal" results. From Table 4.3 we see that the real eigenvalues of the random examples before and after the reordering are equal to machine precision.

Example 2. The examples with conjugate pairs of eigenvalues represent well-conditioned to moderately ill-conditioned problems. Table 3 shows the backward stability quantities. For each of the three examples we show two lines of results corresponding to (A, B) and $P(A,B)Q$, respectively, where P and Q are orthogonal random matrices. Notice, $E_{A,B}$ is much larger than 1, especially, for the last example. However, Table 4 and Table 5 show that the changes of individual eigenvalues are much less than the backward error $E_{A,B}$. This can be explained by the fact that the largest perturbation is in the $(1,2)$-block which

Before reordering	After reordering
8.970109206562917e-01	8.970109206562918e-01
2.785408684158501e-01	2.785408684158499e-01
4.996834562911003e-01	4.996834562911004e-01
9.299846243696946e-01	9.299846243696948e-01
-1.411724350161915e+00	-1.411724350161915e+00
1.582713553370696e-01	1.582713553370699e-01
7.902102449959900e+00	7.902102449959905e+00
2.072270448556391e-01	2.072270448556391e-01
1.285924657614034e+00	1.285924657614035e+00
1.354708319927192e+00	1.354708319927192e+00

Table 2: *Example 1: Eigenvalues before and after reordering*

m	n	$\|A_{21}\|_F$	$\|B_{21}\|_F$	$E_{A,B}$	E_Q	E_Z
2	2	5.68229e-15	2.2672e-16	31.7612	2.5496e+00	3.2874e+00
2	2	3.10502e-15	2.12698e-16	62.0392	1.7505e+00	5.3791e-01
2	2	9.09359e-16	2.81096e-16	2.0661e+03	4.0920e+00	3.3829e+00
2	2	3.35798e-16	2.270947e-16	1.3214e+03	1.1737e+00	2.2499e+00
2	2	1.03369e-14	1.44669e-15	7.1872e+04	2.2364e+00	2.6987e+00
2	2	3.55696e-14	2.49753e-14	1.1188e+06	1.5015e+00	1.9141e+00

Table 3: *Example 2: Backward errors of the direct swapping*

Before reordering
2.000000000000000e+00 - 2.085665361461421e+01i
2.000000000000000e+00 + 2.085665361461422e+01i
1.000000000000000e+00 - 2.017424100183202e+01i
1.000000000000000e+00 + 2.017424100183202e+01i
1.999999999999989e+00 - 2.085665361461421e+01i
2.000000000000005e+00 + 2.085665361461421e+01i
9.999999999999991e-01 - 2.017424100183201e+01i
9.999999999999990e-01 + 2.017424100183201e+01i
1.000000000000000e+00 - 1.732050807568877e+00i
9.999999999999999e-01 + 1.732050807568877e+00i
1.001000000000000e+00 - 1.732916616574496e+00i
1.001000000000000e+00 + 1.732916616574496e+00i
1.000000000000000e+00 - 1.732050807568878e+00i
1.000000000000000e+00 + 1.732050807568878e+00i
1.001000000000000e+00 - 1.732916616574495e+00i
1.001000000000000e+00 + 1.732916616574496e+00i
1.000000000000000e+00 - 1.000000000000000e+00i
9.999999999999999e-01 + 1.000000000000000e+00i
1.001000000000000e+00 - 1.000000000000000e+00i
1.001000000000000e+00 + 1.000000000000000e+00i
9.999999999999926e-01 - 9.999999999999611e-01i
9.999999999999876e-01 + 9.999999999999613e-01i
1.000999999999997e+00 - 1.000000000000077e+00i
1.001000000000003e+00 + 1.000000000000077e+00i

Table 4: *Example 2: Eigenvalues before reordering*

After reordering
2.000000000000002e+00 - 2.085665361461422e+01i
1.999999999999999e+00 + 2.085665361461421e+01i
1.000000000000001e+00 + 2.017424100183201e+01i
9.999999999999999e-01 - 2.017424100183201e+01i
2.000000000000005e+00 + 2.085665361461421e+01i
2.000000000000007e+00 - 2.085665361461421e+01i
9.999999999999942e-01 + 2.017424100183201e+01i
1.000000000000000e+00 - 2.017424100183201e+01i
1.000000000000000e+00 + 1.732050807568877e+00i
9.999999999999999e-01 - 1.732050807568877e+00i
1.001000000000000e+00 - 1.732916616574496e+00i
1.001000000000000e+00 + 1.732916616574496e+00i
9.999999999999999e-01 + 1.732050807568877e+00i
9.999999999999998e-01 - 1.732050807568877e+00i
1.001000000000000e+00 - 1.732916616574496e+00i
1.001000000000000e+00 + 1.732916616574497e+00i
9.999999999999999e-01 + 1.000000000000001e+00i
1.000000000000001e+00 - 1.000000000000001e+00i
1.001000000000000e+00 + 1.000000000000000e+00i
1.001000000000001e+00 - 1.000000000000001e+00i
1.000000000000397e+00 + 9.999999999992505e-01i
1.000000000000398e+00 - 9.999999999992510e-01i
1.001000000000052e+00 - 9.999999999997093e-01i
1.001000000000053e+00 + 9.999999999997096e-01i

Table 5: *Example 2: Eigenvalues after reordering*

m	n	$\|(L,R)\|_F$	$\sigma_{min}(Z)$	Φ	Ψ	$\frac{\Phi}{\Psi}$
2	2	8.2412e+04	1.2937e-02	9.7582e+03	2.9887e+03	3.2651e+00
2	2	8.2412e+04	1.2937e-02	9.7582e+03	2.9887e+03	3.2651e+00
2	2	3.5688e+06	2.5952e-04	2.8845e+04	6.6800e+03	4.3181e+00
2	2	3.5688e+06	2.5952e-04	2.8845e+04	6.6800e+03	4.3181e+00
2	2	7.0469e+07	9.9968e-08	2.0011e+09	1.7806e+07	1.1238e+02
2	2	7.0469e+07	9.9968e-08	2.0011e+09	1.7806e+07	1.1238e+02

Table 6: *Example 2: Condition numbers of the generalized Sylvester equation*

m	n	$\|(R_1, R_2)\|_F$	$\frac{\|(R_1,R_2)\|_F}{(\alpha+\beta)\|(L,R)\|_F+\gamma}$	$\|\mathcal{H}^+ \cdot r\|_2$	$\eta(\bar{L}, \bar{R})$	$\mu(\bar{L}, \bar{R})$
2	2	1.1665e-10	1.1212e-17	5.3351e-17	1.6102e-16	1.4361e+01
2	2	1.8805e-10	1.8075e-17	1.8266e-16	2.5958e-16	1.4361e+01
2	2	4.6770e-10	1.7507e-17	1.2164e-16	1.1039e-16	6.3055e+00
2	2	4.0125e-10	1.5019e-17	9.8135e-17	9.4704e-17	6.3055e+00
2	2	1.4903e-08	1.0572e-18	4.9669e-18	3.6009e-15	3.4060e+03
2	2	3.4193e-07	2.4256e-17	2.8805e-14	8.2616e-14	3.4060e+03

Table 7: *Example 2: Residuals and backward error bounds for the generalized Sylvester equation*

m	n	$\|\tilde{A}_{21}\|_F$	$\|\tilde{B}_{21}\|_F$	$E_{A,B}$	E_Q	E_Z
2	1	1.24127e-16	1.14439e-16	1.0848	3.5355e-01	1.1605e+00
2	2	5.04205e-16	2.53192e-16	1.6862	2.8422e+00	1.5743e+00
1	2	1.67111e-16	1.0968e-16	2.9654	1.5000e+00	2.5062e+00
2	2	1.35974e-16	9.77005e-17	1.4615	1.4604e+00	4.0265e+00
2	2	1.75542e-16	2.42524e-16	1.6814	1.7139e+00	2.3237e+00

Table 8: *Example 3: Backward errors of the direct swapping*

Before reordering	After reordering
1.0	1.000000000000000e+00
2.0	2.000000000000000e+00
∞	∞
∞	∞
∞	∞
1.0	1.000000000000000e+00
1.0	9.999999999999998e-01
∞	∞
1.0	1.000000009978619e+00
1.0	9.999999900213808e-01
∞	-1.769999161545647e+16
∞	∞
1.0	1.000000000000000e+00
2.0	2.000000000000000e+00
∞	2.363027832467060e+08
∞	-2.363027816953079e+08
1.0	1.000000000000000e+00 + 7.619266321069252e-09i
1.0	1.000000000000000e+00 - 7.619266321069254e-09i

Table 9: *Example 3: Eigenvalues before and after reordering*

m	n	α	$\|A_{21}\|_F$	$\|B_{21}\|_F$	$E_{A,B}$	E_Q	E_Z
2	2	10^{-3}	1.40258e-16	6.97943e-14	405.6742	1.3822e+00i	1.1397e+00
2	2	10^{-6}	1.67183e-16	7.4277e-11	4.1034e+05	1.4108e+00	2.3321e+00
2	2	10^{-3}	2.33032e-14	2.46847e-11	1.5589e+09	2.1116e+00	2.4819e+00
2	2	10^{-6}	3.85274e-10	0.000385389	8.8997e+17	1.6202e+00	2.4421e+00

Table 10: *Example 4: Backward errors of the direct swapping*

Before reordering
0.
0.
-1.000500000000104e-03 - 9.999998750657075e-04i
-1.000499999999913e-03 + 9.999998750657075e-04i
0.
0.
-1.000000499989864e-06 - 1.000074695365655e-06i
-1.000000499883942e-06 + 1.000074695365655e-06i
0.
0.
-1.000500000000104e-03 - 9.999998750657075e-04i
-1.000499999999913e-03 + 9.999998750657075e-04i
0.
0.
-1.000000499989864e-06 - 1.000074695365655e-06i
-1.000000499883942e-06 + 1.000074695365655e-06i

Table 11: *Example 4: Eigenvalues before reordering*

After reordering
-3.243337697283541e-10
3.243339037288756e-10
-1.000500000000012e-03 - 9.999998750238749e-04i
-1.000500000000013e-03 + 9.999998750238747e-04i
-1.826601743707151e-12
1.826604202442095e-12
-1.000000499923316e-06 - 1.000025162120888e-06i
-1.000000499923316e-06 + 1.000025162120888e-06i
-1.103093753272721e-08
1.103088413341525e-08
-1.000500000013035e-03 - 9.999998750287687e-04i
-1.000500000013132e-03 + 9.999998750287685e-04i
2.924391626724574e-11 + 7.335718815521697e-09i
2.924391626776018e-11 - 7.335718815521695e-09i
-9.997381110559232e-07 + 9.994584988932529e-07i
-9.997381110093625e-07 - 9.994584988932536e-07i

Table 12: *Example 4: Eigenvalues after reordering*

does not directly affect the eigenvalues.

In Table 6 and Table 7 quantities that reflect the conditioning of the associated generalized Sylvester equation are displayed. Besides $\|(\bar{L}, \bar{R})\|_F$, $\sigma_{min}(Z)$, a new structure preserving condition number Ψ and the condition number Φ based on the separation between the two diagonal blocks are displayed ($\Psi \leq \Phi$) [15]. In Table 7 computed residuals and corresponding backward error bounds are displayed. More precisely, $\|(R_1, R_2)\|_F$, the relative residual $\frac{\|(R_1, R_2)\|_F}{(\alpha+\beta)\|(L, R)\|_F + \gamma}$, the exact backward error $\|\mathcal{H}^+ \cdot r\|_2$, an upper bound on the backward error $\eta(\bar{L}, \bar{R})$, and the growth factor $\mu(\bar{L}, \bar{R})$ that measures by how much the backward error, at most, can be greater than the relative residual (for more details see [15]). The last line in Table 7 illustrates a case when the relative backward error is larger than the relative residual. The exact relations between these results and the change of the eigenvalues after a swap are not yet known.

Example 3. These five examples, including finite as well as infinite defective eigenvalues, illustrate ill-conditioned generalized eigenvalue problems $Ax = \lambda Bx$. The results in Table 8 show that the backward errors after the swapping are all at the machine precision level. However, in Table 9 we see that the defective eigenvalues have changed "drastically" but in accordance with known perturbation theory. As expected, the mean value of the eigenvalues corresponding to finite defective eigenvalues are accurate to machine precision. We also see that the defective infinite eigenvalue "spreads out" similarly (cases four and five).

Example 4. These four examples illustrate problems where the associated generalized Sylvester equation is (extremely) ill-conditioned. The eigenvalues are 0 (of multiplicity 2) and a complex conjugate pair (as a function of $\alpha, 0 < \alpha < 1$). The eigenvalues before and after the reordering are displayed in Table 11 and Table 12, respectively.

Table 10 shows the backward stability quantities for $\alpha = 10^{-3}$ and 10^{-6}. Especially, the results in the last two lines of Table 10, where $col(A_{12})$ is chosen as the left singular vector of the smallest singular value of $I_n \otimes A_{11} - A_{22}^T \otimes I_m$, signal the ill-conditioning of the (reordering) problem. The large backward error for $\alpha = 10^{-6}$ is directly related to the associated generalized Sylvester equation which is almost singular with a large-normed solution ($\sigma_{min}(Z) = O(10^{-16})$, $\Psi = O(10^{12})$), $\|(\bar{L}, \bar{R})\|_F = O(10^{18})$, and $\|(R_1, R_2)\|_F = O(10^1)$).

5 The direct reordering algorithm in practice

The error analysis of the direct reordering method and the numerical experiments suggest that a practical implementation should reject a swap if it would result in too large backward errors (i.e. instability). The test for stability can be performed directly and to a small extra cost. In the following, we formulate a direct algorithm with controlled backward error for swapping two diagonal blocks of a regular matrix pair (A, B) in generalized real Schur form. Here all quantities denote "computed" quantities.

Direct Swapping Algorithm

Step 1 Copy A and B to S and T, respectively.

$$S \leftarrow A, \quad T \leftarrow B.$$

Step 2 Solve for (L, R) in the generalized Sylvester equation:

$$S_{11}R - LS_{22} = -S_{12}\gamma,$$
$$T_{11}R - LT_{22} = -T_{12}\gamma.$$

Use Gaussian elimination with complete pivoting to solve the corresponding linear system and a scaling factor γ to prevent against overflow as in [2], [1].

Step 3 Compute an orthogonal matrix Q:

$$Q^T \begin{bmatrix} L \\ \gamma I_n \end{bmatrix} = \begin{bmatrix} T_L \\ 0 \end{bmatrix}.$$

Use Householder transformations to compute a QR-factorization [13], [1].

Step 4 Compute an orthogonal matrix Z:

$$\begin{bmatrix} \gamma I_m & -R \end{bmatrix} Z = \begin{bmatrix} 0 & T_R \end{bmatrix}.$$

Use Householder transformations to compute an RQ-factorization [13], [1].

Step 5 Perform the swapping tentatively:

$$S \leftarrow Q^T S Z \equiv \begin{bmatrix} S_{11} & S_{12} \\ S_{21} & S_{22} \end{bmatrix},$$

$$T \leftarrow Q^T T Z \equiv \begin{bmatrix} T_{11} & T_{12} \\ T_{21} & T_{22} \end{bmatrix}.$$

Weak stability test: $\|(S_{21}, T_{21})\|_F \leq O(\epsilon \|(S, T)\|_F)$

Strong stability test: $\|(A - Q S Z^T, B - Q T Z^T)\|_F \leq tol$, where *tol* is a user-supplied tolerance.

Step 6 If the swap is accepted ("weakly" or "strongly"), apply the equivalence transformation to (A, B):

$$A \leftarrow S, \quad B \leftarrow T$$

Set the $(1, 2)$-blocks to zero.

Step 7 Standardize existing 2×2 blocks.

Use QZHES (1 rotation) and QZVAL which standardizes and separates 2×2 blocks further [11].

After step 2 it would be possible to compute an optimal block-diagonalizing equivalence transformation that minimizes the condition numbers of the transformation matrices [5],[17]. Since the scaling factors (which possibly are large numbers) will show up in the S_{ij} and T_{ij} blocks, we do not expect any substantial improvements in performing this block-diagonal scaling. Our experiments in Matlab confirm this statement too.

The weak stability test in step 5 only guarantees that the $(2, 1)$-blocks are of size $O(\epsilon)$, while the strong stability test takes all backward errors into account. By choosing *tol* of the size $O(\epsilon \|(A, B)\|_F)$ and rejecting the swap if the error is larger than *tol*, we obtain guaranteed backward stability. One could argue that the $(1, 2)$-blocks could be omitted in the test since they do not explicitly affect the eigenvalues. However, they do implicitly. In the perturbation theory for eigenspaces of matrix pairs (A, B) both the $(1, 2)$- and $(2, 1)$-blocks appear in the condition that restricts the size of the perturbations such that the perturbed and unperturbed matrix pairs have an eigenspace and an associated complementary eigenspace of the same sizes [26], [9]. Further, the extra cost of the complete strong stability test is only marginal.

If we allow $m, n > 2$, the above algorithm will reorder two blocks or clusters of eigenvalues. Then, in general, we have to make a call to QZIT [11] (which implements the implicit double-shift QZ iteration [22]) in step 7 to recover a generalized real Schur form.

6 Conclusions and future work

The error analysis and the numerical experiments show the following characteristics of the direct reordering method:

- It is numerically stable and accurate except for "extremely" ill-conditioned problems. Typically, these problems are related to ill-conditioned, large-normed solutions of the associated generalized Sylvester equation [15].

- The numerical stability can be guaranteed and controlled by computing the size of the backward error and rejecting the swap if it exceeds a certain threshold.

- We can expect "large" changes in individual eigenvalues for ill-conditioned $Ax = \lambda Bx$ problems even if the backward error after the swapping is at the level of machine precision.

  Examples include defective eigenvalues, notably at infinity. This type of inherited ill-conditioning cannot be "cured" by any reordering method. However, one possible remedy is to start to deflate the infinite eigenvalues with a staircase type of algorithm (e.g., see [7, 8]) and then perform the required reordering of the finite eigenvalues. The placement of the infinite cluster could be made either to the $(1,1)$-block or to the $(2,2)$-block.

The future work will be focused on software development based on the direct reordering algorithm with controlled backward error and our goal is to produce reliable and robust software for swapping of diagonal blocks and reordering of eigenvalues in the generalized real Schur form of a regular matrix pair in the style of the routines _TREXC and _TRSEN in LAPACK [1], which perform the corresponding tasks for the standard eigenvalue problem.

Acknowledgements

I am grateful to Nick Higham for reading and commenting on the manuscript.

Financial support has been received from NUTEK under contract 89-02578P.

References

[1] E. Anderson, Z. Bai, C. Bischof, J. Demmel, J. Dongarra, J. Du Croz, A. Greenbaum, S. Hammarling, A. McKenney, S. Ostrouchov, and D. Sorensen. *LAPACK Users' Guide*. Society for Industrial and Applied Mathematics, Philadelphia, 1992.

[2] Z. Bai and J. Demmel. On a direct algorithm for computing invariant subspaces with specified eigenvalues. Computer Science Dept. Technical Report CS-91-138, University of Tennessee,, Knoxville, TN, November 1991. (LAPACK Working Note #38).

[3] R. H. Bartels and G. W. Stewart. Solution of the equation $AX + XB = C$. *Comm. Assoc. Comput. Mach.*, 15, pp 820–826, 1972.

216

[4] A. Bojanczyk, G. Golub, and P. Van Dooren. The periodic Schur decomposition. Algorithms and applications. SCCM Inter. Rept. NA-92-07, Stanford University, August 1992.

[5] J. Demmel. The condition number of equivalence transformations that block diagonalize matrix pencils. *SIAM J. Num. Anal.*, **20**(3), pp 599–610, 1983.

[6] J. Demmel. Computing stable eigendecompositions of matrices. *Lin. Alg. Appl.*, **79**, pp 163–193, 1986.

[7] J. Demmel and B. Kågström. The generalized Schur decomposition of an arbitrary pencil $A - \lambda B$: robust software with error bounds and applications. Part I: Theory and Algorithms. To appear in *ACM Trans. Math. Software*. Also published as Report UMINF-91.22, 1991.

[8] J. Demmel and B. Kågström. The generalized Schur decomposition of an arbitrary pencil $A - \lambda B$: robust software with error bounds and applications. Part II: Software and Applications. To appear *in ACM Trans. Math. Software*. Also published as Report UMINF-91.23, 1991.

[9] J. Demmel and B. Kågström. Computing stable eigendecompositions of matrix pencils. *Lin. Alg. Appl.*, **88/89**, pp 139–186, 1987.

[10] J. Dongarra, S. Hammarling, and J. Wilkinson. Numerical considerations in computing invariant subspaces. *SIAM J. Matrix Anal. Appl.*, **13**(1), pp 145–161, 1992.

[11] B. S. Garbow, J. M. Boyle, J. J. Dongarra, and C. B. Moler. *Matrix Eigensystem Routines – EISPACK Guide Extension*, volume 51 of *Lecture Notes in Computer Sciences*. Springer-Verlag, Berlin, 1977.

[12] G. Golub, S. Nash, and C. Van Loan. A Hessenberg-Schur method for the matrix problem AX+XB=C. *IEEE Trans. Autom. Contr.*, **AC-24**, pp 909–913, 1979.

[13] G. Golub and C. Van Loan. *Matrix Computations*. Second Edition. Johns Hopkins University Press, Baltimore, MD, 1989.

[14] N. J. Higham. Perturbation theory and backward error for $AX - XB = C$. Numerical Analysis Report No. 211, Department of Mathematics, University of Manchester, Manchester M13 9PL, England, 1992. To appear in *BIT*.

[15] B. Kågström. A Perturbation Analysis of the Generalized Sylvester Equation. Report UMINF-92.17, Institute of Information Processing, University of Umeå, S-901 87 Umeå, Sweden, 1992.

[16] B. Kågström and A. Ruhe. ALGORITHM 560: An algorithm for the numerical computation of the Jordan normal form of a complex matrix [F2]. *ACM Trans. Math. Soft.*, **6**(3), pp 437–443, 1980.

[17] B. Kågström and P. Van Dooren. A generalized state-space approach for the additive decomposition of a transfer matrix. *Int. J. Numerical Linear Algebra with Applications*, **1**(2), pp 165–181, 1992.

[18] B. Kågström and L. Westin. GSYLV - Fortran routines for the generalized Schur method with Dif^{-1}-estimators for solving the generalized Sylvester equation. Report UMINF-132.86, Institute of Information Processing, University of Umeå, S-901 87 Umeå, Sweden, 1987. *Also in NAG's SLICOT library, 2nd release 1991.*

[19] B. Kågström and L. Westin. Generalized Schur methods with condition estimators for solving the generalized Sylvester equation. *IEEE Trans. Autom. Contr.*, **34**(4), pp 745–751, 1989.

[20] A. Laub. A Schur method for solving algebraic Riccati equations. *IEEE Trans. Autom. Contr.*, **AC-24**, pp 913–921, 1979.

[21] C. Moler, J. Little, and S. Bangert. *PRO-MATLAB Users' Guide.* The Math Works Incorporation, 1987.

[22] C. Moler and G. Stewart. An algorithm for the generalized matrix eigenvalue problem. *SIAM J. Numer. Anal.*, **10**, pp 241–256, 1973.

[23] K. C. Ng and B. Parlett. Development of an accurate algorithm for Exp(Bt), Part I, Programs to swap diagonal blocks, Part II. Technical Report CPAM-294, University of California, Berkeley, CA, 1989.

[24] J. L. Rigal and J. Gaches. On the compatability of a given solution with the data of a linear system. *J. Assoc. Comput. Mach.*, **14**, pp 543–548, 1967.

[25] A. Ruhe. An algorithm for numerical determination of the structure of a general matrix. *BIT*, **10**, pp 196–216, 1970.

[26] G. W. Stewart. Error and perturbation bounds for subspaces associated with certain eigenvalue problems. *SIAM Review*, **15**(4), pp 727–764, Oct 1973.

[27] G. W. Stewart. Algorithm 406: HQR3 and EXCHNG: Fortran subroutines for calculating and ordering the eigenvalues of a real upper hessenberg matrix. *ACM Trans. Math. Soft.*, **2**, pp 275–280, 1976.

[28] G. W. Stewart. Perturbation bounds for the QR factorization of a matrix. *SIAM J. Num. Anal.*, **14**, pp 509–518, 1977.

[29] G. W. Stewart and J.-G. Sun. *Matrix Perturbation Theory.* Academic Press, New York, 1990.

[30] J.-G. Sun. Perturbation bounds for the Cholesky and QR factorizations. *BIT*, **31**, pp 341–352, 1991.

[31] P. Van Dooren. A generalized eigenvalue approach for solving Riccati equations. *SIAM J. Sci. Stat. Comp.*, **2**, pp 121–135, 1981.

[32] P. Van Dooren. ALGORITHM 590: DSUBSP and EXCHQZ: Fortran routines for computing deflating subspaces with specified spectrum. *ACM Trans. Math. Software*, **8**, pp 376–382, 1982. and (Corrections) Vol. 4, No. 4, p 787, 1983.

218

[33] R. Walker. *Computing the Jordan form for control of dynamic systems*. PhD thesis, Stanford University, Stanford, CA 94305, March 1981. Dept. Aeronautics and Astronautics SUDAAR 528.

[34] J. H. Wilkinson. *The Algebraic Eigenvalue Problem*. Oxford University Press, 1965.

BLOCK SHIFT INVARIANCE AND
EFFICIENT SYSTEM IDENTIFICATION ALGORITHMS

N. KALOUPTSIDIS
University of Athens, Informatics Dept.
Panepistimiopolis
TYPA Buildings 157 71, Athens
Greece
spa64@grathun1.bitnet

ABSTRACT. Various problems in system identification and signal processing are naturally formulated and solved in a multichannel context. Notable examples are identification of multivariable systems and multichannel signal modeling. In other circumstances such as in single input single output system identification, formulation of the problem is carried out in a single channel format, however issues of computational complexity suggest conversion to a multichannel environment, thus paving the ground for an efficient algorithmic realization. This multichannel embedding is applicable provided a certain block shift invariance property prevails. Various applications will be considered. Their main feature is that they involve ℓ_2 optimization via error linearization and as such, they give rise to linear system solvers, which in addition, possess block shift invariance.

KEYWORDS. System identification, signal processing fast algorithms.

1 System identification

Identification is concerned with the determination of a system on the basis of input output data $u(t)$, $y(t)$, $1 \leq t \leq N$, [7], [10], [5]. We assume that the input output sequences are governed by the input output expression

$$y(t) = S(u(t), \eta(t)) \tag{1}$$

The signal η is white noise. The identification task is to design a predictor which will try to reconstuct the data and simultaneously remove the noisy contribution. The predictor is a system that acts on the data sequences $u(t)$, $y(t)$, and produces the output signal $\hat{y}(t)$, a faithful copy of the unknown system's output $y(t)$. Several possibilities exist for the

219

M. S. Moonen et al. (eds.), Linear Algebra for Large Scale and Real-Time Applications, 219–229.

selection of the predictor. The option we shall pick here makes the prediction error white noise. Ideally, this amounts to solving eq.(1) with respect to η

$$\eta(t) = G(u(t), y(t))$$

and then setting

$$\hat{y}(t) = y(t) - G(u(t), y(t)) \tag{2}$$

To come up with affordable identification solutions and efficient algorithmic structures offering the potential of intensive computation handling, we consider the finite parametrization case. Thus we assume that the predictor (2) can be factored as

$$\hat{y}(t) = y(t) - G(u(t), y(t), \theta) \tag{3}$$

where G is a known transformation and θ is a finite dimensional unknown parameter. To determine θ we seek to minimize some measure of the error

$$e(t|\theta) = y(t) - \hat{y}(t|\theta) \tag{4}$$

with respect to θ. A popular performance criterion is the ℓ_2 norm

$$V_N(\theta) = \sum_{t=1}^{N} w_N(t) e^2(t|\theta) \tag{5}$$

The window function $w_N(t)$ reflects the time varying significance of past and recent information. Here we shall stick to the rectangular window $w_N(t) = 1$, $1 \leq t \leq N$, although subsequent analysis is applicable to other window functions, such as the sliding window and the exponential forgetting window.

2   Linear regression and block shift invariance

Minimization of (5) leads to a system of nonlinear equations, unless the predictor depends linearly on θ

$$\hat{y}(t|\theta) = \phi^T(t)\theta \tag{6}$$

This special and important case is referred to as *linear regression*. $\phi(t)$ is called regressor vector and is a known signal. The parameter θ is determined by the linear system of equations

$$R(N)\theta(N) = d(N) \tag{7}$$

where

$$R(N) = \sum_{t=1}^{N} \phi(t)\phi^T(t), \qquad d(N) = \sum_{t=1}^{N} \phi(t)y(t). \tag{8}$$

Next we discuss some cases where the linear regression assumption (6) holds.

2.1 ARX MODELS

The input output relationship of an ARX model is

$$y(t) = \sum_{i=1}^{p} a_i y(t-i) + \sum_{i=1}^{q} b_i u(t-i) + \eta(t)$$

Let the unknown parameter

$$\theta = [a_1, a_2, \cdots, a_p, b_1, b_2, \cdots, b_q]^T$$

The predictor is then linear in θ and the regressor vector is

$$\phi(t) = [y(t-1), \cdots, y(t-p), u(t-1), \cdots, u(t-q)]^T.$$

Besides the linear regression property, ARX models feature another remarkable property. The regressor vector is partitioned into two blocks of lengths p and q consisting of successive shifts. Thus one block is associated with $u(t)$ and its shifts and the other block is associated with the shifts of $y(t)$. We refer to this property as block shift invariance.

Entirely analogous results hold in the multichannel case. The multiinput multioutput ARX system

$$y(t) = \sum_{i=1}^{p} A_i y(t-i) + \sum_{i=1}^{q} B_i u(t-i) + \eta(t)$$

leads to a linear regression with regressor vector

$$\phi(t) = [y^T(t-1), \cdots, y^T(t-p), u^T(t-1), \cdots, u^T(t-q)]^T.$$

We can achieve further flexibility if we allow different memory lengths for each input and output channel. Via a suitable permutation the regressor takes the form

$$\phi(t) = [y_1(t-1) \cdots y_1(t-p_1) \cdots y_2(t-p_2) \cdots u_k(t-1) \cdots u_k(t-q_k)]^T.$$

We observe that multivariable systems lead to linear regressions. Moreover, the regressor vector possesses the block shift invariance property, as is formed from successive samples of a finite number of signals.

2.2 NONLINEAR EQUATION ERROR MODELS

Linear systems is not the only class of systems leadings to linear regressions with block shift invariance. An interesting family of nonlinear systems with analogous properties is the following [6]

$$y(t) + \sum_{i} a_i y(t-i) + \sum_{ij} a_{ij} y(t-i)y(t-j) + \cdots + \sum_{i_1 \ldots i_n} a_{i_1 \ldots i_n} y(t-i_1) \cdots y(t-i_n)$$

$$= \sum_{i} b_i u(t-i) + \sum_{ij} b_{ij} u(t-i)u(t-j) + \cdots + \sum_{i_1 \ldots i_n} b_{i_1 \ldots i_n} u(t-i_1) \cdots u(t-i_n) +$$

$$+ \sum_{i_1 \ldots i_n j_1 \ldots j_m} c_{i_1 \ldots i_n j_1 \ldots j_m} y(t-i_1) \cdots y(t-i_n)u(t-j_1) \cdots u(t-j_m) + \eta(t) \qquad (9)$$

The value of the output at each time instant t is formed from past output values, products of past output values, products of past inputs as well as products of inputs and outputs, all linearly combined. Such systems are approximate finite parametrizations of nonlinear systems of the form

$$y(t) = f(y(t-1), ..., y(t-k), u(t-1), ..., u(t-l))$$

when f is approximated by a finite Taylor series. Alternate parametrizations of the above class using neural network architectures for identification and control purposes, have been discussed in [9].

If the unknown parameter consists of the coefficients associated with the a, b,and c, eq.(9) can be written as a linear regression. Furthermore the regressor vector has the block shift invariance property. To see this, let us for simplicity consider the second order case, where only second order terms appear in the expression. There are 3 second order terms, one consisting of input output products, one consisting of input input products and one consisting of output output products. For each one of them, say the input input term, we form a 2-D array $x^{uu}(i,j)$ whose ij entry is the corresponding product $u(n-i)u(n-j)$

$$x^{uu}(i,j) = u(n-i)u(n-j) \tag{10}$$

We denote by $u_l^{uu}(n)$ the entries of the first row. If we scan the array moving along the diagonals parallel to the main diagonal, we obtain shifts of the first row. Thus if we switch to this indexing we get the block shift invariance property. The new indexing corresponds to a permutation of the regressor vector and clearly does not affect the output computation. It is described by the rule

$$j > i: \qquad (i,j) \to (j-i+1, i)$$

In summary, the regressor vector consists of shifts of the following signals

$$\left\{ \begin{array}{l} y(n), \quad u(n) \\ u_\ell^{yu}(n) = y(n-\ell)u(n), \quad 1 \le \ell \le q \\ u_\ell^{uu}(n) = u(n-\ell)u(n), \quad 1 \le \ell \le p \\ u_\ell^{yy}(n) = y(n-\ell)y(n), \quad 1 \le \ell \le r \end{array} \right\}$$

The input output relationship is then written as

$$\hat{y}(n) = -\sum_{i=1}^{m^y} a(i)y(n-i) - \sum_{i=1}^{m^u} b(i)u(n-i)-$$

$$-\sum_{\ell=1}^{p}\sum_{k=1}^{m_\ell^{yu}} c_{yu}^\ell(k)u_l^{yu}(n-k) - \sum_{\ell=1}^{r}\sum_{k=1}^{m_\ell^{yy}} a_{yy}^\ell(k)u_l^{yy}(n-k)-$$

$$-\sum_{\ell=1}^{q}\sum_{k=1}^{m_\ell^{uu}} b_{uu}^\ell(k)u_l^{uu}(n-k)$$

and the regressor vector has the form

$$\phi(n) = [\mathbf{y}_{m^y}^T(n-1)\mathbf{u}_{m^u}^T(n)\mathbf{u}_{m_1^{yu}}^{yuT}(n)\ldots\mathbf{u}_{m_q^{yu}}^{yuT}(n)$$

$$\mathbf{u}_{m_1^{yy}}^{yyT}(n)\ldots\mathbf{u}_{m_r^{yy}}^{yyT}(n)\mathbf{u}_{m_1^{uu}}^{uuT}(n)\ldots\mathbf{u}_{m_p^{uu}}^{uuT}(n)]^T$$

where

$$\mathbf{u}_{m_\ell^{yu}}^{yu}(n) = [u_\ell^{yu}(n)u_\ell^{yu}(n-1)\dots u_\ell^{yu}(n-m_\ell^{yu}+1)]^T$$

$$\ell = 1, 2\dots q$$

and similarly for the other blocks.

2.3 2-D AND MULTIDIMENSIONAL SYSTEMS

Besides nonlinear systems, 2-D systems also lead to linear regressions with block shift invariance. In fact the nonlinear development of the previous paragraph can be viewed as a special case of the 2-D setup. Indeed, the assignment (10) embedds the nonlinear problem into a 2-D structure. The 2-D counterpart of the ARX system is

$$y(m, n) = - \sum_{(k,l) \in R_y} a_{k,l} y(m-k, n-l)$$

$$- \sum_{(k,l) \in R_u} b_{k,l} u(m-k, n-l) + \eta(m, n)$$

The masks can be fairly general. A convenient description is as a union of strips

$$R_u = \cup_{k=\alpha^u}^{\beta^u} r_k^u, \quad r_k^u = \{(k,l) : -\gamma_\alpha^u \le l \le \delta_\beta^u\}$$

and similarly for R_y.

For each $\alpha^u \le k \le \beta^u$ let $p_k^u = -\gamma_k^u + \delta_k^u + 1$, denote the length of each segment. We scan the array column by column. The basic signal components of the regressor vector are located on the first column. Let

$$u_k(n) = u(m-k, n), \qquad y_k(n) = y(m-k, n)$$

Their shifts are then on the same row. Thus the regressor vector has the form

$$\phi(n) = [\mathcal{Y}_{\mathbf{py}}^t(n) \ \mathcal{U}_{\mathbf{pu}}^t(n)]$$

where

$$\mathcal{U}_{\mathbf{pu}}(n) = [\mathbf{u}_{p_\alpha^u}^t(n)\mathbf{u}_{p_{\alpha+1}^u}^t(n)\dots\mathbf{u}_{p_\beta^u}^t(n)]^t$$

$$\mathcal{Y}_{\mathbf{py}}(n) = [\mathbf{y}_{p_\alpha^y}^t(n)\mathbf{y}_{p_{\alpha y+1}}^t(n)\dots\mathbf{y}_{p_\beta^y}^t(n)]^t$$

and

$$\mathbf{y}_{p_k^y}(n) = [y_k(n-\gamma_k^y) \ y_k(n-\gamma_k^y+1)\dots y_k(n-\delta_k^y)]^t$$

$$\mathbf{u}_{p_k^u}(n) = [u_k(n-\gamma_k^u) \ u_k(n-\gamma_k^u+1)\dots u_k(n-\delta_k^u)]^t$$

Block shift invariance is apparent.

2.4 INSTRUMENTAL VARIABLE METHOD

In the linear regression case the unknown parameter is determined by a linear system of equations where the pertinent matrix R is obtained by a symmetric sum of dyads formed by the regressor vector. Most of the popular fast algorithms that solve the above equations owe their value to the block shift invariance of the regressor vector. A similar situation appears with the instrumental variable methods, albeit the symmetry of the matrix is lost. Indeed, we recall from [7],[10] that the linear regression estimate provided by equations (7),(8) will asymptotically recover the true parameter if the noise signal is white or the system is of the FIR form. If the noise signal is not white, the instrumental variable approach chooses a sequence $\zeta(t)$ obtained by suitable filtering of the data sequence and the parameter θ so that the error is uncorrelated to $\zeta(t)$

$$\frac{1}{N} \sum_{t=1}^{N} \zeta(t) e(t, \theta) = 0$$

This leads to the linear system of equations

$$R_{iv}(N) \theta_{iv}(N) = d_{iv}(N)$$

where

$$R_{iv}(N) = \sum_{t=1}^{N} \zeta(t) \phi^T(t), \qquad d_{iv}(N) = \sum_{t=1}^{N} \zeta(t) y(t)$$

If the unknown system is described by an ARX model for which the noise term is not white, a natural choice of the instrument is

$$\zeta(t) = (\, x(t-1) \quad \ldots \quad x(t-n_x) \quad u(t-1) \quad \ldots \quad u(t-n_u) \,)^T$$

where $x(t)$ is obtained by filtering $u(t)$

$$N(q) x(t) = M(q) u(t)$$

We observe that $\zeta(t)$ is block shift invariant.

Analogous observations hold when multiinput multioutput ARX models, nonlinear and multidimensional systems are considered.

3 Linear system identification and the prediction error method

Identification of realizable linear systems as is carried out by the prediction error method also involves linear system solvers. Furthermore, the underlying "regressor" possesses the block shift invariance property, under general conditions. Let us assume that the data sequences are generated by the linear system

$$y(t) = G(q|\theta) u(t) + H(q|\theta) \eta(t)$$

$\eta(t)$ is a white noise signal. Let us take as predictor the system which will make the difference between its output and the unknown system output white. It is easy to see [7] that it is given by the linear filter

$$\hat{y}(t|\theta) = (1 - H^{-1}(q|\theta)) y(t) + H^{-1}(q|\theta) G(q|\theta) u(t)$$

If the unknown system is rational the predictor is rational as well, and is specified by the equation

$$D(q|\theta) \hat{y}(t|\theta) = N_u(q|\theta) u(t) + N_y(q|\theta) y(t) \tag{11}$$

The predictor must be a stable filter. Therefore we further require that the polynomial $D(q|\theta)$ has its roots inside the unit circle.

If we plug the above expression into the cost (5) and set the derivative equal to zero, we end up with a nonlinear set of equations. Hence we must resort to iterative techniques for the determination of the optimal parameters. Let us assume that $\theta(N)$ has become available. To determine $\theta(N+1)$, we approximate the error by a first order Taylor approximation

$$e(t|\theta) \approx \bar{e}(t|\theta) = y(t) - \hat{y}(t|\theta(N)) - \psi^T(t|\theta(N))(\theta - \theta(N))$$

where

$$\psi^T(t|\theta(N)) = \frac{\partial \hat{y}(t|\theta)}{\partial \theta}|_{\theta=\theta(N)}$$

If we replace the approximate expression for the error into the minimization criterion the latter becomes quadratic in the unknown parameter. Thus $\theta(N+1)$ is approximately given by the linear system of equations

$$R(N+1)\theta(N+1) = r(N+1) \tag{12}$$

$$R(N+1) = \sum_{t=1}^{N+1} \psi(t|\theta(N))\psi^T(t|\theta(N)) \tag{13}$$

$$r(N+1) = \sum_{t=1}^{N+1} \psi(t|\theta(N))[y(t) - \hat{y}(t\theta(N)) + \psi^T(t|\theta(N))\theta(N)] \tag{14}$$

To check whether the above system is amenable to efficient computation we consider the determination of the gradient vector. Differentiation of (11) with respect to θ gives

$$D^2\psi = (D\dot{N}_u - \dot{D}N_u)u + (D\dot{N}_y - \dot{D}N_y)y$$

Thus the gradient is also obtained at the output of a rational filter acting on the data sequence. Morevover since the roots of D are stable, the roots of D^2 are also stable and the predictor gradient is produced by a stable filter.

Next we turn to block shift invariance. One natural assumption that ensures the block shift invariance of the predictor gradient vector, is to assume that the unknown parameter θ is partitioned into k blocks of sizes $m_i \times 1$ each, such that each one of the polynomials associated with the predictor, D, N_u, N_y, becomes a polynomial function of the blocks of θ. For instance,

$$D(q,\theta) = R_D(\theta_1(q),\ldots,\theta_k(q)) \tag{15}$$

and R_D is given by the polynomial

$$R_D(\theta_1(q),\ldots,\theta_k(q)) = \sum_{i_1=0}^{\mu_1}\sum_{i_2=0}^{\mu_2}\sum_{i_k=0}^{\mu_k} a(i_1,i_2,\cdots,i_k)\theta_1^{i_1}(q)\theta_2^{i_2}(q)\cdots\theta_k^{i_k}(q) \tag{16}$$

Similar expressions are valid for the remaining polynomials N_u, and N_y. If we differentiate the predictor dynamics and take into account the above assumption it follows that the predictor gradient possesses the block shift invariance property, [1],[2]. To achieve acceptable levels of computational complexity, all μ_i must be small integers, say, 0,1, or 2. As an example, let us take the generalized Box Jenkins model

$$A(q)y(t) = \frac{B(q)}{F(q)}u(t) + \frac{C(q)}{D(q)}\eta(t)$$

with unknown parameters the corresponding coefficients

$$\theta = (A(q) - 1 \quad B(q) \quad F(q) \quad C(q) - 1 \quad D(q) - 1) = (\theta_1(q) \quad \cdots \quad \theta_5(q))$$

Note that the constant term of the polynomials A, C, and D are known and equal to 1. The predictor is given by

$$C(q)F(q)\hat{y}(t) = [C(q) - A(q)D(q)]F(q)y(t) + B(q)D(q)u(t)$$

Expressions (13), (14) are valid with

$$D(q, \theta) = (1 + \theta_4)\theta_3$$
$$N_u(q, \theta) = \theta_2(q) + \theta_5(q)\theta_2(q)$$
$$N_y(q, \theta) = (\theta_4 - \theta_1 - \theta_5 - \theta_1\theta_5)\theta_3$$

4 Identification of certain nonlinear systems

The previous discussion can be extended to a certain class of nonlinear systems. To do so it will be useful to write models such as (9) in an algebraic fashion. Let i and j be integers. The $\delta_i\delta_j$ shift is an operator that acts on a pair of causal sequences $y(t)$, $u(t)$ and produces the product sequence

$$\delta_i\delta_j[y(t), u(t)] = y(t - i)u(t - j)$$

For multiindices $\underline{i_m} = [i_1 i_2 \ldots i_m]$ and $\underline{j_n} = [j_1 j_2 \ldots j_n]$, we define the mixed shift as

$$\delta_{\underline{i_m}}\delta_{\underline{j_n}}[y(t), u(t)] = y(t - i_1) \cdots y(t - i_m)u(t - j_1) \cdots u(t - j_n)$$

We also consider simple shifts, $\delta_{\underline{i_m}}$ acting on a single sequence as follows

$$\delta_{\underline{i_m}}y(t) = y(t - i_1) \cdots y(t - i_m)$$

Simple shifts can be recovered from mixed shifts through the relation

$$\delta_{\underline{i_m}}\delta_e(y, u) = \delta_{\underline{i_m}}y(t)$$

where δ_e assigns every sequence to the constant sequence taking the value 1. A mixed series is a formal expression of the form

$$\underline{S} = \sum_{m=0}^{\infty} \sum_{n=0}^{\infty} \sum_{(i,j) \in I_n \times J_m} c_{ij}\delta_i\delta_j \qquad I_n = N^n, \qquad J_m = N^m$$

N is the set of natural numbers and Indices i and j run over the set of all multiindeces.

The mixed series becomes a polynomial if only finitely many coefficients c_{ij} are nonzero. Returning to the nonlinear expression (9),we can write it in terms of δ polynomials as

$$A(\delta)y = B(\delta)u + M(\delta, \delta)(y, u) + v \qquad (17)$$

$A(\delta)$ is a δ polynomial containing all single y terms, $B(\delta)$ is a δ polynomial containing all single u terms. Finally $M(\delta,\delta)$ is a mixed polynomial involving both input output terms. The disturbance signal v is obtained by filtering white noise through a linear rational filter

$$Dv = C\eta$$

Both D and C are linear polynomials in the shift

$$D(\delta) = 1 + \sum_i d_i \delta_i, \qquad C(\delta) = 1 + \sum_i c_i \delta_i$$

A wide sense stationary process having power spectral density can be obtained as the output of a LTI stable system driven by white noise. Since the transfer function of this system can be approximated by a rational function, modeling the disturbance by (16) is fairly general.

To determine a predictor filter for the above system we make use of the star product operation between series, [6]. The star product is closely related to the Volterra series expansion of the tandem connection of two Volterra series. Let us consider a single series G and a mixed series S

$$G = \sum_{\theta=0}^{\infty} \sum_{k \in K_\theta} g_k \delta_k \qquad \underline{S} = \sum_{m=0}^{\infty} \sum_{n=0}^{\infty} \sum_{(i,j) \in I_n \times J_m} c_{ij} \delta_i \delta_j$$

Their star product is given by the expression

$$G * \underline{S} = \sum_{\theta=0}^{\infty} \sum_{k \in K_\theta} \sum_{(i_1,i_2,\ldots,i_\theta) \in (\cup I_n \times \cup J_m)^\theta} g_\theta c_{i_1} c_{i_2} \cdots c_{i_\theta} \delta_{i_{11} \oplus k_1} \delta_{i_{12} \oplus k_1} \delta_{i_{21} \oplus k_2}$$

$$\delta_{i_{22} \oplus k_2} \cdots \delta_{i_{\theta 1} \oplus k_\theta} \delta_{i_{\theta 2} \oplus k_\theta}$$

The star product has the following properties

(i) $[A + B] * \underline{C} = A * \underline{C} + B * \underline{C}$

(ii) $A * (B * \underline{C}) = (A * B) * \underline{C}$

(iii) If A linear then $A * [\underline{B} + \underline{C}] = A * \underline{B} + A * \underline{C}$.

With the aid of the star product we next show that the nonlinear filter

$$C\hat{y}(t) = [C - D * A]y(t) + D * Bu(t) + D * M(y(t), u(t))$$

will make the prediction error white. Indeed, we substitute eq.(15) into the latter equation to obtain

$$C\hat{y} = Cy - D * [Bu + M(y,u) + v] + D * Bu + D * M(y,u)$$

or

$$C\hat{y} = Cy - C\eta$$

Since C is linear we infer that the prediction error is white. Comparison with eq.(11) shows that the predictor has the form

$$D(\delta, \theta)\hat{y}(t) = N_y(\delta, \theta)y(t) + N_u(\delta, \theta)u(t) + N_{yu}(\delta, \delta, \theta)(y(t), u(t)) \tag{18}$$

where $D(\delta, \theta)$, $N_y(\delta, \theta)$ and $N_u(\delta, \theta)$ are δ polynomials and $N_{yu}(\delta, \delta, \theta)$ is a mixed δ polynomial. In particular $D(q, \theta)$ is a linear polynomial in δ.

To determine the dynamics of the predictor gradient we differentiate the above equation with respect to θ (for simplicity we denote by *dot* the derivative with respect to θ).

$$\dot{D}(\delta,\theta)\hat{y}(t) + D(\delta,\theta)\dot{\hat{y}}(t) = \dot{N}_y(\delta,\theta)y(t) + \dot{N}_u(\delta,\theta)u(t) + \dot{N}_{yu}(\delta,\delta,\theta)(y(t),u(t))$$

Since D is linear, it commutes with respect to star product with all δ polynomials. Thus we take the star product of D with the above expression and using (16) we have

$$D(\delta,\theta)^2\psi(t) = (D(\delta,\theta)\dot{N}_y(\delta,\theta) - \dot{D}(\delta,\theta)N_y(\delta,\theta))y(t)$$

$$+(D(\delta,\theta)\dot{N}_u(\delta,\theta) - \dot{D}(\delta,\theta)N_u(\delta,\theta))u(t)$$

$$+(D(\delta,\theta)\dot{N}_{yu}(\delta,\delta,\theta) - \dot{D}(\delta,\theta)N_{yu}(\delta,\delta,\theta))(y(t),u(t)) \qquad (19)$$

The predictor dynamics are governed by the linear polynomial D^2. Let us assume that the roots of the polynomial D are stable. Then the predictor dynamics can be written as the tandem connection of a nonlinear system and a linear system as follows

$$v(t) = N_y(\delta,\theta)y(t) + N_u(\delta,\theta)u(t) + N_{yu}(\delta,\delta,\theta)(y(t),u(t))$$

$$D(\delta,\theta)\hat{y}(t) = v(t)$$

The nonlinear system is BIBO stable. Indeed, if the data sequences are bounded, their shifts remain bounded and finite sums and products are bounded as well. Thus $v(t)$ is bounded. The linear system is BIBO stable because the polynomial D is stable by assumption. We thus conclude that the predictor is BIBO stable. In much the same way, the predictor gradient is BIBO stable as well.

Let us take the unknown parameter to be the vector of the unknown coefficients

$$\theta = (A \quad B \quad M \quad C \quad D) \qquad (20)$$

Proceeding as in section 3 we can determine estimates of the unknown parameters through the computation of the family of the linear system of equations (12a)-(12c). If θ is selected as in (18), it can be shown with arguments similar to those discussed in the context of eq. (13),(14), that the predictor gradient has the block shift invariance property.

5 Fast algorithms

We demonstrated in the previous sections that several important identification cases, once formulated as ℓ_2 optimization problems, lead to linear systems of equations whose parameters possess a block shift invariance structure. The block shift invariance property gives rise to a nesting structure which in turn paves the way for the development of fast algorithms. An outline of these algorithms is briefly summarized next. A detailed description is provided in [1],[2].

Several algorithmic variants are available in batch and sequential form. Batch algorithms include Levinson type and Schur type families. Both have multichannel versions involving block submodules such as matrix by matrix multiplications and linear system solvers, and scalar algorithms, involving ordinary real operations. A common feature of all these schemes is that they utilize three basic functions, the primary module, the secondary module and the management scheme. The primary module is the basic Levinson or Schur type building block. It would suffice if the blocks of the regressor vector had the same size. The secondary module is activated in the general case of unequal blocklengths. It provides all necessary recursions to effect the interface between consecutive runs of the primary module. The management scheme controls the flow of operations.

The Schur type algorithms are derived from lattice realizations, once they are excited by appropriate correlation and crosscorrelation sequences. These algorithms are highly parallelizable. Locally recursive forms can be obtained by the canonical mapping methodology. The various reflection coefficients are broadcast as transmittent variables. The above algorithms can be used as adaptive algorithms once they are executed at each instant.

Alternate adaptive schemes result from fast realizations of the recursive least squares algorithm supplemented with a suitable mechanism to account for the numerical instabilities. Adaptive lattice algorithms are also applicable.

References

[1] G. Glentis and N. Kalouptsidis. Efficient Order Recursive Algorithms for Multichannel LS Filtering. *IEEE Trans. on Signal Processing.* July 1992.

[2] G. Glentis and N. Kalouptsidis. Fast adaptive algorithms for multichannel filtering and system identification. *IEEE Trans. on Signal Processing,* July 1992.

[3] G. Glentis and N. Kalouptsidis. Efficient algorithms for the solution of block linear systems with Toeplitz entries. Submitted to *Linear Algebra.*

[4] G. Glentis and N. Kalouptsidis. Efficient adaptive LS algorithms for multichannel Two-Dimentional Filtering and System Identification. Presented in ICASSP-91, Torondo, Canada.

[5] N. Kalouptsidis and S. Theodoridis. *Efficient Signal Processing and System Identification Algorithms.* Prentice Hall, 1992.

[6] S. Kotsios and N. Kalouptsidis. The model matching problem for a certain class of nonlinear systems. To appear in *International Journal of Control.*

[7] L. Ljung. *System Identification- Theory for the user.* Prentice Hall, 1987.

[8] J. V. Mathews. Adaptive polynomial filter *IEEE Signal Processing magazine,* July 91.

[9] K.S. Narendra and K. Parthasarathy. Identification and Control of Dynamical Systems Using Neural Networks, *IEEE Trans. on Neural Networks,* 1(1), pp 4-28, 1990.

[10] T. Söderström and P. Stoica. *System Identification.* Prentice Hall, 1989.

COMPUTING THE SINGULAR VALUE DECOMPOSITION ON A FAT-TREE ARCHITECTURE

T.J. LEE
School of Electrical Engineering
Cornell University
Ithaca, New York 14853, USA
tjlee@ee.cornell.edu

F.T. LUK
Department of Computer Science
Rensselaer Polytechnic Institute
Troy, New York 12180, USA
luk@cs.rpi.edu

D.L. BOLEY
Department of Computer Science
University of Minnesota
Minneapolis, Minnesota 55455, USA
boley@cs.umn.edu

ABSTRACT. The Singular Value Decomposition (SVD) is a matrix tool that plays a critical role in many applications; for example, in signal processing, it is often necessary to calculate the SVD in real time. We present here a new technique for computing the SVD on a parallel architecture whose processors are connected via a fat-tree. We tested our idea on the Connection Machine CM-5, and achieved efficiency up to 40% even for moderately sized matrices.

KEYWORDS. Singular value decomposition, parallel Jacobi algorithm, fat-tree, CM-5.

M. S. Moonen et al. (eds.), Linear Algebra for Large Scale and Real-Time Applications, 231–240.
© 1993 Kluwer Academic Publishers.

1 Introduction

Let A be a real $m \times n$ matrix. Its singular value decomposition (SVD) is given by

$$A = U\Sigma V^T,$$

where U and V are respectively $m \times m$ and $n \times n$ orthogonal matrices and Σ is an $m \times n$ diagonal matrix. The best approach to parallel SVD computation is apparently one of the Jacobi type; see, e.g., [1], [2], [4], [5], [7], [11], [12]. In this paper, we will discuss the efficient implementation of a Jacobi method on a parallel computer with a fat-tree interconnection network. We will propose a new Jacobi ordering for a fat-tree and analyze its behavior both theoretically and experimentally (on a Connection Machine CM-5).

This paper is organized as follows. In the next two subsections, we present the fat-tree architecture and Jacobi algorithm. Section 2 introduces a new fat-tree ordering, and provides some kernel programs. We analyze communication costs on a fat-tree network in Section 3, and discuss implementation results on the CM-5 in Section 4.

1.1 FAT-TREE ARCHITECTURE

The fat-tree was introduced by Leiserson [10] as a novel approach to interconnect the processors of a general-purpose parallel supercomputer. This communication structure can also be seen in the distributed computing environment, such as a network of workstations.

The routing network of the Connection Machine CM-5 [14] is based on the fat-tree. This parallel machine consists of up to $544 \ (= 512 + 32)$ nodes for the model at the Army High Performance Computing Research Center (AHPCRC) at the University of Minnesota, and 32 nodes at the Northeast Parallel Architectures Center (NPAC) at Syracuse University. Each node of the CM-5 is a SPARC chip which runs at 32 MHz and delivers 22 Mips and 5 Mflops. There is a 64 Kbyte instruction and data cache and a 16 Mbyte memory in each node. All the nodes are synchronized. In October of 1992, two vector units will be installed in each processing node; each vector unit is capable of 64 Mflops peak and 40 Mflops sustained [9]. The control and data networks are connected via a *skinny* fat-tree structure. By *skinny*, we mean that the bandwidth does not increase proportionately to the number of nodes; in particular, the bandwidth is 20Mbyte/sec per node in a group of four processors, 10 Mbyte/sec per node in a group of sixteen, and 5Mbyte/sec overall. So data contention may severely degrade performance when all nodes need to access a large set of data from other nodes through the top level of the tree.

1.2 JACOBI ALGORITHM

The one-sided Jacobi method [8] generates an orthogonal matrix V such that the columns of the matrix W, given by $W = AV$, are mutually orthogonal. The matrix V can be generated by a sequence of plane rotations $V^{(1)}, V^{(2)}, \ldots$, where each $V^{(k)}$ is an identity matrix except for four entries: $v_{ii}^{(k)} = \cos\theta$, $v_{ij}^{(k)} = -\sin\theta$, $v_{ji}^{(k)} = \sin\theta$ and $v_{jj}^{(k)} = \cos\theta$, where (i,j) represents the index pair of the columns of A that $V^{(k)}$ orthogonalizes. The SVD computation requires $O(mn^2)$ operations for an $m \times n$ matrix A. For a limited number of processors, i.e., up to $n/2$ processors, an efficient way is to configure them as a linear array along the horizontal dimension. Columns can be distributed either in blocks or in a wraparound fashion. Note from the above derivation that each column-pair can be orthogonalized independently, so that we may transform up to p pairs concurrently, where p denotes the number of processors. This method was used for computing the SVD on special machines, e.g., parallel computers such as the Illiac IV [11] and vector processors such as the CYBER 205 [3]. The one-sided Jacobi method is composed of these major steps:

1. Compute the norm of each column.

2. Compute plane rotations to orthogonalize paired columns.

3. Apply the plane rotations to update the columns and the column norms.

4. Permute the columns in a pre-chosen order to generate the next column pairs, and repeat the process from step 2 .

If the column pairs are distributed to different processors, then step 4 requires communication. In the case of a two-dimensional mesh (as in the ILLIAC IV), each column is itself distributed among different processors and step 3 requires that the rotation parameters be transmitted to all the processors containing each given column pair. In the case of a one-dimensional array, each column pair is stored entirely in one processor and significant speedup is possible if vector units are present within each processor.

In this paper, we use the one-dimensional array, with each processor storing two blocks of columns. That is, we use a *block* Jacobi algorithm, in which the column blocks are circulated according to a given ordering to be defined, and the *cyclic-by-rows* ordering [6] is used within each block.

2 New SVD algorithm

In the past, when the hypercube interconnection topology was in vogue, several Jacobi ordering schemes were proposed [1], [4], [7] to utilize the hypercube structure. Here, for a one-dimensional array of processors with no wraparound, a chess-tournament ordering [2] may be chosen because it does not waste processing power or memory space. However, communication requires a two-way transmission of columns between adjacent processors. An alternative is a ring ordering [4] which uses only one-way transmission, but it requires a wraparound connection. To develop an ideal ordering for a fat-tree, we aim to minimize the total path length by using the extra bandwidth of a fat-tree.

2.1 FAT-TREE ORDERING

It is easiest to describe this ordering by an example. In Figure 1 we show the case for sixteen columns and eight processors. For pedagogic reasons, we use a base 8 numbering of the indices and so A=8, B=9, ..., H=15. The XOR (exclusive-or) column is the binary XOR of the column indices: at each step, the XOR value of each index pair is the same, and from one step to the next this quantity follows the Gray code. The *cost-to-this-step* column denotes the maximum number of levels up the tree the messages must travel to reach their destinations from the previous step. In general, if there are p processors and two columns per processor, then a sweep requires $2p - 2$ steps. We save one step per sweep because the last step of sweep i can be included as the first step for sweep $i + 1$.

Step	Ordering of Index Pairs	XOR	Cost to This Step
0.	(01)(23)(45)(67)(AB)(CD)(EF)(GH)	0001	NA
	Forward Sweep		
1.	(03)(12)(47)(56)(AD)(BC)(EH)(FG)	0011	1
2.	(02)(13)(46)(57)(AC)(BD)(EG)(FH)	0010	1
3.	(06)(17)(24)(35)(AG)(BH)(CE)(DF)	0110	2
4.	(07)(16)(25)(34)(AH)(BG)(CF)(DE)	0111	1
5.	(05)(14)(27)(36)(AF)(BE)(CH)(DG)	0101	2
6.	(04)(15)(26)(37)(AE)(BF)(CG)(DH)	0100	1
7.	(0E)(1F)(2G)(3H)(4A)(5B)(6C)(7D)	1100	3
8.	(0F)(1E)(2H)(3G)(4B)(5A)(6D)(7C)	1101	1
9.	(0H)(1G)(2F)(3E)(4D)(5C)(6B)(7A)	1111	2
10.	(0G)(1H)(2E)(3F)(4C)(5D)(6A)(7B)	1110	1
11.	(0C)(1D)(2A)(3B)(4G)(5H)(6E)(7F)	1010	3
12.	(0D)(1C)(2B)(3A)(4H)(5G)(6F)(7E)	1011	1
13.	(0B)(1A)(2D)(3C)(4F)(5E)(6H)(7G)	1001	2
14.	(0A)(1B)(2C)(3D)(4E)(5F)(6G)(7H)	1000	1
	Backward Sweep		
13.	(0B)(1A)(2D)(3C)(4F)(5E)(6H)(7G)	1001	1
12.	(0D)(1C)(2B)(3A)(4H)(5G)(6F)(7E)	1011	2
11.	(0C)(1D)(2A)(3B)(4G)(5H)(6E)(7F)	1010	1
10.	(0G)(1H)(2E)(3F)(4C)(5D)(6A)(7B)	1110	3
9.	(0H)(1G)(2F)(3E)(4D)(5C)(6B)(7A)	1111	1
8.	(0F)(1E)(2H)(3G)(4B)(5A)(6D)(7C)	1101	2
7.	(0E)(1F)(2G)(3H)(4A)(5B)(6C)(7D)	1100	1
6.	(04)(15)(26)(37)(AE)(BF)(CG)(DH)	0100	3
5.	(05)(14)(27)(36)(AF)(BE)(CH)(DG)	0101	1
4.	(07)(16)(25)(34)(AH)(BG)(CF)(DE)	0111	2
3.	(06)(17)(24)(35)(AG)(BH)(CE)(DF)	0110	1
2.	(02)(13)(46)(57)(AC)(BD)(EG)(FH)	0010	2
1.	(03)(12)(47)(56)(AD)(BC)(EH)(FG)	0011	1
0.	(01)(23)(45)(67)(AB)(CD)(EF)(GH)	0001	1
	Forward Sweep		
1.	(03)(12)(47)(56)(AD)(BC)(EH)(FG)	0011	1
	...		

Figure 1: *Fat-tree ordering based on the Gray code*
(eight processors and sixteen columns).

2.2 KERNEL PROGRAMS

To see how to write a simple node program to generate the fat-tree ordering, we use the following observations from the example in Figure 1. To simplify the presentation, we consider only the forward sweep. At each step, each processor must communicate with a remote processor whose label differs in one bit. The basis for our kernel presented here is to compute a mask such that the *exclusive-or* of the mask with the current processor label yields the remote processor label. When using the Gray code, this mask can be computed using only the step number – it is independent of the processor label.

We also use the following observations. First, we use the fact that the XOR's follow the Gray code. Second, we observe that during the second half of the forward sweep (steps 7-14), the lower half of the columns (numbers $0, \ldots, 7$ in Figure 1) remain fixed in the processor with the same number. Hence the location of the remaining columns is fixed entirely by the Gray code. Third, we observe that the first half of the steps (steps 0-6) amount to doing a Gray code fat-tree ordering on each half of the processor array separately. The only remaining step is the transition from the first half to the second half (step 6 to step 7). Hence we can define the ordering for these steps recursively from the smaller cases.

We can summarize the steps for the forward sweep in the following procedure, in a pseudo-MATLAB notation assuming for the sake of simplicity of the presentation that the sends do not block.

```
% Node program for processor ProcNo for one forward sweep using an array of
% NProcs processors.  Assume Column(1) and Column(2) are the head and tail
% columns, respectively, in the local memory.

Orthogonalize_Individual_Column_Blocks % (within each block);

for StepNo = 1:2*NProcs-2,

    Pairwise_Orthogonalize_Column_Blocks;

    %% for each processor, figure where the data goes to and send it.

    [Mask,ColumnSwitch] = MakeMask(StepNo,ProcNo,NProcs);
    RemoteProcNo = XOR(ProcNo,Mask);

    Send Column(2) to remote processor RemoteProcNo;
    if ColumnSwitch == rotate,
       Column(2) = Column(1);
       Column(1) = receive_from(RemoteProcNo);
    else
       Column(2) = receive_from(RemoteProcNo);
    end;

end;
```

```
function [Mask,ColumnSwitch]=MakeMask(StepNo,ProcNo,NoProcs);

% ColumnSwitch indicates which column of the pair is to be sent/received.

% Mask is the XOR Mask so that RemoteProcNo = XOR(ProcNo,Mask).
% The Mask is computed independent of the processor label ProcNo.

% Handle first 2 steps as special cases to start recursion
if StepNo <= 2,
   Mask=1;
   ColumnSwitch = tail;
   if rem(ProcNo,2) == 1 & StepNo == 1, ColumnSwitch = rotate; end;

% First half of sweep: pretend this is a separate fat tree sweep on each
% half of the processor array.
else if StepNo < NoProcs-1,
   [Mask,ColumnSwitch] = MakeMask(StepNo,rem(ProcNo,NoProcs/2),NoProcs/2);

% Middle of sweep: here is first exchange through top of tree.
else if StepNo == NoProcs-1,
   Mask = NoProcs/2;
   ColumnSwitch = tail;
   if ProcNo >= NoProcs/2, ColumnSwitch = rotate; end;

% Last half of sweep: only tail columns move, figure Mask using Gray codes.
else if StepNo > NoProcs-1,
   Mask = xor(gray(StepNo),gray(StepNo+1));
   ColumnSwitch = tail;

end;
```

2.3 TEST OF CONVERGENCE

For a fat-tree ordering, any consecutive $2p-2$ (or even $2p-1$) steps may not constitute one sweep. We must complete a sweep, either forward or backward, to ensure that all column pairs have been orthogonalized. The convergence test is simple. We maintain a one-bit counter in every processor. The counter is reset at the beginning of every sweep, and is set whenever a column pair needs to be orthogonalized. At the end of the sweep, a global or operation is performed and convergence is achieved if no bit has been set.

3 Analysis on a binary fat-tree network

We consider a binary fat-tree with p processors, and assume that the communication time from one processor to another is determined by the number of links a message has to traverse and the capacity of these links. Our assumption is supported by experimental results reported in [13]. Define a channel to be the communication link between any two adjacent nodes; here a node can be a processor or an internal switching element. The capacity of a channel equals the number of parallel wires in the channel, and thus the maximum number of simultaneous bit-serial messages it can support [10]. Denote the capacity of the channels at the bottom level by γ. Label the levels from bottom up as level $1, 2, \ldots$, so that the capacity of the channels at level l is given by $2^{l-1}\gamma$. Let us ignore start-up and latency costs. Within a single problem, all the messages have the same size and thus we measure the cost of multiple message transmission using *path length*.

For the ring ordering, at each step a message always goes through the top level and the maximum path length equals $2\log p$ (unless otherwise stated, we use base 2 logarithms). Since there is at most one message at each channel, congestion never occurs and it takes $2p - 1$ steps to complete one sweep. The total path length equals $(4p - 2)\log p$.

The fat-tree ordering does not cause congestion on a fat-tree network. Hence it suffices to count the number of times that each level is used. Denote that count by $c(p, l)$. Consider the forward sweep. We see from Figure 1 that with $p = 8$ processors, the top level is used in two transition steps, the middle level in six steps and the bottom in fourteen steps. The first six steps correspond to the fat-tree ordering for the first four processors, and also for the second four processors. In the general case of p processors, there are $2p - 2$ steps using $\log p$ levels, of which the first $p - 2$ steps amount to the ordering for $p/2$ processors. When the number of processors doubles to $2p$, we add a new top level and the first $2p - 2$ steps correspond exactly to the p processor ordering. There are an extra $2p$ steps, of which two use the new top level, four use the next level (the old top level), eight use the following level, etc. Formally, we get the recurrence

$$c(2p, l) = c(p, l) + 4(p/2^l) \qquad \text{for } l = 1, \ldots, \log p,$$

starting with $c(p, \log p) = 2$ and $c(p, l) = 0$ for $l > \log p$. Therefore, $c(p, l) = 4p/2^l - 2$, and the total path length is given by

$$2\sum_{l=1}^{\log p} c(p, l) = 2[(2p - 2) + (p - 2) + \ldots + 14 + 6 + 2] = 8p - 4\log p - 8.$$

For a large p, the path length ratio of the two orderings grows like $\log p/2$, a very attractive result for our new ordering.

4 Connection Machine CM-5

Although the CM-5 network is a 4-way tree, the analysis on 2-way trees is applicable. We take a 4-way tree and expand every interior node into a binary tree consisting of that node with two new children each connected to two of the four former children. The number of levels as well as the path length are doubled. However, the CM-5 is *skinny* and the capacity only doubles at every level. Hence it becomes a *skinny* 2-way tree in which the capacity goes up by $\sqrt{2}$ at each level.

To simplify our analysis, we concentrate on the 32-processor model. So $p = 32$ and there are three tree levels because $\lceil \log_4 p \rceil = 3$. The dominating communication cost for the CM-5 is the overhead time that is spent on address calculation, buffer space management, and so on. Let t_{or} and t_{of} represent the cost of such overhead in each step for the ring and the fat-tree ordering, respectively. Let t_{cf} be the overhead cost for resolving contention in the channels of the CM-5 network when applying the fat-tree ordering, and let t_e be the time for traversing an edge in the network. We note that $t_e < t_{cf} < t_{oh}$, where $t_{oh} \in \{t_{or}, t_{of}\}$, $t_{cf} \approx t_{oh}$, and $t_e \in (t_{oh}/10^3, t_{oh}/10^2)$. The overheads t_{or} and t_{of} depend on the data size and are of equal magnitude.

We proceed to compute the coefficient for t_e, which we assume to equal the number of messages that traverse the channels in one sweep. For the ring ordering, there is no congestion in the networks. So the coefficient for t_e is $2 \cdot 63 \cdot 3$ (=378), and the total cost equals $63\, t_{or} + 378\, t_e$. For the fat-tree ordering, we observe that level 1 is visited 62 times, level 2 fourteen times, and level 3 two times. We model the resolution of the contention by sending messages in batches. Messages through level 2 must be sent in two batches and messages through level 3 in four batches, in order to avoid contention. Hence we account for the thinness of the CM-5 network by assigning a weight of two to level 2 and a weight of four to level 3. The total path length is $2(62 + 2 \cdot 14 + 4 \cdot 2) = 196$ and the total cost equals $62\, t_{of} + 196\, t_e + t_{cf}$. Thus, on the CM-5 the fat-tree ordering may not outperform the ring ordering because of the extra cost associated with message contention.

4.1 EXPERIMENTAL RESULTS

In Table 1 we present implementation results on a 32-node CM-5 for random $n \times n$ matrices with n ranging from 64 to 1024. The program was written in Fortran and each experiment repeated ten times. We measured the overall and computation (by disabling communication) costs for one sweep, and estimated the communication cost by subtracting the latter from the former. Our results show that, despite the message congestion that it causes on the CM-5, the fat-tree ordering gets more competitive as n grows, justifying our effort to minimize the total message path length (see also [13]). The mflops (million floating-point operations per second) figures in Table 2 are computed based on the count that $8n^3$ flops are required for one sweep. We conjecture that the *compute* performance deteriorates when n gets beyond 512 because the cache is no longer large enough to hold the huge column blocks. Nonetheless, our implementation results shows how, as the message size increases (hence t_e increases [13]), the fat-tree ordering quickly becomes competitive.

	n	64	128	256	512	1024
Overall	Ring	$7.595\,\mathrm{e}^{-2}$	$3.229\,\mathrm{e}^{-1}$	2.628	$1.794\,\mathrm{e}^{1}$	$1.380\,\mathrm{e}^{2}$
	Fat-tree	$8.134\,\mathrm{e}^{-2}$	$3.481\,\mathrm{e}^{-1}$	2.237	$1.795\,\mathrm{e}^{1}$	$1.361\,\mathrm{e}^{2}$
Compute	Ring	$3.013\,\mathrm{e}^{-2}$	$2.320\,\mathrm{e}^{-1}$	1.871	$1.493\,\mathrm{e}^{1}$	$1.309\,\mathrm{e}^{2}$
	Fat-tree	$3.436\,\mathrm{e}^{-2}$	$2.420\,\mathrm{e}^{-1}$	1.878	$1.493\,\mathrm{e}^{1}$	$1.310\,\mathrm{e}^{2}$
Communicate	Ring	$4.582\,\mathrm{e}^{-2}$	$0.909\,\mathrm{e}^{-1}$	0.757	3.010	7.110
	Fat-tree	$4.698\,\mathrm{e}^{-2}$	$1.061\,\mathrm{e}^{-1}$	0.359	3.020	5.140

Table 1: *CPU time (seconds) of ring and fat-tree orderings*

	n	64	128	256	512	1024
Overall	Ring	27.61	51.96	51.07	59.85	62.25
	Fat-tree	25.78	48.20	60.00	59.82	63.11
Compute	Ring	69.60	72.32	71.74	71.92	65.62
	Fat-tree	61.03	69.33	71.47	71.92	65.57

Table 2: *Mflops rates of ring and fat-tree orderings*

Acknowledgements

The work of T. J. Lee and F. T. Luk was supported in part by the Joint Services Electronics Program under contract F49620-90-C-0039 at Cornell University; F. T. Luk was also supported by start-up funds at the Rensselaer Polytechnic Institute. The authors thank the AHPCRC and NPAC for time on the CM-5, and Richard Brent and Lennart Johnsson for valuable discussions on CM-5 communication and hardware issues.

References

[1] C. H. Bischof. The two-sided block Jacobi method on a hypercube. In : M. T. Heath (Ed.), *Hypercube Multiprocessors*, SIAM, pp 612–618, 1988.

[2] R. P. Brent and F. T. Luk. The solution of singular-value and symmetric eigenvalue problems on multiprocessor arrays. *SIAM J. Sci. Statist. Comput.*, 6, pp 69–84, 1985.

[3] P. P. M. de Rijk. A one-sided Jacobi algorithm for computing the singular value decomposition on a vector computer. *SIAM J. Sci. Statist. Comput.*, 10, pp 359–371, 1989.

[4] P. J. Eberlein and H. Park. Efficient implementation of Jacobi algorithms and Jacobi sets on distributed memory architectures. *J. Par. Distrib. Comput.*, 8, pp 358–366, 1990.

[5] L. M. Ewerbring and F. T. Luk. Computing the singular value decomposition on the Connection Machine. *IEEE Trans. Computers*, 39, pp 152–155, 1990.

[6] G. E. Forsythe and P. Henrici. The cyclic Jacobi method for computing the principal values of a complex matrix. *Trans. Amer. Math. Soc.*, 94, pp 1–23, 1960.

[7] G. R. Gao and S. J. Thomas. An optimal parallel Jacobi-like solution method for the singular value decomposition. *Internat. Conf. Parallel Proc.*, pp 47–53, 1988.

[8] M. R. Hestenes. Inversion of matrices by biorthogonalization and related results. *J. Soc. Indust. Appl. Math.*, 6, pp 51–90, 1958.

[9] S. L. Johnsson. Private communication, September 1992.

[10] C. E. Leiserson. Fat-trees: Universal networks for hardware-efficient supercomputing. *IEEE Trans. Computers*, c-34, pp 892–901, 1985.

[11] F. T. Luk. Computing the singular-value decomposition on the ILLIAC IV. *ACM Trans. Math. Softw.*, 6, pp 524–539, 1980.

[12] ——, A triangular processor array for computing singular values. *Lin. Alg. Appl.*, 77, pp 259–273, 1986.

[13] R. Ponnusamy, A. Choudhary, and G. Fox. Communication overhead on CM5: an experimental performance evaluation. in *Frontier 92, Fourth Symp. on the Frontiers of Massively Parallel Computation*, IEEE, pp 108–115, 1992.

[14] Thinking Machines Corporation. The Connection Machine CM-5 Technical Summary. October 1991.

A NEW MATRIX DECOMPOSITION FOR SIGNAL PROCESSING

F.T. LUK
Department of Computer Science
Rensselaer Polytechnic Institute
Troy, New York 12180, USA
luk@cs.rpi.edu

S. QIAO
Communications Research Laboratory
McMaster University
Hamilton, Ontario L8S 4K1, Canada
qiao@maccs.dcss.mcmaster.ca

ABSTRACT. We extend the generalized singular value decomposition to a new decomposition that can be updated at a low cost. In addition, we show how a forgetting factor can be incorporated in our decomposition.

KEYWORDS. ULV decomposition, generalized SVD, updating, forgetting factor, signal processing.

1 Problem definition

A recurring matrix problem in signal processing concerns generalized eigenvalues:

$$A^H A x = \lambda B^H B x,$$

where A is $n \times p$, $n \geq p$, B is $m \times p$, and $m \geq p$. We assume further that the matrix B has full column rank. Often, the generalized eigenvalues, call them d_j^2's, satisfy this property:

$$d_1^2 \geq d_2^2 \geq \cdots \geq d_{p-k}^2 >> d_{p-k+1}^2 \approx \cdots \approx d_p^2 . \tag{1}$$

The k-dimensional subspace spanned by the eigenvectors corresponding to the k smallest generalized eigenvalues is called the *noise* subspace. We are interested in the following problem.

Noise Subspace Problem. Compute an orthonormal basis for the noise subspace.

241

M. S. Moonen et al. (eds.), Linear Algebra for Large Scale and Real-Time Applications, 241–247.

2 ULLV decomposition

This problem has a known solution for the *special* case where $B = I_p$, where I_p denotes a $p \times p$ identity matrix. Compute a singular value decomposition (SVD) of A:

$$A = U D_A V^H,$$

where U is $n \times p$ and *orthonormal*, i.e., $U^H U = I_p$, V is $p \times p$ and unitary, D_A is diagonal and $D_A = \text{diag}(d_1, \ldots, d_p)$. From (1) we get

$$d_1 \geq d_2 \geq \cdots \geq d_{p-k} >> d_{p-k+1} \approx \cdots \approx d_p \geq 0, \tag{2}$$

and the desired orthonormal basis is given by the last k columns of V. However, the SVD is not amenable to efficient updating when a new row is added to A. A clever procedure was devised by Stewart [2] in the form of the ULV decomposition (ULVD):

$$A = U L_A V^H,$$

where U is orthonormal and V unitary as in the SVD, but the middle matrix L_A is lower triangular and essentially block diagonal. In particular,

$$L_A = \begin{pmatrix} \check{L}_A & 0 \\ E & K \end{pmatrix}, \tag{3}$$

where

(i) $\check{L}_A$ and K are lower triangular and $\check{L}_A$ is $(p-k) \times (p-k)$;

(ii) $\sigma_{\min}(\check{L}_A) \approx d_{p-k}$ and $\|E\|^2 + \|K\|^2 \approx d_{p-k+1}^2 + \cdots + d_p^2$.

By $\sigma_{\min}(M)$ we mean the smallest singular value of a matrix M, and by $\|M\|$ we refer to the Frobenius norm of the matrix M. Essentially, Stewart showed that to separate out the noise subspace from the signal subspace, it suffices to reduce A to the 2×2 block lower triangular form L_A, where both E and K are very small in norm. The last k columns of V provide an orthonormal basis for the noise subspace.

In this paper we consider the noise subspace problem for the *general* case where $B \neq I_p$. First, the problem may be solved via the generalized SVD (GSVD):

$$A = U_A D_A L V^H \quad \text{and} \quad B = U_B L V^H,$$

where U_A is $n \times p$ and orthonormal, U_B is $m \times p$ and orthonormal, V is $p \times p$ and unitary, L is $p \times p$ and lower triangular, and $D_A = \text{diag}(d_1, \ldots, d_p)$. If the generalized singular values d_j's satisfy (2), then the last k columns of V provide a basis for the noise subspace. We propose here a generalized ULVD (ULLVD):

$$A = U_A L_A L V^H \quad \text{and} \quad B = U_B L V^H, \tag{4}$$

where U_A, U_B, V and L are just as in the GSVD. The new middle matrix L_A has the same form as in (3) and the desired orthonormal basis is given by the last k columns of V.

3 Forgetting factor and updating the ULLVD

We examine how to update the ULLVD efficiently when new rows are added to A or B. As in signal processing, we incorporate a forgetting factor β, where $0 < \beta \leq 1$. Assume that we get the row data vectors $x_1^T, x_2^T, \ldots$, for A, and $y_1^T, y_2^T, \ldots$, for B. Adding a superscript for clarity, we get

$$A^{(n)} = \begin{pmatrix} \beta^{n-1} x_1^T \\ \vdots \\ \beta x_{n-1}^T \\ x_n^T \end{pmatrix} \quad \text{and} \quad B^{(n)} = \begin{pmatrix} \beta^{n-1} y_1^T \\ \vdots \\ \beta y_{n-1}^T \\ y_n^T \end{pmatrix}.$$

Thus,

$$A^{(n+1)} = \begin{pmatrix} \beta A^{(n)} \\ x_{n+1}^T \end{pmatrix} \quad \text{and} \quad B^{(n+1)} = \begin{pmatrix} \beta B^{(n)} \\ y_{n+1}^T \end{pmatrix}.$$

We consider the following problem.

Updating Problem. Given a ULLVD of A and B as in (4), and given $\hat{A}$ and $\hat{B}$ as *simple* updates of A and B, find a ULLVD of $\hat{A}$ and $\hat{B}$.

Case A. Add a Row to A: $A = A^{(n)}$, $\hat{A} = A^{(n+1)}$ and $B = \hat{B} = B^{(n)}$.

Case B. Add a Row to B: $A = \hat{A} = A^{(n)}$, $B = B^{(n)}$ and $\hat{B} = B^{(n+1)}$.

How do we begin? There are many possibilities. We propose to initialize A to a $p \times p$ zero matrix and B to a $p \times p$ matrix $B^{(p)}$; recall that B is required to have full column rank. Let $B = LV^H$ be an LQ decomposition. Then we get $U_A = I_p$, $L_A = 0$, and $U_B = I_p$.

3.1 ADDING A ROW TO A

Step A0. Suppose we have a ULLVD of A and B, and a new row a^T is added to A. That is,

$$\hat{A} = \begin{pmatrix} \beta A \\ a^T \end{pmatrix} \quad \text{and} \quad \hat{B} = B.$$

We write down the initial decompositions as

$$\hat{A} = \begin{pmatrix} U_A & 0 \\ 0^T & 1 \end{pmatrix} \begin{pmatrix} \beta L_A & 0 \\ 0^T & 1 \end{pmatrix} \begin{pmatrix} L \\ a^T V \end{pmatrix} V^H \quad \text{and} \quad \hat{B} = U_B (I_p \quad 0) \begin{pmatrix} L \\ a^T V \end{pmatrix} V^H. \quad (5)$$

Our goal is to transform the matrix factors of (5) into the proper forms of (4).

Step A1. We eliminate all but the first element of the row $a^T V$ using rotations from the right. For $i = p, p-1, \ldots, 2$, we annihilate the i-th element by postmultiplying with a rotation in the $(i-1, i)$ plane. This rotation will be incorporated into V. However, it introduces a nonzero $(i-1, i)$ entry in L, which is eliminated by applying from the left a rotation to rows $i-1$ and i of L. This rotation must be propagated to the left in both decompositions in (5). In the decomposition of B, this rotation can be directly incorporated into U_B because only the I_p part of the second factor is affected. In the decomposition of A, this rotation introduces a nonzero $(i-1, i)$ entry in the second factor, which is annihilated by applying a rotation from the left to its $(i-1)$-st and i-th rows. The rotation can be incorporated into U_A and consequently we get

$$\hat{A} = \begin{pmatrix} \tilde{U}_A & 0 \\ 0^T & 1 \end{pmatrix} \begin{pmatrix} \bar{L}_A & 0 \\ 0^T & 1 \end{pmatrix} \begin{pmatrix} \bar{L} \\ \xi e_1^T \end{pmatrix} \bar{V}^H \quad \text{and} \quad \hat{B} = \bar{U}_B (I_p \;\; 0) \begin{pmatrix} \bar{L} \\ \xi e_1^T \end{pmatrix} \bar{V}^H.$$

Step A2. We zero out the vector ξe_1^T using a transformation from the left. We apply a scaled rotation, say Y, of the form

$$Y = \begin{pmatrix} c^2 & & & & -c\bar{s} \\ & 1 & & & \\ & & 1 & & \\ & & & 1 & \\ cs & & & & c^2 \end{pmatrix},$$

from the left and rotate against the $(1,1)$ element of $\bar{L}$. Note that

$$Y^{-1} = \begin{pmatrix} 1 & & & & \bar{s}/c \\ & 1 & & & \\ & & 1 & & \\ & & & 1 & \\ -s/c & & & & 1 \end{pmatrix}.$$

When the transformation Y^{-1} is propagated to the left, it creates many nonzero elements in the last column of the second factor in both decompositions. The following figure shows the resultant nonzero structures:

$$\begin{pmatrix} \bar{L}_A & 0 \\ 0^T & 1 \end{pmatrix} Y^{-1} = \begin{pmatrix} \times & & & & \times \\ \times & \times & & & \times \\ \times & \times & \times & & \times \\ \times & \times & \times & \times & \times \\ \times & & & & \times \end{pmatrix} \quad \text{and} \quad (I_p \;\; 0) Y^{-1} = \begin{pmatrix} 1 & & & & \times \\ & 1 & & & \\ & & 1 & & \\ & & & 1 & \end{pmatrix}.$$

Fortunately, the last column of these two matrices can be dropped because the third factor now has a zero row at the bottom. Hence

$$\hat{A} = \begin{pmatrix} \tilde{U}_A & 0 \\ 0^T & 1 \end{pmatrix} \begin{pmatrix} \tilde{L}_A \\ \eta e_1^T \end{pmatrix} \hat{L} \hat{V}^H \quad \text{and} \quad B = \hat{U}_B \hat{L} \hat{V}^H.$$

Step A3. We zero out the vector ηe_1^T using a rotation from the left in the $(1, p+1)$ plane. Although the rotation creates many nonzeros in the last column of the first factor, this column can be dropped because the second factor now has a zero bottom row. Finally, we obtain the updated decompositions :

$$\hat{A} = \hat{U}_A \hat{L}_A \hat{L} \hat{V}^H \quad \text{and} \quad \hat{B} = \hat{U}_B \hat{L} \hat{V}^H.$$

Note that in Steps A1 to A3 we have taken special care to arrange the annihilation sequence so that $\hat{L}_A$ may satisfy (3) as much as possible.

Step A4. The rank of $\hat{L}_A$ matrix may increase by one over that of L_A. A deflation followed by possibly an iterative refinement (Stewart [2]) is applied to $\hat{L}_A$ to determine its rank.

3.2 ADDING A ROW TO B

Step B0. When a new row b^T is added to B, i.e.,

$$\hat{A} = A \quad \text{and} \quad \hat{B} = \begin{pmatrix} \beta B \\ b^T \end{pmatrix},$$

we write

$$\hat{A} = U_A(\beta^{-1}L_A \quad 0)\begin{pmatrix} \beta L \\ b^T V \end{pmatrix} V^H \quad \text{and} \quad \hat{B} = \begin{pmatrix} U_B & 0 \\ 0^T & 1 \end{pmatrix}\begin{pmatrix} \beta L \\ b^T V \end{pmatrix} V^H. \tag{6}$$

Again, our goal is to transform the matrix factors of (6) into the proper forms of (4). Unlike (5), here we need to associate the forgetting factor β with L.

Step B1. We eliminate all but the first element of the row $b^T V$ using rotations from the right. Proceeding as in Step A1, we get

$$\hat{A} = \hat{U}_A(\bar{L}_A \quad 0)\begin{pmatrix} \bar{L} \\ \xi e_1^T \end{pmatrix}\hat{V}^H \quad \text{and} \quad \hat{B} = \begin{pmatrix} \bar{U}_B & 0 \\ 0^T & 1 \end{pmatrix}\begin{pmatrix} \bar{L} \\ \xi e_1^T \end{pmatrix}\hat{V}^H.$$

Step B2. We zero out the vector ξe_1^T using a rotation from the left in the $(1, p+1)$ plane. When the rotation is propagated to the left, it creates a nonzero vector right after $\bar{L}_A$ (and also $\bar{U}_B$). Fortunately, this new vector can be dropped since vector ξe_1^T has been annihilated. Consequently,

$$\hat{A} = \hat{U}_A\hat{L}_A\hat{L}\hat{V}^H \quad \text{and} \quad \hat{B} = \hat{U}_B\hat{L}\hat{V}^H.$$

As in Case A, we have arranged the sequence of transformations so that $\hat{L}_A$ will satisfy (3) as much as possible.

Step B3. The rank of $\hat{L}_A$ matrix may increase by one over that of L_A. A deflation followed by possibly iterative refinement (Stewart [2]) is applied to $\hat{L}_A$ to determine its rank.

3.3 COMPUTATIONAL COST

Since we are interested in an orthonormal basis for the noise subspace, the matrices U_A and U_B are not needed. That is, we update only the three $p \times p$ matrices L_A, L and V, giving us a very efficient method. First, consider Case A. The initialization step costs $1.5p^2$ multiplies for forming βL_A and $a^T V$. Step A1 requires approximately $12p^2$ multiplies. Step A2 costs $O(p)$ multiplies, and Step A3 $O(1)$ multiplies. The cost of Step A4 depends on the rank of $\hat{L}_A$ (see [2]), which is usually less than p. Second, for Case B, the initialization step costs $2p^2$

multiplies because we need to form $\beta^{-1}L_A$, βL and $b^T V$. Step B1 requires approximately $12p^2$ multiplies, and Step B2 $O(p)$ multiplies. The cost of Step B3 is dependent on the rank of $\hat{L}_A$.

4 Example

Our new ULLVD algorithm was programmed in MATLAB and run on a SUN3/80 at Mc-Master University; the machine precision equals 2×10^{-16}. The results show that this adaptive algorithm computes the noise subspace just as accurately as a stable GSVD algorithm implemented along the lines described in Van Loan [3] and Luk and Qiao [1]. In our example, B is a random 9×5 matrix. We let

$$A^{(140)} = \begin{pmatrix} A_0 \\ A_1 \\ A_2 \\ A_3 \end{pmatrix},$$

where A_0 is a 5×5 null matrix, A_1 a 20×5 random matrix of rank 3, A_2 a 60×5 random matrix of rank 2, and A_3 a 60×5 random matrix of rank 3. Hence $A^{(140)}$ is a 145×5 matrix whose top five rows are zeros. Define $A^{(n)}$ by

$$A^{(n)} = \begin{pmatrix} A_0 \\ \beta^{n-1}x_1^T \\ \vdots \\ \beta x_{n-1}^T \\ x_n^T \end{pmatrix},$$

for $n = 1, 2, \ldots, 140$. The forgetting factor β is set to 0.7. The relative accuracy of the noise subspace computed by the adaptive ULLVD algorithm is determined as follows. Let

$$A^{(n)} = U^{(n)} L_A^{(n)} L^{(n)} V^{(n)H}$$

be a ULLVD of $A^{(n)}$, and

$$A^{(n)} = \bar{U}^{(n)} D_A^{(n)} \bar{L}^{(n)} \bar{V}^{(n)H}$$

be a GSVD of $A^{(n)}$. Also, partition

$$V^{(n)} = \begin{pmatrix} V_S^{(n)} & V_N^{(n)} \end{pmatrix} \quad \text{and} \quad \bar{V}^{(n)} = \begin{pmatrix} \bar{V}_S^{(n)} & \bar{V}_N^{(n)} \end{pmatrix},$$

so that $V_N^{(n)}$ and $\bar{V}_N^{(n)}$ represent the bases for the noise subspaces computed by the two different algorithms. The parameter $\theta^{(n)} = \|(V_N^{(n)})^H \bar{V}_S^{(n)}\|_2$ approximates the angle between the two noise subspaces. Our experimental results are presented in Table 1. First, we initialized $A^{(0)} = A_0$ and $L_A^{(0)} = 0$, and performed an LQ decomposition on the top five rows of B to get $L^{(0)}$ and $V^{(0)}$. Next, we added the first five data rows x_1^T to x_5^T. As a little detour, we increased the row dimension of B from five to nine; the detected rank of $L_A^{(n)}$ stayed at three and $\theta^{(n)}$ varied between 1.4×10^{-15} and 3.7×10^{-15}. Finally, we added the remaining data rows x_6^T to x_{140}^T. We were pleased that our ULLVD algorithm computed a noise subspace very close to the one given by the GSVD. However, a minor loss in accuracy occurred in steps 52 to 58 (i.e., right after the rank of $L_A^{(n)}$ decreased from four to three)

n	1	2	3	4	5	6	15	20	21	22	$\cdots$	41	42
$\mathrm{rank}(L_A^{(n)})$	1	2	3	3	3	3	3	3	4	5	$\cdots$	5	4
$\theta^{(n)}(\times 10^{-15})$	2.1	2.4	4.2	2.4	1.9	3.5	1.8	2.4	1.5	NA	$\cdots$	NA	2.0

n	50	51	52	53	54	55	56	57	58	59
$\mathrm{rank}(L_A^{(n)})$	4	3	3	3	3	3	3	3		2
$\theta^{(n)}(\times 10^{-12})$	1.6×10^{-3}	2.4×10^{-3}	2.6	2.8	2.6	3.4	3.1	3.3	3.1	1.5×10^{-3}

n	70	80	81	82	83	$\cdots$	103	104	108	109	125	140
$\mathrm{rank}(L_A^{(n)})$	2	2	3	4	5	$\cdots$	5	4	4	3	3	3
$\theta^{(n)}(\times 10^{-15})$	1.3	0.8	1.9	1.4	NA	$\cdots$	NA	1.4	2.6	3.0	1.2	1.8

Table 1: *Accuracy of ULLVD algorithm in noise subspace estimation*

because the condition estimator got confused when the smallest singular values were closely clustered.

Acknowledgements

F. T. Luk was supported in part by startup funds at the Rensselaer Polytechnic Institute, and S. Qiao by the Natural Sciences and Engineering Research Council of Canada under grant OGP0046301.

References

[1] F. T. Luk and S. Qiao. Computing the CS-decomposition on systolic arrays. *SIAM J. Sci. Statist. Comput.*, **7**, pp 1121-1125, 1986.

[2] G. W. Stewart. Updating a rank-revealing ULV decomposition. Report 2627, Department of Computer Science, University of Maryland, College Park, Maryland, 1991.

[3] C. F. Van Loan. Computing the CS and generalized singular value decomposition. *Numer. Math.*, **46**, pp 479-492, 1985.

THE LINEAR ALGEBRA OF PERFECT RECONSTRUCTION FILTERING

M. STEWART
Department of Electrical and Computer Engineering
University of Illinois
Urbana, IL 61820, U.S.A.

G. CYBENKO
Thayer School of Engineering
Dartmouth College
Hanover, NH 03755-8000, U.S.A.

ABSTRACT. There has been considerable recent interest in perfect reconstruction filter banks, wavelets and multiresolution analysis. The basic building blocks for these techniques are special families of convolutions and decimation/interpolation operations. When viewed as linear algebraic operations, those building blocks are unitary operators and related projections. We have developed a general approach to perfect reconstruction and multiresolution analysis that uses polynomial functions of unitary operators and closely related projection operations. This general algebraic framework is powerful in that it easily captures much of the known theory as special cases and gives a simple mechanism for constructing arbitrary perfect reconstruction families of operators. Using the general framework, we derive necessary and sufficient conditions for the perfect reconstruction property, even for higher dimensional nonseparable sampling lattices and operators. The necessary conditions, although explicitly stated, are not computable however.

1 Introduction

Perfect reconstruction filtering can be easily understood in terms of unitary operators acting on sequences. A simple example serves to illustrate the basic principles. Let x_j be the value of a discrete-time signal at time j and let U be an $n \times n$ unitary matrix. From the signal, x, and the unitary operator, U, define a new signal as follows: break the signal x up into blocks of n consecutive samples, say $X_k = (x_{kn+1}, x_{kn+2}, ..., x_{(k+1)n})^T$ and define $Y_k = UX_k$. Now we define the transformed signal, y_j, by $(y_{kn+1}, y_{kn+2}, ..., y_{(k+1)n})^T = Y_k$. Because U is unitary, there is a simple way to invert this operation - merely, apply U^* to the signal y_j in the same way we applied U to x_j. So quite trivially, $X_k = U^*Y_k = U^*UX_k$.

M. S. Moonen et al. (eds.), Linear Algebra for Large Scale and Real-Time Applications, 249–274.
© 1993 *Kluwer Academic Publishers.*

This is perhaps the simplest example of a perfect reconstruction filter bank and the fact that the unitary operator acts on blocks and those block are nonoverlapping makes the whole setup trivial. Another way to obtain y_{kn+m}, with $0 < m < n + 1$, is to note that y_{kn+m} is defined by convolving the whole signal x_j with the mth row of U and then taking every nth sample. The action of taking every nth sample is called *decimation by n*. The recovery process is seen to be *interpolation by n*, namely inserting $n - 1$ zeroes between y_{kn+m} and $y_{(k+1)n+m}$, then convolving the mth interpolated signal obtained this way with the reversed and conjugated mth row of U. Addition of these terms gives us the original signal. At the heart of this example are the processes of convolution, decimation, and interpolation.

The key idea here is that a signal can be decomposed into a bank of signals using convolution and decimation operations. Those signals contain different spectral information depending on the choice of unitary transformation, U. Further processing of these signals, using interpolation and filtering reconstructs the original signal exactly. The example is trivial because the unitary operator acts on nonoverlapping blocks. The nontrivial aspects of the technique begin by looking at convolutions involving banks that have more than n coefficients in each filter. Then the nonoverlapping property goes away and we need conditions other than the simple ones above to ensure perfect reconstruction.

Classically, perfect reconstruction filter banks are designed to split a signal into two frequency bands through filtering and decimation so as to be able to exactly reconstruct the original signal from the filtered and decimated versions. If the filter operators are given by $H^{(0)}$ and $H^{(1)}$, attention will be focused on the case in which the recovery process is upsampling (that is insertion of zeroes between samples or the adjoint operation of decimation on $l^2(\mathcal{Z})$), operating on the channels with $(H^{(0)})^*$ and $(H^{(1)})^*$ (filtering with the time reverse or conjugate filter), and summation of the two channels. With the decimation operator given by D, this process is shown in Figure 1. The operator mapping the input to the output is

$$(H^{(0)})^* D^* D (H^{(0)}) + (H^{(1)})^* D^* D (H^{(1)}). \tag{1}$$

For the output to equal the input, which is the perfect reconstruction property, the operator in (1) must equal the identity operator.

The traditional filtering operators are just polynomials in the right shift operator S : $l^2(\mathcal{Z}) \to l^2(\mathcal{Z})$,

$$H^{(0)} = \sum_j h_j^{(0)} S^j,$$

and

$$H^{(1)} = \sum_j h_j^{(1)} S^j.$$

Normally conditions are found on the coefficients $h_j^{(0)}$ and $h_j^{(1)}$ that allow perfect reconstruction. This paper will show that these conditions can be generalized to the case in which S and D are replaced by more general operators U and Π, provided that U and Π satisfy certain relations. Those relations are abstractions of specific properties of S and

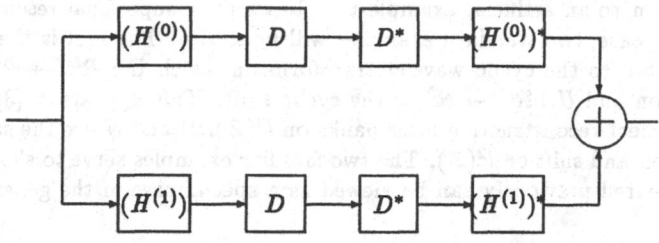

Figure 1: *The Two Channel Filter Bank*

D making it possible to derive a more general class of perfect reconstruction systems. We present conditions on the more general U and Π through a set of operator equations. Later, when actually constructing examples of such operators, the relations will be shown to be met by a class of operators with a similarity structure like shift and decimation on $l^2(\mathcal{Z})$. In fact, the existence of a similarity structure for U which cycles between two particular orthogonal subspaces will be shown to be equivalent to the original operator equations. The two subspaces will turn out to be the range of $\Pi^*\Pi$ and its orthogonal complement. Although interpretations in terms of the orthogonal subspaces and the cyclic structure of U will be noted at relevant points, the development of the conditions on the coefficients will be carried out through use of the equations imposed on the operators. The cyclic structure of U will be introduced as an equivalent formulation only after the perfect reconstruction conditions have been established and specific examples are being given. This is done because the equation based approach seems to be the more concise of the two for deriving the perfect reconstruction conditions, but the similarity structure characterization provides a simple way to construct examples of specific operators.

Several terms which strictly speaking should only be applied to the case in which the banks use operators S and D will be applied by analogy to the general case for U and Π. $H^{(0)}$ and $H^{(1)}$ will be called "filters" even when they are polynomials in U or Π instead of S and D. Similarly, the structure shown in Figure 1 will be referred to as a filter bank even when S and D are replaced by U and Π.

In extending the result to the multidimensional case, sufficient conditions for perfect reconstruction will be given involving the multidimensional Fourier transforms of the filter coefficients on the cosets of the sampling lattice. These will be analogous to conditions arising from a polyphase analysis of the filter bank, as given in [4] and [8]. The conditions will be in the form of equations in the Fourier transforms which must be satisfied on a given set of frequencies. It will also be shown that there is a subset of this set on which satisfaction of the equations is both necessary and sufficient, but, unfortunately, the characterization of this subset will be non-constructive.

Examples of different operators U and Π for which the results are satisfied will be

252

given. In addition to an artificial example that bears little superficial resemblance to the shift/decimation case, two standard examples will be shown. The first is the finite dimensional case, related to the cyclic wavelet transform, in which $\Pi : \Re^N \to \Re^{N/2}$ is a finite length decimation and $U : \Re^N \to \Re^N$ is the cyclic shift. This appears in [3]. The second is the case of perfect reconstruction filter banks on $l^2(\mathcal{Z})$. Π and U are the same as D and S, the decimation and shift on $l^2(\mathcal{Z})$. The two familiar examples serve to show that results which have appeared previously can be viewed as a special case of the general framework presented here.

In this paper we treat the simple case of a single operator separately from the general multidimensional case in order to illustrate the basic concepts. Some results are stronger in the single operator case as well. Section 2 introduces the basic properties and develops a few relationships that result from the assumptions. Section 3 develops the perfect reconstruction conditions for the single operator case. Section 4 contains results on bases constructable from the perfect reconstruction operators. Section 5 identifies some special cases that result from the general theory. Section 6 moves onto the multidimensional case which is handled more succinctly given the development of the single operator theory in the preceding sections. Section 7 is a summary.

2 The operators and their properties

First we give the characterization of the operators U and Π in terms of two equations. Assume two Hilbert spaces H_1 and H_2 and operators $\Pi : H_1 \to H_2$ and unitary $U : H_1 \to H_1$ such that the following two assumptions hold:

Assumption 1 $\qquad \Pi\Pi^* = I_{H_2}$

and

Assumption 2 $\qquad \Pi^*\Pi + U^*\Pi^*\Pi U = I_{H_1}.$

Since U is unitary, an equivalent form of assumption (2) is:

Assumption 3 $\qquad \Pi^*\Pi + U\Pi^*\Pi U^* = I_{H_1}.$

Assumption (1) implies that the operator $\Pi^*\Pi$ is an orthogonal projection, since $(\Pi^*\Pi)^2 = \Pi^*\Pi$ and $(\Pi^*\Pi)^* = \Pi^*\Pi$. It will be seen in section 4 that assumption (2) implies that the projection for the subspace $(R(\Pi^*\Pi))^\perp$ is $U^*\Pi^*\Pi U$. Intuitively, assumption (2) shows that the Hilbert space H_1 can be broken up into two complementary orthogonal subspaces: The range of $\Pi^*\Pi$ and the range of $U^*\Pi^*\Pi U$. The rest of this section will algebraically derive properties of the operators from assumptions (1) and (2), but several of them can be justified by the idea of breaking the Hilbert space into two complementary orthogonal subspaces. This notion will be treated more carefully in section 5.3, when the cyclic structure of U is shown explicitly: U cycles between these two orthogonal subspaces.

The assumptions (1) and (2) are intended to generalize important facts that are clearly true for the case in which the operators U and Π are S and D on $l^2(\mathcal{Z})$. In this special case, the first condition merely says that upsampling followed by decimation does not change the signal, while the second shows how even and odd samples can be used to reconstruct the original signal. The two complementary subspaces mentioned previously are then the subspace of sequences which are zero on odd samples and the subspace of sequences which are zero on even samples, with the operator S shifting between them. U's property of successively mapping between complementary subspaces $R(\Pi^*\Pi)$ and $R(U^*\Pi^*\Pi U)$ is common to all operators satisfying assumptions (1) and (2).

The property to be derived is a generalization of the fact that upsampling, shift, and decimation give zero:

Lemma 1 *For Π and unitary U which satisfy assumptions (1) and (2),*

$$\Pi U \Pi^* = 0.$$

This can be shown by multiplication of assumption (2) by Π on the left and by Π^* on the right to get

$$\Pi\Pi^*\Pi\Pi^* + (\Pi U \Pi^*)^*(\Pi U \Pi^*) = \Pi\Pi^*.$$

Since $\Pi\Pi^* = I_{H_2}$,

$$(\Pi U \Pi^*)^*(\Pi U \Pi^*) = 0,$$

which can only happen if $\Pi U \Pi^* = 0$. This equality can be expressed in terms of the projection $\Pi^*\Pi$: By assumption (1), $\Pi U \Pi^* = 0$ if and only if $\Pi^*\Pi U \Pi^*\Pi = 0$, so that U transforms any vector in the range of the projection $\Pi^*\Pi$ to a vector in its orthogonal complement. With this fact it is simple to see that $U^*\Pi^*\Pi U$ is the projection for $R(\Pi^*\Pi)^\perp$.

If D and S are decimation and shift, then we clearly would have

$$S = DS^2D^*.$$

Upsampling, two shifts, and decimation are the same as performing one shift. It turns out that this equation does not hold in general for U and Π, but the following can be shown:

Lemma 2 *The operator $U_2 = \Pi U^2 \Pi^*$ is unitary.*

This follows from multiplication of assumption (2) by ΠU^* on the left and $U \Pi^*$ on the right to get

$$(\Pi U \Pi^*)^*(\Pi U \Pi^*) + (\Pi U^2 \Pi^*)^*(\Pi U^2 \Pi^*) = \Pi U^*U \Pi^* = I_{H_2}.$$

Since $\Pi U \Pi^* = 0$, we see that $U_2^*U_2 = I_{H_2}$. The fact that $U_2 U_2^* = I_{H_2}$ can be established through a similar operation on assumption (3). In the special case of shift and decimation on $l^2(\mathcal{Z})$, U_2 is also the shift on l^2, while in the finite dimensional case in which $U: \Re^N \to \Re^N$ is a circular shift and $\Pi\Re^N \to \Re^{N/2}$ is a decimation, U_2 is the shift operator on the space $\Re^{N/2}$. In general, it is not always possible to relate the action of U_2 to the action U. This is an issue which comes up when trying to apply the filtering process recursively, as in a wavelet transform: What is the natural operator from which to form the filters at the next level of the transform? An attempt will be made to come to grips with this problem in section 4.

The next result is a generalization of the fact that upsampling, shift by $2n$ and decimation is the same as shifting by n. The result is somewhat different, due to the fact that the identity $S = DS^2D^*$ does not directly apply to U and Π. The correct form of the identity is:

Lemma 3 $\qquad \Pi U^{2n}\Pi^* = (\Pi U^2\Pi^*)^n = U_2^n$

for all integers n.

This result follows for positive n by induction. It is clearly true for $n = 0$. Assume its truth for $n-1$

$$
\begin{aligned}
(\Pi U^2\Pi^*)^n &= (\Pi U^2\Pi^*)^{n-1}(\Pi U^2\Pi^*) \\
&= \Pi U^{2(n-1)}\Pi^*\Pi U^2\Pi^* \\
&= \Pi U^{2(n-1)}(I_{H_1} - U\Pi^*\Pi U^*)U^2\Pi^* \\
&= \Pi U^{2n}\Pi^* - \Pi U^{2n-1}\Pi^*\Pi U\Pi^* \\
&= \Pi U^{2n}\Pi^*.
\end{aligned}
$$

Since U and U_2 are unitary, the identity follows for negative n by taking the adjoint of the identity established for positive n. This result shows the relationship between the action of U_2 and U. It also suggests that U_2 might be a natural candidate for an operator with which to define filters for a second transform on H_2 — a possibility which is discussed in section 4.

The next result extends $\Pi U\Pi^* = 0$ to all odd powers of U.

Lemma 4 $\qquad \Pi U^{(2n+1)}\Pi^* = 0.$

for all integer n.

It can be shown by multiplying assumption (2) by $\Pi(U^*)^{2n}$ on the left and $U^{2n}\Pi^*$ on the right to get

$$
(U_2^*)^n(U_2)^n + (\Pi U^{2n+1}\Pi^*)^*(\Pi U^{2n+1}\Pi^*) = \Pi(U^*)^{2n}(U)^{2n}\Pi^* = I_{H_2},
$$

or, since $U_2^*U_2 = I_{H_2}$, $(\Pi U^{2n+1}\Pi^*)^*(\Pi U^{2n+1}\Pi^*) = 0$, which can only occur if $\Pi U^{2n+1}\Pi^* = 0$. Again, this result has an interpretation in terms of the projection $\Pi^*\Pi$. It is equivalent to $\Pi^*\Pi U^{2n+1}\Pi^*\Pi = 0$, showing that an odd power of U maps anything in the range of $\Pi^*\Pi$ to the orthogonal complement. This is an intuitive result in light of the previous discussion of the cyclic nature of U.

In this section, several important properties of operators satisfying assumptions (1) and (2) have been established. To summarize: for Π and unitary U satisfying assumptions (1) and (2),

$$\Pi U^{2n}\Pi^* = (\Pi U^2\Pi^*)^n = U_2^n, \tag{2}$$

$$\Pi U^{(2n+1)}\Pi^* = 0, \tag{3}$$

and $U_2 = \Pi U^2\Pi^*$ is unitary.

These results will be used to establish necessary and sufficient conditions on the sequences $(h_j^{(0)})$ and $(h_j^{(1)})$ corresponding to the usual perfect reconstruction and orthogonality conditions. They are the properties of the shift and decimation operators which allow such conditions to be established and they are implicit in assumptions (1) and (2).

3 Perfect reconstruction conditions

Define the filters $H^{(0)}$ and $H^{(1)}$ in Figure 1 by

$$H^{(0)} = \sum_j h_j^{(0)} U^j, \tag{4}$$

and

$$H^{(1)} = \sum_j h_j^{(1)} U^j. \tag{5}$$

Replace the operator D in Figure 1 with Π. Let U be unitary and let Π and U jointly satisfy assumptions (1) and (2). We seek conditions guaranteeing the perfect reconstruction property, as in [1],

$$(H^{(0)})^*\Pi^*\Pi(H^{(0)}) + (H^{(1)})^*\Pi^*\Pi(H^{(1)}) = I_{H_1}. \tag{6}$$

An additional condition is usually required for orthogonality,

$$\Pi(H^{(0)})(H^{(1)})^*\Pi^* = 0. \tag{7}$$

The intuition behind (6) is clear from the two channel filter bank as shown in Figure (1). It will be shown in section 4 that the perfect reconstruction property implies that $(H^{(0)})^*\Pi^*\Pi(H^{(0)})$ and $(H^{(1)})^*\Pi^*\Pi(H^{(1)})$ are projections. The orthogonality condition merely expresses the orthogonality of the ranges of these two projections. Since assumption (2) divides H_1 into two subspaces, and two filters are used, this case corresponds to maximal sampling. With maximal sampling, the orthogonality condition will be shown to be redundant. It is implied by the perfect reconstruction property. For now, though, we work with the orthogonality condition as separate from the perfect reconstruction condition.

The first step in using the properties of U and Π to establish conditions on $h_j^{(0)}$ and $h_j^{(1)}$ will be a lemma concerning U and Π.

Lemma 5 *If U and Π satisfy assumptions (1) and (2) then for any self adjoint $X : H_1 \rightarrow H_1$, $X = I_{H_1}$ if and only if*

1. $\Pi X \Pi^* = I_{H_2}$

2. $\Pi U^* X U \Pi^* = I_{H_2}$

3. $\Pi X U \Pi^* = 0$ *or equivalently,* $\Pi U^* X \Pi^* = 0$.

Proof: Define an operator $V : H_2 \times H_2 \rightarrow H_1$ by

$$V = \begin{bmatrix} \Pi^* & U\Pi^* \end{bmatrix}$$

. Note that

$$VV^* = \Pi^*\Pi + U\Pi^*\Pi U^* = I_{H_1}$$

and

$$V^*V = \begin{bmatrix} \Pi\Pi^* & \Pi U\Pi^* \\ \Pi U^*\Pi^* & \Pi\Pi^* \end{bmatrix} = I_{H_2 \times H_2}.$$

This means that $X = I_{H_1}$ if and only if $V^*XV = I_{H_2 \times H_2}$. Since

$$V^*XV = \begin{bmatrix} \Pi X \Pi^* & \Pi X U \Pi^* \\ \Pi U^* X \Pi^* & \Pi U^* X U \Pi^* \end{bmatrix},$$

$X = I_{H_1}$ if and only if the three conditions stated in the lemma are satisfied. $\square$

Letting X in the lemma equal $(H^{(0)})^*\Pi^*\Pi(H^{(0)}) + (H^{(1)})^*\Pi^*\Pi(H^{(1)})$, we see that the perfect reconstruction property,

$$(H^{(0)})^*\Pi^*\Pi(H^{(0)}) + (H^{(1)})^*\Pi^*\Pi(H^{(1)}) = I_{H_1},$$

is satisfied if and only if

$$\Pi(H^{(0)})^*\Pi^*\Pi(H^{(0)})\Pi^* + \Pi(H^{(1)})^*\Pi^*\Pi(H^{(1)})\Pi^* = I_{H_2},$$

$$\Pi U^*(H^{(0)})^*\Pi^*\Pi(H^{(0)})U\Pi^* + \Pi U^*(H^{(1)})^*\Pi^*\Pi(H^{(1)})U\Pi^* = I_{H_2},$$

and

$$\Pi(H^{(0)})^*\Pi^*\Pi(H^{(0)})U\Pi^* + \Pi(H^{(1)})^*\Pi^*\Pi(H^{(1)})U\Pi^* = 0.$$

We have transformed the condition on an operator in H_1 to three equivalent conditions on operators in H_2.

The term $\Pi(H^{(0)})\Pi^*$ can be written, using equations (3) and (4),

$$\begin{aligned} \Pi(H^{(0)})\Pi^* &= \sum_j h_j^{(0)} \Pi U^j \Pi^* \\ &= \sum_j h_{2j}^{(0)} \Pi U^{2j} \Pi^* \\ &= \sum_j h_{2j}^{(0)} U_2^j \end{aligned}$$

and similarly,

$$\begin{aligned} \Pi(H^{(0)})U\Pi^* &= \sum_j h_j^{(0)} \Pi U^{j+1} \Pi^* \\ &= \sum_j h_{2j-1}^{(0)} \Pi U^{2j} \Pi^* \\ &= \sum_j h_{2j-1}^{(0)} U_2^j. \end{aligned}$$

There are corresponding relations for $H^{(1)}$:

$$\Pi(H^{(1)})\Pi^* = \sum_j h_{2j}^{(1)} U_2^j,$$

and

$$\Pi(H^{(1)})U\Pi^* = \sum_j h_{2j-1}^{(1)} U_2^j.$$

If we define the polynomial

$$A_k^{(j)}(U_2) = \sum_n h_{2n-k}^{(j)} U_2^n,$$

for $k, j \in \{0, 1\}$, then the perfect reconstruction conditions become

$$[A_0^{(0)}(U_2)]^*[A_0^{(0)}(U_2)] + [A_0^{(1)}(U_2)]^*[A_0^{(1)}(U_2)] = I_{H_2},$$

$$[A_1^{(0)}(U_2)]^*[A_1^{(0)}(U_2)] + [A_1^{(1)}(U_2)]^*[A_1^{(1)}(U_2)] = I_{H_2},$$

and

$$[A_0^{(0)}(U_2)]^*[A_1^{(0)}(U_2)] + [A_0^{(1)}(U_2)]^*[A_1^{(1)}(U_2)] = 0.$$

The orthogonality condition (7) can be expressed in terms of these polynomials by noting:

$$
\begin{aligned}
\Pi(H^{(0)})(H^{(1)})^*\Pi^* &= \Pi H(\Pi^*\Pi + U\Pi^*\Pi U^*)(H^{(1)})^*\Pi^* \\
&= \Pi(H^{(0)})\Pi^*\Pi(H^{(1)})^*\Pi^* + \Pi(H^{(0)})U\Pi^*\Pi U^*(H^{(1)})^*\Pi^* \\
&= [A_0^{(0)}(U_2)][A_0^{(1)}(U_2)]^* + [A_1^{(0)}(U_2)][A_1^{(1)}(U_2)]^* = 0.
\end{aligned}
$$

These conditions can be expressed in an even simpler form by letting the spectral decomposition (see [2]) of U_2 be

$$U_2 = \int_0^{2\pi} e^{i\lambda}\, dP(\lambda).$$

where P is a resolution of the identity. So that the full conditions for perfect reconstruction are

$$\int_0^{2\pi} [\overline{A_0^{(0)}(e^{i\lambda})}A_0^{(0)}(e^{i\lambda}) + \overline{A_0^{(1)}(e^{i\lambda})}A_0^{(1)}(e^{i\lambda})]\, dP(\lambda) = I_{H_2}, \tag{8}$$

$$\int_0^{2\pi} [\overline{A_1^{(0)}(e^{i\lambda})}A_1^{(0)}(e^{i\lambda}) + \overline{A_1^{(1)}(e^{i\lambda})}A_1^{(1)}(e^{i\lambda})]\, dP(\lambda) = I_{H_2}, \tag{9}$$

$$\int_0^{2\pi} [\overline{A_0^{(0)}(e^{i\lambda})}A_1^{(0)}(e^{i\lambda}) + \overline{A_0^{(1)}(e^{i\lambda})}A_1^{(1)}(e^{i\lambda})]\, dP(\lambda) = 0, \tag{10}$$

and for orthogonality

$$\int_0^{2\pi} [A_0^{(0)}(e^{i\lambda})\overline{A_0^{(1)}(e^{i\lambda})} + A_1^{(0)}(e^{i\lambda})\overline{A_1^{(1)}(e^{i\lambda})}]\, dP(\lambda) = 0. \tag{11}$$

Since $P(\lambda)$ is increasing only for λ such that $e^{i\lambda}$ is in the spectrum of U_2, it is sufficient for the integrands of equations (8) and (9) to be equal to one and the integrands of equations (10) and (11) to be equal to zero on the spectrum of U_2. This is also necessary, since for any continuous function f on the unit circle, the spectrum of $f(U_2)$ is

$$\sigma(f(U_2)) = \{f(\lambda) : \lambda \in \sigma(U_2)\}.$$

So that $f(U_2) = I_{H_2}$ implies that

$$\sigma(f(U_2)) = \{1\} = \{f(\lambda) : \lambda \in \sigma(U_2)\},$$

or $f(\lambda) = 1$ for all $\lambda \in \sigma(U_2)$. Similarly, if $f(U_2) = 0$, then $f(\lambda) = 0$ for all $\lambda \in \sigma(U_2)$. It is both necessary and sufficient for the integrands of equations (8) and (9) to be equal to one and the integrands of equations (10) and (11) to be equal to zero on the spectrum of U_2. This establishes the following theorem:

Theorem 1 *Given $\Pi : H_1 \to H_2$ and unitary $U : H_1 \to H_1$ so that*

1. $\Pi\Pi^* = I_{H_2}$

2. $\Pi^*\Pi + U^*\Pi^*\Pi U = I_{H_1}$

and operators

$$(H^{(0)}) = \sum_j (h_j^{(0)})U^j,$$

and

$$(H^{(1)}) = \sum_j (h_j^{(1)})U^j,$$

define $U_2 = \Pi U^2 \Pi^*$ *and*

$$A_k^{(j)}(e^{i\lambda}) = \sum_n h_{2n-k}^{(j)} e^{in\lambda}.$$

Then the perfect reconstruction property

$$(H^{(0)})^*\Pi^*\Pi(H^{(0)}) + (H^{(1)})^*\Pi^*\Pi(H^{(1)}) = I_{H_1}$$

is satisfied if and only if

$$|A_0^{(0)}(e^{i\lambda})|^2 + |A_0^{(1)}(e^{i\lambda})|^2 = 1, \tag{12}$$

$$|A_1^{(0)}(e^{i\lambda})|^2 + |A_1^{(1)}(e^{i\lambda})|^2 = 1, \tag{13}$$

and

$$\overline{A_0^{(0)}(e^{i\lambda})}A_1^{(0)}(e^{i\lambda}) + \overline{A_0^{(1)}(e^{i\lambda})}A_1^{(1)}(e^{i\lambda}) = 0, \tag{14}$$

hold for all $e^{i\lambda} \in \sigma(U_2)$. *The orthogonality condition* $\Pi(H^{(0)})(H^{(1)})^*\Pi^* = 0$ *is satisfied if and only if*

$$A_0^{(0)}(e^{i\lambda})\overline{A_0^{(1)}(e^{i\lambda})} + A_1^{(0)}(e^{i\lambda})\overline{A_1^{(1)}(e^{i\lambda})} = 0 \tag{15}$$

for all $e^{i\lambda} \in \sigma(U_2)$.

Equations (12) through (15) are actually the standard condition on the coefficients $h_j^{(0)}$ and $h_j^{(0)}$ for perfect reconstruction and orthogonality. What is new is that these equations must be satisfied on the spectrum of the operator U. In the case of shift and decimation on $l^2(\mathcal{Z})$, the spectrum is the entire unit circle.

Since these equations are standard, we can use a standard alternate form: Perfect reconstruction and orthogonality are achieved if and only if

$$|A_0^{(0)}(e^{i\lambda})|^2 + |A_1^{(0)}(e^{i\lambda})|^2 = 1, \tag{16}$$

$$A_0^{(1)}(e^{i\lambda}) = V(e^{i\lambda})\overline{A_1^{(0)}(e^{i\lambda})}, \tag{17}$$

and

$$A_1^{(1)}(e^{i\lambda}) = -V(e^{i\lambda})\overline{A_0^{(0)}(e^{i\lambda})} \tag{18}$$

for some $|V(e^{i\lambda})|$ for all $e^{i\lambda} \in \sigma(U_2)$. A proof of the equivalence of these conditions with those in the theorem can be found in the paper by Grossman and Poor [3].

4 An orthonormal basis for H_2

It is natural to ask if equations (12) through (15) contain some sort of redundancy, and if so, is there some more concise form. It turns out that in the two channel case using assumption (2), the perfect reconstruction conditions (12) through (14) imply the orthogonality condition (15). This can be seen by restating the conditions in a matrix form: equations (12) through (14) are satisfied if and only if the matrix

$$M = \begin{bmatrix} A_0^{(0)}(e^{i\lambda}) & A_0^{(1)}(e^{i\lambda}) \\ A_1^{(0)}(e^{i\lambda}) & A_1^{(1)}(e^{i\lambda}) \end{bmatrix}$$

satisfies $MM^H = I$ for all $e^{i\lambda} \in \sigma(U_2)$. Such an M is called paraunitary. The equivalence can be seen by comparing the elements of MM^H with equations 12 through 14.

Assuming that the equations (12) through (14) are satisfied, we get that $MM^H = I$ and hence that $M^H M = I$ for all $e^{i\lambda} \in \sigma(U_2)$. This is a result of the fact that M is a square matrix: $MM^H = I$ implies $M^H M = I$. The fact that M is square is equivalent to maximal sampling, with the number of filters equaling the number of terms in assumption (2). Looking at the elements of the matrix $M^H M$ gives the three new relations:

$$|A_0^{(0)}(e^{i\lambda})|^2 + |A_1^{(0)}(e^{i\lambda})|^2 = 1, \tag{19}$$

$$|A_0^{(1)}(e^{i\lambda})|^2 + |A_1^{(1)}(e^{i\lambda})|^2 = 1, \tag{20}$$

and

$$A_0^{(0)}(e^{i\lambda})\overline{A_0^{(1)}(e^{i\lambda})} + A_1^{(0)}(e^{i\lambda})\overline{A_1^{(1)}(e^{i\lambda})} = 0, \tag{21}$$

for all $e^{i\lambda} \in \sigma(U_2)$.

Equation (21) is the same as (15). This is the dependence mentioned previously: The perfect reconstruction conditions imply the orthogonality condition.

The two new equations (19) and (20) are equivalent to $\Pi(H^{(0)})(H^{(0)})^*\Pi^* = I_{H_2}$ and $\Pi(H^{(1)})(H^{(1)})^*\Pi^* = I_{H_2}$ respectively. This can be easily shown. Take the case involving $(H^{(0)})$:

$$\begin{aligned} \Pi(H^{(0)})(H^{(0)})^*\Pi^* &= \Pi(H^{(0)})(\Pi^*\Pi + U\Pi^*\Pi U^*)(H^{(0)})^*\Pi^* \\ &= \Pi(H^{(0)})\Pi^*\Pi(H^{(0)})^*\Pi^*\Pi(H^{(0)})U\Pi^*\Pi U^*(H^{(0)})^*\Pi^* \\ &= (A_0^{(0)}(U_2))(A_0^{(0)}(U_2))^* + (A_1^{(0)}(U_2))(A_1^{(0)}(U_2))^*, \end{aligned}$$

so that

$$\Pi(H^{(0)})(H^{(0)})^*\Pi^* = I_{H_2} \tag{22}$$

if and only if (19) is satisfied. Similarly,

$$\Pi(H^{(1)})(H^{(1)})^*\Pi^* = I_{H_2} \tag{23}$$

if and only if (20) is satisfied. Furthermore, these equalities are achieved, along with the orthogonality condition $\Pi(H^{(0)})(H^{(1)})^*\Pi^* = 0$, by satisfaction of the perfect reconstruction conditions (12) through (14). These new relations will be particularly useful in interpreting the transform in terms of bases for H_1 and H_2.

It was mentioned earlier without proof that the operators

$$(H^{(0)})^*\Pi^*\Pi(H^{(0)}),$$

and

$$(H^{(1)})^*\Pi^*\Pi(H^{(1)})$$

are projections. This follows easily from (22) and (23). Further, the ranges of these projections are orthogonal by the fact that $\Pi(H^{(0)})(H^{(1)})^*\Pi^* = 0$. This means that the outputs of the two separate channels before summation are projections of the original signal onto these two orthogonal complementary subspaces. These subspaces should not be confused with the subspaces $R(\Pi^*\Pi)$ and its complement $R(U^*\Pi^*\Pi U)$.

There is a previously unmentioned relation between U and Π which will also be useful in constructing bases:

$$U^2\Pi^* = \Pi^*U_2. \tag{24}$$

Verification of this identity is simple: assumption (2) implies that

$$\Pi^*\Pi U\Pi^* + U^*\Pi^*\Pi U^2\Pi^* = U\Pi^*.$$

or since $\Pi^*\Pi U\Pi^* = 0$ and U is unitary,

$$\Pi^*\Pi U^2\Pi^* = U^2\Pi^*.$$

which is the desired result with $U_2 = \Pi U^2\Pi^*$.

The new identities established in this section lead to the following result:

Theorem 2 *Assume that U, Π, $(H^{(0)})$ and $(H^{(1)})$ satisfy the perfect reconstruction conditions. If $\{U_2^n\delta\}_n$ is an orthonormal basis for H_2 for some $\delta \in H_2$, then*

$$\{U^{2n}(H^{(0)})^*\Pi^*\delta\}_n \cup \{U^{2n}(H^{(1)})^*\Pi^*\delta\}_n$$

forms an orthonormal basis for H_1.

Proof: The orthonormality follows from the orthonormality of $\{U_2^n\delta\}_n$ and the following:

$$
\begin{aligned}
\langle U^{2n}(H^{(0)})^*\Pi^*\delta, U^{2m}(H^{(0)})^*\Pi^*\delta\rangle &= \langle (H^{(0)})^*\Pi^*U_2^n\delta, (H^{(0)})^*\Pi^*U_2^m\delta\rangle \\
&= \langle \Pi(H^{(0)})(H^{(0)})^*\Pi^*U_2^n\delta, U_2^m\delta\rangle \\
&= \langle U_2^n\delta, U_2^m\delta\rangle.
\end{aligned}
$$

Similarly

$$\langle U^{2n}(H^{(1)})^*\Pi^*\delta, U^{2m}(H^{(1)})^*\Pi^*\delta\rangle = \langle U_2^n\delta, U_2^m\delta\rangle,$$

and

$$\langle U^{2n}(H^{(0)})^*\Pi^*\delta, U^{2m}(H^{(1)})^*\Pi^*\delta\rangle = \langle \Pi(H^{(1)})(H^{(0)})^*\Pi^*U_2^n\delta, U_2^m\delta\rangle = 0.$$

This shows orthonormality.

The fact that the set is a basis follows since for any $v \in H_1$ there exists a representation of $\Pi(H^{(0)})v$ and $\Pi(H^{(1)})v$ as

$$\Pi(H^{(0)})v = \sum_n x_n U_2^n\delta,$$

and

$$\Pi(H^{(1)})v = \sum_n y_n U_2^n \delta$$

where $x_n = \langle \Pi(H^{(0)})v, U_2^n \delta \rangle$ and $y_n = \langle \Pi(H^{(1)})v, U_2^n \delta \rangle$. Using the perfect reconstruction property:

$$\begin{aligned}
v &= (H^{(0)})^* \Pi^* \Pi (H^{(0)}) v + (H^{(1)})^* \Pi^* \Pi (H^{(1)}) v \\
&= (H^{(0)})^* \Pi^* \left(\sum_n x_n U_2^n \delta \right) + (H^{(1)})^* \Pi^* \left(\sum_n y_n U_2^n \delta \right) \\
&= \sum_n x_n U^{2n} (H^{(0)})^* \Pi^* \delta + y_n U^{2n} (H^{(1)})^* \Pi^* \delta.
\end{aligned}$$

So any $v \in (H^{(0)})_1$ can be represented as a combination of

$$U^{2n} (H^{(0)})^* \Pi^* \delta,$$

and

$$U^{2m} (H^{(1)})^* \Pi^* \delta. \quad \square$$

In the normal case of perfect reconstruction on $l^2(\mathcal{Z})$ using shift and decimation, δ is defined to be the unit pulse sequence, and with additional constraints, $h_j^{(0)}$ and $h_j^{(1)}$ are wavelet filters. In the case of a wavelet transform we are interested in carrying out the analysis part of the filter bank recursively. Is it possible to carry such a recursive transform with operators other than shift or decimation?

If the process is to be carried out recursively then we would probably desire to use the same coefficients at each level. If $H_1 \neq H_2$, however, then we can't even use the same operators U and Π. For that reason, we will consider only the case of $H_1 = H_2$.

Consider what happens in one analysis stage. Suppose that the input is v. The representation of the output of the $H^{(0)}$ channel with respect to the basis $u_2^n \delta$ is:

$$\langle \Pi H^{(0)} v, U_2^n \delta \rangle = \langle v, (H^{(0)})^* \Pi^* U_2^n \delta \rangle = \langle v, U^{2n} (H^{(0)})^* \Pi^* \delta \rangle.$$

So that the representation of the output of the $H^{(0)}$ channel with respect to the basis $U_2^n \delta$ is the same as that of the input with respect to the basis $U^{2n} (H^{(0)})^* \Pi^* \delta$. An obvious, similar result applies to the $H^{(1)}$ channel. It might be important that the operator used to construct the filters at each stage be the same as those which generate the basis with δ. To achieve this, it might be desirable to require $U = U_2$. These are speculations, the issue of exactly how to define a recursive transform, and its interpretation in terms of basis functions, have not been resolved.

5 Examples

Conditions (12) through (15) cover several special cases. Among them are standard perfect reconstruction on $l^2(\mathcal{Z})$ with shift and decimation, perfect reconstruction on $\Re^N$ using the cyclic shift and decimation, and perfect reconstruction based on operators other than shift and decimation.

5.1 THE CASE OF $l^2(\mathcal{Z})$

Let $H_1 = H_2 = l^2(\mathcal{Z})$. Let Π be the decimation operator, and U be the right shift. This means that Π^* will be the upsampling operator and U^* will be the left shift. U is unitary, and clearly these operators satisfy the assumptions (1) and (2).

The operator U in $l^2(\mathcal{Z})$ has a spectrum consisting of the entire unit circle. This means that the conditions (12) through (15) must be satisfied for all $\lambda \in [0, 2\pi]$.

This is the condition in terms of Fourier transforms that is usually applied to perfect reconstruction filter banks. It is also applied to wavelet transforms on $l^2(\mathcal{Z})$ (along with other conditions), [1].

5.2 PERFECT RECONSTRUCTION ON $\Re^N$

Take $H_1 = \Re^N$ and $H_2 = \Re^{N/2}$, Π is the $N/2 \times N$ decimation matrix and U is the $N \times N$ right circular shift matrix. As an example, for $N = 4$

$$\Pi = \begin{bmatrix} 1 & 0 & 0 & 0 \\ 0 & 0 & 1 & 0 \end{bmatrix}$$

and

$$U = \begin{bmatrix} 0 & 0 & 0 & 1 \\ 1 & 0 & 0 & 0 \\ 0 & 1 & 0 & 0 \\ 0 & 0 & 1 & 0 \end{bmatrix}.$$

Again, these operators clearly satisfy assumptions (1) and (2). Further, since $U^N = I_{H_1}$, it is only necessary to consider finite length sequences $h_j^{(0)}$ and $h_j^{(1)}$:

$$H^{(0)} = \sum_{j=0}^{N-1} h_j^{(0)} U^j,$$

and

$$H^{(1)} = \sum_{j=0}^{N-1} h_j^{(1)} U^j.$$

The filter operators $H^{(0)}$ and $H^{(1)}$ perform circular convolution of the input vector with the vectors $h_j^{(0)}$ and $h_j^{(1)}$ respectively.

The matrix $U_2 = \Pi U^2 \Pi^*$ is just the right circular shift matrix on $\Re^{N/2}$. It has eigenvalues

$$(e^{i\frac{2\pi}{N/2}})^k$$

for $k = 0, 1, ..., N/2-1$, so that the points at which the polynomials $A_0^{(0)}$, $A_1^{(0)}$, $A_0^{(1)}$, and $A_1^{(1)}$ must satisfy equations (12) through (15) are those points on the unit circle corresponding to the DFT's of the sequences $h_{2j}^{(0)}$, $h_{2j-1}^{(0)}$, $h_{2j}^{(1)}$, and $h_{2j-1}^{(1)}$. It is therefore necessary and sufficient that the DFT's of $h_{2j}^{(0)}$, $h_{2j-1}^{(0)}$, $h_{2j}^{(1)}$, and $h_{2j-1}^{(1)}$ satisfy the equations (12) through

(15). These are also conditions which can be applied to a cyclic wavelet transform as derived in [3].

5.3 A CONSTRUCTIVE CHARACTERIZATION OF U AND Π

Although there are other possibilities for U and Π than the shift and decimation operators, it is not immediately obvious how to construct these from the assumptions (1) and (2). The following theorem provides a characterization which is more useful in constructing examples:

Theorem 3 *Given Π so that $\Pi\Pi^* = I_{H_2}$, U is unitary and the condition $\Pi^*\Pi + U^*\Pi^*\Pi U$ are satisfied if and only if there exists $V : H_1 \to H_2 \times H_2$ such that*

$$V = \begin{bmatrix} \Pi \\ \Pi' \end{bmatrix},$$

$V^*V = I_{H_1}$, $VV^* = I_{H_2 \times H_2}$ *and*

$$U = V^* \begin{bmatrix} 0 & B \\ A & 0 \end{bmatrix} V = V^*WV$$

for unitary A and B.

Proof: If U is unitary and $\Pi^*\Pi + U^*\Pi^*\Pi U = I_{H_1}$, then it is simple to verify, using results from previous sections that $\Pi' = \Pi U$, $A = \Pi U^2 \Pi^*$, and $B = I_{H_2}$ satisfy all the conditions in the theorem.

If there exist such Π', A and B, then we need to check if U is unitary. We know that

$$V^*V = \Pi^*\Pi + \Pi'^*\Pi' = I_{H_1}$$

and also that $VV^* = I_{H_2 \times H_2}$ implies that $\Pi\Pi'^* = I_{H_2}$ and $\Pi\Pi'^* = 0$ so,

$$\begin{aligned}
U^*U &= (\Pi^*B\Pi' + \Pi'^*A\Pi)^*(\Pi^*B\Pi' + \Pi'^*A\Pi) \\
&= \Pi^*\Pi + \Pi'^*\Pi' \\
&= I_{H_1}.
\end{aligned}$$

Similarly it can be shown $UU^* = I_{H_1}$.

We also need to check the validity of assumption (2):

$$\begin{aligned}
\Pi^*\Pi + U^*\Pi^*\Pi U &= \Pi^*\Pi + V^*W^* \begin{bmatrix} \Pi \\ \Pi' \end{bmatrix} \Pi^*\Pi \begin{bmatrix} \Pi^* & \Pi'^* \end{bmatrix} WV \\
&= \Pi^*\Pi + V^*W^* \begin{bmatrix} I_{H_2} \\ 0 \end{bmatrix} \begin{bmatrix} I_{H_2} & 0 \end{bmatrix} WV \\
&= \Pi^*\Pi + V^* \begin{bmatrix} 0 \\ B^* \end{bmatrix} \begin{bmatrix} 0 & B \end{bmatrix} V \\
&= \Pi^*\Pi + V^* \begin{bmatrix} 0 & 0 \\ 0 & I_{H_2} \end{bmatrix} V \\
&= \Pi^*\Pi + \Pi'^*\Pi' \\
&= I_{H_1}.
\end{aligned}$$

This establishes the theorem. □

If we are given an operator Π and can find an operator Π', theorem 3 gives a means for constructing unitary operators U which will satisfy assumption (2). In the finite dimensional case of $\Pi : \mathbf{C}^N \to \mathbf{C}^{N/2}$, it is clear how to construct Π' given Π: The rows of Π' form an orthonormal basis for the orthogonal complement of the span of the rows of Π. We will use this fact to construct a finite dimensional example.

Let $H_1 = \mathbf{C}^4$ and $H_2 = \mathbf{C}^2$. Define

$$\Pi = \begin{bmatrix} 1 & 2 & 3 & 0 \\ 0 & -3 & 2 & 1 \end{bmatrix} / \sqrt{14},$$

$$\Pi' = \begin{bmatrix} 2 & -1 & 0 & -3 \\ 3 & 0 & -1 & 2 \end{bmatrix} / \sqrt{14},$$

and

$$W = \begin{bmatrix} 0 & 0 & 1/\sqrt{2} & -1/\sqrt{2} \\ 0 & 0 & 1/\sqrt{2} & 1/\sqrt{2} \\ 0 & 1 & 0 & 0 \\ 1 & 0 & 0 & 0 \end{bmatrix}.$$

Clearly Π, Π', and W satisfy the conditions of theorem 3. Therefore, $U = V^*WV$ can be computed

$$U = \begin{bmatrix} 3 - 1/\sqrt{2} & -1/\sqrt{2} & 13 + 1/\sqrt{2} & 2 - 5/\sqrt{2} \\ -17/\sqrt{2} & 3 + 1/\sqrt{2} & -2 + 5/\sqrt{2} & -1 - 7/\sqrt{2} \\ -1 + 7/\sqrt{2} & -2 - 5/\sqrt{2} & -3 + 1/\sqrt{2} & -17/\sqrt{2} \\ 2 + 5/\sqrt{2} & 13 - 1/\sqrt{2} & -1/\sqrt{2} & -3 - 1/\sqrt{2} \end{bmatrix} / 14.$$

It is simple to verify that U is unitary, and $\Pi^*\Pi + U^*\Pi^*\Pi U = I_{H_1}$. Using these operators it is possible to get coefficients using theorem 1.

With U as above, U_2 is

$$\Pi U^2 \Pi^* = \begin{bmatrix} -1/\sqrt{2} & 1/\sqrt{2} \\ 1/\sqrt{2} & 1/\sqrt{2} \end{bmatrix}$$

with eigenvalues e^{i0} and $e^{i\pi}$.

Define the sequences $\hat{a}_0^{(0)}(j) = \{1,0\}$, $\hat{a}_0^{(0)}(j) = \{0,1\}$, $\hat{a}_0^{(1)}(j) = \{0,1\}$, and $\hat{a}_1^{(1)}(j) = \{-1,0\}$ for $j = 0,1$. Clearly

$$|\hat{a}_0^{(0)}(j)|^2 + |\hat{a}_0^{(1)}(j)|^2 = 1,$$

$$|\hat{a}_1^{(0)}(j)|^2 + |\hat{a}_1^{(1)}(j)|^2 = 1,$$

$$\overline{\hat{a}_0^{(0)}(j)}\hat{a}_1^{(0)}(j) + \overline{\hat{a}_0^{(1)}(j)}\hat{a}_1^{(1)}(j) = 0,$$

and

$$\hat{a}_0^{(0)}(j)\overline{\hat{a}_0^{(1)}(j)} + \hat{a}_1^{(0)}(j)\overline{\hat{a}_1^{(1)}(j)} = 0$$

for $j = 0,1$.

If we demand that $\hat{a}_0^{(0)}(j)$, $\hat{a}_1^{(0)}(j)$, $\hat{a}_0^{(1)}(j)$, and $\hat{a}_1^{(1)}(j)$ be values of $A_0^{(0)}(e^{i\lambda_j})$, $A_1^{(0)}(e^{i\lambda_j})$, $A_0^{(1)}(e^{i\lambda_j})$, and $A_1^{(1)}(e^{i\lambda_j})$ respectively for $\lambda_0 = 0$ and $\lambda_1 = \pi$, then finding h_{2j}, h_{2j+1}, g_{2j}, and g_{2j+1} becomes a frequency sampling design problem:

$$\begin{bmatrix} h_0 \\ h_2 \end{bmatrix} = \begin{bmatrix} (e^{i\lambda_0})^0 & (e^{i\lambda_0})^1 \\ (e^{i\lambda_1})^0 & (e^{i\lambda_1})^1 \end{bmatrix}^{-1} \begin{bmatrix} \hat{a}(0) \\ \hat{a}(1) \end{bmatrix}$$

and

$$\begin{bmatrix} h_1 \\ h_3 \end{bmatrix} = \begin{bmatrix} (e^{i\lambda_0})^1 & (e^{i\lambda_0})^2 \\ (e^{i\lambda_1})^1 & (e^{i\lambda_1})^2 \end{bmatrix}^{-1} \begin{bmatrix} \hat{b}(0) \\ \hat{b}(1) \end{bmatrix}.$$

This gives

$$(h_j^{(0)}) = \{1/2, -1/2, 1/2, 1/2\},$$

and similarly,

$$(h_j^{(1)}) = \{1/2, -1/2, -1/2, -1/2\}.$$

So if,

$$(H^{(0)}) = \sum_{j=0}^{3} (h_j^{(0)}) U^j,$$

and

$$(H^{(1)}) = \sum_{j=0}^{3} (h_j^{(1)}) U^j,$$

then $(H^{(0)})^* \Pi^* \Pi (H^{(0)}) + (H^{(1)})^* \Pi^* \Pi (H^{(1)}) = I$ and $\Pi(H^{(0)})(H^{(1)})^* \Pi^* = 0$. These operators will satisfy both perfect reconstruction and orthogonality conditions.

6 Extension to the multidimensional case

It turns out that similar results can be derived for the D dimensional M channel case. The generalization involves a number of unitary operators $U_k : H_1 \rightarrow H_1$, $1 \leq k \leq D$, corresponding to shifts along the various dimensions of the lattice. The notation required is more complex. To help matters, let a D dimensional vector of commuting unitary operators

$$\mathbf{U} = \begin{bmatrix} U_1 & U_2 & \cdots & U_D \end{bmatrix}^T,$$

raised to an integer vector

$$\mathbf{j} = \begin{bmatrix} j_1 & j_2 & \cdots & j_D \end{bmatrix}^T \in \mathcal{Z}^D,$$

denote

$$\mathbf{U}^{\mathbf{j}} = \prod_{k=1}^{D} U_k^{j_k}.$$

As the U_k commute, the ordering of the product is not important. Similarly, the adjoint of an operator of vectors raised to a vector power, is defined to be

$$(\mathbf{U}^*)^{\mathbf{j}} = \prod_{k=1}^{D} (U_k^*)^{j_k}.$$

We also write the componentwise product of two integer vectors as

$$\mathbf{i} \circ \mathbf{j} = \begin{bmatrix} i_1 j_1 & i_2 j_2 & \cdots & i_D j_D \end{bmatrix}^T.$$

This is similar to the notation used in [4] and in [8]. We also use the absolute value of an integer vector to mean

$$|\mathbf{j}| = \begin{bmatrix} |j_1| & |j_2| & \cdots & |j_D| \end{bmatrix}^T.$$

Throughout the development, we will assume a correspondence between the operators and shifts on $\mathcal{Z}^D$. We assume that we are given a matrix $\mathbf{N}$,

$$N = \begin{bmatrix} \mathbf{n}_1 & \mathbf{n}_2 & \cdots & \mathbf{n}_D \end{bmatrix}.$$

with full rank which will be used in equations which define the properties of the operators. The columns of the matrix can be interpreted as the *generators* of some lattice $L \subset \mathcal{Z}^D$. That is, $\mathbf{l} \in L$ if and only if $\mathbf{l} = \mathbf{N}\mathbf{i} = \sum_i \mathbf{n}_i x_i$ for some $\mathbf{i} \in \mathcal{Z}^D$. Any set of $\mathbf{n}_i$ which can be used in this manner to produce L is called a set of *generators* for the lattice L. We will assume that we are given the matrix $\mathbf{N}$ which will be used to define both L and the equations which we will require the operators U_k to satisfy. Both the initial assumptions and the results which are derived from them will be expressed as relations which must be satisfied for all $\mathbf{Ni}$, $\mathbf{i} \in \mathcal{Z}^D$. This is equivalent to the relation being satisfied for all $\mathbf{l} = \mathbf{Ni} \in L$. From this it is clear that the results will be dependent only on the lattice L and not on the particular generators chosen.

A *unit cell* of the lattice generated by $\mathbf{N}$ will be defined as a minimal set of vectors $J = \{\mathbf{j}_i : \in \mathcal{Z}^D\}$ such that any $\mathbf{x} \in \mathcal{Z}^D$ can be represented as

$$\mathbf{x} = \mathbf{Ni} + \mathbf{j}$$

for some $\mathbf{j} \in J$ and $\mathbf{i} \in \mathcal{Z}^D$. It is simple to show that the requirements that J be minimal and that N have full rank imply that the representation of $\mathbf{x}$ in terms of $\mathbf{i}$ and $\mathbf{j}$ must be unique.

We will denote the number of elements in the set J by $|J|$.

As an example of these definitions, let

$$N = \begin{bmatrix} 1 & 1 \\ 1 & -1 \end{bmatrix}.$$

The lattice produced by these generators is shown in Figure 2. A valid unit cell is

$$J = \{ \begin{bmatrix} 1 & 0 \end{bmatrix}^T \}.$$

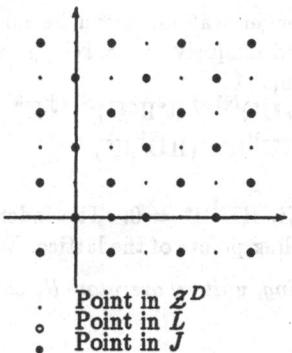

Point in Z^D
Point in L
Point in J

Figure 2: *An Example of a Lattice*

As stated before, we require that the operators U_k be unitary and commute with each other. Using this new notation, we make the following assumptions

Assumption 4 $\Pi\Pi^* = I_{H_2}$,

Assumption 5 $\sum_{j \in J} (U^*)^j \Pi^* \Pi (U)^j = I_{H_1}$,

and

Assumption 6 $(U^*)^{Ni} \Pi^* \Pi (U)^{Ni} = \Pi^* \Pi$

for integer vectors i.

Nothing analogous to assumption (6) appears in the one dimensional case. A similar equation does follow from assumption (1), since by (24)

$$(U^{2i})^* \Pi^* \Pi (U^{2i}) = \Pi^*(U_2^i)^*(U_2^i)\Pi = \Pi^*\Pi.$$

It seems that the periodicity is implicit in assumption (1) and (2) in the single dimensional case, but it must be assumed in the multidimensional case.

As with the single dimensional case, we can get a relation $\Pi U^j \Pi^* = 0$ except for certain j. To specify which j, multiply assumption (5) by Π on the left and Π^* on the right to get

$$\sum_{j \in J} \Pi(U^*)^j \Pi^* \Pi (U)^j \Pi^*,$$

so that because of assumption (4)

$$\sum_{\substack{j \in J \\ j \neq 0}} \Pi(U^*)^j \Pi^* \Pi (U)^j \Pi^* = 0.$$

Since this is a sum of positive semidefinite terms, each term must be zero. This means that

$$\Pi U^j \Pi^* = 0 \tag{25}$$

for $j \in J, j \neq 0$.

It turns out that a much stronger statement can be made when $\Pi U^{\mathbf{j}} \Pi^*$ is zero. Any integer vector $\mathbf{j}$ can be represented uniquely as $\mathbf{j} = N\mathbf{i} + \mathbf{j}'$, where $\mathbf{i}$ is an integer vector and $\mathbf{j}' \in J$. This means that for $\mathbf{j}' \neq 0$,

$$
\begin{aligned}
(\Pi U^{\mathbf{j}} \Pi^*)^*(\Pi U^{\mathbf{j}} \Pi^*) &= \Pi (U^*)^{N\mathbf{i}+\mathbf{j}'} \Pi^* \Pi (U)^{N\mathbf{i}+\mathbf{j}'} \Pi^* \\
&= (\Pi U^{\mathbf{j}'} \Pi^*)^*(\Pi U^{\mathbf{j}'} \Pi^*) \\
&= 0.
\end{aligned}
$$

So that for $\mathbf{j}$ not of the form $N\mathbf{i}$, $\Pi U^{\mathbf{j}} \Pi^* = 0$. The index $\mathbf{j}$ for which the expression is nonzero corresponds to the sampling points of the lattice. We restate this fact as a theorem

Theorem 4 *For Π and commuting, unitary operators U_k satisfying assumptions (4) through (6),*

$$
\Pi U^{\mathbf{m}} \Pi^* = 0
$$

unless $\mathbf{m} = N\mathbf{i}$ *for some integer vector* $\mathbf{i}$.

As in the single dimensional case, the condition $\Pi \Pi^* = I_{H_2}$ implies that $\Pi^* \Pi$ is a projection. Also $\Pi U^{\mathbf{m}} \Pi^* = 0$ if and only if $\Pi^* \Pi U^{\mathbf{m}} \Pi^* \Pi = 0$. Theorem 4 states that $U^{\mathbf{m}}$ maps any vector in $R(\Pi^* \Pi)$ into $(R(\Pi^* \Pi))^{\perp}$ unless $\mathbf{m} = N\mathbf{i}$ for some integer vector $\mathbf{i}$. This idea can be extended by noting that the operators

$$
(U^*)^{\mathbf{j}} \Pi^* \Pi (U)^{\mathbf{j}}
$$

for $\mathbf{j} \in J$ are projections. We look at how the operator $U^{\mathbf{k}}$ shifts between the ranges of these projection

$$
(U^*)^{\mathbf{i}} \Pi^* \Pi (U)^{\mathbf{i}} (U^{\mathbf{k}})(U^*)^{\mathbf{j}} \Pi^* \Pi (U)^{\mathbf{j}}
$$

and see that this expression is nonzero only when $\mathbf{i} + \mathbf{k} = N\mathbf{l} + \mathbf{j}$ for some integer vector $\mathbf{l}$. The operators U_k have exactly the same structure with respect to the subspaces

$$
R((U^*)^{\mathbf{j}} \Pi^* \Pi (U)^{\mathbf{j}})
$$

as the shifts do with subspaces of multidimensional sequences which are nonzero only on the different cosets of the lattice generated by N. This is analogous to the cyclic structure of U in the one dimensional case.

The next result shows how expressions of the form $\Pi U^{N\mathbf{i}} \Pi^*$ can be broken up into a product of powers of the D operators $\Pi U^{\mathbf{n}_i} \Pi^*$. It turns out that these operators will be unitary, and that the perfect reconstruction conditions will be expressed in terms of their spectra.

Theorem 5 *For Π and commuting, unitary operators U_k satisfying assumptions (4) through (6)*

$$
\Pi U^{N\mathbf{i}} \Pi^* = \prod_{k=1}^{D} (\Pi U^{\mathbf{n}_k} \Pi^*)^{i_k}.
$$

Proof: The result follows from the fact that

$$
\begin{aligned}
\Pi U^{N\mathbf{i}} \Pi^* \Pi U^{N\mathbf{j}} \Pi^* &= \Pi U^{N(\mathbf{i}+\mathbf{j})} (U^{-N\mathbf{j}} \Pi^* \Pi U^{N\mathbf{j}}) \Pi^* \\
&= \Pi U^{N(\mathbf{i}+\mathbf{j})} \Pi^* \Pi \Pi^*
\end{aligned}
$$

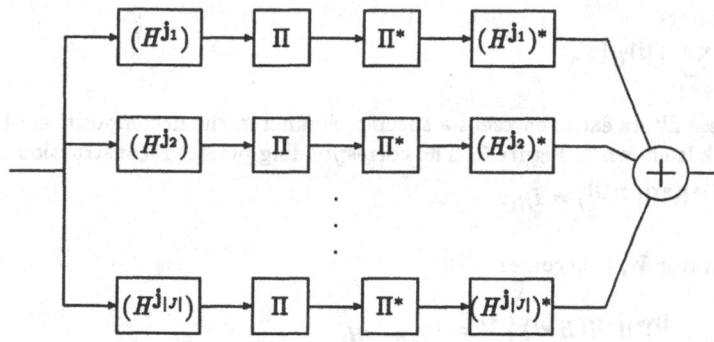

Figure 3: *The General Filter Bank*

$$= \Pi U^{N(i+j)} \Pi^*.$$

In particular,

$$\Pi U^{n_i} \Pi^* \Pi U^{n_j} \Pi^* = \Pi U^{n_i + n_j} \Pi^*.$$

The theorem follows from application of this equation. □

Theorem 6 *For* Π *and commuting, unitary operators* U_k *satisfying assumptions (4) through (6) the operators*

$$\Pi U^{Ni} \Pi^*$$

are unitary and commutative for integer vectors **i**.

Proof: The theorem follows trivially since by assumption (6),

$$\begin{aligned}
(\Pi U^{Ni} \Pi^*)^* (\Pi U^{Ni} \Pi^*) &= \Pi((U^*)^{Ni} \Pi^* \Pi(U)^{Ni}) \Pi^* \\
&= \Pi \Pi^* \Pi \Pi^* \\
&= I_{H_1}.
\end{aligned}$$

It can be similarly shown that

$$(\Pi U^{Ni} \Pi^*)(\Pi U^{Ni} \Pi^*)^* = I_{H_1},$$

and it follows that $\Pi U^{N \circ i} \Pi^*$ is unitary. Commutativity is implicit in the proof of theorem 5. □

At this point we define an operator $V : H_2 \times \cdots H_2$ similar to that defined for the case of the single operator U. In particular, let

$$V = \left[\ (U^*)^{j_1} \Pi^* \quad (U^*)^{j_2} \Pi^* \quad \cdots \quad (U^*)^{j_{|J|}} \Pi^* \ \right]$$

where j_k is any ordering of the elements of J. As before $VV^* = I_{H_1}$ by assumption (5). From the theorems just proven showing when terms of the form $\Pi U^j \Pi^*$ are zero, it is simple to see that $V^*V = I_{H_2} \times \cdots \times H_2$.

Define the filters

$$H^{(j)} = \sum_{i \in \mathbb{Z}^D} h_i^{(j)} U^i$$

for $j \in J$. These filters each represent a specific channel in the decomposition of the signal. The filter bank is shown in Figure 3. The corresponding perfect reconstruction condition is

$$\sum_{j \in J} (H^{(j)})^* \Pi^* \Pi (H^{(j)}) = I_{H_1}.$$

Using the operator V, it becomes

$$V^* \left(\sum_{j \in J} (H^{(j)})^* \Pi^* \Pi (H^{(j)}) \right) V = I_{H_2 \times \cdots \times H_2}.$$

Looking at each term of this matrix operator gives

$$\sum_{j \in J} \Pi(U)^k (H^{(j)})^* \Pi^* \Pi (H^{(j)}) (U^*)^l \Pi^* = \begin{cases} I_{H_2} & \text{if } k = l \\ 0 & \text{otherwise} \end{cases}$$

for $k, l \in J$. The terms of the form $\Pi U^k (H^{(j)})^* \Pi^*$ can be simplified:

$$\Pi U^k (H^{(j)})^* \Pi^* = \sum_{i \in \mathbb{Z}^D} h_i^{(j)} \Pi U^{i+k} \Pi^*$$

$$= \sum_{i \in \mathbb{Z}^D} h_{Ni-k}^{(j)} \Pi U^{Ni} \Pi^*$$

$$= \sum_{i \in \mathbb{Z}^D} h_{Ni-k}^{(j)} \left(\prod_{m=1}^{D} (\Pi U^{n_m} \Pi^*)^{i_m} \right).$$

Define the operators $T_m = \Pi U^{n_m} \Pi^*$ and

$$T = \begin{bmatrix} T_1 & T_2 & \cdots & T_D \end{bmatrix}^T.$$

Define the functions

$$A_k^{(j)}(T_1, \ldots, T_D) = A_k^{(j)}(T) = \sum_{i \in \mathbb{Z}^D} h_{Ni-k}^{(j)} T^i$$

for $j \in J$. The perfect reconstruction conditions can then be expressed as

$$\sum_{j \in J} (A_k^{(j)}(T))(A_l^{(j)}(T))^* = \begin{cases} I_{H_2} & \text{if } k = l \\ 0 & \text{otherwise} \end{cases}$$

for $k, l \in J$.

To develop these conditions further, we will make use of a theorem which can be found in [6]:

Theorem 7 *For any sequence of mutually *-commutative resolvable operators $T_1, T_2, \ldots,$ there exists a Hermitian operator C so that the class of functions of the operators $T_1, T_2, \ldots,$ is identical to the class of functions of C.*

*-commutativity of T_i means that

$$T_i T_j = T_j T_i$$

and

$$T_j^* T_i = T_i T_j^*.$$

The class of resolvable operators corresponds to the class of Normal operators of which unitary operators are a subset.

In particular the operators T_m can be represented as continuous functions of some Hermitian C:

$$T_m = f_m(C).$$

for $1 \leq m \leq D$. In terms of C, the perfect reconstruction construction conditions become

$$\sum_{j \in J} (A_k^{(j)}(f_1(C), \ldots f_D(C)))(A_l^{(j)}(f_1(C), \ldots, f_D(C)))^* = \begin{cases} I_{H_2} & \text{if } k = 1 \\ 0 & \text{otherwise} \end{cases}.$$

for $k, l \in J$. Letting the spectral decomposition of C be

$$C = \int_{-\infty}^{\infty} \mu \, dP(\mu),$$

the conditions become

$$\sum_{j \in J} A_k^{(j)}(f_1(\mu), \ldots f_D(\mu)) \overline{A_l^{(j)}(f_1(\mu), \ldots, f_D(\mu))} = \begin{cases} 1 & \text{if } k = 1 \\ 0 & \text{otherwise} \end{cases}. \tag{26}$$

for all $\mu \in \sigma(C)$, for $k, l \in J$.

Because $f_m(\mu) \in \sigma(T_m)$ for all $\mu \in \sigma(C)$, it is sufficient to require that

$$\sum_{j \in J} A_k^{(j)}(e^{i\lambda_1}, \ldots, e^{i\lambda_D}) \overline{A_l^{(j)}(e^{i\lambda_1}, \ldots, e^{i\lambda_D})} = \begin{cases} 1 & \text{if } k = 1 \\ 0 & \text{otherwise} \end{cases}.$$

whenever $e^{i\lambda_m} \in \sigma(T_m)$ for all $1 \leq m \leq D$. This means that sufficient conditions for perfect reconstruction can be expressed in terms of the multidimensional Fourier Transforms

$$\hat{H}_k^{(j)}(\Lambda) = \sum_{i \in Z^D} h_{Ni-k}^{(j)} e^{i\Lambda^T n},$$

where Λ is a vector with components λ_m, as

$$\sum_{j \in J} \hat{H}_k^{(j)}(\Lambda) \overline{\hat{H}_l^{(j)}(\Lambda)} = \begin{cases} 1 & \text{if } k = 1 \\ 0 & \text{otherwise} \end{cases}, \tag{27}$$

for $k, l \in J$ and for Λ such that $\lambda_m \in \sigma(T_m)$.

It is sufficient to satisfy equation (27) on the set of Λ specified. Equation (26) shows that there is a subset of this set of Λ on which the conditions are both necessary and sufficient. Unfortunately, the D-dimensional curve traced out by $f_1(\mu), \ldots, f_D(\mu)$ forms a Peano curve which is of no use in identifying this subset. There doesn't seem to be a way to use this approach to construct such a set — only to show its existence.

It is possible to propose orthogonality conditions similar to those used in the one dimensional case. We require that

$$\Pi(H^{(\mathbf{k})})(H^{(\mathbf{l})})^*\Pi^* = 0 \tag{28}$$

for all $\mathbf{k} \neq \mathbf{l}$, $\mathbf{k}, \mathbf{l} \in J$. By assumption (5) this is equivalent to

$$
\begin{aligned}
\Pi(H^{(\mathbf{k})})(H^{(\mathbf{l})})^*\Pi^* &= \Pi(H^{(\mathbf{k})})\left(\sum_{\mathbf{j} \in J}(\mathbf{U}^{\mathbf{j}})^*\Pi^*\Pi(\mathbf{U}^{\mathbf{j}})\right)(H^{(\mathbf{l})})^*\Pi^* \\
&= \sum_{\mathbf{j} \in J}\Pi(H^{(\mathbf{k})})(\mathbf{U}^{\mathbf{j}})^*\Pi^*\Pi(\mathbf{U}^{\mathbf{j}})(H^{(\mathbf{l})})\Pi^* \\
&= 0.
\end{aligned}
$$

This takes a form similar to the perfect reconstruction conditions, and by the same type of argument, the orthogonality condition (28) is satisfied if and only if:

$$\sum_{\mathbf{j} \in J}\overline{A_{\mathbf{j}}^{(\mathbf{k})}(f_1(\mu), \ldots, f_D(\mu))}A_{\mathbf{j}}^{(\mathbf{l})}(f_1(\mu), \ldots, f_D(\mu)) = 0 \tag{29}$$

for all $\mathbf{k} \neq \mathbf{l}, \mathbf{k}, \mathbf{l} \in J$ and for $\mu \in \sigma(C)$.

The corresponding sufficient conditions for orthogonality are

$$\sum_{\mathbf{j} \in J}\overline{\hat{H}_{\mathbf{j}}^{(\mathbf{k})}(\Lambda)}\hat{H}_{\mathbf{j}}^{(\mathbf{l})}(\Lambda) = 0 \tag{30}$$

for all Λ for which $\lambda_m \in \sigma(T_m)$ whenever $\mathbf{k} \neq \mathbf{l}$, $\mathbf{k}, \mathbf{l} \in J$.

At this point, there has been no justification for using $|J|$ filters $H^{(\mathbf{j})}$. We will concentrate on the solution of equation (27) for a single value of Λ. In the single dimensional case only two such filters were used. In fact, it would be perfectly legitimate to write expressions of the form (27) and (30) for two filters: merely replace $\sum_{\mathbf{j} \in J}$ with $\sum_{j=1}^2$. The filters would then be $H^{(1)}$ and $H^{(2)}$. It would, however, be meaningless to do so, because it can be shown that $|J|$ filters are required.

Instead of indexing the transforms $A_{\mathbf{k}}^{(\mathbf{j})}(\Lambda)$ by vector indices, we use scalars: $A_k^{(j)}(\lambda)$, $1 \leq j \leq P$, and $1 \leq k \leq |J|$. The exact ordering in the correspondence between the vector and scalar versions is not important. Is it possible for the perfect reconstruction and orthogonality conditions to be satisfied if $P < |J|$? A simple restatement of equation (27) in terms of matrix multiplication shows that this is impossible.

Define

$$
M = \begin{bmatrix}
\hat{H}_1^1(\Lambda) & \hat{H}_1^2(\Lambda) & \cdots & \hat{H}_1^P(\Lambda) \\
\hat{H}_2^1(\Lambda) & \hat{H}_2^2(\Lambda) & \cdots & \hat{H}_2^P(\Lambda) \\
\vdots & \vdots & & \vdots \\
\hat{H}_{|J|}^1(\Lambda) & \hat{H}_{|J|}^2(\Lambda) & \cdots & \hat{H}_{|J|}^P(\Lambda)
\end{bmatrix}.
$$

The perfect reconstruction conditions (27) are then equivalent to $MM^H = I$ since

$$(MM^H)_{kl} = \sum_{j=1}^P (M)_{kj}(M^H)_{jl}$$

$$= \sum_{j=1}^{P} (M)_{kj} \overline{(M)_{lj}}$$

$$= \sum_{j=1}^{P} \hat{H}_k^j(\Lambda) \overline{\hat{H}_l^j(\Lambda)}.$$

M cannot have orthonormal rows if $P < |J|$, and hence for the condition $MM^H = I$ to be satisfied $P \geq |J|$ is necessary.

In a similar manner, the orthogonality condition boils down to requiring the off diagonal elements of $M^H M$ to be zero. This means that the columns of M must be orthogonal, not necessarily orthonormal. Ignoring the possibility of any zero columns, leads to the conclusion that the diagonal matrix $M^H M$ has full rank. This is not possible if $P > |J|$. We conclude that $P = |J|$ is the only interesting possibility for which the perfect reconstruction and orthogonality conditions can be satisfied. This justifies the original use of the set J for indexing $H^{(j)}$.

It is interesting to note that if $P = |J|$, then $MM^H = I$ implies that $M^H M = I$. Under the condition that there are $|J|$ filters, satisfaction of the perfect reconstruction conditions implies satisfaction of the orthogonality conditions.

In [4] and [8] sufficient conditions for perfect reconstruction in terms of the product of the analysis and synthesis polyphase matrices equaling the identity. In the case considered here the synthesis polyphase matrix is just the adjoint of the analysis polyphase matrix. They correspond to M and M^H respectively, so that the sufficient conditions given here in matrix form correspond to the standard conditions. What is new is the fact that these conditions apply to operators other than shift and decimation, and the fact that there is a set of frequencies on which the conditions are both necessary and sufficient.

7 Conclusions

This paper has shown that conditions usually derived for the coefficients of filters in perfect reconstruction filter banks can be applied to coefficients of polynomials in unitary operators more general than the shift to give "filter" operators which have a perfect reconstruction property:

$$(H^{(0)})^*\Pi^*\Pi(H^{(0)}) + (H^{(1)})^*\Pi^*\Pi(H^{(1)}).$$

The conditions on the operator $U : H_1 \to H_1$, of which $H^{(0)}$ and $H^{(1)}$ are functions, and the operator $\Pi : H_1 \to H_2$ can be expressed either in terms of the two equations

$$\Pi\Pi^* = I_{H_2}$$

and

$$\Pi^*\Pi + U^*\Pi^*\Pi U = I_{H_1},$$

or in terms of a cyclic similarity structure for U. The derivation of the conditions on the coefficients was given in terms of the equations, with important results reinterpreted in terms of the cyclic structure of U. Examples of special cases of the result were discussed.

274

An extension of the conditions on U and Π to a case involving multiple operators U_k was discussed. This corresponded to the multidimensional case. The end result was a sufficient condition on the Fourier transforms of the coefficients on the cosets of a particular lattice. A set of frequencies over which satisfaction of these conditions is sufficient to achieve perfect reconstruction was specified. The existence of a subset of this set of frequencies over which the conditions are both necessary and sufficient was shown. Unfortunately, there doesn't seem to be any constructive way to produce this set.

References

[1] I. Daubechies. Orthonormal Bases of Compactly Supported Wavelets. *Comm. Pure Appl. Math.*, **41**, pp 909-996, 1988.

[2] C. DeVito. *Functional Analysis and Linear Operator Theory*. Addison-Wesley, 1990.

[3] R.L. Grossman and H.V. Poor. *Wavelet Transforms Associated with Finite Cyclic Groups*.

[4] J. Kovacevic and M. Vetterli. Nonseparable Multidimensional Perfect Reconstruction Filter Banks and Wavelet Bases for $\Re^n$, *IEEE Trans. on Inform. Theory*, Vol. **38**, pp 533-555, 1992.

[5] S.G. Mallat. Multiresolution Approximations and Wavelet Orthonormal Bases of $L^2(R)$. *Trans. Amer. Math. Soc.*, Vol. **315**, pp 69-87, 1989.

[6] A.I. Plesner. *Spectral Theory of Linear Operators, Vol. II*, Frederick Ungar Publishing Co., 1969.

[7] G. Strang. Wavelets and Dilation Equations: A Brief Introduction, *SIAM Review*, Vol. **31**, pp 614-627, 1989.

[8] E. Viscito and J.P. Allebach. The Analysis and Design of Multidimensional FIR Perfect Reconstruction Filter Banks for Arbitrary Sampling Lattices, *IEEE Trans. on Circuits and Systems*, Vol. **38**, pp 29-41, 1991.

DETERMINING RANK IN THE PRESENCE OF ERROR

G. W. STEWART
Department of Computer Science
University of Maryland
College Park, MD 20752
stewart@cs.umd.edu

ABSTRACT. The problem of determining rank in the presence of error occurs in a number of applications. The usual approach is to compute a rank-revealing decomposition and make a decision about the rank by examining the small elements of the decomposition. In this paper we look at three commonly use decompositions: the singular value decomposition, the pivoted QR decomposition, and the URV decomposition.

KEYWORDS. Rank degeneracy, error, singular value decomposition, QR decomposition, pivoting, URV decomposition.

Introduction

The problem of determining the rank of a matrix has any number of mathematical solutions. For example, if X is an $n \times p$ $(n \geq p)$ of rank k, then X can be reduced by elementary transformations to a row echelon form in which the first k rows of X are linearly independent and the remaining rows are zero. This factorization is perhaps the most widely known example of a *rank-revealing decomposition* — a decomposition in which the rank can be read off from the pattern of zero and nonzero elements. There are, of course, many other rank revealing decompositions; e.g., the singular value decomposition, the QRP decomposition, and a variety of complete orthogonal factorizations.

The problem is far more difficult when the elements of X are contaminated with error, so that instead of X we observe

$$\tilde{X} = X + E,$$

where E is unknown. In this case we must determine the rank k of the original matrix from the contaminated matrix $\tilde{X}$. The usual approach is to compute a rank-revealing decomposition of $\tilde{X}$. Since, in general, $\tilde{X}$ will be of full rank, the decomposition will not reveal the rank by the structure of its zero elements. Instead one looks at the structure of the "small" elements in the hope that they will say something about the rank of X. However, there are several difficulties with this general approach, which we now list briefly.

275

M. S. Moonen et al. (eds.), Linear Algebra for Large Scale and Real-Time Applications, 275–291.
© 1993 *Kluwer Academic Publishers.*

1. If we compute a decomposition $\tilde{Z} = \tilde{U}^T \tilde{X} \tilde{V}$ of $\tilde{X}$ corresponding to the rank-revealing decomposition $Z = U^T X V$ of X, there is no guarantee that $\tilde{Z}$ have small elements in place of the revealing zeros of Z. As we shall see, this is a problem even when the transformations U and V are orthogonal.

2. We must have some knowledge of E to say when elements in a rank-revealing decomposition are "small". Usually this knowledge is in the form of an estimate of a norm of E, or the size of a "typical" element of E, or even a statistical distribution of the elements of E. Whatever form this knowledge takes, it must come from outside sources, i.e. the nature of the application.

3. We also need to know something about X. For example, if $\tilde{X} = \mathrm{diag}(1, 10^{-3})$ and we know that $\|E\| \cong 10^2$ in the spectral norm, we cannot say that X has rank one — only that X being of rank one is not inconsistent with what we know. In order to make a stronger statement we need to know, say, that the smallest nonzero singular value of X is greater than 10^{-2}.

4. The results of a rank determination can vary with the scaling. For example, if the last row of the matrix X in the preceding item of this list is multiplied by 10^3, then X becomes the identity, and the most we can say about E is that $\|E\| \cong 10$. In this case the data are consistent with X being the zero matrix! The usual fix is to attempt to scale $\tilde{X}$ so that the elements of E are roughly the same size. But this is not always easily done.

5. In most applications, rank determination is only a beginning. What is done subsequently often requires a knowledge of the column or null spaces of X or X^T. Thus any rank-revealing decomposition must produce approximations to these spaces.

6. In many applications the rows of X are not fixed but change as rows are added and deleted. Since rank-revealing decompositions are usually too expensive to recompute *ab initio*, our decomposition should be updatable.

Keeping these difficulties in mind, we are going discuss three decompositions that are used to determine the rank of a matrix: the singular value decomposition, the pivoted QR decomposition (also called the QRP decomposition), and an intermediary called the URV decomposition. All three are based on orthogonal transformations, which have the desirable property that they cannot magnify the error in X. In the next three sections we will describe these decompositions and discuss their numerical properties. In the concluding section we will treat the problem of using these decompositions to determine rank in the presence of error.

1 The singular value decomposition

The singular value decomposition is the *crème de la crème* among rank-revealing decomposition. It has the form

$$U^T X V = \begin{pmatrix} \Sigma \\ 0 \end{pmatrix}, \tag{1}$$

where U and V are orthogonal and

$$\Sigma = \mathrm{diag}(\sigma_1, \sigma_2, \ldots, \sigma_p),$$

with

$$\sigma_1 \geq \cdots \geq \sigma_k > 0 = \sigma_{k+1} = \cdots = \sigma_p.$$

Thus the rank k of X is revealed by the fact that its $p - k$ largest singular values are nonzero while its k smallest singular values are zero.

The singular value decomposition easily provides orthonormal bases for range and null spaces associated with X. Specifically, if we partition

$$U = (U_1 \; U_2) \quad \text{and} \quad V = (V_1 \; V_2), \tag{2}$$

then:

1. The columns of U_1 form an orthonormal basis for the column space of X.

2. The columns of U_2 form an orthonormal basis for the null space of X^{T}.

3. The columns of V_1 form an orthonormal basis for the column space of X^{T}.

4. The columns of V_2 form an orthonormal basis for the null space of X.

The singular value decomposition behaves well in the presence of error. Specifically, if

$$\tilde{U}\tilde{X}\tilde{V}^{\mathrm{T}} = \begin{pmatrix} \tilde{\Sigma} \\ 0 \end{pmatrix}$$

is the singular value decomposition of $\tilde{X}$, then it follows from Schmidt's theorem that

$$\tilde{\sigma}_{k+1}^2 + \cdots + \tilde{\sigma}_p^2 \leq \|E\|_{\mathrm{F}}^2, \tag{3}$$

where $\| \cdot \|_{\mathrm{F}}$ denotes the Frobenius norm. Thus the singular value decomposition of $\tilde{X}$ reveals the rank in the sense that the sum of squares of its $p - k$ smallest singular values are bounded by the Frobenius norm of E.[1] Moreover, the spaces spanned by $\tilde{U}_1$, $\tilde{U}_2$, $\tilde{V}_1$, and $\tilde{V}_2$ are approximations to the subspaces listed above that are accurate to about $\sigma_k^{-1}\|E\|$. Thus if σ_k is reasonably large compared to E — i.e., the problem has a favorable signal to noise ratio — the singular value decomposition provides good approximations to the desired subspaces.

The singular value decomposition can be computed in many ways, among which the following three are the most common.

1. Reduce X to bidiagonal form by two sided orthogonal transformations and reduce the bidiagonal form to diagonal form by a variant of the QR algorithm.

2. Reduce X to triangular form by transformations applied on the right (i.e., compute the QR decomposition of X). Then compute the singular value decomposition of the triangular matrix.

---

[1] Actually, each of the $n - k$ smallest singular values are bounded by the spectral norm of E; however, this result is less useful in practice than (3).

3. Compute the eigendecomposition of the symmetric cross-product matrix $X^T X$.

The first method is by far the most expensive, though it is standard for one-shot jobs. The other two approaches have the advantage that the intermediate decompositions can be updated in $O(p^2)$ time, the third trivially and very quickly. It is true that when we update in this way, we loose the ability to compute the matrix U; but in most updating applications U is not required. It is sometimes objected that computing the singular value decomposition via the cross-product matrix is numerically unstable. But these instabilities only become important when σ_k^2/σ_1^2 approaches the rounding unit of the computer arithmetic — something that seldom happens in the presence of errors other than rounding error.

In the updating game, no matter which of the above algorithms is used, one is left with the problem of updating a singular value decomposition or eigendecomposition of a square matrix of order p. Unfortunately, no algorithms that perform these updates in less that $O(p^3)$ time are known, a fact that severely restricts the use of the singular value decomposition in real-time applications. Recent work has focused on maintaining an approximate diagonal form that is good enough for practical purposes. However, an alternative to which we now turn, is to work with more computationally tractable decompositions.

2 QRP decompositions

The pivoted QR decomposition, or the QRP decomposition as it will be called here, is actually a class of decompositions. Specifically, there is a permutation matrix P and an orthogonal matrix Q such that

$$Q^T X P = \begin{pmatrix} R \\ 0 \end{pmatrix}, \tag{4}$$

where

$$R = \begin{pmatrix} R_{11} & R_{12} \\ 0 & 0 \end{pmatrix}$$

with R_{11} an upper triangular matrix of order k having positive diagonal elements. The decomposition is not unique, since P can be any permutation matrix (also called a *pivot matrix*) such that the first k columns of XP are unique. Once P has been determined, however, the matrices R_{11} and R_{12} are uniquely determined, as are the first k columns of Q.

If we partition $XP = (X_1\ X_2)$ and $Q = (Q_1\ Q_2)$, where X_1 and Q_1 have k columns, then

1. The columns of X_1 and Q_1 form an orthonormal basis for the column space of X.

2. The columns of Q_2 form an orthonormal basis for the null space of X^T.

3. The columns of $(R_{11}\ R_{12})^T$ form a (nonorthonormal) basis for the column space of X^T.

4. The columns of $(-R_{12}^T R_{11}^{-T} \ I)^T$ form a (nonorthonormal) basis for the null space of X.

This list illustrates some of the strengths and weaknesses of QRP decompositions. As far as the row space of X and the null space of X^T are concerned, the matrix Q is entirely analogous to the matrix U of the singular value decompositions. Moreover, the first k columns of XP form a basis for the columns space of X; i.e., the QRP decomposition picks out a set of linearly independent columns, unlike the singular value decomposition, which merely furnishes a basis for the column space. Unfortunately, there is no analogue of the matrix V, and the corresponding bases, which must be obtained from R, are not orthonormal. Moreover, the basis for the null space of X requires additional computation for its formation. This is particularly unfortunate, since many applications require an orthonormal basis for this subspace.

In the presence of error, we should like to determine a permutation matrix $\tilde{P}$ and an orthogonal matrix $\tilde{Q}$ such that

$$\tilde{Q}^T \tilde{X} \tilde{P} = \begin{pmatrix} \tilde{R}_{11} & \tilde{R}_{12} \\ 0 & G_{22} \end{pmatrix}, \tag{5}$$

where R_{11} is upper triangular of order k with positive diagonal elements, and G_{22} is a triangular matrix satisfying

$$\|G_{22}\| = O(\|E\|).$$

Note that this amounts to finding a suitable pivot matrix P, since once P is chosen, the rest of the decomposition is essentially unique.

The question of the existence of a rank-revealing QRP decomposition in the sense of the preceding paragraph has only recently been answered in the affirmative, Unfortunately, the proof is not constructive, and the problem of efficiently computing a provably rank-revealing QRP decomposition is still an active area of research (for more see the notes and references at the end of the paper).

For practical purposes, however, almost any sensible strategy will work. The standard algorithm is unitary triangularization with pivoting on the column of largest norm. This procedure is mathematically (though not numerically) equivalent to the Gram-Schmidt algorithm in which the largest projected vector is the next to enter the orthogonalization.

An alternative is the rank-revealing algorithm of T. Chan, which starts from an unpivoted QR decomposition and moves linearly dependent columns of R to the end. The first step of this algorithm is typical. A condition estimator is used to find a vector v of norm one such that $\epsilon = \|Rv\|$ is small. A permutation matrix $\hat{P}$ and an orthogonal matrix $\hat{Q}$ are determined so that

1. the last component of $\hat{v} = \hat{P}^T v$ is the largest,

2. $\hat{R} = \hat{Q}^T R \hat{P}$ is upper triangular.

Since the magnitude of the last component of $\hat{v}$ is not less than $1/\sqrt{p}$ and $\|\hat{R}\hat{v}\| = \epsilon$, it follows that $|\hat{r}_{pp}| \le \sqrt{p}\epsilon$; i.e., $\hat{R}$ reveals the degeneracy in the rank of R. Unfortunately, when this procedure is iterated, the provable bound on the size of the resulting G_{22} grows exponentially, though in practice the algorithms works well enough.

At present there seems to be no efficient algorithm for updating a rank-revealing QRP algorithm.

3 URV and ULV decompositions

Although the simplicity of QRP decompositions makes them attractive, the fact that P must be a permutation matrix is a combinatorial constraint that makes analysis difficult and algorithms hard to come by. In this section we will consider another class of decompositions that mitigates these problems by relaxing the restriction on P.

The decompositions are based on what is sometimes called a complete orthogonal decomposition. If X is exactly of rank k, there are orthogonal matrices U and V such that

$$ U^T X V = \begin{pmatrix} R_{11} & 0 \\ 0 & 0 \end{pmatrix}, $$

where R is an upper triangular matrix of order k having positive diagonal elements. Such a decomposition is not unique: the singular value decomposition is an extreme example. However, by relaxing the restriction that the decomposition be diagonal, we introduce extra degrees of freedom in U and V that make for flexability.

This flexability does not imply a loss of information. If we partition U and V as in (2), then the statements following that equation remain true. In other words, the decomposition provides orthonormal bases for the range and null spaces associated with X.

A rank-revealing URV decomposition of $\tilde{X}$ is a decomposition of the form

$$ \tilde{U}^T \tilde{X} \tilde{V} = \begin{pmatrix} R_{11} & F_{12} \\ 0 & G_{22} \end{pmatrix}, $$

where, as usual, R is an upper triangular matrix of order k having positive diagonal elements and F_{12} and G_{22} are of order $\|E\|$. Unlike the QRP decomposition, the URV decomposition can be made provably rank revealing. The process begins like Chan's method by reducing X to a triangular form R and using a condition estimator to find a vector v of norm one such that $\epsilon = \|Rv\|$ is small. Orthogonal matrices $\hat{U}$ and $\hat{V}$ are determined so that

1. $\hat{V}^T v = e_p$, where e_p is the vector whose last component is one and whose other components are zero,

2. $\hat{R} = \hat{U}^T R \hat{V}$ is upper triangular.

It then follows that $\hat{R}$ has the form

$$ \hat{R} = \begin{pmatrix} \hat{R}_{11} & f_{12} \\ 0 & \gamma_{22} \end{pmatrix}, $$

where $\|(f_{12}^T \ \gamma_{22})\| = \epsilon$. Thus $\hat{R}$ reveals the rank degeneracy in R. However, instead of just the (p,p)-element of $\hat{R}$ being small, as in the QRP decomposition, the entire last column of

$\hat{R}$ is small. This fact allows us to iterate the process on $\hat{R}_{11}$ to get a provably rank-revealing decomposition of X. At each stage, a block variant of the QR algorithm, can be applied to further reduce the size of f_{12}.

An attractive feature of rank-revealing URV decompositions is that they can be updated in $O(p^2)$ time. Moreover, the updating algorithm can be implemented in $O(p)$ time on a linear array of p processors. The updating algorithm can be started with the zero matrix, so that there is no need to compute an initial decomposition.

There is also a rank-revealing ULV decomposition, in which the target matrix is lower triangular. Surprisingly, these decompositions are not mere variants of one another but have different mathematical algorithmic properties — an area for future research.

4 Rank determination

The term "rank-revealing decomposition" is something of a misnomer, since it implies that the decomposition automatically reveals rank. As we indicated in Items 2 and 3 of the list in the introduction, a decomposition alone is never sufficient: we need to know something about the error, and perhaps also about the original matrix. In this section, we will discuss the how to use our three rank-revealing decompositions to determine rank in the presence of errors.

4.1 THE SINGULAR VALUE DECOMPOSITION

Suppose for the moment that we know the rank k of X and desire to estimate the matrix X from $\tilde{X}$. A natural procedure is to try to approximate $\tilde{X}$ by a matrix of rank k in the least squares sense. Otherwise put, our estimate of X is a matrix $\hat{X}$ that satisfies

$$\|\tilde{X} - \hat{X}\|_F = \min_{\text{rank}(Y) \leq k} \|\tilde{X} - Y\|_F.$$

Fischer's theorem says that $\hat{X}$ exists and that

$$\|\tilde{X} - \hat{X}\|_F^2 = \tilde{\sigma}_{k+1}^2 + \cdots + \tilde{\sigma}_n^2 \stackrel{\text{def}}{=} \tilde{\tau}_k^2 \tag{6}$$

For our purposes the most important consequence of Fischer's theorem is the following. Since $\hat{X}$ is minimizing, the right hand side of (6) can only increase when we replace $\hat{X}$ by X. Consequently,

$$\|E\|_F^2 \geq \|X - \hat{X}\|_F^2 = \tilde{\tau}_k^2.$$

The implication is that if $\|E\|_F$ is smaller than $\tilde{\tau}_k^2$ then X could not possibly been of rank k. Thus a natural choice of k is the smallest integer such that

$$\|E\|_F^2 \geq \tilde{\tau}_k^2.$$

This strategy works well, provided the errors are well scaled and σ_k is well above the error level. In this case $\tilde{\tau}_k^2$ remains below $\|E\|_F^2$, but the presence of $\tilde{\sigma}_k$ forces the sum $\tilde{\tau}_{k-1}^2$ to be larger than $\|E\|_F^2$ (recall that from the perturbation theory for singular values, $\tilde{\sigma}_k \geq \sigma_k - \|E\|$, so that $\tilde{\sigma}_k$ is large along with σ_k).

If we are willing to assume more about E, we can refine our procedure. Let us suppose that the elements of E are uncorrelated random variables with mean zero and standard deviation ϵ. Because the matrices U and V in the singular value decomposition (1) of X are orthogonal, the elements of the matrices H_{ij} in

$$U^T \tilde{X} V = U^T X V + U^T E V \equiv \begin{pmatrix} \Sigma_1 + H_{11} & H_{12} \\ H_{21} & H_{22} \end{pmatrix} \tag{7}$$

are also uncorrelated with mean zero and standard deviation ϵ. Now if σ_k is large compared with $\|E\|$, then

$$\tilde{\tau}_k^2 = \|H_{22}\|_F^2 + O(\|E\|^3). \tag{8}$$

It follows that that the average value of $\tilde{\sigma}_{k+1}^2 + \cdots + \tilde{\sigma}_p^2$ is approximated by

$$\mathbf{E}(\tilde{\tau}_k^2) \cong (n-k)(p-k)\epsilon^2.$$

Consequently, we should choose k to be the smallest integer such that

$$\tilde{\tau}_k^2 < \phi \cdot (n-k)(p-k)\epsilon^2. \tag{9}$$

The number ϕ in (9) is a fudge factor that compensates for the fact that $\tilde{\tau}_k^2$ will often be larger than its mean. If it is too small, i.e. too near one, the test will tend to overestimate the rank. If it is too large, the test will tend to underestimate the rank. If we know the distribution of the elements of E and the value of σ_k, we can choose ϕ to trade these errors off against one another. However, it seldom happens in practice that we have such precise information, and the value of ϕ must usually be chosen on the basis of experience.

The statistical assumptions about the elements of E — that they are uncorrelated with mean zero and common standard deviation ϵ — correspond to the equal error scaling mentioned in Item 4 of the list in the introduction. If these assumptions are not satisfied, it may be possible to scale the problem so that they are. For example, if the rows of E are uncorrelated with mean zero and dispersion (variance) matrix Δ, then the elements of $E\Delta^{-\frac{1}{2}}$ are uncorrelated with mean zero and standard deviation one. This process is sometimes called "whitening" the noise. However, it cannot be applied when Δ is singular (e.g., when a column of $\tilde{X}$ is without error). What to do in such situations is imperfectly understood.

Finally, it is important not to lay too much stress on detailed statistical assumptions about the error. Informally all that is needed is for the elements of E to be roughly the same size ϵ and to remain so under unrelated orthogonal transformations. In that case the elements of the matrix $H_{22} = U_2^T E V_2$ will be of roughly of size ϵ and $\|H_{22}\|_F^2$ will be approximately $(n-k)(p-k)\epsilon^2$. This is all that is required for the validity of the test (9).

4.2 QRP DECOMPOSITIONS

Rank determination with QRP decompositions is not as straightforward as it is with the singular value decomposition. In the first place there is no analogue of Fisher's theorem for the decomposition. Moreover, if we attempt to repeat the development that produced the test (9) we run into difficulties.

To see why, let us apply the transformations Q and P to the matrix $\tilde{X}$. The result is

$$U^T \tilde{X} P = \begin{pmatrix} R_{11} + H_{11} & R_{12} + H_{12} \\ H_{21} & H_{22} \end{pmatrix}. \tag{10}$$

Now the matrix H_{22} is quite tractable. Under our assumptions about the distribution of the elements of E, the expectation of $\|H_{22}\|_F^2$ is $(n-k)(p-k)\epsilon^2$. Unfortunately, $\|H_{22}\|_F$ is not an approximation of $\|G_{22}\|_F$, which is what we have to work with. The reason is that $U^T \tilde{X} P$ is not in triangular form, and when we reduce it to triangular form, the matrix G_{22} [c.f. (5)] becomes contaminated by the elements of the large matrix R_{12}. (The same sort of thing does not occur with the singular value decomposition because *both* off-diagonal blocks in (7) are small.)

Fortunately, we can still approximate the expected value of $\|G_{22}\|_F$. Specifically, the norm of G_{22} is the norm last $n-k$ rows of the projection of the last column of (10) onto the orthogonal complement of the space spanned by the first column. Explicitly, the norm of G_{22} is the norm of the matrix

$$H_{22} - H_{21}[(R_{11} + H_{11})^T (R_{11} + H_{11}) + H_{21}^T H_{21}]^{-1}[(R_{11} + H_{11})^T (R_{12} + H_{12}) + H_{21}^T H_{22}].$$

If we ignore higher order terms, we get

$$\|G_{22}\|_F^2 \cong \|H_{22} - H_{21} R_{11}^{-1} R_{12}\|_F^2. \tag{11}$$

Consequently, the expected value of $\|G_{22}\|_F^2$ should be approximately

$$(n-k)\text{trace}[I_{p-k} + R_{12}^T (R_{11}^T R_{11})^{-1} R_{12}]\epsilon^2. \tag{12}$$

Of course we do not know R_{11} and R_{12}; however, the computed matrices $\tilde{R}_{11}$ and $\tilde{R}_{12}$ are small perturbations of the originals and can be used in their place. Thus, our test is to choose k to be the smallest integer such that

$$\|G_{22}\|_F^2 < \phi \cdot (n-k)\text{trace}[I_{p-k} + \tilde{R}_{12}^T (\tilde{R}_{11}^T \tilde{R}_{11})^{-1} \tilde{R}_{12}]\epsilon^2, \tag{13}$$

where ϕ is the usual fudge factor.

4.3 URV DECOMPOSITIONS

The same argument that was used to show (8) can be used to show that for the URV decomposition

$$\tilde{\tau}_k^2 = \|G_{22}\|_F^2 + O(\|E\|^3).$$

Consequently an appropriate test for the URV decomposition is to choose the smallest k such that

$$\|G_{22}\|_F^2 < \phi \cdot (n-k)(p-k)\epsilon^2.$$

5 Notes and references

The problem of rank determination in the presence of error arises in a number of applications: e.g., variable selection in statistics and engineering [7, 47, 61], direction of arrival estimation in signal processing [1, 58, 59], and the projection of ill-conditioned problems onto manifolds where they become well conditioned [51, 21, 22]. In many instances, the original matrix X is not exactly of rank k as we have described it in the introduction. Instead physical approximations or infelicities in the model make X only approximately of rank k, though the deviation must be less than the error for the techniques described here to have approximate validity.

Closely related, but of a different flavor, is the problem of regularizing the ill-posed problems which arise from discretizations of compact or unbounded operators [24, 32, 45, 57, 71, 75].

The fact that something must be known about the errors in order to make statements about rank is a commonplace in areas like signal processing, where the errors are relatively large, or statistics, where there is a vast literature under the heading "errors in the variables" [2, 3, 5, 4, 8, 20, 34, 48, 65], or numerical analysis, where there is a growing literature under the heading "total least squares" [31, 41, 63, 72, 74].

Although the updating of least squares solutions goes back to Gauss [35], the updating of decompositions seems to have arisen in linear program, where the inverse basis matrix must be updated [19]. A closely related problem is that of downdating — the removal of rows from X — a process that is also called windowing. The literature on updating and downdating is too voluminous to survey here.

The singular value decomposition dates to the last half of the nineteenth century (for a history see [67]). The theorem cited here as Schmidt's theorem [60], is often attributed to Eckart and Young [29], who rediscovered it thirty years later. The popularity of the singular value decomposition in numerical analysis is due to Golub and Kahan [38].

Until recently reduction to bidiagonal form followed by a variant of the QR algorithm, due to Golub [40], has been the standard way to compute the decomposition. Recently new algorithms for reducing the bidiagonal matrix have been proposed [25, 30]. The idea of first computing the QR decomposition has been exploited by Chan [12, 11]. Beltrami [6] first established the existence of the singular value decomposition in 1873 by computing the eigendecomposition of the cross-product matrix, and this is still a popular way of doing things in some disciplines. In fact, sometimes the singular value decomposition completely disappears.

Algorithms for updating the singular value decomposition have been given in [10, 18]; however, they require $O(p^3)$ operations, the same as required to compute the decomposition from scratch. Iterative algorithms that maintain an approximate factorization may be found in [54, 55, 56].

Formulas for the discrete version of the Gram-Schmidt algorithm can be found in the first supplement to Laplace's *Théoria Analytique des Probabilities* [52]; however, Laplace was after an expression for the variance of a regression parameter and did not regard his formulas as a computational device. Gram [44] and Schmidt [60] orthogonalized series of functions: Gram by determinantal expressions (hence the Gramian matrix) and Schmidt by the now classic algorithm. The use of orthogonal transformations to compute the decomposition is due to Householder [50], Bogert and Burris [9], and Golub [36]. The last mentioned work also contains the notion of column pivoting and the first updating algorithm for the

QR decomposition. The name QR decomposition is from Francis's QR algorithm [33], which uses the decomposition.

Although pivoting for column size while computing the QR decomposition has long been regarded as a reliable way of determining rank (e.g., see [39, 62]), Chan [13] was the first to give bounds for a rank-revealing decomposition (the descriptive phrase "rank revealing" was coined by him). Unfortunately, the bounds were exponential in the defect $p - k$ in the rank. In fact, only recently have Hong and Pan [49] established the existence of a rank revealing QR decomposition. Although their approach is not constructive, Chandrasekaran and Ipsen [14] have given an algorithm, which unfortunately has combinatorial complexity (this paper is an excellent source for other pivoting strategies that have appeared in the literature). In a personal communication and Pan and Tang have described and algorithm that requires less work.

It is important to distinguish the sense in which the theory of Hong and Pan and the algorithms mentioned above are rank revealing. Both take an integer k and produce a permutation that reveals if there is a gap between the kth and $(k+1)$th singular value. Change k and the permutation changes, so that the rank is not necessarily revealed for all k simultaneously.

The ability to cheaply compute an approximate null vector of a triangular matrix — a topic which goes under the slightly misleading name of "conditions estimation" — is fundamental to some algorithms for computing a rank-revealing QRP decomposition as well as the URV and ULV decompositions. Although the first such algorithm is found in [43], it was LINPACK [27] that popularized the idea. For a survey with references see [46].

URV and ULV decompositions [68, 66] had their genesis in the author's unsuccessful attempt to update a rank-revealing QR factorization. A refinement step, which tends to decrease the size of the off-diagonal elements has been analyzed in [53] (see also [56, 15, 28]). A parallel implementation of the updating algorithm is described in [69].

The methods treated here are not the only ones for revealing rank. For example, methods based on the Lanczos algorithm have been proposed for the case where the rank is small [17, 77, 78].

The perturbation of singular values, including Fischer's theorem, is surveyed in [70]. The relation (8) is a consequence of theorems in [53]. The approach to rank determination followed here is rather crude, suitable for the crude models and data one can expect in practice. However, if one can assume normality, then $\tilde{\tau}_k^2$ is approximately χ^2, a fact that can be used to determine a value for the fudge facter ϕ in (9). More generally the singular values $\sigma_{k+1}^2, \ldots \sigma_p^2$ are approximately the eigenvalues of a Wishart matrix, whose distributions are known (e.g., see [26, 16]).

The problem of poorly scaled errors is closely related to the problem of artificial ill-conditioning, which is discussed in [64]. The equal error scaling advocated there is the equivalent of noise whitening. One solution to the problem of constrained errors is to project the problem onto a submanifold where the errors can be whitened [23, 37, 73].

The results on testing QRP decompositions appear to be new. The consequence of (11) and (12) are that $\|G_{22}\|_F^2$ will tend to be larger than $\tilde{\rho}_k^2$. Comparing (9) and (13), we see that the latter has been increased to compensate for this fact.

References

[1] G. Adams, M. F. Griffin, and G. W. Stewart. Direction-of-arrival estimation using the rank-revealing URV decomposition. *Proceedings of the IEEE International Conference on Acoustics, Speech and Signal Processing,* Washington, DC, IEEE, 1991.

[2] T. W. Anderson. Estimation of linear functional relationships: Approximate distributions and connections with simultaneous equations in econometrics. *Journal of the Royal Statistical Society,* **38**, pp 1–31, 1976.

[3] A. E. Beaton, D. B. Rubin, and J. L. Barone. The acceptability of regression solutions: Another look at computational accuracy. *Journal of the American Statistical Association,* **71**, pp 158–168, 1976.

[4] D. A. Belsley. Assessing the presence of harmful collinearity and other forms of weak data through a test for signal-to-noise. *Journal of Econometrics,* **20**, pp 211–253, 1982.

[5] D. A. Belsley, A. E. Kuh, and R. E. Welsch. *Regression Diagnostics: Identifying Influential Data and Sources of Collinearity.* John Wiley and Sons, New York, 1980.

[6] E. Beltrami. Sulle funzioni bilineari. *Giornale di Matematiche ad Uso degli Studenti Delle Universita,* **11**, pp 98–106, 1873. An English translation by D. Boley is available as University of Minnesota, Department of Computer Science, Technical Report 90–37, 1990.

[7] K. N. Berk. Comparing subset regression procedures. *Technometrics,* **20**, pp 1–6, 1978.

[8] J. Berkson. Are there two regressions. *Journal of the American Statistical Association,* **45**, pp 164–180, 1950.

[9] D. Bogert and W. R. Burris. Comparison of least squares algorithms. Report ORNL-3499, Neutron Physics Division, Oak Ridge National Laboratory. Vol. 1, Sec. 5.5., 1963

[10] J. R. Bunch and C. P. Nielsen. Updating the singular value decomposition. *Numerische Mathematik,* **31**, pp 111–129, 1978.

[11] T. F. Chan. Algorithm 581: An improved algorithm for computing the singular value decomposition. *ACM Transactions on Mathematical Software,* **8**, pp 84–88, 1982.

[12] T. F. Chan. An improved algorithm for computing the singular value decomposition. *ACM Transactions on Mathematical Software,* **8**, pp 72–83, 1982.

[13] T. F. Chan. Rank revealing QR factorizations. *Linear Algebra and Its Applications,* **88/89**, pp 67–82, 1987.

[14] S. Chandrasekaran and I. Ipsen. Perturbation theory for the solution of systems of linear equations. Research Report YALEU/DCS/RR-866, Department of Computer Science, Yale University, 1991.

[15] S. Chandrasekaran and I. Ipsen. Analysis of a QR algorithm for computing singular values. Research Report YALEU/DCS/RR-917, Department of Computer Science, Yale University, 1992.

[16] R. Choudary, Hanumara, and W. A. Thompson Jr. Percentage points of the extreme roots of a Wishart matrix. *Biometrika*, **55**, pp 505–512, 1968.

[17] P. Comon and G. H. Golub. Tracking a few extreme singular values and vectors in signal processing. *Proc. IEEE*, **78**, pp 1327–1343, 1990.

[18] J. J. M. Cuppen. The singular value decomposition in product form. *SIAM Journal on Scientific and Statistical Computing*, 4, pp 216–222, 1983. Cited in [42].

[19] G. B. Dantzig. *Linear Programming and Extensions*. Princeton University Press, Princeton, New Jersey, 1963.

[20] R. B. Davies and B. Hutton. The effects of errors in the independent variables in linear regression. *Biometrika*, **62**, pp 383–391, 1975.

[21] J. Demmel. *A Numerical Analyst's Jordan Canonical Form*. PhD thesis, Department of Computer Science, University of California at Berleley, 1983. Center for Pure and Applied Mathematics Technical Report PAM-156.

[22] J. Demmel. On condition numbers and the distance to the nearest ill-posed problem. *Numerische Mathematik*, **51**, pp 251–290, 1987.

[23] J. Demmel. The smallest perturbation of a submatrix which lowers the rank and constrained total least squares problems. *SIAM Journal on Numerical Analysis*, **24**, pp 199–206, 1987.

[24] J. Demmel and B. Kågström. Accurate solutions of ill-posed problems in control theory. *SIAM Journal on Matrix Analysis and Applications*, **9**, pp 126–145, 1988.

[25] J. Demmel and W. Kahan. Accurate singular values of bidiagonal matrices. *SIAM Journal on Scientific and Statistical Computing*, 11, pp 873–912, 1989.

[26] A. P. Dempster. *Elements of Continuous Multivariate Analysis*. Addison-Wesley, Reading, Massachusetts, 1969.

[27] J. J. Dongarra, J. R. Bunch, C. B. Moler, and G. W. Stewart. *LINPACK User's Guide*. SIAM, Philadelphia, 1979.

[28] E. M. Dowling, L. P. Ammann, and R. D. DeGroat. A tqr-iteration based adaptive svd for real time angle and frequency tracking. Erik Johnson School of Engineering and Computer Science, The University of Texas at Dallas. Manuscript submitted to *IEEE Transactions on Signal Processing*, 1992.

[29] C. Eckart and G. Young. The approximation of one matrix by another of lower rank. *Psychometrika*, **1**, pp 211–218, 1936.

[30] K. V. Fernando and B. Parlett. Accurate singular values and differential qd algorithms. Technical Report PAM–554, Department of Mathematics, University of California, Berkeley, 1992.

[31] R. D. Fierro and J. R. Bunch. Multicollinearity and total least squares. Preprint Series 977, Institute for Mathematics and Its Applications, 1992.

[32] H. E. Fleming. Equivalence of regularization and truncated iteration in the solution of ill-posed image reconstruction problems. *Linear Algebra and Its Applicaations*, **130**, pp 133–150, 1990.

[33] J. G. F. Francis. The QR transformation, parts I and II. *Computer Journal*, **4**, pp 265–271, pp 332–345, 1961, 1962.

[34] W. A. Fuller. *Measurement Error Models*. John Wiley, New York, 1987.

[35] C. F. Gauss. Theoria combinationis observationum erroribus minimis obnoxiae, pars posterior. In : *Werke, IV*, pp 27–53. Königlichen Gesellshaft der Wissenschaften zu Göttingin (1880), 1823.

[36] G. H. Golub. Numerical methods for solving least squares problems. *Numerische Mathematik*, **7**, pp 206–216, 1965.

[37] G. H. Golub, A. Hoffman, and G. W. Stewart. A generalization of the Eckart-Young matrix approximation theorem. *Linear Algebra and Its Applications*, **88/89**, pp 317–327, 1987.

[38] G. H. Golub and W. Kahan. Calculating the singular values and pseudo-inverse of a matrix. *SIAM Journal on Numerical Analysis*, **2**, pp 205–224, 1965.

[39] G. H. Golub, V. Klema, and G. W. Stewart. Rank degeneracy and least squares problems. Technical Report TR-456, Department of Computer Science, University of Maryland, 1976.

[40] G. H. Golub and C. Reinsch. Singular value decomposition and least squares solution. *Numerische Mathematik*, **14**, pp 403–420, 1970. Also in [76, pp.134–151].

[41] G. H. Golub and C. F. Van Loan. An analysis of the total least squares problem. *SIAM Journal on Numerical Analysis*, **17**, pp 883–893, 1980.

[42] G. H. Golub and C. F. Van Loan. *Matrix Computations*. Johns Hopkins University Press, Baltimore, Maryland, 2nd edition, 1989.

[43] W. B. Gragg and G. W. Stewart. A stable variant of the secant method for solving nonlinear equations. *SIAM Journal on Numerical Analysis*, **13**, pp 880–903, 1976.

[44] J. P. Gram. Über die Entwicklung reeler Functionen in Reihen mittelst der Methode der kleinsten Quadrate. *Journal für die reine und angewandte Mathematik*, **94**, pp 41–73, 1883.

[45] P. C. Hansen. Truncated singular value decomposition solutions to discrete ill-posed problems with ill-determined numerical rank. *SIAM Journal on Scientific and Statistical Computing*, **11**, pp 503–518, 1990.

[46] N. J. Higham. A survey of condition number estimation for triangular matrices. *SIAM Review*, **29**, pp 575–596, 1987.

[47] R. R. Hocking. The analysis and selection of variables in linear regression. *Biometrics*, **32**, pp 1–49, 1976.

[48] S. D. Hodges and P. G. Moore. Data uncertainties and least squares regression. *Applied Statistics*, **21**, pp 185–195, 1972.

[49] Y. P. Hong and C.-T. Pan. Rank-revealing QR factorizations and the singular value decomposition. *Mathematics of Computation*, **58**, pp 213–232, 1992.

[50] A. S. Householder. Unitary triangularization of a nonsymmetric matrix. *Journal of the ACM*, **5**, pp 339–342, 1958.

[51] W. Kahan. Conserving confluence curbs ill-conditioning. Technical Report 6, Computer Science Department, University of California, Berkeley, 1972.

[52] P. S. Laplace. *Théoria analytique des probabilities (3rd ed.) premier supplement: Sur l'application du calcul des probabilités a la philosophie naturelle. Oeuvres, v. 7.* Gauthier-Villars, 1820. Supplement published before 1820.

[53] R. Mathias and G. W. Stewart. A block qr algorithm and the singular value decomposition. Technical Report CS-TR 2626, Department of Computer Science, University of Maryland, College Park, 1992. To appear in *Linear Algebra adn Its Applications*.

[54] M. Moonen. *Jacobi-Type Updating Algorithms for Signal Processing, Systems Identification and Control.* PhD thesis, Katholieke Universiteit Leuven, 1990.

[55] M. Moonen, P. Van Dooren, and J. Vandewalle. Combined Jacobi-type algorithms in signal processing. In : R. J. Vaccaro (Ed.), *SVD and Signal Processing, II,* pp 177-188, Amsterdam, 1991. Elsevier Science Publishers.

[56] M. Moonen, P. Van Dooren, and F. Vanpoucke. On the QR algorithm and updating the SVD and URV decomposition in parallel. Report ESAT Katholieke Universiteit Leuven, 1992.

[57] D. P. O'Leary and J. A. Simmons. A bidiagonalization-regularization procedure for large scale discretizations of ill-posed problems. *SIAM Journal on Scientific and Statistical Computing*, **2**, pp 474–489, 1981.

290

[58] S. Prasac and B. Chandna. Direction-of-arrival estimation using rank revealing QR factorization. *IEEE Transactions on Signal Processing*, **39**, pp 1224–1229, 1991. Citation communicated by Per Christian Hansen.

[59] R. Roy and T. Kailath. ESPRIT–estimation of signal parameters via rotational invariance techniques. In : F. A. Grünbaum, J. W. Helton, and P. Khargonekar (Eds), *Signal Processing Part II: Control Theory and Applications*, pp 369–411, New York, 1990. Springer.

[60] E. Schmidt. Zur Theorie der linearen und nichtlinearen Integralgleichungen. I Teil. Entwicklung willkürlichen Funktionen nach System vorgeschriebener. *Mathematische Annalen*, **63**, pp 433–476, 1907.

[61] G. N. Stenbakken, T. M. Souders, and G. W. Stewart. Ambiguity groups and testability. *IEEE Transactions on Instrumentation*, **38**, pp 941–947, 1989.

[62] G. W. Stewart. The efficient generation of random orthogonal matrices with an application to condition estimators. *SIAM Journal on Numerical Analysis*, **17**, pp 403–404, 1980.

[63] G. W. Stewart. On the invariance of perturbed null vectors under column scaling. *Numerische Mathematik*, **44**, pp 61–65, 1984.

[64] G. W. Stewart. Rank degeneracy. *SIAM Journal on Scientific and Statistical Computing*, **5**, pp 403–413, 1984.

[65] G. W. Stewart. Collinearity and least squares regression. *Statistical Science*, **2**, pp 68–100, 1987.

[66] G. W. Stewart. Updating a rank-revealing ULV decomposition. Technical Report CS-TR 2627, Department of Computer Science, University of Maryland, 1991.

[67] G. W. Stewart. On the early history of the singular value decomposition. Technical Report CS-TR-2848, Department of Computer Science, University of Maryland, College Park, 1992.

[68] G. W. Stewart. An updating algorithm for subspace tracking. *IEEE Transactions on Signal Processing*, **40**, pp 1535–1541, 1992.

[69] G. W. Stewart. Updating URV decompositions in parallel. Technical Report CS-TR-2880, Department of Computer Science, University of Maryland, College Park, 1992.

[70] G. W. Stewart and J.-G. Sun. *Matrix Perturbation Theory*. Academic Press, Boston, 1990.

[71] A. N. Tihonov. Regularization of incorrectly posed problems. *Soviet Mathematics*, **4**, pp 1624–1627, 1963.

[72] S. Van Huffel. *Analysis of the Total Least Squares Problem and Its Use in Parameter Estimation*. PhD thesis, Katholeike Universiteit Leuven, 1987.

[73] S. Van Huffel and J. Vandervalle. Analysis and properties of the generalized total least squares problem $AX \approx B$ when some or all columns in A are subject to error. *SIAM Journal on Matrix Analysis and Applications*, **10**, pp 294–315, 1989.

[74] S. Van Huffel and J. Vandewalle. *The Total Least Squares Problem: Computational Aspects and Analysis*. SIAM, Philadelphia, 1991.

[75] J. M. Varah. A practical examination of some numerical methods for linear discrete ill-posed problems. *SIAM Review*, **21**, pp 100–111, 1979.

[76] J. H. Wilkinson and C. Reinsch. *Handbook for Automatic Computation. Vol. II Linear Algebra*. Springer, New York, 1971.

[77] G. Xu and T. Kailath. Fast signal-subspace decomposition — Part I: Ideal covariance matrices. Manuscript submitted to ASSP. Information Systems Laboratory, Stanford University, 1990.

[78] G. Xu and T. Kailath. Fast signal-subspace decomposition — Part II: Sample covariance matrices. Manuscript submitted to ASSP. Information Systems Laboratory, Stanford University, 1990.

APPROXIMATION WITH KRONECKER PRODUCTS

C.F. VAN LOAN and N. PITSIANIS
Department of Computer Science
Cornell University
Ithaca, New York 14853, U.S.A.
cv@cs.cornell.edu

ABSTRACT. Let A be an m-by-n matrix with $m = m_1 m_2$ and $n = n_1 n_2$. We consider the problem of finding $B \in \mathbb{R}^{m_1 \times n_1}$ and $C \in \mathbb{R}^{m_2 \times n_2}$ so that $\| A - B \otimes C \|_F$ is minimized. This problem can be solved by computing the largest singular value and associated singular vectors of a permuted version of A. If A is symmetric, definite, non-negative, or banded, then the minimizing B and C are similarly structured. The idea of using Kronecker product preconditioners is briefly discussed.

KEYWORDS. Kronecker product, preconditioners, block matrices.

1 Background

Suppose $A \in \mathbb{R}^{m \times n}$ with $m = m_1 m_2$ and $n = n_1 n_2$. This paper is about the minimization of

$$\phi_A(B, C) = \| A - B \otimes C \|_F^2$$

where $B \in \mathbb{R}^{m_1 \times n_1}$, $C \in \mathbb{R}^{m_2 \times n_2}$, and " $\otimes$ " denotes the Kronecker product.

Our interest in this problem stems from preliminary experience with Kronecker product preconditioners in the conjugate gradient setting. Suppose $A \in \mathbb{R}^{n \times n}$ with $n = n_1 n_2$ and that M is the preconditioner. For this solution process to be successful, the preconditioner should "capture" the essence of A as much as possible subject to the constraint that a linear system $Mz = r$ is "easy" to solve. In our context, we capture A through the minimization $\phi_A(B, C)$ with $B \in \mathbb{R}^{n_1 \times n_1}$ and $C \in \mathbb{R}^{n_2 \times n_2}$. Systems of the form $Mz \equiv (B \otimes C)z = r$ are easy to solve because only $O(n^{3/2})$ flops are required if $n_1 \approx n_2 \approx \sqrt{n}$. To appreciate this point, observe that $(B \otimes C)z = r$ is equivalent to

$$CZB^T = R \tag{1}$$

M. S. Moonen et al. (eds.), *Linear Algebra for Large Scale and Real-Time Applications*, 293–314.

where Z and R are n_2-by-n_1 matrices whose columns are segments of the vectors z and r respectively:

$$Z(:,k) = z((k-1)n_2 + 1 : kn_2)$$
$$k = 1:n_1 .$$
$$R(:,k) = r((k-1)n_2 + 1 : kn_2)$$

(At this point the reader may wish to review the algebra of Kronecker products. See [21] or [29].) If B and C are nonsingular and we apply Gaussian elimination with partial pivoting to produce the factorizations $P_1 B = L_1 U_1$ and $P_2 C = L_2 U_2$, then $2(n_1^3 + n_2^3)/3$ flops are required. The ensuing multiple triangular system solves involve an additional $2(n_1^2 n_2 + n_1 n_2^2)$ flops. If $n = n_1^2 = n_2^2$, then a total of $16n^{3/2}/3$ flops are needed.

An instructive way to look at the above solution process is to recognize that

$$(P_1 \otimes P_2)(B \otimes C) = (L_1 \otimes L_2)(U_1 \otimes U_2)$$

is an LU (with partial pivoting) factorization of $B \otimes C$. This illustrates the adage that *a given factorization of $B \otimes C$ can usually be obtained by taking the Kronecker product of the corresponding B and C factorizations* :

Cholesky: $\begin{aligned} B &= L_1 L_1^T \\ C &= L_2 L_2^T \end{aligned}$ $\Rightarrow$ $(B \otimes C) = (L_1 \otimes L_2)(L_1 \otimes L_2)^T$

QR: $\begin{aligned} B &= Q_1 R_1 \\ C &= Q_2 R_2 \end{aligned}$ $\Rightarrow$ $(B \otimes C) = (Q_1 \otimes Q_2)(R_1 \otimes R_2)^T$

SVD: $\begin{aligned} B &= U_1 \Sigma_1 V_1^T \\ C &= U_2 \Sigma_2 V_2^T \end{aligned}$ $\Rightarrow$ $(B \otimes C) = (U_1 \otimes U_2)(\Sigma_1 \otimes \Sigma_2)(V_1 \otimes V_2)^T$

Schur: $\begin{aligned} B &= U_1 D_1 U_1^H \\ C &= U_2 D_2 U_2^H \end{aligned}$ $\Rightarrow$ $(B \otimes C) = (U_1 \otimes U_2)(D_1 \otimes D_2)(U_1 \otimes U_2)^H$

Here we are exploiting the fact that

$$\text{Kronecker products of} \left\{ \begin{array}{c} \text{orthogonal} \\ \text{triangular} \\ \text{diagonal} \end{array} \right\} \text{matrices are} \left\{ \begin{array}{c} \text{orthogonal} \\ \text{triangular} \\ \text{diagonal} \end{array} \right\}.$$

For a practical illustration of Kronecker product factorizations, see [10] where the idea is applied with QR in a photogrammetry application.

Some factorizations are not "preserved" when Kronecker products are taken:

- A real Schur decomposition of $B \otimes C$ is not obtained by taking the Kronecker product of the real Schur decompositions of B and C because the 2-by-2 bumps in the factors can create "block bumps" in the product. The computational ramifications of this fact are discussed in [3, 13].

- If QR with column pivoting is used to produce the factorizations $B\Pi_1 = Q_1 R_1$ and $C\Pi_2 = Q_2 R_2$, then $(B \otimes C)(\Pi_1 \otimes \Pi_2) = (Q_1 \otimes Q_2)(R_1 \otimes R_2)$ is *not* the factorization rendered by the same algorithm applied to $B \otimes C$.

Despite these anomalies, it is clear that the solution of Kronecker product systems is a nice problem with much structure to exploit. Not only are $O(n^{3/2})$ solution procedures available, but the form of (1) suggests opportunities for using the level-3 BLAS and parallel processing.

The act of finding good preconditioners through an appropriately constrained minimization of $\| A - M \|_F$ is not new. For example, [5] derives a useful class of preconditioners for the case when A is Toeplitz by solving

$$\min_{M \text{ circulant}} \| A - M \|_F .$$

Generalizations of this for matrices with Toeplitz blocks are discussed in [6].

Our presentation is organized as follows. First, we characterize the optimum Kronecker factors B and C in terms of the singular value decomposition of a permuted version of A. Algorithms for determining B and C are discussed is §3 and §4. The important cases when A is banded, non-negative, symmetric, and definite are handled in §5 along with some additional specially structured examples. In §6 we briefly examine the use of Kronecker product preconditioners.

We conclude this section with a few pointers to related work. The Kronecker product has a long history in mathematics and an excellent review is offered in [18]. Computational aspects of the operation are detailed in [25, 9].

Kronecker products arise in a number of applied areas. See [1, 28, 4, 17, 26] for Kronecker product discussions of generalized spectra, higher order statistics, systems theory, image processing, and photogrammetry.

In recent years there have been a number of developments that point to an increased role of the Kronecker product in the area of high performance matrix computations. In [22] is developed a parallel programming methodology that revolves around the Kronecker product. See also [23]. In [27, 29] is shown how the organization of fast transforms is clarified through the "language" of Kronecker products.

2 The rank-1 approximation

Consider the uniform blocking of an $m_1 m_2$-by-$n_1 n_2$ matrix A.

$$A = \begin{bmatrix} A_{11} & A_{12} & \cdots & A_{1,n_1} \\ A_{21} & A_{22} & \cdots & A_{2,n_1} \\ \vdots & \vdots & \ddots & \vdots \\ A_{m_1,1} & A_{m_1,2} & \cdots & A_{m_1,n_1} \end{bmatrix}, \qquad A_{ij} \in \mathbb{R}^{m_2 \times n_2} . \tag{2}$$

Using Matlab colon notation, the (i, j) block is given by

$$A_{ij} = A((i-1)m_2 + 1{:}im_2, (j-1)n_2 + 1{:}jn_2),$$

the submatrix defined by rows $(i-1)m_2 + 1$ to im_2 and columns $(j-1)n_2 + 1$ to jn_2. It is not hard to show using the definition of the Kronecker product that

$$\phi_A(B, C) = \sum_{i=1}^{m_1} \sum_{j=1}^{n_1} \| A_{ij} - b_{ij}C \|_F^2. \tag{3}$$

By keeping the B matrix "intact," we also have

$$\phi_A(B, C) = \sum_{i=1}^{m_2} \sum_{j=1}^{n_2} \| \hat{A}_{ij} - c_{ij}B \|_F^2, \tag{4}$$

where

$$\hat{A}_{ij} = A(i{:}m_2{:}m, j{:}n_2{:}n)$$

is the m_1-by-n_1 submatrix defined by rows $i, i + m_2, i + 2m_2, \ldots, i + (m_1 - 1)m_2$ and columns $j, j + n_2, j + 2n_2, \ldots, j + (n_1 - 1)n_2$. Thinking of matrices at the block level is the key to high performance matrix computations. See [15].

To proceed further with the analysis of $\phi_A(B, C)$, we require the *vec* operation, which is a way of turning matrices into vectors by "stacking" the columns:

$$X \in \mathbb{R}^{p \times q} \Rightarrow vec(X) = \begin{bmatrix} X(1{:}p, 1) \\ X(1{:}p, 2) \\ \vdots \\ X(1{:}p, q) \end{bmatrix} \in \mathbb{R}^{pq}.$$

It turns out that the *vec* operator can be used to express the minimization of $\| A - B \otimes C \|_F^2$ as a rank-1 approximation problem. The idea is to rearrange A into another matrix $\tilde{A}$ so that the sum of squares that arise in $\| A - B \otimes C \|_F^2$ is exactly the same as the sum of squares that arise in $\| \tilde{A} - vec(B)vec(C)^T \|_F^2$. For example, in a 4-by-4 problem with 2-by-2 blocks,

$$\| A - B \otimes C \|_F = \left\| \begin{bmatrix} a_{11} & a_{21} & a_{12} & a_{22} \\ a_{31} & a_{41} & a_{32} & a_{42} \\ a_{13} & a_{23} & a_{14} & a_{24} \\ a_{33} & a_{43} & a_{34} & a_{44} \end{bmatrix} - \begin{bmatrix} b_{11} \\ b_{21} \\ b_{12} \\ b_{22} \end{bmatrix} \begin{bmatrix} c_{11} & c_{21} & c_{12} & c_{22} \end{bmatrix} \right\|_F.$$

Refer to the above permuted version of A as $\tilde{A}$. Note that $\tilde{A}$ is *not* of the form PAQ where P and Q are permutation matrices. Indeed, in our example

- the four rows of $\tilde{A}$ are *vec*'s of the 2-by-2 blocks of A:

$$A = \begin{bmatrix} A_{11} & A_{12} \\ A_{21} & A_{22} \end{bmatrix} \quad \Rightarrow \quad \tilde{A} = \begin{bmatrix} vec(A_{11})^T \\ vec(A_{21})^T \\ vec(A_{12})^T \\ vec(A_{22})^T \end{bmatrix}.$$

- the *vec*'s of the 2-by-2 blocks of $\tilde{A}^T$ are columns of A:

$$\tilde{A} = \begin{bmatrix} \tilde{A}_{11} & \tilde{A}_{12} \\ \tilde{A}_{21} & \tilde{A}_{22} \end{bmatrix} \quad \Rightarrow \quad A = \left[\, vec(\tilde{A}_{11}^T) \mid vec(\tilde{A}_{12}^T) \mid vec(\tilde{A}_{21}^T) \mid vec(\tilde{A}_{22}^T) \, \right].$$

In general, if $m = m_1 m_2$, $n = n_1 n_2$, $A \in \mathbb{R}^{m \times n}$, and we have the blocking (2), then we define the *rearrangement* of A (relative to the blocking parameters m_1, m_2, n_1, and n_2) by

$$\mathcal{R}(A) = \begin{bmatrix} A_1 \\ \vdots \\ A_{n_1} \end{bmatrix}, \qquad A_j = \begin{bmatrix} vec(A_{1,j})^T \\ \vdots \\ vec(A_{m_1,j})^T \end{bmatrix}, \qquad j = 1{:}n_1. \tag{5}$$

Note that $\mathcal{R}(A)$ has $m_1 n_1$ rows and $m_2 n_2$ columns. Thus, $\mathcal{R}(A)$ need not be the same size as A. For example, if $m = m_1 m_2 = 2 \cdot 2$ and $n = n_1 n_2 = 3 \cdot 2$, then A is 4-by-6 but

$$\mathcal{R}(A) = \left[\begin{array}{cc|cc} a_{11} & a_{21} & a_{12} & a_{22} \\ a_{31} & a_{41} & a_{32} & a_{42} \\ \hline a_{13} & a_{23} & a_{14} & a_{24} \\ a_{33} & a_{43} & a_{34} & a_{44} \\ \hline a_{15} & a_{25} & a_{16} & a_{26} \\ a_{35} & a_{45} & a_{36} & a_{46} \end{array} \right].$$

We are now set to establish a key result that connects the problem of minimizing $\phi_A(B, C)$ with the problem of approximating $\tilde{A}$ with a rank-1 matrix.

Theorem 1 *Assume that $A \in \mathbb{R}^{m \times n}$ with $m = m_1 m_2$ and $n = n_1 n_2$. If $B \in \mathbb{R}^{m_1 \times n_1}$ and $C \in \mathbb{R}^{m_2 \times n_2}$, then*

$$\| A - B \otimes C \|_F = \| \mathcal{R}(A) - vec(B) vec(C)^T \|_F.$$

Proof. By applying the *vec* operator in (3) we get:

$$\begin{aligned}
\| A - B \otimes C \|_F^2 &= \sum_{j=1}^{n_1} \sum_{i=1}^{m_1} \| vec(A_{ij}) - b_{ij} vec(C) \|_2^2 \\
&= \sum_{j=1}^{n_1} \sum_{i=1}^{m_1} \| vec(A_{ij})^T - b_{ij} vec(C)^T \|_2^2 \\
&= \sum_{j=1}^{n_1} \| A_j - B(:,j) vec(C)^T \|_F^2 \\
&= \| \mathcal{R}(A) - vec(B) vec(C)^T \|_F^2. \quad \square
\end{aligned}$$

The approximation of a given matrix by a rank-1 matrix has a well-known solution in terms of the singular value decomposition.

Corollary 2 *Assume that $A \in \mathbb{R}^{m \times n}$ with $m = m_1 m_2$ and $n = n_1 n_2$. If $\tilde{A} = \mathcal{R}(A)$ has singular value decomposition*

$$U^T \tilde{A} V = \Sigma = \mathrm{diag}(\sigma_i)$$

where σ_1 is the largest singular value, and $U(:,1)$ and $V(:,1)$ are the corresponding singular vectors, then the matrices $B \in \mathbb{R}^{m_1 \times n_1}$ and $C \in \mathbb{R}^{m_2 \times n_2}$ defined by $vec(B) = \sigma_1 U(:,1)$ and $vec(C) = V(:,1)$ minimize $\| A - B \otimes C \|_F$.

Proof. See [15, p. 73]. $\square$

The definition (5) of $\mathcal{R}(A)$ is in terms of the blocks A_{ij} in (2). An alternative characterization can be obtained in terms of the columns of A. In particular, we show that

$$
\mathcal{R}(A) = \begin{bmatrix} \tilde{A}_{11} & \cdots & \tilde{A}_{1,n_2} \\ \vdots & \ddots & \vdots \\ \tilde{A}_{n_1,1} & \cdots & \tilde{A}_{n_1,n_2} \end{bmatrix}.
\tag{6}
$$

where $\tilde{A}_{ij} \in \mathbb{R}^{m_1 \times m_2}$ is defined by

$$vec(\tilde{A}_{ij}^T) = A(:,(i-1)n_2 + j) \qquad 1 \leq i \leq n_1,\ 1 \leq j \leq n_2.$$

In view of (5) we need only confirm that

$$
A_i = \begin{bmatrix} vec(A_{1,i})^T \\ \vdots \\ vec(A_{m_1,i})^T \end{bmatrix} = \begin{bmatrix} \tilde{A}_{i,1} \mid \tilde{A}_{i,2} \mid \cdots \mid \tilde{A}_{i,n_2} \end{bmatrix}.
\tag{7}
$$

For $s = 1{:}m_2$, $p = 1{:}n_2$, and $q = 1{:}m_2$ we have

$$[A_i]_{s,(p-1)m_2+q} = \left[vec(A_{s,i})^T \right]_{(p-1)m_2+q} = A((s-1)m_2 + q, (i-1)n_2 + p).$$

But (7) immediately follows because we also have

$$\left[\tilde{A}_{i,1} \mid \tilde{A}_{i,2} \mid \cdots \mid \tilde{A}_{i,n_2} \right]_{s,(p-1)m_2+q} = \left[\tilde{A}_{i,p} \right]_{sq} = A((s-1)m_2 + q, (i-1)n_2 + p).$$

3 SVD framework

The Golub-Reinsch SVD algorithm can be used for computing the largest singular value and corresponding singular vectors of $\mathcal{R}(A)$. However, in view of the potentially large dimension of $\tilde{A} = \mathcal{R}(A)$ in some applications, it may be more appropriate to use the SVD Lanczos process described in [12]. Here is how to proceed with the computation of $B \in \mathbb{R}^{m_1 \times n_1}$ and $C \in \mathbb{R}^{m_2 \times n_2}$:

Framework 1.

C = initial guess.
$v_1 \leftarrow vec(C)/\| C \|_F$
$p_0 \leftarrow v_1$; $\beta_0 \leftarrow 1$; $j \leftarrow 0$; $u_0 \leftarrow 0$
while $\beta_j \neq 0$ (or some other less stringent criteria.)
 $v_{j+1} \leftarrow p_j/\beta_j$
 $j \leftarrow j + 1$
 $r_j \leftarrow \tilde{A}v_j - \beta_{j-1}u_{j-1}$
 $\alpha_j \leftarrow \| r_j \|_2$
 $u_j \leftarrow r_j/\alpha_j$
 $p_j \leftarrow \tilde{A}^T u_j - \alpha_j v_j$;
 $\beta_j \leftarrow \| p_j \|_2$
 end {while}
Compute the largest singular value σ_1 and associated left and right
 singular vectors u_B and v_B of the bidiagonal matrix with diagonal
 $\alpha_1, \ldots, \alpha_j$ and upper diagonal $\beta_1, \ldots, \beta_{j-1}$.
Define B by $vec(B) = \sigma_1[u_1, \ldots, u_j]u_B$ and C by $vec(C) = [v_1, \ldots, v_j]v_B$

There are many subtleties associated with the Lanzcos process and we refer the reader to [8] or [15, p. 98ff] for details.

Our only implementation discussion concerns the matrix-vector products $\tilde{A}x$ and $\tilde{A}^T x$ that are required by the iteration. The explicit formation of $\mathcal{R}(A) = \tilde{A}$ is *not* necessary. For example, working with the characterization (5), here is a dot product formulation for $y \leftarrow \tilde{A}x$:

```
for j = 1:n₁
    for i = 1:m₁
        y((j − 1)m₁ + i) ← vec(Aᵢⱼ)ᵀx
    end
end
```

A saxpy-based procedure for $y \leftarrow \tilde{A}^T x$ proceeds as follows:

```
y(1:m₂n₂) ← 0
for j = 1:n₁
    for i = 1:m₁
        y ← y + x((j − 1)m₁ + i)vec(Aᵢⱼ)
    end
end
```

By working with (6) we have the following alternative block formulation for $y \leftarrow \tilde{A}x$:

```
y(1:m₁n₁) ← 0
for i = 1:n₁
    rows = (i − 1)m₁ + 1:im₁
    for j = 1:n₂
        Define Z ∈ ℝ^(m₁×m₂) by vec(Zᵀ) = A(:,(i − 1)n₂ + j)
        y(rows) ← y(rows) + Zx((j − 1)m₂ + 1:jm₂)
    end
end
```

```
        end
    end
```

Likewise, we can formulate a procedure for $y \leftarrow \tilde{A}^T x$ that is based upon (6):

```
y(1:m₂n₂) ← 0
for i = 1:n₂
    rows = (i − 1)m₂ + 1:im₂
    for j = 1:n₁
        Define Z ∈ ℝ^{m₂×m₁} by vec(Zᵀ) = A(:, (j − 1)n₂ + i)
        y(rows) ← y(rows) + Zᵀx((j − 1)m₁ + 1:jm₁)
    end
end
```

Each of these products requires $2m_1 n_1 m_2 n_2 = 2mn$ flops assuming that $\tilde{A}$ is treated as a dense matrix.

4 The separable least squares framework

Note that if we fix C, then the problem of minimizing $\phi_A(B, C) = \| A - B \otimes C \|_F$ is a linear least squares problem with unknowns b_{ij}. Likewise, if B is fixed, then the minimization of ϕ_A is a linear least squares problem in the c_{ij}. The following theorem specifies the solution to these linear least squares problems and requires the concept of matrix trace:

$$X \in \mathbb{R}^{q \times q} \Rightarrow tr(X) = \sum_{i=1}^{q} x_{ii} .$$

Theorem 3 *Suppose* $m = m_1 m_2$, $n = n_1 n_2$, *and* $A \in \mathbb{R}^{m \times n}$. *If* $C \in \mathbb{R}^{m_2 \times n_2}$ *is fixed, then the matrix* $B \in \mathbb{R}^{m_1 \times n_1}$ *defined by*

$$b_{ij} = \frac{tr(A_{ij}^T C)}{tr(C^T C)} \qquad 1 \le i \le m_1,\ 1 \le j \le n_1 \tag{8}$$

minimizes $\| A - B \otimes C \|_F$ *where* $A_{ij} = A((i-1)m_2 + 1:im_2, (j-1)n_2 + 1:jn_2)$. *Likewise, if* $B \in \mathbb{R}^{m_1 \times n_1}$ *is fixed, then the matrix* $C \in \mathbb{R}^{m_2 \times n_2}$ *defined by*

$$c_{ij} = \frac{tr(\hat{A}_{ij}^T B)}{tr(B^T B)} \qquad 1 \le i \le m_2,\ 1 \le j \le n_2 \tag{9}$$

minimizes $\| A - B \otimes C \|_F$ *where* $\hat{A}_{ij} = A(i:m_2:m, j:n_2:n)$.

Proof. Since

$$\begin{aligned} \| A_{ij} - b_{ij} C \|_F^2 &= tr((A_{ij} - b_{ij}C)^T(A_{ij} - b_{ij}C)) \\ &= \| A_{ij} \|_F^2 - 2b_{ij} tr(C^T A_{ij}) + b_{ij}^2 \| C \|_F^2 \end{aligned}$$

it follows from (2.2) that

$$\frac{\partial \phi_A(B, C)}{\partial b_{ij}} = -2\, tr(C^T A_{ij}) + 2b_{ij} \| C \|_F^2 .$$

Setting these partials to zero defines the required matrix B. The proof of (9) is similar. ☐

The above result suggests that we can compute B and C by taking the *separable least squares* described in [2]. The idea is to minimize $\phi_A(B,C)$ by alternately improving the B and C matrices through a sequence of linear least squares optimizations:

Framework 2.

$\quad C = C_0$ (given starting matrix)
$\quad$**Repeat:**
$\quad\quad \gamma \leftarrow tr(C^T C)$
$\quad\quad$**for** $i = 1{:}m_1$
$\quad\quad\quad$**for** $j = 1{:}n_1$
$\quad\quad\quad\quad b_{ij} \leftarrow tr(C^T A_{ij})/\gamma$
$\quad\quad \beta \leftarrow tr(B^T B)$
$\quad\quad$**for** $i = 1{:}m_2$
$\quad\quad\quad$**for** $j = 1{:}n_2$
$\quad\quad\quad\quad c_{ij} \leftarrow tr(B^T \hat{A}_{ij})/\beta$

This process requires $4m_1 n_1 m_2 n_2 = 4mn$ flops per iteration, the same as Framework 1. Other methods for nonlinear least squares problems with variables that separate are discussed in [14, 24].

Framework 2 amounts to a power method for the largest singular value of $\tilde{A} = \mathcal{R}(A)$. To see this we switch to "tilde-space" and observe that if

$$\phi(b,c) = \| \tilde{A} - bc^T \|_F^2 \qquad b \in \mathbb{R}^{m_1 n_1}, \; c \in \mathbb{R}^{m_2 n_2},$$

then the gradient is given by

$$\nabla\phi(b,c) = -2 \left[\begin{array}{c} \tilde{A}c - (c^T c)b \\ \tilde{A}^T b - (b^T b)c \end{array} \right].$$

If b is fixed, then the minimizing c is obtained by setting $c = \tilde{A}^T b / b^T b$ for then the c-partials are all zero. Likewise, if c is fixed, then the minimizing b is given by $b = \tilde{A}c/c^T c$. After k passes through the iteration

$\quad c = c_0$ (given starting vector)
$\quad$**Repeat:**
$\quad\quad b \leftarrow \tilde{A}c/c^T c$
$\quad\quad c \leftarrow \tilde{A}^T b/b^T b$

the vector c is in the direction of $(\tilde{A}^T \tilde{A})^k c_0$ and the vector b is in the direction of $(\tilde{A}\tilde{A}^T)^{k-1}\tilde{A}c_0$.

The practical implementation of this framework involves all the subtleties that are associated with the power method. See [30] for a discussion.

5 Structured problems

As we alluded to in §1, the Kronecker product of two structured matrices is usually structured in the same way:

$$\text{If } B \text{ and } C \text{ are } \left\{ \begin{array}{l} \text{banded} \\ \text{non-negative} \\ \text{symmetric} \\ \text{positive definite} \\ \text{stochastic} \\ \text{orthogonal} \end{array} \right\}, \text{ then } B \otimes C \text{ is } \left\{ \begin{array}{l} \text{banded} \\ \text{non-negative} \\ \text{symmetric} \\ \text{positive definite} \\ \text{stochastic} \\ \text{orthogonal} \end{array} \right\}.$$

We are interested in the structure of the solution to the Kronecker approximation problem given that A is structured. In the following subsections we use Corollary 2 and Theorem 3 to establish a number results about structured problems.

5.1 BANDEDNESS

We first show how bandedness in A "shows up" in B and C.

Theorem 4 *Suppose* $n = n_1 n_2$, $A \in \mathbb{R}^{n \times n}$ *has bandwidth* pn_2, *and that each block in*

$$A = \left[\begin{array}{ccc} A_{11} & \cdots & A_{1,n_1} \\ \vdots & \ddots & \vdots \\ A_{n_1,1} & \cdots & A_{n_1 n_1} \end{array} \right] \qquad A_{ij} \in \mathbb{R}^{n_2 \times n_2}$$

has bandwidth q *or less. If* $B \in \mathbb{R}^{n_1 \times n_1}$ *and* $C \in \mathbb{R}^{n_2 \times n_2}$ *minimize* $\| A - B \otimes C \|_F$, *then* B *has bandwidth* p *and* C *has bandwidth* q.

Proof. Since A has bandwidth pn_2, it follows that $A_{ij} = 0$ if $|i - j| > p$. From (3) we have $b_{ij} = 0$ whenever $|i - j| > p$. Since each A_{ij} has bandwidth q, it follows that the minimization of $\| A_{ij} - b_{ij} C \|_F$ requires setting c_{rs} to zero whenever $|r - s| > q$. Thus, a minimizing C must have bandwidth q. $\square$

5.2 NON-NEGATIVITY

We first show that if A and C are non-negative, then the B that minimizes $\phi_A(B, C)$ is also non-negative.

Theorem 5 *If* $m = m_1 m_2$, $n = n_1 n_2$, $A \in \mathbb{R}^{m \times n}$, *and* $C \in \mathbb{R}^{m_2 \times n_2}$ *are non-negative, then there exists a non-negative* $B \in \mathbb{R}^{m_1 \times n_1}$ *that minimizes* $\| A - B \otimes C \|_F$.

Proof. Using the non-negativity of C and Theorem 3,

$$b_{ij} = \frac{tr(A_{ij}^T C)}{tr(C^T C)} \geq 0$$

for $i = 1:m_1$ and $j = 1:n_1$. $\square$

In the same way, we can show that if A and B are non-negative, then the C that minimizes $\| A - B \otimes C \|$ is also non-negative. Thus, if we start with a non-negative C in Framework 2, then all subsequent B and C matrices are non-negative. The following theorem shows that this restriction poses no difficulty because the optimum B and C are also non-negative.

Theorem 6 *If $m = m_1 m_2$, $n = n_1 n_2$, and $A \in \mathbb{R}^{m \times n}$ is non-negative, then there exist non-negative matrices $B \in \mathbb{R}^{m_1 \times n_1}$ and $C \in \mathbb{R}^{m_2 \times m_2}$ such that $\| A - B \otimes C \|_F$ is minimized.*

Proof. Note that $\tilde{A} = \mathcal{R}(A)$ has non-negative entries and let σ_1 be its largest singular value. Peron-Frobenius theory tells us that there exist non-negative $u \in \mathbb{R}^{m_1 n_1}$ and $v \in \mathbb{R}^{m_2 n_2}$ so that $\tilde{A}^T \tilde{A} v = \sigma_1^2 v$ and $\tilde{A} \tilde{A}^T u = \sigma_1^2 u$. (See [20, p.503].) But u and v are the right and left singular vectors of $\tilde{A}$ and so the matrices B and C as specified in Corollary 2 are non-negative. $\square$

5.3 SYMMETRY

Turning next to the issue of symmetry, we show that if A and C are symmetric, then a symmetric B can be found to minimize $\phi_A(B, C)$.

Theorem 7 *If $n = n_1 n_2$, $A \in \mathbb{R}^{n \times n}$ and $C \in \mathbb{R}^{n_2 \times n_2}$ are symmetric, then there exists a symmetric $B \in \mathbb{R}^{n_1 \times n_1}$ that minimizes $\| A - B \otimes C \|_F$.*

Proof. Since A is symmetric, $A_{ji} = A_{ij}^T$. Using elementary properties of the trace we have

$$b_{ij} = \frac{tr(A_{ij}^T C)}{tr(C^T C)} = \frac{tr(A_{ji} C)}{tr(C^T C)} = \frac{tr(C A_{ji})}{tr(C^T C)} = \frac{tr(A_{ji}^T C)}{tr(C^T C)} = b_{ji}$$

for all $1 \leq i, j \leq n_1$. It follows that B is symmetric. $\square$

It is equally straightforward to establish that a symmetric C can be found to minimize $\| A - B \otimes C \|_F$ is A and B are symmetric.

Analogous results are applicable if the "frozen factor" is skew-symmetric:

Theorem 8 *If $n = n_1 n_2$, $A \in \mathbb{R}^{n \times n}$ is symmetric and $C \in \mathbb{R}^{n_2 \times n_2}$ is skew-symmetric, then there exists a skew-symmetric $B \in \mathbb{R}^{n_1 \times n_1}$ that minimizes $\| A - B \otimes C \|_F$.*

Proof.

$$b_{ij} = \frac{tr(A_{ij}^T C)}{tr(C^T C)} = -\frac{tr(A_{ji}^T C)}{tr(C^T C)} = -b_{ji}. \quad \square$$

The optimum Kronecker approximation of a symmetric matrix may have skew-symmetric factors as consideration of the following example shows:

$$A = \begin{bmatrix} 0 & 0 & 0 & 1 \\ 0 & 0 & -1 & 0 \\ 0 & -1 & 0 & 0 \\ 1 & 0 & 0 & 0 \end{bmatrix} = \begin{bmatrix} 0 & 1 \\ -1 & 0 \end{bmatrix} \otimes \begin{bmatrix} 0 & 1 \\ -1 & 0 \end{bmatrix}.$$

For this particular A, it is not possible to find symmetric B and C for which we have $A = B \otimes C$. The following theorem summarizes the situation.

Theorem 9 *Suppose $n = n_1 n_2$ and $A \in \mathbb{R}^{n \times n}$ is symmetric. If $\| A - B \otimes C \|_F$ cannot minimized by symmetric matrices $B \in \mathbb{R}^{n_1 \times n_1}$ and $C \in \mathbb{R}^{n_2 \times n_2}$, then it can be minimized by skew-symmetric matrices $B \in \mathbb{R}^{n_1 \times n_1}$ and $C \in \mathbb{R}^{n_2 \times n_2}$.*

Proof. For any positive integer q, define the following orthogonal subpaces of $\mathbb{R}^{q^2}$:

$$S_+^{(q)} = \{x \in \mathbb{R}^{q^2} : x = vec(X) \text{ for some symmetric } X \in \mathbb{R}^{q \times q}\}$$

$$S_-^{(q)} = \{x \in \mathbb{R}^{q^2} : x = vec(X) \text{ for some skew-symmetric } X \in \mathbb{R}^{q \times q}\}$$

Note that $\mathbb{R}^{q^2} = S_+^{(q)} \oplus S_1^{(q)}$.

Now suppose that $y = \mathcal{R}(A)x$ and that $X \in \mathbb{R}^{n_2 \times n_2}$ and $Y \in \mathbb{R}^{n_1 \times n_1}$ are defined by $x = vec(X)$ and $y = vec(Y)$, respectively. From (2)) we know that

$$[Y]_{ij} = vec(A_{ij})^T x = tr(A_{ij}^T X) \qquad 1 \le i, j \le n_1.$$

If $x \in S_+^{(n_2)}$, then since A is symmetric we have

$$[Y]_{ij} - [Y]_{ji} = tr((A_{ij}^T - A_{ji}^T)X) = tr((A_{ij}^T - A_{ij})X) = vec(A_{ij}^T - A_{ij})^T x = 0$$

since $vec(A_{ij}^T - A_{ij}) \in S_-^{(n_2)}$. Thus,

$$x \in S_+^{(n_2)} \Rightarrow \mathcal{R}(A)x \in S_+^{(n_1)}$$

Likewise,

$$x \in S_1^{(n_2)} \Rightarrow \mathcal{R}(A)x \in S_1^{(n_1)}$$

and so $(S_+^{(n_2)}, S_+^{(n_1)})$ and $(S_-^{(n_2)}, S_-^{(n_1)})$ are singular subspace pairs for $\mathcal{R}(A)$. It follows that the largest singular value and corresponding singular vectors must be associated with one of these pairs. $\square$

Theorem 9 can also be established by observing that if A is symmetric, then

$$P_{n_1} \mathcal{R}(A) P_{n_2}^T = \mathcal{R}(A)$$

where P_q designates the *vec permutation matrix* on $\mathbb{R}^{q^2}$:

$$P_q vec(X) = vec(X^T) \qquad X \in \mathbb{R}^{q \times q}.$$

This permutation connects the *vec* of a matrix and the *vec* of its transpose. See [19] for further details.

5.4 POSITIVE DEFINITENESS

We first show that if the initial guess matrix in Framework 2 is positive definite, then all subsequent B and C iterates are positive definite.

Theorem 10 *If $n = n_1^2$, $A \in \mathbb{R}^{n \times n}$ and $C \in \mathbb{R}^{n_2 \times n_2}$ are symmetric positive definite, then there exists a symmetric positive definite $B \in \mathbb{R}^{n_1 \times n_1}$ that minimizes $\phi_A(B, C)$. Likewise, if $B \in \mathbb{R}^{n_1 \times n_1}$ is symmetric positive definite, then there exists a symmetric positive definite $C \in \mathbb{R}^{n_2 \times n_2}$ that minimizes $\phi_A(B, C)$.*

Proof. If each entry b_{ij} in $B \in \mathbb{R}^{n_1 \times n_1}$ satisfies $b_{ij} = tr(C^T A_{ij})/tr(C^T C)$, and if $y \in \mathbb{R}^{n_1}$, then using the linearity of the trace we have

$$y^T B y = \sum_{i=1}^{n_1} \sum_{j=1}^{n_1} b_{ij} y_i y_j = \sum_{i=1}^{n_1} \sum_{j=1}^{n_1} y_i y_j tr(C^T A_{ij})/tr(C^T C) = tr(C^T \hat{A}) \qquad (10)$$

where

$$\hat{A} = \sum_{i=1}^{n_1} \sum_{j=1}^{n_1} y_i y_j A_{ij}.$$

The matrix $\hat{A}$ is positive definite because for any $z \in \mathbb{R}^{n_1}$ we have

$$0 < (z \otimes y)^T A (z \otimes y) = \left[z_1 y^T \mid \cdots \mid z_{n_1} y^T \right] [A_{ij}] \begin{bmatrix} z_1 y \\ \vdots \\ z_{n_1} y \end{bmatrix} = z^T \hat{A} z.$$

Since C is positive definite, it has a Cholesky factorization $C = LL^T$. From (10) and the fact that the trace is invariant under similarity transformations, gives

$$y^T B y = tr(C^T \hat{A}) = tr(LL^T \hat{A}) = tr(L^{-1}(LL^T \hat{A})L) = tr(L^T \hat{A} L) > 0.$$

The proof that C is positive definite when B is given is similar. $\square$

The next result shows that if A is symmetric and positive definite, then the same can be said about the optimum B and C.

Theorem 11 *If $n = n_1 n_2$ and $A \in \mathbb{R}^{n \times n}$ is symmetric positive definite, then there exists symmetric positive definite $B \in \mathbb{R}^{n_1 \times n_1}$ and $C \in \mathbb{R}^{n_2 \times n_2}$ that minimize $\phi_A(B, C)$.*

Proof. From Theorem 9 we may select the optimum B and C to be either both skew-symmetric or both symmetric. We first show that the latter must be the case.

If B is skew-symmetric, then there exists a real orthogonal U_B such that

$$U_B^T B U_B = B_1 \qquad (11)$$

where B_1 is a direct sum of 1-by-1 and 2-by-2 skew-symmetric blocks. The 1-by-1's are (of course) zero and the 2-by-2's have the form

$$M = \begin{bmatrix} 0 & m \\ -m & 0 \end{bmatrix}$$

and correspond to the complex conjugate eigenpairs of B. The decomposition (11) is just the real Schur decomposition. Note that the unitary matrix

$$Z = \frac{1}{\sqrt{2}} \begin{bmatrix} 1 & i \\ i & 1 \end{bmatrix}$$

diagonalizes M:

$$Z^H M Z = \begin{bmatrix} im & 0 \\ 0 & -im \end{bmatrix}.$$

Let V_B be the unitary matrix that has copies of Z on the diagonal which correspond to the 2-by-2 blocks in B_1, and which is the identity elsewhere. It follows that

$$V_B^H U_B^T B U_B V_B = D_B$$

is diagonal. Let us refer to this decomposition as the *structured Schur decomposition* of B. Assume that C is also skew-symmetric and let

$$V_C^H U_C^T C U_C V_C = D_C$$

be its structured Schur decomposition. For a matrix H, let $|H|$ be the matrix obtained by taking the absolute values of each entry. Since

$$Z \left| \begin{bmatrix} im & 0 \\ 0 & -im \end{bmatrix} \right| Z^H = |m| I_2$$

it is easy to check that the matrices

$$\begin{aligned} B_+ &= U_B V_B |D_B| V_B^H U_B^T \\ C_+ &= U_C V_C |D_C| V_C^H U_C^T \end{aligned}$$

are real and symmetric.

Let $Q = Q_B \otimes Q_C$ where $Q_B = U_B V_B$ and $Q_C = U_C V_C$. Define the *off* operation on matrices as follows:

$$\text{off}(M) = \sum_{i \neq j} m_{ij}^2.$$

Setting D_A to be the diagonal part of $Q^H A Q$, we see that

$$\begin{aligned} \| A - B_+ \otimes C_+ \|_F^2 &= \| Q^H A Q - |D_B| \otimes |D_C| \|_F^2 \\ &= \text{off}(Q^H A Q) + \| D_A - |D_B| \otimes |D_C| \|_F^2 \end{aligned}$$

while

$$\begin{aligned} \| A - B \otimes C \|_F^2 &= \| Q^H A Q - D_B \otimes D_C \|_F^2 \\ &= \text{off}(Q^H A Q) + \| D_A - D_B \otimes D_C \|_F^2. \end{aligned}$$

Since $Q^H A Q$ is positive definite, D_A has positive diagonal entries. Moreover, $D_B \otimes D_C$ is a real diagonal matrix with some negative diagonal entries. It follows that

$$\| D_A - |D_B| \otimes |D_C| \|_F^2 < \| D_A - D_B \otimes D_C \|_F^2.$$

and so

$$\| A - B_+ \otimes C_+ \|_F < \| A - B \otimes C \|_F.$$

This shows that a skew-symmetric pair cannot minimize $\phi_A(B, C)$.

Knowing now that the optimizing B and C are symmetric, it remains for us to show that they are both positive definite. Suppose

$$Q_1^T B Q_1 = D_1 = \text{diag}(\lambda_1, \ldots, \lambda_{n_1})$$
$$Q_2^T C Q_2 = D_2 = \text{diag}(\mu_1, \ldots, \mu_{n_2})$$

are Schur decompositions. Set $Q = Q_1 \otimes Q_2$ and let $D = \text{diag}(d_1, \ldots, d_n)$ be the diagonal part of $F = Q^T A Q$. Thus,

$$\| A - B \otimes C \|_F^2 = \| Q^T (A - B \otimes C) Q \|_F^2$$
$$= \| F - D_1 \otimes D_2 \|_F^2 = \| D - D_1 \otimes D_2 \|_F^2 + \text{off}(F).$$

Note that

$$\| D - D_1 \otimes D_2 \|_F^2 = \sum_{i=1}^{n_1} \sum_{j=1}^{n_2} = (d_{(i-1)n_2+j} - \lambda_i \mu_j)^2.$$

Since D has positive diagonal entries and

$$(d_{(i-1)n_2+j} - \lambda_i \mu_j)^2 - (d_{(i-1)n_2+j} - |\lambda_i \mu_j|)^2 = |\lambda_i|^2 |\mu_j|^2 - \lambda_i^2 \mu_j^2, > 0,$$

it follows that the λ_i and μ_j should all have the same sign. Otherwise, B and C will not render the minimum sum of squares. Since $\phi_A(-B, -C) = \phi_A(B, C)$, we may assume without loss of generality that this sign is positive. This implies that symmetric positive definite B and C may be chosen to be minimize $\phi_A(B, C)$. $\square$

5.5 SUMS OF KRONECKER PRODUCTS

Next, we consider the situation when the matrix A to be approximated is a sum of Kronecker products:

$$A = \sum_{i=1}^{p} (G_i \otimes F_i).$$

Assume that each G_i is m_1-by-n_1 and each F_i is m_2-by-n_2. It follows that if $f_i = \text{vec}(F_i)$ and $g_i = \text{vec}(G_i)$, then

$$\tilde{A} = \mathcal{R}(A) = \sum_{i=1}^{p} \mathcal{R}(G_i \otimes F_i) = \sum_{i=1}^{p} g_i f_i^T$$

is a rank-p matrix. This has two important ramifications. First, it means that matrix-vector products of the form $\tilde{A}x$ and $\tilde{A}^T x$ cost $O((m+n)p)$ flops where $m = m_1 m_2$ and $n = n_1 n_2$. Second, it means that the optimum B and C are linear combinations of the G_i and F_i:

$$B = \alpha_1 G_1 + \cdots + \alpha_p G_p$$
$$C = \beta_1 F_1 + \cdots + \beta_p F_p$$

The problem of approximating matrices of the form $(I \otimes F) + (G \otimes I)$ is discussed further in §6.

5.6 APPROXIMATION WITH LINEAR HOMOGENEOUS CONSTRAINTS

Consider the problem of approximating A with a Kronecker product $B \otimes C$ that has a prescribed structure. If the constraints on B and C are linear and homogeneous, then we are looking at a problem with the following form:

$$\min_{\substack{S_1^T vec(B) = 0 \\ S_2^T vec(C) = 0}} \| A - B \otimes C \|_F. \tag{12}$$

Here, $A \in \mathbb{R}^{m \times n}$, $m = m_1 m_2$, $n = n_1 n_2$, $B \in \mathbb{R}^{m_1 \times n_1}$, $C \in \mathbb{R}^{m_2 \times n_2}$, $S_1 \in \mathbb{R}^{m_1 n_1 \times p_1}$, $S_2 \in \mathbb{R}^{m_2 n_2 \times p_2}$, and we assume that S_1 and S_2 have full column rank. By choosing these constraint matrices properly, we can force B and C to take on any prescribed sparsity pattern. Circulant, Toeplitz, Hankel, and Hamiltonian structures can also be imposed.

To solve the constrained problem we follow the techniques espoused in [11] where various modified eigenvalue problems are discussed. Let $b = vec(B)$, $c = vec(C)$, and assume that we have the QR factorizations

$$S_1 = Q_1 \begin{bmatrix} R_1 \\ 0 \end{bmatrix} \qquad S_2 = Q_2 \begin{bmatrix} R_2 \\ 0 \end{bmatrix} \tag{13}$$

where R_1 and R_2 are square. If

$$Q_1^T \mathcal{R}(A) Q_2 = \begin{bmatrix} \tilde{A}_{11} & \tilde{A}_{12} \\ \tilde{A}_{21} & \tilde{A}_{22} \end{bmatrix}, \qquad Q_1^T b = \begin{bmatrix} b_1 \\ b_2 \end{bmatrix}, \qquad Q_2^T c = \begin{bmatrix} c_1 \\ c_2 \end{bmatrix}$$

are partitioned conformably with (13), then (12) transforms to the problem of minimizing

$$\left\| \begin{bmatrix} \tilde{A}_{11} & \tilde{A}_{12} \\ \tilde{A}_{21} & \tilde{A}_{22} \end{bmatrix} - \begin{bmatrix} b_1 \\ b_2 \end{bmatrix} \begin{bmatrix} c_1 \\ c_2 \end{bmatrix}^T \right\|_F$$

subject to the constraints

$$\begin{bmatrix} R_1^T | 0 \end{bmatrix} \begin{bmatrix} b_1 \\ b_2 \end{bmatrix} = 0, \qquad \begin{bmatrix} R_2^T | 0 \end{bmatrix} \begin{bmatrix} c_1 \\ c_2 \end{bmatrix} = 0.$$

It follows that b_1 and c_1 are both zero and that the optimum b_2 and c_2 can be obtained by solving the unconstrained problem

$$\min \| \tilde{A}_{22} - b_2 c_2^T \|_F.$$

Collecting results, we see that B and C are prescribed by

$$vec(B) = Q_1 \begin{bmatrix} 0 \\ b_2 \end{bmatrix}, \qquad vec(C) = Q_2 \begin{bmatrix} 0 \\ c_2 \end{bmatrix}.$$

5.7 STOCHASTIC AND ORTHOGONAL PROBLEMS

The non-negative matrix $A \in \mathbb{R}^{n \times n}$ is *stochastic* if $e_n^T A = e_n^T$ where e_n is the n-vector of ones. If $n = n_1 n_2$ and $B \in \mathbb{R}^{n_1 \times n_1}$ and $C \in \mathbb{R}^{n_2 \times n_2}$ minimize $\phi_A(B, C)$, then it does *not* follow that B and C are stochastic. For example, if

$$A = \begin{bmatrix} .1 & .5 & .2 & .6 \\ .4 & .1 & .1 & .2 \\ .2 & .0 & .3 & .1 \\ .3 & .4 & .4 & .1 \end{bmatrix},$$

then, after normalizing B and C so that $b_{11} + b_{21} = 1$ we have

$$B = \begin{bmatrix} .6228 & .5939 \\ .3772 & .4298 \end{bmatrix} \quad C = \begin{bmatrix} .3610 & .6657 \\ .5560 & .3512 \end{bmatrix}.$$

Note that B and C are not quite stochastic. Thus, to get the best stochastic Kronecker product approximation we must apply a constrained nonlinear least squares solver to the problem

$$\min_{\substack{e_{n_1}^T B = e_{n_1}^T,\, B \geq 0 \\ e_{n_2}^T C = e_{n_2}^T,\, C \geq 0}} \| A - B \otimes C \|_F$$

Another structured problem that is not solvable by our SVD framework is the case when A is orthogonal and we insist that the optimizing B and C be orthogonal. It does *not* follow that orthogonal B and C minimize $\phi_A(B, C)$. Thus, we are led to another constrained nonlinear leasts squares problem:

$$\min_{\substack{B^T B = I_{n_1} \\ C^T C = I_{n_2}}} \| A - B \otimes C \|_F.$$

A reasonable initial guess (B_0, C_0) in this setting is to set B_0 and C_0 to be the closest orthogonal matrices to the B and C that minimize $\phi_A(B, C)$.

6 Kronecker product preconditioners

To acquire some intuition about the use of Kronecker products as pre-conditioners, consider the $Ax = b$ problem where

$$A = a_1(I_{n_1} \otimes I_{n_2}) + a_2(I_{n_1} \otimes J_{n_2}) + a_2(J_{n_1} \otimes I_{n_2}) + a_3(J_{n_1} \otimes J_{n_2}), \tag{14}$$

$n = n_1 n_2$, and J_m is the m-by-m symmetric tridiagonal matrix

$$J_m = \begin{bmatrix} 0 & 1 & 0 & \cdots & 0 \\ 1 & 0 & 1 & \cdots & 0 \\ \vdots & \vdots & \ddots & & \vdots \\ 0 & 0 & \cdots & 0 & 1 \\ 0 & 0 & \cdots & 1 & 0 \end{bmatrix}.$$

Matrices with this structure arise in many applications. For example, the usual discretization of Poisson's equation on a rectangle with the Dirichlet stencil

a_3	a_2	a_3		0	-1	0
a_2	a_1	a_2	$=$	-1	4	-1
a_3	a_2	a_3		0	-1	0

leads to

$$A = (2I_{n_1} - J_{n_1}) \otimes I_{n_2} + I_{n_1} \otimes (2I_{n_2} - J_{n_2}). \tag{15}$$

In computer vision, the Laplace stencil defined by

a_3	a_2	a_3		-1	-4	-1
a_2	a_1	a_2	$=$	-4	20	-4
a_3	a_2	a_3		-1	-4	-1

is frequently used and this leads to

$$A = (2I_{n_1} - J_{n_1}) \otimes (5I_{n_2} - \frac{1}{2}J_{n_2}) + (5I_{n_1} - \frac{1}{2}J_{n_1}) \otimes (2I_{n_2} - J_{n_2}).$$

In either case, if we define the constants

$$\alpha_1 = 2, \qquad \alpha_2 = 2\left(a_2 - \sqrt{a_2^2 - a_1 a_3}\right)/a_1,$$

$$\beta_1 = a_1/4, \quad \beta_2 = a_1 a_3\left(\left(a_2 - \sqrt{a_2^2 - a_1 a_3}\right)\right)/4,$$

then it can be shown that

$$A = (\alpha_1 I_{n_1} + \alpha_2 J_{n_1}) \otimes (\beta_1 I_{n_2} + \beta_2 J_{n_2}) + (\beta_1 I_{n_1} + \beta_2 J_{n_1}) \otimes (\alpha_1 I_{n_2} + \alpha_2 J_{n_2}).$$

Thus, A is the sum of two Kronecker products and the remarks made in §5.5 apply. Since the rank of $\tilde{A}$ is two, the singular vectors that define the optimal B and C can be computed in $O(n)$ flops. These matrices are tridiagonal, symmetric, and positive definite in view of the discussions in §5.

Let us focus on the case when A is given by (15). For simplicity, define the 1-2-1 tridiagonal matrix

$$T_m = 2I_m - J_m$$

and note that

$$A = T_{n_1} \otimes I_{n_2} + I_{n_1} \otimes T_{n_2}.$$

From §5.5 we know that the optimizing B and C have the form

$$B = b_1 I_{n_1} + b_2 T_{n_1}$$
$$C = c_1 I_{n_2} + c_2 T_{n_2}.$$

The matrix T_m has known eigenvalues:

$$Q_m^T T_m Q_m = D_m = \text{diag}(\lambda_1^{(m)}, \ldots, \lambda_m^{(m)}), \qquad \lambda_j^{(m)} = 4\sin^2\left(\frac{j\pi}{2(m+1)}\right).$$

Using this result, it can be shown that the Kronecker approximation problem involves choosing b_1, b_2, c_1, and c_2 so that

$$
\begin{aligned}
\| A - B \otimes C \|_F^2 &= \| (T_{n_1} \otimes I_{n_2} + I_{n_1} \otimes T_{n_2}) - (b_1 I_{n_1} + b_2 T_{n_1}) \otimes (c_1 I_{n_2} + c_2 T_{n_2}) \|_F^2 \\
&= \| (D_{n_1} \otimes I_{n_2} + I_{n_1} \otimes D_{n_2}) - (b_1 I_{n_1} + b_2 D_{n_1}) \otimes (c_1 I_{n_2} + c_2 D_{n_2}) \|_F \\
&= \sum_{i=1}^{n_1} \sum_{j=1}^{n_2} \left[(\lambda_i^{(n_1)} + \lambda_j^{(n_2)}) - (b_1 + b_2 \lambda_i^{(n_1)})(c_1 + c_2 \lambda_j^{(n_2)}) \right]^2
\end{aligned}
$$

is minimized. The eigenvalue distribution of $M^{-1}A$, which is crucial to the success of $M = B \otimes C$ as a preconditioner, can also be examined in closed form once b_1, b_2, c_1, and c_2 are known:

$$\lambda_{ij}(M^{-1}A) = \frac{\lambda_i^{(n_1)} + \lambda_j^{(n_2)}}{(b_1 + b_2\lambda_i^{(n_1)})(c_1 + c_2\lambda_j^{(n_2)})}. \tag{16}$$

We ran some experiments in the square case $n_1 = n_2 = \sqrt{n}$. It can be shown that about $10n$ flops are required to solve a system of the form $Mz = r$ assuming that the LDL^T factorizations of B and C are available. By way of comparison, about $9n$ flops are involved when an incomplete Cholesky (IC) preconditioner is used. In the following table we compare these two preconditioners:

$\sqrt{n}$	IC Iterations	Kronecker Iterations
16	14	19
32	23	33
64	39	56
128	51	74
256	66	93

Table 1. *Comparison of Preconditioners on Model Problem*

Random right hand sides were used with termination criteria $r^T A r \le 10^{-6}$ where $r = b - Ax$ is the residual of the approximate solution. We have no "proof" why reasonable convergence occurs before $\sqrt{n}$ steps. A plot of the spectrum of $M^{-1}A$ using (16) reveals that many eigenvalues of $M^{-1}A$ are clustered about 1. However, the clustering is not definitive enough to suggest that $O(\sqrt{n})$ convergence is provable.

The Kronecker preconditioner applied to the above model problem compares favorably with many of the other block preconditioners that are reported in [7]. In a distributed memory environment, we suspect that the Kronecker approach may be very attractive because the preconditioner equation $CZB^T = R$ is structured perfectly for parallel computation–but that is the subject of ongoing research.

312

Acknowledgements

Nick Trefethen of Cornell University suggested the connection between Kronecker and rank-1 approximation. Roger Horn of the University of Utah enhanced our understanding of the rearrangement operator and pointed the way to the proof of Theorem 9.

This research was partially supported by the Cornell Theory Center, which receives major funding from the National Science Foundation and IBM Corporation, with additional support from New York State and members of its Corporate Research Institute.

References

[1] H.C. Andrews and J. Kane. Kronecker Matrices, Computer Implementation, and Generalized Spectra. *J. Assoc. Comput. Mach.*, **17**, pp 260–268, 1970.

[2] R.H. Barham and W. Drane. An Algorithm for Least Squares Estimation of Nonlinear Parameters when Some of the Parameters are Linear. *Technometrics*, **14**, pp 757–766, 1972.

[3] R.H. Bartels and G.W. Stewart. Solution of the Equation $AX + XB = C$. *Comm. ACM*, **15**, pp 820–826, 1972.

[4] J.W. Brewer. Kronecker Products and Matrix Calculus in System Theory. *IEEE Trans. on Circuits and Systems*, **25**, pp 772–781, 1978.

[5] T.F. Chan. An Optimal Circulant Preconditioner for Toeplitz Systems. *SIAM J. Sci. Stat. Comp.*, **9**, pp 766–771, 1988.

[6] R. Chan and X-Q Jin. A Family of Block Preconditioners for Block Systems. *SIAM J. Sci. Stat. Comp.*, **13**, pp 1218–1235, 1992.

[7] P. Concus, G.H. Golub, and G. Meurant. Block Preconditioning for the Conjugate Gradient Method. *SIAM J. Sci. Stat. Comp.*, **6**, pp 220–252, 1985.

[8] J. Cullum and R.A. Willoughby. *Lanczos Algorithms for Large Sparse Symmetric Eigenvalue Computations, Volume I (Theory) and II (Programs)*, Birkhauser, Boston, 1985.

[9] C. de Boor. Efficient Computer Manipulation of Tensor Products. *ACM Trans. Math. Software*, **5**, pp 173–182, 1979.

[10] D.W. Fausett and C. Fulton. Large Least Squares Problems Involving Kronecker Products. *SIAM J. Matrix Analysis*, to appear, 1992.

[11] G.H. Golub. Some Modified Eigenvalue Problems. *SIAM Review*, **15**, pp 318-344, 1973.

[12] G.H. Golub, F. Luk, and M. Overton. A Block Lanczos Method for Computing the Singular Values and Corresponding Singular Vectors of a Matrix. *ACM Trans. Math. Soft.*, **7**, pp 149–169, 1981.

[13] G.H. Golub, S. Nash, and C. Van Loan. A Hessenberg-Schur Method for the Matrix Problem $AX + XB = C$. *IEEE Trans. Auto. Cont.*, **AC-24**, pp 909-913, 1979.

[14] G.H. Golub and V. Pereya. The Differentiation of PseudoInverses and Nonlinear least Squares Problems Whose Variables Separate. *SIAM J. Numer. Analysis*, **10**, pp 413–432, 1973.

[15] G.H. Golub and C. Van Loan. *Matrix Computations, 2nd Ed.*, Johns Hopkins University Press, Baltimore, MD, 1989.

[16] A. Graham. *Kronecker Products and Matrix Calculus with Applications*, Ellis Horwood Ltd., Chichester, England, 1981.

[17] S.R Heap and D.J. Lindler. Block Iterative Restoration of Astronomical Images with the Massively Parallel Processor. *Proc. of the First Aerospace Symposium on Massively Parallel Scientific Computation*, pp 99-109, 1986.

[18] H.V. Henderson, F. Pukelsheim, and S.R. Searle. On the History of the Kronecker Product. *Linear and Multilinear Algebra*, **14**, pp 113–120, 1983.

[19] H.V. Henderson and S.R. Searle. The Vec-Permutation Matrix, The Vec Operator and Kronecker Products: A Review. *Linear and Multilinear Algebra*, **9**, pp 271–288, 1981.

[20] R.A. Horn and C.A. Johnson. *Matrix Analysis*, Cambridge University Press, New York, 1985.

[21] R.A. Horn and C.A. Johnson. *Topics in Matrix Analysis*, Cambridge University Press, New York, 1991.

[22] C-H Huang, J.R. Johnson, and R.W. Johnson. Multilinear Algebra and Parallel Programming. *J. Supercomputing*, **5**, pp 189–217, 1991.

[23] J. Johnson, R.W. Johnson, D. Rodriguez, and R. Tolimieri. A Methodology for Designing, Modifying, and Implementing Fourier Transform Algorithms on Various Architectures. *Circuits, Systems, and Signal Processing*, **9**, pp 449–500, 1990.

[24] L. Kaufman. A Variable Projection Method for Solving Separable Nonlinear Least Squares Problems. *BIT*, **15**, pp 49–57, 1975.

[25] V. Pereyra and G. Scherer. Efficient Computer Manipulation of Tensor Products with Applications to Multidimensional Approximation. *Mathematics of Computation*, **27**, pp 595–604, 1973.

[26] U.A. Rauhala. Introduction to Array Algebra. *Photogrammetric Engineering and Remote Sensing*, **46**(2), pp 177–182, 1980.

[27] P.A. Regalia and S. Mitra. Kronecker Products, Unitary Matrices, and Signal Processing Applications. *SIAM Rev.*, **31**, pp 586–613, 1989.

[28] A. Swami and J. Mendel. Time and Lag Recursive Computation of Cumulants from a State-Space Model. *IEEE Trans. Auto. Cont.*, **35**, pp 4–17, 1990.

314

[29] C. Van Loan. *Computational Frameworks for the Fast Fourier Transform*, SIAM Publications, Philadelphia, PA, 1992.

[30] J.H. Wilkinson. *The Algebraic Eigenvalue Problem*. Oxford University Press, New York, 1965.

SOME LINEAR ALGEBRA ISSUES IN LARGE-SCALE OPTIMIZATION

M.H. WRIGHT
AT&T Bell Laboratories
600 Mountain Avenue, Room 2C-462
Murray Hill, New Jersey 07974
U.S.A.
mhw@research.att.com

ABSTRACT. Solving linear systems is a fundamental ingredient in numerical optimization. Particularly in large-scale optimization, these systems encompass a surprisingly wide variation not only in form and representation of matrices, but also in the level of solution accuracy. A broad and often vaguely specified concept in the latter context is "inexact" solution. For unconstrained, linearly constrained and nonlinearly constrained optimization in turn, we discuss selected features of large-scale problems and of the associated forms of inexactness. Various unresolved issues in each problem category are mentioned, along with recent areas of research activity stressing linear algebra.

KEYWORDS. Large-scale optimization, inexactness, active-set methods, SQP methods, interior methods.

1 Introduction

An optimization problem involves finding the "best": given a specified measure of goodness (the *objective function*), we try to determine its optimal value by adjusting a set of variables that may be subject to restrictions or limitations (*constraints*). Linear algebra and optimization are closely related in many ways, of which we mention two:

- Optimization methods include linear algebraic subproblems (mostly, linear systems). Hence, the needs of optimization methods indicate areas of linear algebra in which improvements would be helpful.

- Efficient linear algebra techniques inspire optimization methods that use them. If researchers in optimization know that a certain kind of linear system can be solved quickly and stably, it is natural to develop optimization methods that include such

315

M. S. Moonen et al. (eds.), Linear Algebra for Large Scale and Real-Time Applications, 315–337.
© 1993 *Kluwer Academic Publishers.*

systems. For example, the simplex method for linear programming and quasi-Newton methods for nonlinear optimization rely on the ability to perform low-rank updates to matrix factorizations.

A *large-scale* optimization problem cannot be given a precise general definition, since the notion of "size" depends on context. Broadly speaking, however, a "large" problem is one in which the *number of variables* (or constraints) noticeably affects our ability to solve the problem in the given computing environment. Confusion in discussing large-scale optimization arises because some authors restrict their definition of "large" problems to those in which the associated linear systems are *sparse*. Although sparsity does indeed characterize many, perhaps even a majority, of large-scale problems solved today, it is *not* a requirement. The recent surveys [6, 10] discuss many issues in large-scale optimization, and provide extensive bibliographies.

It is undeniable that larger and larger optimization problems are being solved successfully. In certain areas, such as linear programming, problems considered impossibly large only a few years ago are solved almost routinely in practical applications—sometimes in real time. The state of the art has advanced in part because of hardware: enormous gains in computer speed and storage, and increasing availability of high-performance computing environments in a variety of architectures. Progress in algorithms, however, has been even more significant.

For any particular large problem, issues such as efficient use of storage, organization of calculations, and adaptation to specialized architectures must be faced—for example, determining the best way to represent and solve partial differential equations that define an objective function. In addition, the same issues appear in a more general form as they apply to common linear algebraic procedures within optimization algorithms. The exigencies of linear algebra accordingly dictate attention to the inner workings of large-scale optimization methods.

Two themes will recur throughout this paper:

1. The special, interesting or important features of large-scale optimization problems and methods, particularly as they relate to linear algebra;

2. The role of *inexactness* in solving linear algebraic subproblems. (Although inexactness is frequently mentioned in the optimization literature, it is rarely carefully defined.)

2 Formulation

2.1 PROBLEM STATEMENT

The objective function, which we assume is minimized, will be denoted by $f(x)$, where x is a real n-vector. Constraints occur in two varieties: equalities of the form $c_j(x) = 0$, and inequalities of the form $c_j(x) \geq 0$. (For convenience, we have arbitrarily chosen to express inequalities with a "greater than" relation.) The constraints may include a mixture of equalities and inequalities. The objective and constraint functions are assumed to be twice-continuously differentiable.

The gradient and Hessian matrix of $f(x)$ will be denoted by $g(x)$ and $H(x)$. The n-vector $a_j(x)$ is the gradient of $c_j(x)$, and $H_j(x)$ is its Hessian. The matrix $A(x)$ is defined as

$$A(x) = \begin{pmatrix} a_1^T(x) \\ \vdots \\ a_m^T(x) \end{pmatrix},$$

and is the usual Jacobian of the set of m constraints $\{c_j(x)\}$. (Note that the jth row of $A(x)$ is $a_j^T(x)$.) The generic solution of an optimization problem will be denoted by x^*, which is assumed to be only a local solution. For simplicity and brevity, we assume for each problem category that standard second-order sufficiency conditions hold at x^*. Details of these conditions are given in most textbooks on optimization—for example, [9, 10, 15, 24].

Constraint functions are often specialized beyond the general form above. A *linear constraint* has the form $a_j^T x = b_j$ or $a_j^T x \geq b_j$, where a_j is a constant vector and b_j is a scalar. The Jacobian matrix of a set of linear constraints is thus a constant matrix, and the Hessian of a linear constraint is zero. The simplest and probably the most frequent form of linear constraint imposes lower and/or upper bounds on a variable, i.e. $l_j \leq x_j \leq u_j$.

For large-scale problems, it can be advantageous to "remove" inequality constraints by transforming to so-called *standard form*, in which the only inequality constraints are bounds on the variables. To produce standard form, inequalities are replaced by equalities that include extra variables subject to bounds, so that the inequality $c_j(x) \geq 0$ is replaced by the equality $c_j(x) - y_j = 0$, where the new "slack" variable y_j must satisfy $y_j \geq 0$. See [15, 24] for a discussion of some of the benefits of standard form. The main advantage is that, for some algorithms, standard form ensures that certain matrices remain fixed in dimension throughout the optimization, thereby simplifying the linear algebra.

2.2 RELEVANT PROBLEM FEATURES

We have already mentioned that optimization problems are characterized by features such as the presence or absence of constraints, the linearity or nonlinearity of constraints, and the options of inequality or standard form for the constraints. Beyond these features, a major difference between small and large problems is the attention paid to variations connected with linear algebra.

For small optimization problems, the linear algebra costs for an iteration can typically be estimated from the problem dimensions, which bound the size of matrices involved. Consider unconstrained minimization using a Newton-based method for small n (see Section 3.1), which requires a Cholesky factorization of the dense $n \times n$ Hessian matrix at a cost of approximately $n^3/6$ flops. Even if a small problem has features that could be exploited to speed up the factorization (say, many elements of the Hessian are known to be zero or one), it would not generally be considered worthwhile to expend the effort to develop a special-purpose routine unless the problem were to be solved an extremely large number of times.

In contrast, research concerned with gaining speed in solving large-scale problems must focus on linear algebra issues that may be quite specialized. For large problems, linear algebraic choices are not only more varied than for small problems, but also can exert a dominant influence on solution speed. The tradeoffs are difficult to quantify and analyze, since shortcuts in linear algebra during one part of an algorithm necessarily lead to changes in the algorithm's overall behavior. For example, obtaining an approximate solution to a linear system using a "cheap" method (rather than factorizing a large matrix to solve the system exactly) may increase the number of iterations required to solve the problem, which means that more systems need to be solved. Depending on the increase in the number of iterations and the relative costs per iteration, the cheaper method for solving the system may or may not be worthwhile. An informed decision about the preferred strategy is not straightforward, and is likely to be problem-dependent. (This phenomenon, although unsettling, is familiar in the field of sparse linear algebra.)

Some features of large-scale optimization problems—those that dictate the nature of the linear systems—have an obvious and direct effect on linear algebra. Many large optimization problems arise from modeling of physical systems. When continuous problems are represented in discretized form by piecewise interpolants, for example, the optimal value of any given variable tends to be influenced by only a relatively small number of nearby variables. Such limited and localized interactions often imply sparsity in the resulting Jacobian and Hessian matrices.

More generally, the problem functions may be "partially separable", in the sense that the objective function can be expressed as the sum of element functions on which small subsets of the variables have disjoint effects. For a discussion of the general concept of group-partial separability, see [29]. We note later that partially-separable structure in the problem allows an efficient representation of curvature information; see Section 3.2.

Another feature, less well understood but equally important, of some large problems is that they are highly *structured*—*not* the same as sparse. For example, in a problem involving modeling of closely related physical phenomena, the Jacobian or Hessian may contain different multiples of the same (dense) matrix, or of matrices that differ only in minor ways. Enormous gains in efficiency can result if linear algebraic procedures are designed to exploit this high commonality. For an example of such improvements, see [33].

In addition to sparsity and structure, certain properties of the solution that are usually not known in advance can significantly influence algorithm performance, although they do not appear at first glance relevant to linear algebra. Such properties include: the proportion of nonlinear constraints; the fraction of constraints active at the solution; and the nearness of the constraints to linearity. To illustrate how these affect linear algebra, consider the MINOS code [39], a widely-used software package for large-scale optimization. The algorithm of MINOS explicitly stores and performs rank-one modifications on the Cholesky factor of a dense matrix whose dimension is typically close to the number of degrees of freedom at the solution (roughly speaking, the number of variables minus the number of active constraints). Hence, although MINOS is highly reliable on general problems, it is inherently most effective on problems where the fraction of active constraints is relatively large.

It should be clear that no single algorithm will be "the best" for all large-scale problems. Most optimization algorithms are linked to a (frequently unstated) particular *model of the underlying problem*. Characterization of the underlying problem can be crucial to understanding why certain algorithmic choices are made, such as the form of the linear systems to be solved when several mathematically equivalent alternatives are available. A large-scale algorithm will be most effective when applied to an actual problem that closely matches its underlying problem.

3 Unconstrained optimization

3.1 SMALL PROBLEMS

It has long been considered that Newton's method is the "ideal" optimization method. Newton's method for unconstrained optimization is based on approximating the variation of f near the kth iterate x_k by a local quadratic model derived from the Taylor-series expansion:

$$f(x_k + p) \approx f(x_k) + g_k^T p + \tfrac{1}{2} p^T H_k p, \tag{1}$$

where g_k and H_k are the gradient and Hessian evaluated at x_k. Our theme of inexactness makes its first appearance in this most classical of methods, since the quadratic formulation is not in general an exact representation of $f(x_k + p)$.

If H_k is positive definite, the step p_N from x_k to the (unique) minimizer of the local model is called the *Newton direction*, and satisfies the *Newton equations*:

$$H_k p_N = -g_k. \tag{2}$$

When $H(x^*)$ is positive definite and x_k is sufficiently close to x^*, a "pure" Newton method in which the next iterate is $x_k + p_N$ displays quadratic convergence to x^* (see, e.g., [9, 10, 15]). Otherwise, Newton's method is reliable only if safeguards are included to ensure convergence from an arbitrary starting point and to treat non-positive-definiteness of the Hessian.

Newton-based methods tend to be categorized as either line search or trust region methods; see [9, 10, 15]. We consider only line search methods; linear algebraic issues analogous to many of those raised for line search methods also apply to trust region methods. In a line search method, the new iterate x_{k+1} satisfies the relation

$$x_{k+1} = x_k + \alpha_k p_k. \tag{3}$$

The n-vector p_k, the search direction, is defined using curvature information; its calculation will be discussed later. The scalar α_k is a positive number chosen during a line search to produce a suitable decrease in the value of f compared to that at x_k. A pure Newton method corresponds to $p_k = p_N$ and $\alpha_k = 1$.

Within Newton's method, linear algebra is invariably described as playing the role of "solving the Newton equations". However, this seemingly innocuous statement is vastly oversimplified and can even be misleading. In fact, it may be *undesirable* to solve the Newton equations exactly, for at least three reasons. First, for all problems, small and large, the current local model may be inadequate—for example, the Hessian may be indefinite or singular. With an indefinite Hessian, it is probably undesirable to take a step along the Newton direction; when the Hessian is singular, the Newton direction may not exist. In either case, an alternative procedure is used to define p_k. For large problems, two other factors contribute to the undesirability of solving the Newton equations exactly: the cost of evaluating and storing the exact Hessian may be too high; or the linear algebra needed to solve the Newton equations may be too expensive.

The alert reader should pause at this point to observe an implicit and highly crucial assumption underlying the last two justifications of the preceding paragraph. Asserting that it is "too expensive" to solve the Newton equations, yet assuming that we shall nonetheless succeed in solving the original problem, implies that we do not actually *need* to solve the Newton equations exactly. (This conclusion is so easy for optimizers to accept that it is often taken for granted, thereby leading to confusion.) To belabor the point for emphasis: if solving the Newton equations were essential, failing to solve them would not be acceptable. But, since solving the Newton equations is only a means for reaching the desired end of minimizing f, a cheaper procedure can be used if it will achieve the same result. Consequently the luxury of inexactness, which we shall describe in various forms in the rest of the paper, is permitted as long as it leads to an effective method for solving the given problem.

Although the search direction p_k need not solve the Newton equations exactly, it must satisfy other criteria. In particular, standard line search methods require p_k to be a descent direction for f at x_k, i.e. $p_k^T g_k < 0$. When p_k is a descent direction, then, for all sufficiently small positive steps α, it can be guaranteed that $f(x_k + \alpha p_k) < f(x_k)$. The property that p_k is a descent direction will hold automatically if p_k satisfies the linear system

$$M_k p_k = -g_k, \tag{4}$$

for any positive definite matrix M_k. We assume henceforth that p_k always satisfies this equation, and much of our discussion will concern the possible choices for M_k.

A second, more elusive, criterion is that p_k should lead to a "good" reduction in f, and to eventual fast convergence with $\alpha_k = 1$ in (3). What this means (speaking impressionistically) is that, if p_k satisfies $M_k p_k = -g_k$, then M_k must contain useful curvature information, i.e. information that somehow represents the important properties of the exact Hessian. It is impossible to be precise about this characterization, since the essential Hessian information needed in M_k depends on the problem.

For small problems, a technique widely used to deal with possible indefiniteness or singularity of the Hessian is to apply a *modified Cholesky factorization* to H_k; the two most popular factorizations of this type are described in [24] and [48]. Modified Cholesky factorizations fulfill two purposes simultaneously: they check for positive-definiteness of H_k while calculating the Cholesky factors of a matrix M_k used to define the search direction from (4). If H_k is sufficiently positive definite, M_k is taken as H_k, and no modification is required. Otherwise, the Cholesky factors of a *modified* matrix M_k are calculated. The matrix M_k is guaranteed to be positive definite, and is given by $H_k + E_k$, with E_k a nonnegative diagonal matrix (*not* a multiple of the identity). The standard modified Cholesky factorizations always exist, and can be shown to satisfy a bound on the size of $\|E_k\|$.

If second derivatives are too expensive but first derivatives are available, *quasi-Newton methods* are both reliable and efficient. These methods, popularized initially in the 1960's, store an *approximation* M_k to the Hessian, calculate p_k from (4), and then perform updates of low rank (one or two) to M_k after each iteration. The usual choice for the initial approximation M_0 is the identity, so that the first step is equivalent to steepest descent. Quasi-Newton methods have been shown to have many desirable theoretical and computational properties, including superlinear convergence. With respect to the theme of inexactness, notice that a quasi-Newton matrix M_k is only an approximation to the Hessian. For small problems, a factorization of M_k is typically stored, and the equations (4) are solved exactly. Details about quasi-Newton methods may be found in, for example, [9, 10, 15, 24].

Modified Newton and quasi-Newton approaches allow substantial inexactness for small problems. When all goes well, however, the level of inexactness is expected to decrease (or even disappear) as the solution is approached. In a Newton-based method involving a modified Cholesky factorization, the Hessian should not be modified near x^*. With a quasi-Newton method, the changes to M_k should eventually become small in norm as well as in rank, indicating that M_k has become a good representative of the true Hessian.

3.2 LARGE PROBLEMS

Once we move to large-scale unconstrained problems, the third issue mentioned previously—the high linear algebra cost of solving the dense Newton equations exactly—becomes significant. In some instances, efficiency can be maintained simply by switching from a dense to a sparse factorization. If exact second derivatives are available and H_k is sparse, its sparse Cholesky factorization can be calculated; see [11]. Of course, care must be taken to perform the modifications needed to treat indefiniteness without impairing efficiency of the sparse factorization; see [48]. When exact second derivatives are expensive but the Hessian has a known sparsity pattern, clever techniques have been developed for choosing vectors that allow a finite-difference approximation to the Hessian to be calculated with a small number of gradient evaluations; see [3].

In some situations, storage limitations or lack of sparsity mean that it is undesirable to factorize the Hessian in any form. If, however, matrix-vector products involving the Hessian can be calculated efficiently, *truncated Newton* methods introduce yet another form of inexactness. The idea is to "solve" the Newton equations inexactly by performing steps of an *iterative method* such as the conjugate-gradient method (customarily with preconditioning), but stopping before the system has been solved to full accuracy. It was shown in [8] that, if the accuracy with which the Newton equations are solved increases appropriately with the iteration number, the desired quadratic convergence rate of Newton's method can be retained. Broadly speaking, quadratic convergence sets in only near the solution, which is thus where the Newton equations need to be solved accurately. At early iterations, a substantial degree of inaccuracy can be tolerated without sacrificing the desired properties of descent and ultimate fast convergence. In an effective truncated Newton method, preconditioned conjugate-gradient iterations are "truncated" (stopped) using a suitable criterion. A preconditioning strategy that accumulates good curvature information is essential to retain fast Newton-like convergence near the solution. Without preconditioning, a large number of conjugate-gradient iterations would be required to solve the Newton equations accurately.

We do not treat the *nonlinear* conjugate-gradient method (see, e.g., [9, 10, 15, 24]), which performs unconstrained minimization by storing only a few n-vectors, because its interesting issues are not primarily linear algebraic.

If calculation of second derivatives is infeasible for a large problem, our discussion of small problems (Section 3.1) suggests that the method of choice would be a quasi-Newton method. For several years in the 1970's, researchers studied sparse quasi-Newton methods in which each quasi-Newton approximation M_k would retain the prescribed sparsity pattern of the true Hessian. Unfortunately, it turned out to be impossible to preserve positive-definiteness of M_k, a specified sparsity pattern, and satisfaction of the standard quasi-Newton condition; see [49]. Hence this idea has been mostly abandoned, at least for the moment, in favor of other closely related but viable alternatives.

When storing an explicit $n \times n$ quasi-Newton approximation is difficult, a *limited-memory* quasi-Newton method can be considered. With these methods, the low-rank updates are represented by storing the n-vectors that define them rather than by performing explicit updates to an approximate Hessian M_k (or its inverse). As usual, the search direction satisfies (4), but the updated matrix is not stored or factorized; p_k is actually calculated by performing a sequence of inner products with the saved vectors. Only a small number of the most recent vectors (say, 5 or 10) are stored, and earlier updates are discarded. The theoretical properties of limited-memory methods are not nearly as attractive as those of standard quasi-Newton methods, but these methods have nonetheless been successful in practice on many problems. Recent papers on this topic include, for example, [34, 40].

Although apparently unachievable in general, a quasi-Newton method tailored to structured problems can be realized when f is partially separable. In this case, small, separate, explicit quasi-Newton approximations to "pieces" of the Hessian can be developed and carefully combined to produce a low-storage quasi-Newton approximation M_k containing the most important curvature information. For details, see [28, 29, 7].

In terms of inexactness, note that with a truncated Newton method, a limited-memory quasi-Newton method, or a partially-separable quasi-Newton method, x^* may be found efficiently without ever calculating the exact solution of any system remotely resembling the Newton equations! This observation—that inexactness may be considerable without apparent damage to efficiency—leads to an important point: *many large-scale problems are very easy*. To see how well-behaved they are, consider the famous two-dimensional Rosenbrock's function with its steeply curving valley, which appears in essentially all standard optimization textbooks. An efficient Newton-based method typically requires around 30 iterations to minimize Rosenbrock's function, starting from the usual initial point. In contrast, it is not uncommon for researchers to report successful solution of problems with thousands of variables in 30–50 iterations. The difficulty associated with Rosenbrock's function has obviously not "scaled up" for these large problems, which means that they are special in some way not yet fully understood. (This phenomenon appears to be similar to the fact that real-world linear programs solved by the simplex method never require anything like the known worst-case number of iterations.) Thus, fortunately, the judicious introduction of inexactness appears to be acceptable for many practical large-scale problems.

4 Linear constraints

We now consider minimizing $f(x)$ subject to a set of linear constraints, possibly including bounds on the variables. Because of space limitations, we mention only a small selection of issues in large-scale linearly constrained optimization—those associated with linear algebra in general, and inexactness in particular. See [15, 24] for further discussion.

4.1 ACTIVE-SET METHODS

As in Section 3, we consider line search methods in which the new iterate x_{k+1} is given by $x_k + \alpha_k p_k$. The constraints are presumed to be in the form $Ax \geq b$; equality constraints and bounds can be handled in a conceptually straightforward way.

Our attention in this paper is limited to the class of *active-set methods*. Active-set methods have been criticized in recent years for their inefficiency, mostly for the "classical" or "one-at-a-time" strategy used to modify the working set, which we describe later. Various alternatives have been proposed, primarily for the case of convex quadratic programming with bound constraints; see, for example, [38]. Despite some developments in non-active-set software, no clear "winner" has emerged for practical solution of general linearly constrained problems; furthermore, active-set methods provide a useful paradigm for understanding and developing new algorithms. The linear algebra issues associated with active-set methods are sufficiently close to those for newer strategies so that this discussion is likely to remain relevant even if active-set methods are superseded.

The motivation for active-set methods derives from the following observations. Suppose that the set of constraints active (i.e., that hold with equality) at x^* is known; let $\hat{m}$ denote the number of active constraints, $\hat{A}$ the Jacobian of the active constraints, and $\hat{b}$ the associated subvector of b. (We assume for simplicity that $\hat{A}$ has full row rank.) Finally, assume that the current iterate x_k is feasible and satisfies $\hat{A}x_k = \hat{b}$. Let Z denote a basis for the null space of $\hat{A}$, so that $\hat{A}Z = 0$, and define p_k as Zp_Z for some $(n - \hat{m})$-vector p_Z. Then

$$\hat{A}(x_k + \alpha p_k) = \hat{A}x_k + \alpha \hat{A}Zp_Z = \hat{b}.$$

By moving only in the null space of $\hat{A}$, subsequent iterates will automatically stay "on" the active constraints.

If, as in the unconstrained case, a local quadratic model is defined of the variation in f (see (1)), but restricted to directions in the null space of $\hat{A}$, we have

$$f(x_k + Zp_Z) \approx f_k + p_Z^T Z^T g_k + \tfrac{1}{2} p_Z^T Z^T H_k Z p_Z. \tag{5}$$

The number of degrees of freedom in (5) is the dimension of p_Z, and we may apply unconstrained techniques within a reduced subspace to choose a "good" p_Z. The vector $Z^T g_k$ and the matrix $Z^T H_k Z$ are called the reduced gradient and reduced Hessian. If p_Z satisfies the reduced Newton equations:

$$Z^T H_k Z p_Z = -Z^T g, \tag{6}$$

then $p_k = Z p_Z$ is the exact analogue of the Newton direction p_N for an unconstrained problem; compare (2).

In general, of course, the true active set is *unknown*, and must be predicted. An active-set method is based on maintaining a prediction of the active set (called the *working set*) that is used to define the search direction. At any given iteration, constraints in the working set are treated as equalities for the purpose of defining Z and the reduced gradient/Hessian.

Assume that the Jacobian of constraints in the current working set is $\hat{A}_k$, and that $\hat{A}_k Z = 0$. When $Z^T g_k \neq 0$, the search direction is defined as $Z p_Z$, where p_Z is chosen so that p_k is a descent direction for f; for instance, p_Z could solve some version of the reduced Newton equations (6). After a suitable p_k is determined, a steplength $\alpha_k > 0$ is chosen such that $x_k + \alpha_k p_k$ is feasible and $f(x_k + \alpha_k p_k)$ is sufficiently less than $f(x)$. If the step α_k "hits" a constraint not currently in the working set, this constraint is added to the working set before beginning the next iteration.

If $Z^T g_k = 0$, tests are performed to see whether x_k is optimal. Such tests involve the signs of the Lagrange multipliers with respect to the working set, and the inertia of the current reduced Hessian. If the reduced Hessian is positive definite and a Lagrange multiplier is negative, the associated constraint is deleted from the working set. Other, more complicated, situations arise that will not be considered here. Note that, according to this description, constraints are added to or deleted from the working set one by one, so that each change in the working set is a *rank-one modification* to $\hat{A}_k$ (adding or deleting a row). This strategy for altering the working set is *not* essential for an active-set method (in fact, other approaches have been suggested), but its abandonment leads to more complicated linear algebra, especially for general linear constraints.

4.2 DEFINING A BASIS FOR THE NULL SPACE

Based on the reduced Newton equations (6) and the required form $Z p_Z$ for p_k, a straightforward transcription of an active-set method would involve calculation of a null-space basis Z. For small problems, an explicit Z is typically obtained by processing the orthogonal transformations used to compute a QR factorization of $\hat{A}_k$; this produces an orthonormal basis for the null space, which has several numerical advantages. The linear algebra involved in a dense active-set method might thus include updating the QR factorization as $\hat{A}$ changes, and updating the chosen representation of the reduced Hessian.

For large problems in which the Jacobian of the constraints is sparse, techniques for computing and updating an orthonormal null-space basis have been studied by several researchers (see, e.g., [4]). Unfortunately, such calculations remain relatively expensive, and research is still proceeding. A popular and long-standing alternative is to use the "reduced-gradient form of Z", which is closely related to procedures associated with the simplex method for standard-form linear programming. If the working set matrix $\hat{A}_k$ is expressed as $(B \quad S)$, where B is square and nonsingular, the columns of the following matrix Z constitute a (generally non-orthonormal) basis for the null space of $\hat{A}_k$:

$$ Z = \begin{pmatrix} -B^{-1}S \\ I \end{pmatrix} . \tag{7} $$

The reduced-gradient Z is *not* formed explicitly. Rather, products of Z and Z^T with vectors are computed by solving linear systems involving B and B^T. These systems are typically solved using a sparse LU factorization of B, which is updated as $\hat{A}_k$ changes.

The reduced-gradient form of Z is highly efficient for many large problems, and features in, for example, the MINOS code [39]. However, it is possible to remain in the null space without explicit reference to Z; a technique for doing so is considered in Section 4.4.

4.3 THE REDUCED NEWTON EQUATIONS

In addition to ensuring that p_k lies in the null space of $\hat{A}$, an effective active-set method must somehow define a "good" direction p_z. By direct analogy with the unconstrained case, p_z might be computed by solving the reduced Newton equations (6) or a quasi-Newton version thereof, or by inexactly solving such a system.

A serious difficulty is that, despite availability and sparsity of H_k, the reduced Hessian $Z^T H_k Z$ is not necessarily sparse. Furthermore, calculating the reduced Hessian from scratch can be quite expensive, even with the reduced-gradient form of Z (7). Several options have been proposed when the exact Hessian is available, the most obvious of which involve application of a truncated Newton method; an open question is the choice of preconditioner.

If second derivatives are expensive and the dimension of Z is reasonably small, a quasi-Newton approximation of the *reduced Hessian* can be maintained, as in MINOS [39]. If the working set remains fixed during a given iteration, the approximate reduced Hessian can be updated as usual. If the working set is altered, Z changes dimension, and the effect of a modified Z must somehow be reflected in the reduced Hessian approximation. An open research area today is the issue of representing and updating an approximate reduced Hessian when f is partially separable or group-partially separable.

In the formulation described thus far of an active-set method for linear constraints, inexactness plays two different roles. Although it is clearly possible, perhaps even desirable, to solve the reduced Newton equations inexactly, it is *unacceptable* to use an inexact Z. With linear constraints, an active-set method depends on staying in the null space, which means that equations enforcing this property must be solved accurately. To paraphrase George Orwell's *Animal Farm*, "Some equations are more equal than others".

4.4 THE KARUSH-KUHN-TUCKER SYSTEM

We now consider whether the search direction can be forced to remain in the null space without explicitly invoking a null-space basis Z. An increasingly popular approach, especially well suited for sparse problems, is to calculate p_k by solving a larger system whose result is mathematically equivalent to the desired effect. Although this may seem contradictory, it can be faster to factorize a sparse but much larger system (see (9), below) from scratch than to perform a single quasi-Newton update on a dense (albeit reduced) Hessian.

The vector p_k satisfying

$$p_k = Z p_z, \quad \text{with} \quad Z^T H_k Z p_z = -Z^T g_k, \tag{8}$$

is also part of the solution of the $(n + \hat{m})$-dimensional symmetric indefinite *Karush-Kuhn-Tucker* (KKT) equations:

$$\begin{pmatrix} H_k & \hat{A}_k^T \\ \hat{A}_k & 0 \end{pmatrix} \begin{pmatrix} p_k \\ -\lambda_k \end{pmatrix} = K \begin{pmatrix} p_k \\ -\lambda_k \end{pmatrix} = \begin{pmatrix} -g_k \\ 0 \end{pmatrix}. \tag{9}$$

Because of the last $\hat{m}$ equations, the vector p_k must satisfy $\hat{A}_k p_k = 0$, and hence lies by definition in the null space of $\hat{A}_k$. With exact arithmetic, solving the KKT system produces the identical direction as the formulation (8) in which Z appears.

The eigenvalue structure (inertia) of K is closely related to that of the reduced Hessian; see, for example, [23]. In particular, when $\hat{A}_k$ has full row rank and the reduced Hessian is positive definite, K is nonsingular. We shall see later that the KKT system plays a significant role in SQP methods for nonlinearly constrained optimization and in interior methods; see Section 5.1.

For problems in which the Hessian H_k and Jacobian $\hat{A}_k$ are sparse, the KKT system would most logically be solved by calculating a sparse symmetric indefinite factorization of K; see [11] for a discussion of sparse factorization techniques. When standard software of this kind was first applied extensively to KKT systems (in the mid- to late-1980's), a loss of efficiency was observed compared to its usual performance. To explain this phenomenon, we note that the software conducts an initial "analyze" phase devoted to constructing a reordering based on the sparsity pattern, and then executes a numerical "factorize" phase in which interchanges are performed as necessary for stability. Because most matrices to which the software had been applied arose from the context of partial differential equations, the analyze phase was based on the presumption that the diagonal elements were nonzero (which is clearly not true for KKT systems); because of this assumption, many more interchanges than expected were necessitated in the factorization phase.

Since then, new codes have been developed that do not suffer from this difficulty. In [12], a general sparse symmetric indefinite factorization code is presented that uses several new strategies for promoting sparsity and retaining stability. A sparse indefinite solver developed for interior methods (see Section 5.2) is described in [17] that guards against poor orderings by combining the analysis and factorization phases. A more specialized strategy is based on the observation that the sparsity patterns of H_k and $\hat{A}_k$ in the KKT system are related in certain large-scale problems, and it may be possible to calculate the factorization more efficiently by exploiting this special structure; see, e.g., [22].

The solution of KKT systems is an example where an application driven by optimization has led to a change in strategy in a popular "black-box" linear algebra package. Several interesting research questions remain open concerning the KKT approach. One of the most pressing involves development of reliable and effective strategies for coping with an "inappropriate" Hessian, which in this context is the reduced Hessian. The basic difficulties of detecting and treating indefiniteness and singularity are the same as those mentioned in our earlier discussion of unconstrained optimization (Section 3.1), greatly exacerbated because the relevant matrix (the reduced Hessian) is *not* available. The linear system (9) being solved does not directly involve the reduced Hessian; rather, the full Hessian H_k is embedded in the larger KKT matrix. The best way to detect and modify indefiniteness or singularity of the reduced Hessian while factorizing K is unclear, as is the most effective strategy for ensuring that p_k will be a descent direction for f. See, for example, [16].

As with solving the reduced Newton equations (6), it is tempting to suggest that the KKT system should be solved iteratively, and perhaps inexactly. Since the KKT matrix is not positive definite, a special form of the conjugate-gradient method must be used, such as SYMMLQ [43]. Specialized preconditioners for KKT systems are also advisable; see [18]. If a substantial level of inexactness is allowed, the crucial relation $\hat{A}_k p_k = 0$ that keeps p_k in the null space may not be satisfied. Whether this is acceptable depends on the situation. In some instances, remaining in the null space is essential; in others, however, a certain level of inaccuracy is tolerable, such as when KKT systems are solved as subproblems within interior methods for nonlinearly constrained problems (see Section 5.2).

4.5 QUADRATIC PROGRAMMING

Much attention has been devoted to active-set methods for linear and quadratic programming. (In fact, the simplex method for linear programming is the most famous active-set method.) Rather than attempt to cover linear programming, we refer the reader to the recent survey [25], which discusses both simplex and interior methods.

The literature on dense quadratic programming (QP) is vast; see, for example, [15, 23]. Both [22] and [27] describe large-scale QP methods based on solving the KKT system (9), with highly specialized techniques that exploit the quadratic nature of the objective function. In the Schur-complement QP method [22], a sparse factorization is calculated of the KKT matrix defined with a suitable initial working set $\hat{A}_0$. Although change in the working set leads to a theoretically altered version of K, the needed quantities can be calculated using the initial (fixed) factorization and a dense factorization of the Schur complement, a matrix whose dimension is equal to the number of QP iterations. If the number of QP iterations is reasonably small—say, in the hundreds—this technique can be quite efficient.

One reason for interest in quadratic programming is that QP subproblems occur in the extremely effective class of sequential quadratic programming (SQP) methods for nonlinearly constrained optimization (see Section 5.1). Progress in large-scale quadratic programming is therefore likely to have a major effect on large-scale nonlinearly constrained problems as well.

5   Nonlinear constraints

Consider the generic nonlinear inequality-constrained problem

$$\underset{x \in \mathcal{R}^n}{\text{minimize}} \ f(x) \quad \text{subject to} \quad c_j(x) \geq 0, \ j = 1, \ldots, m, \tag{10}$$

where for simplicity of exposition we include only inequality constraints. We have omitted detailed consideration of nonlinear equality constraints because of space limitations and because their treatment is less focused on linear algebra, not because they are "easy".

The terminology in discussing (10) is slightly more complicated than for the linearly constrained case. Assume that $\hat{m}$ constraints are active at the solution x^*; for any point x, $\hat{A}(x)$ means the $\hat{m} \times n$ Jacobian of the constraints active at x^*, evaluated at x. Let $Z(x)$ denote a matrix whose columns form a basis for the null space of $\hat{A}(x)$, so that $\hat{A}(x)Z(x) = 0$. In general $Z(x)$ is not unique, but may be defined as a smooth function in regions where $\hat{A}(x)$ has full rank (see [5]). Let Z^* denote a basis for the null space of $\hat{A}(x^*)$, and Y^* a basis for the range space of $\hat{A}^T(x^*)$.

To simplify our discussion, it is convenient to assume that the following conditions (second-order sufficiency as well as strict complementarity) hold at x^*:

(i) $g(x^*) = A^T(x^*)\lambda^*$, where λ^* is called the Lagrange multiplier vector;

(ii) $\lambda_j^* c_j(x^*) = 0$ for $j = 1, \ldots, m$;

(iii) $\lambda_j^* > 0$ if constraint j is active;

(iv) the gradients of the active constraints at x^* are linearly independent, i.e. $\hat{A}(x^*)$ has full row rank;

(v) $Z^{*T}W^*Z^*$ is positive definite, where $W^* = H(x^*) - \sum_{j=1}^{m} \lambda_j^* H_j(x^*)$; the matrix W^* is the Hessian of the Lagrangian function evaluated at x^*.

For large-scale problems with nonlinear constraints, the most popular solution technique for many years has been the method of MINOS [39]. More recently, the code LANCELOT [7], which is based on trust-region and augmented Lagrangian methods, has become widely available. Both of these provide reliable and soundly-based software for general large-scale problems.

In this section, we treat two other approaches that are less well established for large problems, but involve some interesting linear algebra issues.

5.1 SEQUENTIAL QUADRATIC PPROGRAMMING (SQP) METHODS

For small nonlinearly constrained problems with inequality constraints, arguably the most popular technique for at least ten years has been that of sequential quadratic programming (SQP); detailed discussions of SQP methods are given in, for example, [45, 15, 24]. Several suggestions have been made for applying trust-region techniques to nonlinear constraints, but most of these thus far have been limited to equality constraints; see [2, 50].

A line search SQP method (the only kind to be considered here) chooses the search direction p_k as the solution of a *quadratic programming subproblem* in which the quadratic objective represents the local variation in the *Lagrangian function*. As in (3), the new iterate is taken as $x_k + \alpha_k p_k$, where the step α_k is usually chosen to yield a sufficient decrease in a merit function (a suitable combination of the objective and constraint functions).

In developing SQP methods, a range of strategies has been implemented to balance the gain in speed from a correct early identification of the active set against the loss of efficiency caused by an incorrect prediction. By analogy with the linearly constrained case, a so-called "EQP" strategy can be applied to define a QP subproblem containing only equality constraints corresponding to linearizations of constraints in a specified working set (a prediction of constraints active at the solution). The associated QP subproblem is

$$\text{minimize} \quad \tfrac{1}{2}p^T W_k p + p^T g_k \tag{11}$$

$$\text{subject to} \quad \hat{A}_k p = -\hat{c}_k,$$

where W_k is an approximation of the Hessian of the Lagrangian, g_k is the gradient of f, $\hat{A}_k$ is the Jacobian of constraints in the working set at x_k, and $\hat{c}_k$ is the vector of constraints in the working set. Let Y be a basis for the range space of $\hat{A}_k{}^T$, and Z a basis for the null space of $\hat{A}_k$.

The solution of (11) can be expressed as

$$p = Y p_Y + Z p_z, \tag{12}$$

where p_Y and p_z satisfy the following equations:

$$\hat{A}_k Y p_Y = -\hat{c}_k; \tag{13}$$

$$Z^T W_k Z p_z = -Z^T g_k. \tag{14}$$

The benefits of an EQP approach mainly involve simplification in the computational and informational requirements for calculating p_k. In particular, no iterations are required to solve the QP subproblem (11), and only the reduced Hessian of the Lagrangian function appears in (14). A distinction from the equations (6) for the linearly constrained case is that, if $\hat{c}_k \neq 0$, p_k has a non-zero portion in the range space of $\hat{A}_k{}^T$ (see (13)), and hence a basis Y for this subspace enters the calculations. The major difficulty with an EQP approach is that a poor choice of working set may lead to inefficiency; unfortunately, iron-clad strategies for predicting the active set do not exist.

At the other extreme is a philosophy that, in effect, lets the QP subproblem determine its own active set, and includes linearizations of all the constraints as inequalities:

$$\text{minimize} \quad \tfrac{1}{2}p^T W_k p + p^T g_k \tag{15}$$

$$\text{subject to} \quad A_k p \geq -c_k,$$

where A_k is the complete Jacobian and c_k is the vector of constraint values. A full-blown "IQP" strategy assumes exact solution of (15), which not only involves an unpredictable number of QP iterations, but also requires a representation of the complete matrix W_k rather than the reduced matrix $Z^T W_k Z$. On a typical nonlinearly constrained problem, solving (15) is more costly than solving (11), but the total number of "outer" iterations (QP subproblems) needed to solve the original problem is smaller. It is reasonably well established that neither strategy is consistently better on all problems, and many researchers have sought a compromise between the two.

Techniques based on availability of only a reduced Hessian have been examined for several years; see [42, 30]. In effect, the reduced Hessian is "unprojected" back into the full space so that the QP iterations can proceed; however, the solution obtained in this way may be the exact solution of a QP different from (15). The recent algorithm in [13] is specifically designed for large sparse problems, and uses a combination of dense and sparse techniques, including special sparse forms for the needed subspace bases.

A somewhat different perspective for small problems is applied in [46], which begins with the subproblem (15), but does not necessarily solve it to completion. This strategy provides yet another variant on the concept of an "inexact" solution. We stress that "partially solving" (15) by performing only a few iterations of a standard active-set QP method does not ensure that the solution so found is close in any sense to the exact solution. On the other hand, the iterations performed generate information that may be used to construct a descent direction for a suitable SQP merit function.

Whatever the choice of QP subproblem, all SQP algorithms "solve" (in some sense) KKT systems of the form (9), or a sequence of such systems. In some cases the $(2,2)$ block may be nonzero; for instance, the $S\ell_2 QP$ method [45] involves solving a KKT-like system in which the $(2,2)$ block is a diagonal matrix converging to zero.

Substantial progress in the development of large-scale SQP methods has already occurred, and is likely to continue, perhaps with a proliferation of mixed techniques. For example, finite-differencing can be used to approximate products of the Hessian with selected vectors, such as columns of Z, the null-space basis; this highly effective technique has been applied within a combined SQP-interior method (see [47]). In [1], an SQP method is described in which the constraint Jacobian is stored in sparse form, the subproblems are solved using a Schur-complement QP method (see Section 4.5), and the QP Hessian is represented by limited-memory quasi-Newton updates (see Section 3.2), modified by a trust-region strategy to ensure positive-definiteness. Second-derivative SQP methods are likely to be especially effective on sparse problems once QP-related issues involving the proper treatment of indefiniteness and singularity are resolved; see Section 4.4.

5.2 INTERIOR METHODS

In 1984, Karmarkar announced his discovery of a polynomial-time method for linear programming [32], along with numerical tests showing that it out-performed the simplex method on large problems. In [20], this method was shown to have a formal connection with classical barrier methods for nonlinear programming. Since then, enormous activity on barrier and other interior methods has continued, focused mainly on complexity rather than numerical results. The classic reference on interior methods is [14]; more recent surveys include [26] and [52].

Interior methods deal with inequality constraints and, in their original form, construct a sequence of strictly feasible iterates (hence the designation "interior"). Given the inequality-constrained problem (10), the associated logarithmic barrier function is

$$B(x,\mu) = f(x) - \mu \sum_{i=1}^{m} \log c_i(x), \qquad (16)$$

where the positive parameter μ is called the *barrier parameter*. Let x_μ denote an unconstrained minimizer of $B(x,\mu)$. Under reasonably mild assumptions, it can be shown that $\lim_{\mu \to 0} x_\mu = x^*$. Here, for simplicity of exposition we have assumed second-order sufficiency and strict complementarity. Under these added conditions, the sequence $\{x_\mu\}$ defines a differentiable trajectory converging to x^* nontangentially to the active constraint normals. For details, see, e.g., [52].

It would appear that using an interior method to solve a constrained problem simply involves solving a sequence of unconstrained problems, namely minimizing $B(x, \mu)$ for a suitable selection of barrier parameters converging to zero. This was indeed a large part of the motivation for barrier methods in the 1960's, when techniques for unconstrained optimization were far more developed than those for constrained problems. However, barrier methods fell from favor for a variety of reasons, and were in fact dismissed as outmoded for many years. One reason for their loss of popularity was a perceived (and observed) inefficiency compared with alternative methods, such as augmented Lagrangian methods (on which LANCELOT [7] is based) and SQP methods. Another reason involves linear algebra, which we now discuss in some detail.

The barrier Hessian with respect to x, denoted by H_B, is given by

$$H_B(x, \mu) = H(x) - \sum_{j=1}^{m} \frac{\mu}{c_j(x)} H_j(x) + \mu A^T(x) C^{-2}(x) A(x), \qquad (17)$$

where $C^2(x)$ is the matrix $\operatorname{diag}(c_j^2(x))$. We define the m-vector $d(x, \mu)$ as

$$d_j(x, \mu) = \frac{\mu}{c_j(x)}, \quad j = 1, \ldots, m. \qquad (18)$$

It follows from our assumptions of second-order sufficiency and strict complementarity that, at points x_μ on the barrier trajectory, the value of d_j is within $O(\mu)$ of λ_j^*, the optimal Lagrange multiplier; see [14, 52].

(In stating results about the barrier Hessian, we use the standard notation $O(\cdot)$, $\Omega(\cdot)$ and $\Theta(\cdot)$ from complexity theory; see, for example, [44]. If ϕ is a function of a positive variable h, p is fixed, and there exists a constant $\kappa_u > 0$ such that $|\phi| \leq \kappa_u h^p$ for all sufficiently small h, then $\phi = O(h^p)$. If there exists a constant $\kappa_l > 0$ such that $|\phi| \geq \kappa_l h^p$ for all sufficiently small h, then $\phi = \Omega(h^p)$. If there exist constants $\kappa_l > 0$ and $\kappa_u > 0$ such that $\kappa_l h^p \leq |\phi| \leq \kappa_u h^p$ for all sufficiently small h, then $\phi = \Theta(h^p)$.)

It can be shown that, near x^*, the barrier Hessian exhibits a special structure in its eigenvalues and invariant subspaces. In particular, except in special cases the barrier Hessian is ill-conditioned in an entire neighborhood of x^*. It is shown in [53] that, for all points sufficiently near x^*, under appropriate assumptions,

$$\operatorname{cond}(\mu H_B) = \operatorname{cond}(H_B) = d_{\max} \, \Omega(1/\mu),$$

where $d_{\max}$ is the maximum element of the vector d of (18). The value of $d_{\max}$ can be unboundedly large if we permit points very close to the boundary of the feasible region, for which a constraint is inordinately small relative to μ. For points close to x^* with the additional property that c_j is $\Theta(\mu)$ for all j, so that $d_{\max} = \Theta(1)$, then the barrier Hessian H_B has simple invariant subspaces that are "close to" (within $O(\mu)$ of) those defined by Z^* and Y^*. For a detailed discussion, see [53].

These results can be interpreted as a statement that, as the iterates of a barrier method converge, the invariant subspace associated with the unbounded ("large") eigenvalues of the barrier Hessian is close to the range space of $\hat{A}^T$, and the invariant subspace associated with the bounded ("small") eigenvalues is close to the null space of $\hat{A}$. For the scaled matrix

μH_B, one might say informally that its "small space" (the subspace of vectors v for which $\|\mu H_B v\|$ is small relative to $\|v\|$), is close to the null space of $\hat{A}$.

Because the performance of Newton-like methods is affected adversely by ill-conditioning in the Hessian, this feature of the barrier function causes difficulties for "black-box" unconstrained methods. Fortunately, the structure of the ill-conditioning is known, and need not be a fatal flaw if appropriate care is taken. In fact, the analysis sketched above shows that barrier Hessian itself reveals approximate subspace information about the active constraint Jacobian at x^*.

The saving grace for barrier methods is that, at points near the barrier trajectory, a highly accurate approximation to the step to x^* is given by $p = Y p_Y + Z p_Z$, where Y and Z are bases for range and null spaces associated with the current active constraint Jacobian, and p_Y and p_Z satisfy

$$\hat{A} Y p_Y = v_1, \qquad Z^T W Z p_Z = Z^T v_2, \tag{19}$$

for vectors v_1 and v_2 closely related to those of (13) and (14). The crucial point is that the condition of both linear systems in (19) reflects that of the problem (10) itself, and is not large simply as an artifact of applying a barrier transformation. The ability to approximate a step to x^* while solving well-conditioned systems motivated the barrier trajectory algorithm of [51], and has appeared more recently in the algorithm proposed in [41].

Unfortunately, approaches based on solving the equations (19) must make an estimate of the unknown active set $\hat{A}$. Although an incorrect prediction of $\hat{A}$ in this context has only a local effect on the algorithm, an important feature of interior methods is that they do not make an explicit prediction of the active set. To resolve this apparent dilemma, we need to ask whether there are linear algebraic techniques that allow us to cope with asymptotic ill-conditioning whose structure is known in advance, but which is revealed only implicitly. Note that the barrier Hessian will in general remain nonsingular even close to x^*; it is singular only in the limit.

For small problems, a strategy has been developed [53] based on the rank-revealing Cholesky factorization, which is analyzed in detail in [31]. While factorizing the barrier Hessian H_B, a decision is made about its nearness to singularity. If H_B is safely nonsingular, the Cholesky factors can be used in the standard way to calculate a Newton-based direction; see Section 3.1. If H_B is deemed numerically rank-deficient, it is "replaced" by an exactly rank-deficient matrix constructed from the Cholesky factors, which can be transformed to provide approximate bases Y and Z for the needed range and null spaces. In complete contrast to the case of linear constraints, here Y and Z are permitted to be inexact, since it is not necessary (or even possible) to remain in the null space. Near x^*, when the inevitable ill-conditioning strikes, approximate subspace information can thus be deduced that allows calculation of a good approximation to the Newton direction, without solving the ill-conditioned Newton equations and without explicitly predicting the active set. Inexactness cannot be too extreme, since the decision about rank of the barrier Hessian must be correct to obtain the desired Newton-like behavior near x^*. In any case, a satisfactory descent direction for the barrier function can be calculated.

The work of [41], an interior method for large-scale problems with bound constraints, calculates the search direction in separate range- and null-space components based on a working set, and solves the Newton-like equations in (19) inexactly using a preconditioned truncated conjugate-gradient method. For *general* large-scale nonlinearly constrained problems, however, serious application of interior methods is not particularly advanced.

An approach suggested in [19] in the context of linear and quadratic programming is to solve a larger, KKT-like system in which the known ill-conditioning can be dealt with more directly. Linear systems involving H_B can be solved via systems containing the larger symmetric matrices

$$\begin{pmatrix} -C_k^2/\mu & A_k \\ A_k^T & W_k \end{pmatrix} \quad \text{and} \quad \begin{pmatrix} \mu C_k^{-2} & 0 & I \\ 0 & W_k & A_k^T \\ I & A_k & 0 \end{pmatrix}, \tag{20}$$

where C_k is the diagonal matrix of constraint values at x_k. The ill-conditioning in these matrices is revealed in the $(1,1)$ block, and may be treated by a special interchange strategy for calculating the symmetric indefinite factorizations.

Another highly promising approach, which remains to be explored and implemented, is the primal-dual technique proposed in [35], which avoids ill-conditioning by abandoning symmetry. In this method, the systems to be solved contain matrices of the form

$$\begin{pmatrix} W_k & A_k^T \\ \Lambda_k A_k & -C_k \end{pmatrix}, \tag{21}$$

where Λ_k is a diagonal matrix of Lagrange multiplier estimates for all the constraints, including active and inactive. Because of strict complementarity (relation (iii) at the beginning of this section), the constraints converging to zero correspond to strictly positive Lagrange multipliers, and inactive constraints correspond to Lagrange multipliers converging to zero. The matrix (21) does *not* become ill-conditioned as x^* is approached, which is why its application might be effective. (There is, of course, an obvious relationship (via an ill-conditioned transformation) between (21) and the matrices of (20).)

6 Conclusions

Linear algebra in large-scale optimization comes in many flavors, ranging from the simplest "vanilla" to a complicated mixture of ingredients involving low-rank updates, representation of matrices in compact form, and special techniques for detecting and treating ill-conditioning. Despite enormous progress, many large optimization problems cannot be solved satisfactorily today. Happily for researchers in linear algebra and optimization as well as practitioners with real problems to solve, these difficulties are being ameliorated by continuing improvements in both fields.

References

[1] J. T. Betts and P. D. Frank. A sparse nonlinear optimization algorithm. Report AMS-

TR-173, Boeing Computer Services, Seattle, Washington, 1991.

[2] R. H. Byrd, R. A. Tapia and Y. Zhang. An SQP augmented Lagrangian BFGS algorithm for constrained optimization. *SIAM J. Optim.*, **2**, pp 210–241, 1992.

[3] T. F. Coleman and J. J. Moré. Estimation of sparse Hessian matrices and graph coloring problems. *SIAM J. Numer. Anal.*, **28**, pp 243–270, 1984.

[4] T. F. Coleman and A. Pothen. The null space problem I. Complexity. *SIAM J. Alg. Disc. Meth.*, **7**, pp 527–537, 1986.

[5] T. F. Coleman and D. C. Sorensen. A note on the computation of an orthogonal basis for the null space of a matrix. *Math. Prog.*, **29**, pp 234–242, 1984.

[6] A. R. Conn, N. Gould and P. L. Toint. Large-scale nonlinear constrained optimization. Report 92-02, Facultés Universitaires de Namur, Namur, Belgium, 1992.

[7] A. R. Conn, N. Gould and P. L. Toint. LANCELOT: *a Fortran package for large-scale nonlinear optimization.* Springer-Verlag, Berlin and New York, 1992.

[8] R. S. Dembo, S. C. Eisenstat and T. Steihaug. Inexact Newton methods. *SIAM J. Numer. Anal.*, **19**, pp 400–408, 1982.

[9] J. E. Dennis, Jr. and R. B. Schnabel. *Numerical methods for unconstrained optimization and nonlinear equations.* Prentice-Hall, Englewood Cliffs, New Jersey, 1983.

[10] J. E. Dennis, Jr. and R. B. Schnabel. A view of unconstrained optimization. In : G. L. Nemhauser, A. H. G. Rinnooy Kan and M. J. Todd (Eds.), *Optimization*, Elsevier Science Publishers, North-Holland, Amsterdam, pp 1–72, 1989.

[11] I. S. Duff, A. M. Erisman and J. K. Reid. *Direct methods for sparse matrices.* Oxford University Press, London, 1986.

[12] I. S. Duff, N. I. M. Gould, J. K. Reid, J. A. Scott and K. Turner. The factorization of sparse symmetric indefinite matrices. Report RAL-90-066, Rutherford Appleton Laboratory, Oxon, United Kingdom, 1990.

[13] S. K. Eldersveld. Large-scale sequential quadratic programming algorithms. Report SOL 92-02, Department of Operations Research, Stanford University, Stanford, California, 1992.

[14] A. V. Fiacco and G. P. McCormick. *Nonlinear programming: sequential unconstrained minimization techniques.* John Wiley and Sons, New York, 1968. Republished by SIAM, Philadelphia, 1990.

[15] R. Fletcher. *Practical methods of optimization* (second edition). John Wiley and Sons, Chichester and New York, 1987.

[16] A. L. Forsgren and W. Murray. Newton methods for large-scale linear equality-constrained minimization. Report SOL 90-6, Department of Operations Research, Stanford University, Stanford, California, 1990.

[17] R. Fourer and S. Mehrotra. Performance of an augmented system approach for solving least-squares problems in an interior-point method for linear programming. *Committee on Algorithms Newsletter*, 19, pp 26–31, 1991.

[18] P. E. Gill, W. Murray, D. B. Ponceleón, and M. A. Saunders. Preconditioners for indefinite systems arising in optimization. *SIAM J. Matrix Anal. Appl.*, 13, pp 292–311, 1992.

[19] P. E. Gill, W. Murray, D. B. Ponceleón, and M. A. Saunders. Solving reduced KKT systems in barrier methods for linear and quadratic programming. Report SOL 91-7, Department of Operations Research, Stanford University, Stanford, California, 1991.

[20] P. E. Gill, W. Murray, M. A. Saunders, J. A. Tomlin and M. H. Wright. On projected Newton barrier methods for linear programming and an equivalence to Karmarkar's projective method. *Math. Prog.*, 36, pp 183–209, 1986.

[21] P. E. Gill, W. Murray, M. A. Saunders and M. H. Wright. Constrained nonlinear programming. In : G. L. Nemhauser, A. H. G. Rinnooy Kan and M. J. Todd (Eds.), *Optimization*, Elsevier Science Publishers, North-Holland, Amsterdam, pp 171–210, 1989.

[22] P. E. Gill, W. Murray, M. A. Saunders and M. H. Wright. A Schur-complement method for sparse quadratic programming. In : M. G. Cox and S. J. Hammarling (Eds.), *Reliable numerical computation*, Clarendon Press, Oxford, pp 113–138, 1990.

[23] P. E. Gill, W. Murray, M. A. Saunders and M. H. Wright. Inertia-controlling methods for general quadratic programming. *SIAM Review*, 33, pp 1–36, 1991.

[24] P. E. Gill, W. Murray and M. H. Wright. *Practical optimization*. Academic Press, London, 1981.

[25] D. Goldfarb and M. J. Todd. Linear programming. In : G. L. Nemhauser, A. H. G. Rinnooy Kan and M. J. Todd (Eds.), *Optimization*, Elsevier Science Publishers, North-Holland, Amsterdam, pp 73–170, 1989.

[26] C. C. Gonzaga. Path-following methods for linear programming. *SIAM Review*, 34, pp 167–224, 1992.

[27] N. I. M. Gould. An algorithm for large-scale quadratic programming. *IMA J. Numer. Anal.*, 11, pp 299–324, 1991.

[28] A. Griewank and P. L. Toint. Partitioned variable metric updates for large sparse optimization problems. *Numer. Math.*, 39, pp 119–137, 1982.

[29] A. Griewank and P. L. Toint. Numerical experiments with partially separable optimization problems. In : D. F. Griffiths (Ed.), *Numerical analysis: proceedings Dundee 1983*, Lecture notes in mathematics 1066, Springer-Verlag, Berlin, pp 203–220, 1984.

[30] C. B. Gurwitz and M. L. Overton. SQP methods based on approximating a projected Hessian matrix. *SIAM J. Sci. Stat. Comp.*, 10, pp 631–653, 1989.

[31] N. J. Higham. Analysis of the Cholesky decomposition of a semi-definite matrix. In : M. G. Cox and S. J. Hammarling (Eds.), *Reliable numerical computation*, Clarendon Press, Oxford, pp 161–185, 1990.

[32] N. K. Karmarkar. A new polynomial time algorithm for linear programming. *Combinatorica*, **4**, pp 373–395, 1984.

[33] L. Kaufman and G. Sylvester. Separable nonlinear least-squares with multiple right-hand sides. *SIAM J. Matrix Anal. Appl.*, **13**, pp 68–89, 1992.

[34] D. C. Liu and J. Nocedal. On the limited memory BFGS method for large scale optimization. *Math. Prog.*, **45**, pp 503–528, 1989.

[35] G. P. McCormick. The superlinear convergence of a nonlinear primal-dual algorithm. Report T-550/91, Department of Operations Research, George Washington University, Washington, DC, 1991.

[36] J. J. Moré. Recent developments in algorithms and software for trust region methods. In : A. Bachem, M. Grötschel and B. Korte (Eds.), *Mathematical programming: the state of the art*, Springer-Verlag, Berlin and New York, pp 258–287, 1983.

[37] J. J. Moré and D. C. Sorensen. Newton's method. In : G. H. Golub (Ed.), *Studies in numerical analysis*, Mathematical Association of America, Providence, Rhode Island, pp 29–82, 1984.

[38] J. J. Moré and G. Toraldo. Algorithms for bound-constrained quadratic programming. *Numer. Math.*, **55**, pp 377-400, 1989.

[39] B. A. Murtagh and M. A. Saunders. MINOS 5.1 user's guide. Report SOL 83-20R, Department of Operations Research, Stanford University, Stanford, California, 1987.

[40] S. G. Nash and J. Nocedal. A numerical study of the limited memory BFGS method and the truncated-Newton method for large-scale optimization. *SIAM J. Optim.*, **1**, pp 358–372, 1991.

[41] S. G. Nash and A. Sofer. A barrier method for large-scale constrained optimization. Report 91-10, Department of Operations Research and Applied Statistics, George Mason University, Fairfax, Virginia, 1991.

[42] J. Nocedal and M. L. Overton. Projected Hessian updating algorithms for nonlinearly constrained optimization. *SIAM J. Numer. Anal.*, **22**, pp 821–850, 1985.

[43] C. C. Paige and M. A. Saunders. Solutions of sparse indefinite systems of linear equations. *SIAM J. Numer. Anal.*, **12**, pp 617–629, 1975.

[44] C. R. Papadimitriou and K. Steiglitz. *Combinatorial optimization: algorithms and complexity*. Prentice-Hall, Englewood Cliffs, New Jersey, 1982.

[45] M. J. D. Powell. Variable metric methods for constrained optimization. In : A. Bachem, M. Grötschel and B. Korte (Eds.), *Mathematical programming: the state of the art*, Springer-Verlag, Berlin and New York, pp 288–311, 1983.

[46] F. J. Prieto Sequential quadratic programming algorithms for optimization. Ph. D. thesis, Department of Operations Research, Stanford University, 1989.

[47] U. T. Ringertz. Optimal design of nonlinear shell structures. Report FFA TN 91-18, Aeronautical Research Institute of Sweden, 1991.

[48] R. B. Schnabel and E. Eskow. A new modified Cholesky factorization. *SIAM J. Sci. Stat. Comp.*, **11**, pp 1136–1158, 1990.

[49] D. C. Sorensen. Collinear scaling and sequential estimation in sparse optimization algorithms. *Math. Prog. Study*, **18**, pp 135–159, 1982.

[50] K. A. Williamson. A robust trust region algorithm for nonlinear programming. Report TR 90-22, Department of Mathematical Sciences, Rice University, Houston, Texas, 1990 (revised 1991).

[51] M. H. Wright. *Numerical methods for nonlinearly constrained optimization*. Ph. D. thesis, Computer Science Department, Stanford University, 1976.

[52] M. H. Wright. Interior methods for constrained optimization. In : A. Iserles (Ed.) *Acta numerica 1992*, Cambridge University Press, New York, pp 341–407, 1992.

[53] M. H. Wright. Determining subspace information from the Hessian of a barrier function. Report NAM 92-02, AT&T Bell Laboratories, Murray Hill, New Jersey, 1992.

CONTRIBUTED LECTURES

DIRECT AND INVERSE UNITARY EIGENPROBLEMS IN SIGNAL PROCESSING: AN OVERVIEW

G.S. AMMAR
Department of Mathematical Sciences
Northern Illinois University
DeKalb, IL 60115, U.S.A.
ammar@math.niu.edu

W.B. GRAGG
Department of Mathematics
Naval Postgraduate School
Monterey, CA 93940, U.S.A.
gragg@guinness.math.nps.navy.mil

L. REICHEL
Department of Mathematics and Computer Science
Kent State University
Kent, OH 44242, U.S.A.
reichel@mcs.kent.edu

KEYWORDS. Unitary Hessenberg matrices, Szegö polynomials, frequency estimation, updating and downdating least-squares approximants, eigenvalue computations, inverse eigenvalue problem.

Let $\mathcal{H}_n$ denote the set of unitary upper Hessenberg matrices of order n with nonnegative subdiagonal elements. These matrices bear many similarities with real symmetric tridiagonal matrices, both in terms of their structure, their underlying connections with orthogonal polynomials, and the existence of efficient algorithms for solving eigenproblems for these matrices. The *Schur parameterization* of $\mathcal{H}_n$ provides the means for the development of efficient algorithms for finding eigenvalues and eigenvectors of these matrices. These algorithms include the QR algorithm for unitary Hessenberg matrices [9], an algorithm for solving the orthogonal eigenproblem using two half-size singular value decompositions [1], a divide-and-conquer method [10, 5], an approach based on matrix pencils [6], and a unitary analog of the Sturm sequence method [7]. Aspects of inverse eigenproblems for unitary

341

M. S. Moonen et al. (eds.), Linear Algebra for Large Scale and Real-Time Applications, 341–343.
© 1993 *Kluwer Academic Publishers.*

342

Hessenberg matrices are considered in [3] and efficient algorithms for constructing a unitary Hessenberg matrix from spectral data are presented in [13, 3].

Unitary Hessenberg matrices are fundamentally connected with Szegő polynomials; i.e., with polynomials orthogonal with respect to a measure on the unit circle in the complex plane [8]. In particular, the Schur parameters of a unitary Hessenberg matrix $H \in \mathcal{H}_n$ are the recurrence coefficients of the Szegő polynomials determined by a discrete measure on the unit circle. Moreover, the monic Szegő polynomial of degree k is the characteristic polynomial of the leading principal submatrix H_k of $H = H_n$. The zeros of the Szegő polynomial of degree n, which are the eigenvalues of H, are the nodes of the discrete measure. Furthermore, the weights of the measure are proportional to the squared moduli of the first components of the normalized eigenvectors of H. Algorithms for eigenproblems and inverse eigenproblems for unitary Hessenberg matrices are therefore useful in several computational problems in signal processing involving Szegő polynomials.

Eigenproblems for unitary Hessenberg matrices naturally arise in several frequency estimation procedures in signal processing, including Pisarenko's method [2] and the composite sinusoidal modeling method [11].

In [12], an efficient and reliable algorithm is presented for discrete least-squares approximation on the unit circle by algebraic polynomials. This algorithms also applies to the special case of discrete least-squares approximation of a real-valued function given at arbitrary distinct nodes in $[0, 2\pi)$ by trigonometric polynomials. The algorithm is based on the solution of an inverse eigenproblem for unitary Hessenberg matrices presented in [3]. This algorithm is an *updating* procedure in that the least-squares approximant is obtained by incorporating the nodes of the inner product one at a time. Numerical experiments show that the algorithm produces consistently accurate results that are often better than those obtained by general QR decomposition methods for the least-squares problem.

In [4] an algorithm is presented for *downdating* the Szegő polynomials and given least-squares approximant when a node is deleted from the inner product. This scheme uses the unitary Hessenberg QR algorithm [9] with the node to be deleted as a shift. This algorithm can be combined with the fast Fourier transform (FFT) when the given nodes are a subset of equispaced points. This situation arises, for example, when an FFT is performed on a data set with some missing or spurious values. Moreover, the updating and downdating procedures, based on solving inverse eigenvalue problems and eigenvalue problems, can be combined to yield a sliding window scheme, in which one node is replaced by another.

References

[1] G.S. Ammar, W.B. Gragg and L. Reichel. On the eigenproblem for orthogonal matrices. *Proc. of the 25th Conference on Decision and Control*, IEEE, New York, pp 1063–1066, 1986.

[2] G.S. Ammar, W.B. Gragg and L. Reichel. Determination of Pisarenko frequency estimates as eigenvalues of an orthogonal matrix. In : F.T. Luk (Ed.), *Advanced Algorithms and Architectures for Signal Processing II*, Proc. SPIE, Vol. 826, pp 143–145, 1987.

[3] G.S. Ammar, W.B. Gragg and L. Reichel. Constructing a unitary Hessenberg matrix from spectral data. In : G.H. Golub and P. Van Dooren (Eds.), *Numerical Linear Algebra, Digital Signal Processing and Parallel Algorithms*, Springer, New York, pp 385–396, 1991.

[4] G.S. Ammar, W.B. Gragg and L. Reichel. Downdating of Szegö polynomials and data fitting applications. *Lin. Alg. Appl.*, **172**, pp 315–336, 1992.

[5] G.S. Ammar, L. Reichel, and D.C. Sorensen. An implementation of a divide and conquer algorithm for the unitary eigenproblem. *ACM Trans. Math. Software,* to appear.

[6] A. Bunse-Gerstner and L. Elsner. Schur parameter pencils for the solution of the unitary eigenproblem. *Lin. Alg. Appl.*, **154–156**, pp 741–778, 1991.

[7] A. Bunse-Gerstner and C. He. A Sturm sequence of polynomials for unitary Hessenberg matrices, preprint.

[8] W.B. Gragg. Positive definite Toeplitz matrices, the Arnoldi process for isometric operators, and Gaussian quadrature on the unit circle (in Russian). In : E.S. Nikolaev (Ed.), *Numerical Methods in Linear Algebra*, Moscow University Press, Moscow, pp 16-32, 1982.

[9] W.B. Gragg. The QR algorithm for unitary Hessenberg matrices. *J. Comput. Appl. Math.*, **16**, pp 1–8, 1986.

[10] W.B. Gragg and L. Reichel. A divide and conquer method for unitary and orthogonal eigenproblems. *Numer. Math.*, **57**, pp 695–718, 1990.

[11] L. Reichel and G.S. Ammar. Fast approximation of dominant harmonics by solving an orthogonal eigenvalue problem. In : J. McWhirter et al. (Eds.), *Proc. Second IMA Conference on Mathematics in Signal Processing*, Oxford University Press, pp 575–591, 1990.

[12] L. Reichel, G.S. Ammar and W.B. Gragg. Discrete least squares approximation by trigonometric polynomials. *Math. Comp.*, **57**, pp 273-289, 1991.

[13] T.-L. Wang. *Convergence of the QR Algorithm with Origin Shifts for Real Symmetric Tridiagonal and Unitary Hessenberg Matrices*. Ph.D. Dissertation, Dept. of Mathematics, University of Kentucky, Lexington, KY, 1988.

BLOCK IMPLEMENTATIONS OF THE SYMMETRIC QR AND JACOBI ALGORITHMS

P. ARBENZ and M. OETTLI
ETH Zürich
Institut für Wissenschaftliches Rechnen
8092 Zürich
Switzerland
arbenz@inf.ethz.ch / oettli@inf.ethz.ch

KEYWORDS. Block QR algorithm, Block Jacobi algorithm.

1 Introduction

On modern supercomputers with vector- and/or parallel-processing capabilities, it is very important to avoid unnecessary memory references, as moving data between different levels of memory (registers, cache, main memory) is slow compared to performing arithmetic operations on it. This fact has led to the development of block algorithms based on matrix-matrix operations. In this note we discuss block variants of QR and Jacobi algorithms for computing the spectral decomposition of symmetric matrices and report on numerical tests, which have been performed on an ALLIANT FX/80.

2 Description of the methods

Block QR Algorithm. The starting point of this algorithm is the Bai-Demmel block QR algorithm for real nonsymmetric matrices based on a multishift iteration[2].

The initial reduction to tridiagonal form is as usual done by a sequence of Householder transformations. Each one is chosen to introduce zeros in one column below the subdiagonal and in the corresponding row. To achieve better memory utilization, p consecutive transformations are aggregated to introduce the zeros block-wise.

The tridiagonal QR iteration has been modified to a block method by choosing k shifts simultaneously. This can be considered as a generalization of the double shift QR method used for unsymmetric matrices. The bulge of nonzero elements outside the tridiagonal band, which has to be chased down the diagonal, is now $k \times k$. It can be chased down the diagonal by the same block reduction method used in the initial tridiagonalization.

345

M. S. Moonen et al. (eds.), Linear Algebra for Large Scale and Real-Time Applications, 345–346.
© 1993 *Kluwer Academic Publishers.*

Size	Algorithm	t(sec)	$\max_i \|Tq_i - \lambda_i q_i\|_2$	$\max_i \|[q_1,\ldots,q_n]^T q_i - e_i\|_2$
100	TRED2/TQL2	2.66	0.59093E-12	0.62630E-14
	BTRED/BQR	2.97	0.16983E-11	0.64044E-14
	RUTISH	14.9	0.10199E-12	0.26609E-14
	VESBLK	2.79	0.45009E-12	0.26723E-14
200	TRED2/TQL2	19.4	0.22968E-11	0.10263E-13
	BTRED/BQR	12.7	0.29786E-11	0.81243E-14
	RUTISH	120.0	0.30690E-12	0.36766E-14
	VESBLK	17.1	0.90176E-12	0.31513E-14
400	TRED2/TQL2	153.0	0.45461E-11	0.15090E-13
	BTRED/BQR	71.5	0.96399E-11	0.11459E-13
	RUTISH	1020.0	0.10427E-11	0.48764E-14
	VESBLK	110.0	0.27734E-11	0.25046E-14

Table 1: *Total time and accuracy of results on ALLIANT FX/80.*

Block Jacobi Algorithm. Veselić and Hari [3] proposed a one-sided Jacobi algorithm for symmetric positive definite matrices A: Compute the Cholesky decomposition $A = LL^T$ and set $\tilde{A} = L^T L$. Apply the Jacobi algorithm to $\tilde{A}$, constructing a sequence $\tilde{A}_k = L_k^T L_k = Q_k^T L^T L Q_k$, $(k \to \infty)$. The sequence $\{L_k\}_{k=0}^{\infty}$ ($\tilde{A}$ is not explicitly used) converges to the eigenvector matrix of A. The norms of the columns of this matrix give the eigenvalues.

A parallel version of this one-sided block method is readily derived, if the ordering of the rotations is chosen appropriately. The algorithm can be blocked by aggregating rotations [1].

3   Results

To analyze the performance of our block algorithms, we compared them with TRED2/TQL2 from EISPACK and an implementation of Rutishauser's Jacobi algorithm RUTISH.

In Table 1, the performance on an eight-processor ALLIANT FX/80 is given. While both classical algorithms were executed sequentially, the block QR algorithm BTRED/BQR as well as the block Jacobi algorithm VESBLK exploited parallelism and ran on all eight processors. The timing results show, that BTRED/BQR is fastest. The reason is a high speedup in the tridiagonalization phase. We did not get any speedups over TQL2 with our implementation of the multishift QR iteration. Although, classical Jacobi is not competitive with the QR algorithm, the blocked parallel Jacobi implementation VESBLK is even faster than TRED2/TQL2. In situations where the original matrix is close to diagonal, block Jacobi may be superior to QR.

On a CRAY Y-MP/464 the blocked algorithms performed poorer than the nonblocked ones, on one as well as on four processors [1].

References

[1] P. Arbenz and M. Oettli. Block implementations of the symmetric QR and Jacobi algorithms. Research Report 178, Departement Informatik, ETH Zürich, June 1992.

[2] Z. Bai and J. Demmel. On A block implementation of Hessenberg multishift QR iterations. *Int. J. High Speed Comput.*, 1, pp 97–112, 1989.

[3] K. Veselić and V. Hari. A note on a one-sided Jacobi algorithm. *Numer. Math.*, 56, pp 627–633, 1989.

LINEAR ALGEBRA FOR LARGE-SCALE INFORMATION RETRIEVAL APPLICATIONS

M.W. BERRY
Department of Computer Science
University of Tennessee
107 Ayres Hall
Knoxville, TN 37996-1301, U.S.A.
berry@cs.utk.edu

KEYWORDS. Information retrieval, Lanczos algorithms, singular value decomposition, sparse matrices, numerical software, updating databases.

Currently, most approaches to retrieving textual materials from large databases depend on a lexical match between words in users' requests and those in or assigned to database objects. Because of the tremendous diversity in the words people use to describe the same object, lexical methods are necessarily incomplete and imprecise. Using the singular value decomposition (SVD), one can take advantage of the implicit higher-order structure in the association of terms with text objects by determining the SVD of large sparse term by text-object matrices. Terms and objects represented by dominant singular subspaces are then matched against user queries in the *semantic* space. Preliminary tests ([2]) find this method widely applicable and a promising way to improve users' access to many kinds of textual materials, or to objects and services for which textual descriptions are available.

In collaboration with Susan Dumais from Bell Communications Research (Bellcore) in Morristown, NJ and Dulce Ponceleon from Apple Computer Inc. in Cupertino, CA, we have implemented both a Fortran-77 and ANSI-C numerical library of sparse SVD algorithms (SVDPACK) on machines such as the Cray Y-MP, Convex C3840, IBM RS-6000-550, DEC 5000-100, SUN SPARCstation-2, and Apple Macintosh II/fx for use in a new approach to information retrieval: *latent semantic indexing*. Using Lanczos-based sparse SVD methods, we have been able to efficiently determine lower-rank estimates of very large sparse *term-document* matrices associated with various scientific databases of technical reports, medical abstracts, aeronautics reports, information science documents, magazine articles, and encyclopedia entries (see [1]).

Specifically, for an m by n *term-document* matrix whose m rows and n columns (where m is much larger than n) correspond to terms and documents, respectively, latent semantic indexing seeks the closest (in a least squares sense) rank-k matrix, where k is much smaller than n. This particular rank-k matrix effectively captures the major associational structure in the original term-document matrix and removes obscuring *noise* due to word choice

347

M. S. Moonen et al. (eds.), *Linear Algebra for Large Scale and Real-Time Applications*, 347–348.
© 1993 *Kluwer Academic Publishers.*

(synonymy and polysemy). Since relatively few terms are used as referents to a given document, the rectangular term-document matrix is quite sparse. Each matrix element indicates the frequency with which a term occurs in a particular document. Depending upon the size of the database from which the term-document is generated, such a matrix can have several thousand rows and slightly fewer columns. The largest term-document matrix processed to date is derived from an on-line encyclopedia database of approximately 60000 terms and 30000 entries (documents). Using the Fortran-77 implementation of SVDPACK, a rank-300 approximation to this matrix (which has just under 3 million nonzero elements) required only 16 CPU minutes on a Cray Y-MP/8-64.

By using lower rank approximations to term-document matrices produced via SVD-PACK, minor differences in terminology or word usage are virtually ignored. Moreover, the closeness of documents is determined by the overall pattern of term usage, so documents can be classified together regardless of the precise words that are used to describe them, and their description depends on a consensus of their term meanings, thus dampening the effects of polysemy. As a result, terms that do not actually appear in a document may still be used as referents, if that is consistent with the major patterns of association in the data. Position in the reduced space (dimension k as opposed to n) then serves as a new type of *semantic indexing*.

The optimal number k of singular values and singular vectors of the sparse term-document matrices to use in the lower rank approximation (or reduced model) for latent semantic indexing is an open question. Current research suggests anywhere between 100 to 500 of the largest singular values and corresponding singular vectors are needed for effective indexing. While correlations between the spectra of term-document matrices and the query-matching success rate for latent semantic indexing models need further exploration, we should address effective means for updating the SVD of evolving databases so that the optimal number of factors, k, can be obtained in some iterative fashion.

Suppose we add d documents and want to construct the $(m + t) \times (n + d)$ sparse term-document matrix $\tilde{A}$, where t is the number of *new* terms added to the database (assume global weighting used). Updating the matrix A consists of $(i.)$ appending d document vectors defined by the $m \times d$ sparse matrix D, $(ii.)$ appending new term vectors defined by the $m \times (n + d)$ sparse matrix T, and $(iii.)$ updating j term vectors corresponding to any terms found in the new document vectors using the sparse matrix $A + Y_j Z_j^T$. Hence, we seek the SVD of the resultant matrix $\tilde{A}$, where $\tilde{A} \equiv \left(\dfrac{A + Y_j Z_j^T \,|\, D}{T} \right)$. Our current research involves the development of effective algorithms using SVDPACK to obtain the SVD($\tilde{A}$) from the SVD(A).

References

[1] M. W. Berry. Large Scale Singular Value Computations. *International Journal of Supercomputer Applications*, Vol. 6, No. 1, pp 13-49, 1992.

[2] S. Deerwester, S. Dumais, G. Furnas, T. Landauer and R. Harshman. Indexing by latent semantic analysis. *Journal of the American Society for Information Science*, Vol. 41, No. 6, pp 391-407, 1990.

MATCHED FILTER VS. LEAST-SQUARES APPROXIMATION

L.H.J. BIERENS
TNO Physics and Electronics Laboratory
P.O. Box 96864
2509 JG The Hague
The Netherlands
laurens.bierens@fel.tno.nl

KEYWORDS. Least-squares, least-squares approximation, matched filter.

A well-known problem is the estimation of an unknown n-dimensional wide-sense stationary signal vector x with variance χ^2 which is related to an m-dimensional observed data vector b by the linear relationship

$$b = Ax + n \tag{1}$$

where A is an $(m \times n)$ matrix and n is an m-dimensional vector containing white noise with variance ν^2. The solution of this equation is known as [3]

$$\hat{x} = A^+ b \tag{2}$$

where A^+ is the pseudo-inverse of A. However, when A is bad-conditioned (i.e. the smallest singular values are close zero) then noise is amplified.

The approach that we propose is based on finding a tradeoff between optimum SNR and optimum signal estimation. Let $\hat{x} = Cb$ be an estimation of Hx then we can define the expected error power as

$$P_e = E\{\|\hat{x} - x\|_2^2\} = E\{\|(CA - H)x\|_2^2\} + \nu^2 \|C\|_F^2 \tag{3}$$

where H is a diagonal response matrix containing the diagonal elements of CA. I.e. we assume that CAx is a good estimation of Hx if the off-diagonal elements of CA are small. Let the expected power of the desired signal Hx be given as

$$S = E\{\|Hx\|_2^2\} = \chi^2 \|H\|_F^2 \tag{4}$$

Then the normalized expected error power is given as

$$\frac{P_e}{S} = \frac{E\{\|(CA - H)x\|_2^2\}}{\chi^2 \|H\|_F^2} + \frac{\nu^2 \|C\|_F^2}{\chi^2 \|H\|_F^2} \triangleq \Gamma(C) + \Lambda(C) \tag{5}$$

349

M. S. Moonen et al. (eds.), Linear Algebra for Large Scale and Real-Time Applications, 349–350.

where $\Gamma(\mathbf{C})$ is error component that is related to the signal power and $\Lambda(\mathbf{C})$ the error component related to the noise power. Solving equation (1) is now equivalent to finding a tradeoff between the two problems

1) $\min_{\mathbf{C}} \Gamma(\mathbf{C})$

2) $\min_{\mathbf{C}} \Lambda(\mathbf{C})$

We can now consider the following cases:

- Solve $\min_{\mathbf{C}} \Gamma(\mathbf{C})$: This results in $\mathbf{C} = \mathbf{A}^+$ i.e. the classical leasts-square solution. But we see that the size of $\Lambda(\mathbf{A}^+)$ depends on $\|\mathbf{A}^+\|_F^2 = \sum_{i=1}^r \sigma_i^{-2}$, $r = \text{rank}\{\mathbf{A}\}$ and $\sigma_i \ i = 1, \cdots, r$ the singular values of $\mathbf{A}$ in decreasing order. So if the smallest singular values approach zero then $\Lambda(\mathbf{A}^+)$ will be very large.

- Solve $\min_{\mathbf{C}} \Lambda(\mathbf{C})$: This results in $\mathbf{C} = \mathbf{A}^*$, which makes sense only if $\mathbf{A}$ is almost orthogonal. In linear filter theory this is known as matched filtering [1]. In typical radar and acoustical applications this is the case, however, in general $\Gamma(\mathbf{A}^*)$ will not be small.

- Solve $\min_{\mathbf{C}}\{\Gamma(\mathbf{C}) + \Lambda(\mathbf{C})\}$: In stead of using the full rank pseudo-inverse $\mathbf{A}^+$ we use a reduced rank pseudo-inverse $\mathbf{A}_q^+$, by setting the smallest $r - q$ singular values equal to zero [2]. This is also known as leasts-square approximation. It is obvious that there exists a q such that $\Gamma(\mathbf{A}_q^+) + \Lambda(\mathbf{A}_q^+)$ is minimum. In most cases this solution will give a better estimation of $\mathbf{Hx}$ then leasts-square estimation or matched filter estimation.

The method that we have described can be used to solve arbitrary linear equations of the form of equation (1) . As an example we have used it to deconvolve a radar echo signal which has typically low SNR. The optimum rank solution results in a better estimation then the classical estimation techniques in radar signal processing (matched filtering). Moreover, since in radar applications $\mathbf{A}$ is Toeplitz and band (i.e. equation (1) represents a convolution) we can approximate the reduced rank matrix $\mathbf{A}_q^+$ also by a Toeplitz and band matrix.

References

[1] A. Papoulis. *Probability, random variables, and stochastic processes.* McGraw-Hill, 1984.

[2] L.L. Scharf and D.W. Tufts. Rank reduction for modeling stationary signals. *IEEE Trans. ASSP*, **35(3)**, pp 350-355, 1987.

[3] S.A. Tretter. *Introduction to discrete-time signal processing.* John Wiley & Sons, 1976.

REORDERING DIAGONAL BLOCKS IN REAL SCHUR FORM

A.W. BOJANCZYK
Cornell University
Dept. Electrical Engineering
Ithaca, NY 14853-3801, U.S.A.
adamb@toeplitz.ee.cornell.edu

P. VAN DOOREN
University of Illinois at Urbana-Champaign
Coordinated Science Laboratory
1308 W. Main Str.
Urbana, IL 61801, U.S.A.
vdooren@maggie.csl.uiuc.edu

KEYWORDS. Invariant subspaces, eigenvalues, reordering.

1 Introduction

The problem of reordering eigenvalues of a matrix in real Schur form arises in the computation of the invariant subspaces corresponding to a group of eigenvalues of the matrix. A
basic step in such reordering is to swapp two neighboring 1×1 or 2×2 diagonal blocks by an
orthogonal transformation. Swapping two 1×1 blocks or swapping 1×1 and 2×2 blocks are
well understood [3]. Swapping two 2×2 blocks poses some numerical difficulties. Recently,
Bai and Demmel [1] have proposed an algorithm for swapping two 2×2 blocks which is for
all practical purposes backward stable. In this note we describe an alternative approach
for swapping two 2×2 blocks which is based on an eigenvector calculation. It appears that
the method guarantees small rounding errors in the (2,1) block of the transformed 4×4
matrix even if the two 2×2 blocks have almost the same eigenvalues.

2 Reordering eigenvalues

Assume that A is a 4×4 block triangular matrix,

$$A = \begin{pmatrix} A_{11} & A_{12} \\ 0 & A_{22} \end{pmatrix} = \begin{pmatrix} a_{11} & a_{12} & a_{13} & a_{14} \\ a_{21} & a_{22} & a_{23} & a_{24} \\ 0 & 0 & a_{33} & a_{34} \\ 0 & 0 & a_{43} & a_{44} \end{pmatrix},$$

where A_{11} and A_{22} are 2×2 with pairs of complex conjugate eigenvalues $\lambda_1, \bar{\lambda}_1$ and $\lambda_2, \bar{\lambda}_2$.

351

M. S. Moonen et al. (eds.), Linear Algebra for Large Scale and Real-Time Applications, 351–352.
© 1993 Kluwer Academic Publishers.

We can further assume that A_{11} and A_{22} are in the standard form,

$$A_{11} = \begin{pmatrix} \alpha_1 & \beta_1/k_1 \\ -\beta_1 k_1 & \alpha_1 \end{pmatrix} \quad \text{and} \quad A_{22} = \begin{pmatrix} \alpha_2 & \beta_2/k_2 \\ -\beta_2 k_2 & \alpha_2 \end{pmatrix}.$$

We want to find an orthogonal transformation Q such that

$$\hat{A} \equiv QAQ^T = \begin{pmatrix} \hat{A}_{22} & \hat{A}_{12} \\ 0 & \hat{A}_{11} \end{pmatrix},$$

where $\hat{A}_{11}$ and $\hat{A}_{22}$ are similar to A_{11} and A_{22} respectively.

The standard form implies that $\lambda_2 = \alpha_2 + \beta_2 \cdot i$ is the eigenvalue of A_{22}. Thus $A(\lambda_2) = A - \lambda_2 \cdot I$ is singular as its (2,2) diagonal block has rank 1. Now one can find a sequence of complex Givens rotations such that

$$\begin{pmatrix} a_{11} - \lambda_2 & a_{12} & a_{13} & a_{14} \\ a_{21} & a_{22} - \lambda_2 & a_{23} & a_{24} \\ 0 & 0 & a_{33} - \lambda_2 & a_{34} \\ 0 & 0 & a_{43} & a_{44} - \lambda_2 \end{pmatrix} G_{34}^{(1)} G_{12}^{(2)} G_{23}^{(3)} G_{12}^{(4)} = \begin{pmatrix} 0^{(4)} & \tilde{a}_{12} & \tilde{a}_{13} & \tilde{a}_{14} \\ 0^{(2)} & 0^{(3)} & \tilde{a}_{23} & \tilde{a}_{24} \\ 0 & 0 & 0^{(1)} & \tilde{a}_{34} \\ 0 & 0 & 0^{(1)} & \tilde{a}_{44} \end{pmatrix}$$

where $G_{ij}^{(k)}$ denotes a complex Givens rotation operating in the plane (i,j) introducing zero at the position marked as (k) on the right hand side of the relation. Let $G = G_{34}^{(1)} G_{12}^{(2)} G_{23}^{(3)} G_{12}^{(4)}$. Then $y = u + v \cdot i = Ge_1$, where $u = [u_1, u_2, u_3, u_4]^T$ and $v = [v_1, v_2, v_3, v_4]^T$ are real vectors, is the complex eigenvector corresponding to λ_2. Hence

$$\begin{pmatrix} A_{11} & A_{12} \\ 0 & A_{22} \end{pmatrix} (u \ \ v) = (u \ \ v) \begin{pmatrix} \alpha_2 & \beta_2 \\ -\beta_2 & \alpha_2 \end{pmatrix}.$$

Moreover, because A_{22} is assumed to be in a standard form, $u_4 = v_3 = 0$. The similarity transformation Q can be expressed now as a product of real Givens rotations which triangularizes the matrix $[u \ v]$. More precisely, let $Q = J_{23}^{(4)} J_{34}^{(3)} J_{12}^{(2)} J_{23}^{(1)}$ be such that

$$J_{23}^{(4)} J_{34}^{(3)} J_{12}^{(2)} J_{23}^{(1)} \begin{pmatrix} u_1 & v_1 \\ u_2 & v_2 \\ u_3 & 0 \\ 0 & v_4 \end{pmatrix} = \begin{pmatrix} \tilde{u}_1 & \tilde{v}_1 \\ 0^{(2)} & \tilde{v}_2 \\ 0^{(1)} & 0^{(4)} \\ 0 & 0^{(3)} \end{pmatrix},$$

where $J_{ij}^{(k)}$ denotes the corresponding rotation. Then Q is the desired similarity transformation.

Numerous numerical tests suggest that in the presence of rounding errors the relative error in the (2,1) block of the transformed matrix $\hat{A}$ is proportional to the machine precision. The algorithm can be extended to cover the case of swapping diagonal blocks in the periodic Schur form [2].

References

[1] Z. Bai and J.W. Demmel. On swapping diagonal blocks in real Schur form. Technical Report, IMA, University of Minnesota, 1992.

[2] A. Bojanczyk, G. Golub, P. Van Dooren. The periodic Schur form. Algorithms and Applications, SCCM Intern. Rept. NA-92-07, Stanford University, August 1992.

[3] P. Van Dooren. A generalized eigenvalue approach for solving Riccati equations, *SIAM Sci. & Stat. Comp.*, 2, pp 121-135, 1981.

PLACING ZEROES AND THE KRONECKER CANONICAL FORM

D. L. BOLEY
University of Minnesota
Computer Science Dept.
Minneapolis, MN 55455, U.S.A.
boley@mail.cs.umn.edu

P. VAN DOOREN
University of Illinois
Electrical Eng. Dept.
Urbana-Champaign, IL 61801, U.S.A.
vdooren@uicsl.csl.uiuc.edu

KEYWORDS. Transmission zeroes, control, Kronecker Form, Pencils.

Given a linear time-invariant control model

$$\dot{\mathbf{x}}(t) = A\mathbf{x}(t) + B\mathbf{u}(t) \quad \mathbf{y}(t) = C\mathbf{x}(t) + D\mathbf{u}(t), \tag{1}$$

it is well known that its transmission zeroes are the generalized eigenvalues of the pencil

$$E - \lambda F = \begin{pmatrix} A - \lambda I & B \\ C & D \end{pmatrix} \tag{2}$$

(discounting simple infinite eigenvalues) [1]. We would like to add outputs or inputs to (1) (respectively, append rows or columns to (2)) to place new zeroes in desired locations in the complex plane.

To fix ideas, we analyze how many zeroes we can place by appending rows to (2). We write the Kronecker Canonical Form for (2):

$$P(E - \lambda F)Q = \begin{pmatrix} R(\lambda) & (0) & (0) \\ 0 & J(\lambda) & (0) \\ 0 & 0 & L(\lambda) \end{pmatrix}, \tag{3}$$

where R is the collection of right (short fat) Kronecker blocks (always full row rank for all λ), L is the collections of left (tall thin) Kronecker blocks (always full column rank), and J is the regular part (square and nonsingular except at isolated points). In general P, Q are nonsingular matrices, but if we restrict them to unitary, then the entries marked (0) will

353

M. S. Moonen et al. (eds.), Linear Algebra for Large Scale and Real-Time Applications, 353–354.
© 1993 *Kluwer Academic Publishers.*

be nonzero. We append some number of new rows to obtain

$$\begin{pmatrix} P & 0 \\ 0 & I \end{pmatrix} \left[\begin{pmatrix} E \\ \widehat{Z} \end{pmatrix} - \lambda \begin{pmatrix} F \\ 0 \end{pmatrix} \right] Q = \begin{pmatrix} R(\lambda) & (0) & (0) \\ 0 & J(\lambda) & (0) \\ 0 & 0 & L(\lambda] \\ \widehat{Z}_1 & \widehat{Z}_2 & \widehat{Z}_3 \end{pmatrix}. \tag{4}$$

Only $\widehat{Z}_1$ can be used to place new zeroes. So we need only consider the sub-pencil

$$\begin{pmatrix} R(\lambda) \\ \widehat{Z}_1 \end{pmatrix} = \begin{pmatrix} R_1(\lambda) & & \\ & \ddots & \\ & & R_l(\lambda) \\ Z_1 & \cdots & Z_l \end{pmatrix}, \tag{5}$$

where each R_i is $s_i \times (s_i + 1)$ and represents a single Kronecker block. When appending a single row, every new zero of (5) must be a zero of every individual pencil $(R_i^T(\lambda), Z_i^T)^T$ for every i for which $Z_i \neq 0$. Hence one can place at most s_i zeroes for some i. In order to place those zeroes, one must actually extract the individual Kronecker blocks (the Staircase algorithm [4] is not enough).

To place the zeroes for a single Kronecker block, suppose that $R(\lambda) = (\mathbf{b}, A) - \lambda(0, F)$ is a single right Kronecker block and we add the single row $(\gamma, \mathbf{z}^T)$. Then observe that the finite zeroes of $\begin{pmatrix} R(\lambda) \\ (\gamma, \mathbf{z}^T) \end{pmatrix}$ are exactly the eigenvalues of the pencil $A + \mathbf{b}\gamma^{-1}\mathbf{z}^T - \lambda F$. We can choose $\gamma = 1$ and choose $\mathbf{z}^T$ by standard pole placement techniques [2, 3]. In this case, the new row is generally unique.

To extract the individual Kronecker blocks, we first apply the Staircase algorithm [4] to extract the right Kronecker indices. We then permute the rows and columns to extract the smallest right Kronecker block into the upper left corner and decouple this block from the rest of the pencil. On the remaining collection of right Kronecker blocks we repeat this step to extract the next smallest right Kronecker block, until all the right Kronecker blocks have been extracted. At each step, to decouple the upper left from the lower right, we annihilate the entries in the lower left block – in a very particular order which ends up completely filling in the upper right block. All the transformations applied are unitary transformations, and the result will be a pencil of the form (4) (with (0) nonzero and P, Q unitary), where $s_1 \leq s_2 \leq \cdots \leq s_l$.

References

[1] A. Emami-Naeini and P. Van Dooren. Computation of zeros of linear multivariable systems. *Automatica*, **18**, pp 415–430, 1982.

[2] J. Kautsky, N. K. Nichols, and P. Van Dooren. Robust pole assignment in linear state feedback. *Int. J. Control*, **41**, pp 1129–1155, 1985.

[3] G. Miminis and C. C. Paige. A direct method for pole assignment of time-invariant multi-input linear systems. *Automatica*, **24(3)**, pp 343–356, 1988.

[4] P. Van Dooren. The computation of Kronecker's canonical form of a singular pencil. *Lin. Alg. & Appl.*, **27**, pp 103–141, 1979.

ANALYSIS OF THE RECURSIVE LEAST SQUARES LATTICE ALGORITHM

J.R. BUNCH and R.C. LeBORNE
Department of Mathematics
University of California at San Diego
La Jolla, CA 92093-0112, U.S.A.
jrb@sdna3.ucsd.edu

We address some numerical issues regarding the numerical behavior of a specific type of lattice-based algorithm. Our interest is centered around the relationship between the propagation of numerical errors and the choice for design parameters such as single or double precision or the thresholds for adaptively expanding and contracting the filter order.

Fast Recursive Least Squares Lattice (fast RLSL) algorithms have enjoyed increasing popularity in recent years in fields ranging from biomedical engineering and communications to control, radar, sonar, and seismology. To address concerns of divergence with some of these algorithms, research has been focused on the convergence and stability properties of these fast filters. Direct and indirect updating, a priori and a posteriori forms, and the normalized and unnormalized versions are some of the variations of the fast RLSL that have been comparatively studied for accuracy and stability.

Stochastic analysis has been used to study the stability and accuracy of lattice filters after the filter has reached a steady state condition. Here, the goal has been to determine the long term effects from a single error input into the system by applying the definitions of asymptotic and exponential stability. The a priori version with indirect updating has been shown to be exponentially stable when a single error has been introduced and allowed to propagate in time. Assuming steady state conditions, the issue of accuracy by considering finite precision effects in floating and fixed point representations has been addressed. The error feedback form of the a priori and a posteriori lattice algorithm was introduced as a more numerically robust lattice filter. The analysis used a scalar example and numerical experiments.

The notion of backward consistency and backward stability from numerical analysis has been used to study the effects of roundoff errors that form a computed value which can be interpreted as the exact realization from a perturbed set of inputs. Insights regarding filter initialization (hard and soft starts), minimum realizations, reachability, and regions of stability were the results of this form of analysis.

M. S. Moonen et al. (eds.), *Linear Algebra for Large Scale and Real-Time Applications*, 355–356.
© 1993 *Kluwer Academic Publishers.*

356

We draw from the above analyses to provide new insights into roundoff effects evolving in time and propagating in order for the a posteriori unnormalized version of the recursive least-squares lattice (RLSL) filter. We present error recursions for the forward and backward prediction errors and their respective error-squares (energies).

We show that accumulation errors generated at a local stage are passed to stages of the next order. Local accumulation errors for the forward (backward) prediction errors are inversely related to the modulus of the prediction error-squares, directly related to the accumulated error of the dual backward (forward) prediction error and error-square, and proportional in modulus to the update from the previous stage. Error buildup occurs when the filter is allowed to adaptively expand to an order which propagates the forward and backward error-squares which are small in modulus. Here, the term small means that the modulus of the error-squares are close in modulus to the accumulated errors for the prediction coefficients. Additionally, roundoff errors are amplified when the current state is updated by a quantity that is greater than unity - a condition which often occurs.

The forward and backward energy-squares propagate errors in a slightly different manner than their prediction error counterparts. Errors are again passed forward in order; however, any additional roundoff is proportional to the current update as well as the error accumulated in the dual error-square term, but inversely related to the modulus of the dual error-square.

We conjecture that errors may accumulate at a rate that is proportional to the threshold parameter which allows filter order expansion. The growth rate for error accumulation will be changed if the errors ever become close in modulus to this threshold parameter. This regrowth effect has been observed and controlled by programming all parameters in double precision.

ADAPTIVE CHEBYSHEV ITERATION BASED ON MODIFIED MOMENTS

D. CALVETTI
Department of Pure and Applied Mathematics
Stevens Institute of Technology
Hoboken, NJ 07030, U.S.A.
na.calvetti@na-net.ornl.gov

G.H. GOLUB
Computer Science Department
Stanford University
Stanford, CA 94305, U.S.A.
na.golub@na-net.ornl.gov

L. REICHEL
Department of Mathematics and Computer Science
Kent State University
Kent, OH 44242, U.S.A.
na.reichel@na-net.ornl.gov

KEYWORDS. Iterative method, nonsymmetric linear system, modified moment.

The problem of solving a linear system of equations

$$Ax = b, \qquad A \in \mathbb{R}^{N \times N}, \qquad x, b \in \mathbb{R}^N, \tag{1}$$

with a large, sparse and nonsymmetric matrix A arises in many applications. A Chebyshev iterative method based on scaled Chebyshev polynomials p_n for an interval in the complex plane can be used to solve (1) when the spectrum of A lies in the right half plane. Manteuffel [3] discusses such Chebyshev iterative schemes and shows that the iterations depend on two parameters only, the center d and the focal length c of an ellipse in the complex plane with foci at $d \pm c$. In these schemes, the p_n are Chebyshev polynomials for the interval between the foci, and are scaled so that $p_n(0) = 1$. Let x_0 denote a given initial approximate solution of (1). The three-term recurrence relation for the p_n yields an inexpensive recurrence relation for computing a sequence of approximate solutions $x_1, x_2, \ldots$ of (1). The iterates x_n determined by the Chebyshev iterative method are such that the error vectors $e_n :=$

357

M. S. Moonen et al. (eds.), Linear Algebra for Large Scale and Real-Time Applications, 357–358.
© 1993 *Kluwer Academic Publishers.*

$A^{-1}b - x_n$ satisfy $e_n = p_n(A)e_0$, $n \geq 0$. Let $\lambda(A)$ denote the spectrum of A. If the parameters d and c are chosen so that

$$\max_{z \in \lambda(A)} |p_n(z)|$$

decreases rapidly as n increases, then the Euclidean norm of e_n decreases rapidly as n increases as well. Chebyshev iteration is an attractive solution method if parameters d and c exist, such that there is an ellipse with foci at $d \pm c$ which contains $\lambda(A)$ and is not very close to the origin. Assuming that such an ellipse exists, its center d and focal length c can be determined if $\lambda(A)$ is explicitly known. However, in general, $\lambda(A)$ is neither known nor easy to determine.

In [3] Manteuffel describes algorithms for dynamic estimation of the parameters d and c based on the power method applied to A, or modifications thereof. The parameters d and c are chosen so that $d \pm c$ are the foci of the the smallest ellipse containing available estimates of eigenvalues of A.

The paper [2] describes two adaptive Chebyshev algorithms in which inner products of certain residual vectors $r_n := b - Ax_n$ are interpreted as modified moments. The modified moments are computed and together with the recursion coefficients of the p_n are input to the modified Chebyshev algorithm. This algorithm determines a nonsymmetric tridiagonal matrix, whose eigenvalues are estimates of eigenvalues of A. The tridiagonal matrix obtained is the same tridiagonal matrix that is determined by the nonsymmetric Lanczos algorithm with suitable initial vectors. We note that the nonsymmetric Lanczos algorithm requires multiplication with the transpose of the matrix A in each iteration. This is not necessary in our scheme. The computed eigenvalue estimates are used to compute parameters d and c by determining the smallest ellipse that contains the estimates. From the location of the foci at $d \pm c$ of this ellipse, the parameters d and c can easily be computed. Our schemes require less computer storage than the methods in [3]. Moreover, computed examples indicate that our scheme yields faster convergence than the implementation [1] of the methods [3]. Details and computed examples are presented in [2].

References

[1] S.F. Ashby. CHEBYCODE: a FORTRAN implementation of Manteuffel's adaptive Chebyshev algorithm. Report UIUCDCS-R-85-1203, Department of Computer Science, University of Illinois at Urbana-Champaign, Urbana, IL, 1985.

[2] D. Calvetti, G.H. Golub and L. Reichel. Adaptive Chebyshev iterative methods for nonsymmetric linear systems based on modified moments. Report ICM-9205-30, Institute for Computational Mathematics, Kent State University, Kent, OH 44242, 1992.

[3] T.A. Manteuffel. Adaptive procedure for estimation of parameters for the nonsymmetric Chebyshev iteration. Numer. Math., 31, pp 187–208, 1978.

CONTINUOUS REALIZATION METHODS AND THEIR APPLICATIONS

M.T. CHU
Department of Mathematics
North Carolina State University
Raleigh, NC 27695-8205, U.S.A.
chu@gauss.math.ncsu.edu

KEYWORDS. Continuous realization, isospectral flows, QR-type algorithm, project gradients.

Realization process, in a sense, means any deducible procedure that we use to ra tionalize and solve problems. In mathematics, especially for existence questions, a realization process often appears in the form of an iterative procedure or a differential equation. For years researchers have taken great effort to describe, analyze, and modify realization processes. Nowadays the success is especially evident in discrete numerical algorithms. On the other hand, the use of differential equations to issues in computational mathematics has been found recently to afford fundamental insights into the structure and behavior of existing discrete methods and, sometimes, to suggest new and improved numerical methods.

Continuous realization methods are based on the idea of connecting two abstract problems through a mathematical bridge. Usually one of the abstract problems is a make-up whose solution is trivial while the other is the real problem whose solution is difficult to find. The bridge, if it exists, is regarded as a continuous path in the problem space. Following the path means deforming the underlying abstract problem mathematically. It is hoped that by following the path, the obvious solution will systematically be evolved (flowed) into the solution that we are seeking for.

Obviously, the most important issue in the flow approach is the assurance that a bridge connecting the two abstract problems does exist. The construction of a bridge can be motivated in several different ways: Sometimes an existing discrete numerical method may be extended directly into a continuous model [7, 3], sometimes a differential equation arises naturally with a certain physics background [8], and more often a vector field is constructed with a specific task in mind [1].

Applications of continuous realization methods include nonlinear algebraic systems [7], eigenvalue problems [3, 8], singular value problems [2], least squares approximation problems subject to eigenvalue or singular value constraints [4], inverse eigenvalue problems for Toeplitz matrices or non-negative matrices [6], quadratic programming problems and simultaneous reduction problems [5].

359

M. S. Moonen et al. (eds.), Linear Algebra for Large Scale and Real-Time Applications, 359–360.
© 1993 *Kluwer Academic Publishers.*

The continuous realization approach has the following advantages:
1. Areas of applications of continuous realization process is very broad while many well developed classical results are immediately available for studying the dynamics of continuous systems. The study of continuous system could shed critical insights into the understanding of the dynamics of an existing discrete methods.
2. Continuous realization sometimes unifies different discrete methods as special cases of its discretization and often gives rise to the design of new numerical algorithms.
3. Differential systems results from continuous realization present immediate challenge to the current numerical ODE techniques. Partially this is due to the fact that usually a certain invariant manifold needs to be preserved during the integration. Partially this is due to the fact that matrix differential equations are especially suitable for integration on a massively data-parallel computing system. Thus matrix differential equations may be used as benchmark problems for testing new ODE techniques. Conversely, new ODE techniques may further benefit the numerical computation of matrix differential equations.
4. Many existence problems, seemingly impossible to be tackled by any conventional discrete methods, may be solved by formulating special differential equations that ensure a specific task is taking place continuously. Continuous methods, in a sense, give a smoother control in realization a problem. This, therefore, opens a new field of applications of numerical ODE techniques.
5. In contrast to the local properties of some discrete methods, the continuous approach usually offers a global method for solving the underlying problem.

References

[1] R.W. Brockett. Dynamical systems that sort lists, diagonalize matrices and solve linear programming problems. *Proceedings of the 27th IEEE Conference on Decision and Control*, IEEE, pp 799-803, 1988, and *Lin. Alg. Appl.*, **146**, pp 79-91, 1991.

[2] M.T. Chu. On a differential equation approach to the singular value decomposition of bi-diagonal matrices. *Lin. Alg. Appl.*, **80**, pp 71-79, 1986.

[3] M.T. Chu. On the continuous realization of iterative processes. *SIAM Review*, **30**, pp 375-387, 1988.

[4] M.T. Chu and K.R. Driessel. The projected gradient method for least squares matrix approximations with spectral constraints. *SIAM J. Num. Anal.*, **27**, pp 1050-1060, 1990.

[5] M.T. Chu. A continuous Jacobi-like approach to the simultaneous reduction of real matrices. *Lin. Alg. Appl.*, **147**, pp 75-96, 1991.

[6] M.T. Chu. Matrix differential equations: A continuous realization process for linear algebra problems. *Nonlinear Anal.*, TMA, **18**, pp 1125-1146, 1992.

[7] H.B. Keller. Global homotopies and Newton methods. In : C. de Boor and G. Golub (Eds.), *Recent Advances Numerical Analysis*, Academic Press, New York, pp 73-94, 1978.

[8] W.W. Symes. The QR algorithm and scattering for the finite non-periodic Toda lattice. *Physica*, *4D*, pp 275-280, 1982.

ASYMPTOTIC BEHAVIOR OF ORTHOGONAL POLYNOMIALS

T. DEHN
Institut für Praktische Mathematik
Universität Karlsruhe
D-W-7500 Karlsruhe
Federal Republic of Germany
na.dehn@na-net.ornl.gov

KEYWORDS. Orthogonal polynomials, recurrence coefficients, ratio asymptotics.

We discuss the asymptotic behavior of polynomials q_n of degree n, defined by

$$zq_n(z) = \beta_n q_{n+1}(z) + \alpha_n q_n(z) + \beta_{n-1} q_{n-1}(z) \quad \text{for} \quad n = 0, 1, \ldots \tag{1}$$

with positive β_n, real α_n, and initial values $q_{-1}(z) = 0$ and $q_0(z) = \gamma_0 > 0$. By J. Favard's theorem (see [1] for further references), these polynomials are orthonormal with respect to some nonnegative measure μ with infinite real support E. In many applications, only the moments of μ and the recurrence coefficients can be computed. Hence it is useful to study their influence. The interest in this field was renewed in 1979 by the following theorem of P. Nevai.

Theorem 1. *Let $\alpha_n \to 0$ and $\beta_n \to \frac{1}{2}$. Then $E = [-1, 1] \cup B$, where B is a countable set, which has no limit points outside $[-1, 1]$ and may be empty ([5], Section 3.3). More,*

$$\frac{q_{n+1}(z)}{q_n(z)} \longrightarrow \begin{cases} \Phi(z) & \text{locally uniformly for } z \in \mathbb{C} \setminus E \ ([5], \text{ Theorem 4.1.13}), \\ 1/\Phi(z) & \text{pointwise for } z \in B \ ([5], \text{ Theorem 4.1.18}). \end{cases} \tag{2}$$

Hereby Φ maps $C \setminus [-1, 1]$ conformally onto the exterior of the unit disc.

Parts of Nevai's theorem are contained in old works of H. Poincaré (1885), O. Blumenthal (1898) and E. B. van Vleck (1904). For details see [2] and [3], where new proofs for Nevai's theorem are given. By a theorem proved by E. A. Rakhmanov in 1977, every absolutely continuous measure with support $[-1, 1]$ fulfills the assumptions of Theorem 1. A new proof and a short history of Rakhmanov's theorem are provided in [4].

Geronimo, van Assche and others have generalized Nevai's theorem to orthogonal polynomials with **asymptotically periodic recurrence coefficients**

$$\lim_{n \to \infty} \beta_{nm+k} = b_k \quad \text{and} \quad \lim_{n \to \infty} \alpha_{nm+k} = a_k, \quad k = 0, 1, \ldots, m-1,$$

using Weyl's theorem on compact perturbations of self-adjoint operators [3], [6].

But many asymptotic properties of orthonormal polynomials can be obtained with sim-

361

M. S. Moonen et al. (eds.), Linear Algebra for Large Scale and Real-Time Applications, 361–362.
© 1993 *Kluwer Academic Publishers.*

362

pler methods. Take a look at (2). The convergence is locally uniform for $z \in \mathbb{C} \setminus E$. Hence in this domain the ratios of the orthonormal polynomials form a normal family. The key point is, that from the interlacing property of the zeros it is rather simple to prove directly the normal family property for $z \notin \mathbb{C} \setminus [a,b]$, where $[a,b]$ is any interval containing all zeros. Such an interval can be derived easily from bounds for the recurrence coefficients. We set

$$\mathbf{A_n}(z) := \begin{pmatrix} \frac{z - \alpha_n}{\beta_n} & -\frac{\beta_{n-1}}{\beta_n} \\ 1 & 0 \end{pmatrix},$$

and a simple reformulation of (1) yields

$$\begin{pmatrix} q_{n+1}(z) \\ q_n(z) \end{pmatrix} = \mathbf{A_n}(z) \begin{pmatrix} q_n(z) \\ q_{n-1}(z) \end{pmatrix} = \mathbf{A_n}(z) \cdot \ldots \cdot \mathbf{A_{n-m+1}}(z) \begin{pmatrix} q_{n-m+1}(z) \\ q_{n-m}(z) \end{pmatrix}. \quad (3)$$

In the Nevai case,

$$\lim_{n \to \infty} \mathbf{A_n}(z) = \begin{pmatrix} 2z & -1 \\ 1 & 0 \end{pmatrix},$$

whose dominating eigenvalue is $\Phi(z)$. It is now easy to prove Theorem 1 with the monotonicity of $\{q_n(z)\}$ for large positive z [2]. But more is possible. With the limits of the matrix products in (3), the asymptotic behavior of orthonormal polynomials with asymptotically periodic recurrence coefficients can be related to m **similar** polynomial matrices

$$\mathbf{B_k}(z) = \begin{pmatrix} a_{k,m}(z) & b_{k,m-1}(z) \\ c_{k,m-1}(z) & d_{k,m-2}(z) \end{pmatrix} \quad \text{for} \quad k = 0, 1, \ldots, m-1.$$

The $\mathbf{B_k}(z)$ have determinant 1. E now consists of the union of a countable set B and at most m closed intervals. The endpoints of these intervals are given by $a_{k,m}(z) + d_{k,m-2}(z) \in \{\pm 2\}$. For $z \in \mathbb{C} \setminus E$ the ratios $q_{n+m}(z)/q_n(z)$ converge locally uniformly to the dominating eigenvalue of $\mathbf{B_k}(z)$, independently of k. The points outside of E, in which the zeros of $\{q_{nm+k}\}$ cluster, fulfill $c_{k,m-1}(z) = 0$. For proofs, details and further references see [1].

References

[1] T. Dehn. *Nennerpolynome von Padé und Padé-Typ-Approximationen an Hamburgerfunktionen* (in German). PhD thesis, University of Karlsruhe, 1992.

[2] T. Dehn. A shortcut to asymptotics for orthogonal polynomials. To appear in Proceedings ICCAM 92, *Journal of Computational and Applied Mathematics*.

[3] A. Maté, P. Nevai, and W. van Assche. The support of measures associated with orthogonal polynomials and the spectra of the related self-adjoint operators. *Rocky Mountain Journal of Mathematics*, **21**(1), pp 501–527, 1991.

[4] A. Maté, P. Nevai, and V. Totik. Asymptotics for the Ratio of Leading Coefficients of Orthonormal Polynomials on the Unit Circle. *Constr. Approx.*, **1**, pp 64–72, 1985.

[5] P. Nevai. Orthogonal Polynomials. *Memoirs of the AMS*, **213**, USA, 1979.

[6] W. van Assche. Asymptotics for Orthogonal Polynomials. *Lecture Notes in Mathematics*, **1265**, Springer-Verlag, Berlin and Heidelberg, Germany, 1987.

CADCS AND PARALLEL COMPUTING

F. DUMORTIER, A. VAN CAUWENBERGHE and L. BOULLART
Automatic Control Laboratory
University of Ghent
Grotesteenweg Noord 6
B-9052 Zwijnaarde, Belgium
fd@autoctrl.rug.ac.be

KEYWORDS. CADCS, predictive control, parallel computing.

1 Introduction

In modern control engineering the field of Computer Aided Design in Control Systems has become a major area of research, which involves many topics: simulation, analysis and design of control systems, linear and nonlinear control, multivariable control, frequency-domain and time-domain analysis, state-space systems, continuous-time and discrete-time systems, etc.. Current CADCS software packages cover the whole spectrum of classical and modern control theory. As in many areas of science the arrival of performant micro-computers and minicomputers has boosted the level of fundamental and applied research. In CADCS new advanced and complex numerical control algorithms are continuously being developed and refined. The real-time application of these powerful control methods is however very much limited, especially in time-critical situations, due to the physical speed limits of todays microprocessors. The arrival of new computer architectures, s.a. vector computing and parallel processing, and new technologies, s.a. GaAs chips and optical fibres, can offer a breakthrough to the current computing limitations. It is now generally believed that concurrent processing is the only way to further increase computer performance and to keep the cost/performance ratio within reasonable bounds.

2 Parallel linear algebra routines - multiprocessors

In CADCS numerical control algorithms mainly involve the solution of large sets of linear equations, the solution of partial differential equations over a large data space, or iterative operations. In many cases this comes down to the execution of elementary matrix operations, s.a. matrix/vector summation and multiplication, matrix-inversion, LU-decomposition, etc.. The possible parallelisation of these elementary operations offers a speedup which directly results in a straightforward speedup of the control algorithm. As

363

M. S. Moonen et al. (eds.), *Linear Algebra for Large Scale and Real-Time Applications*, 363–364.
© 1993 *Kluwer Academic Publishers.*

there is currently no generalisation or unification in the field of concurrent processing, and as no automatic compilers do yet exist, it is very important that for each specific parallel hardware platform a basic library of elementary matrix routines is built. This is a major keystone for the succes of parallel computing.

At the Automatic Control Laboratory of the University of Ghent a Sun Sparcstation 370 UNIX workstation is equipped with a Meiko Computing Surface, currently consisting of 32 transputers. This machine is a typical example of a "Multiple Instruction Multiple Data" concurrent computer with distributed memory, which interprocessor communication is based on the principle of message passing. CSBuild end CSTools form the main parallel programming environment, which consists of a collection of C and Fortran callable interprocessor communication and connection routines. The programming language is C. Other available software environments are OCCAM and Express. Using the CSTools parallel programming environment a basic library of elementary matrix-routines has now been realized: matrix/vector summation and multiplication, matrix-inversion, LU-decomposition, triangular system solving have currently been implemented ([1], [2], [3], [4]). Of course this library will be extended with other interesting matrix-routines as time evolves.

3 Parallel computing and predictive control

At the moment this library is applied in CADCS software for long-range predictive control. Long-range predictive control methods are very popular and robust control methods which are based on the following principles. Computing a model prediction of the system output over a prediction horizon and allowing for consecutive control input variable changes over a control horizon, a control law is calculated which minimizes a quadratic cost-criterion comparing the process model output with a reference trajectory. The long-range predictive control methods currently under investigation are Model Algorithmic Control and Dynamic Matrix Control, which are based on a non-parametric process model: the impulse response and step response respectively. To obtain a good precision and numerical robustness a large number of data points has to be included, which results in large computation matrices. Taking into account the strictly non parallelisable part in the code of both algorithms, during the experiments a good (near linear) speedup was obtained. Other long-range predictive control methods and their multivariable counterparts will be tackled in the future.

References

[1] R.H. Bisseling and J.G.G. Van de Vorst. Parallel LU Decomposition on a Mesh Network of Transputers. *Lectures Notes in Computer Science*, **384**, Springer-Verlag, 1991.

[2] R.H. Bisseling and J.G.G. Van de Vorst. Parallel Triangular System Solving on a Mesh Network of Transputers. *SIAM J. Sci. Statist. Comput.*, 1991.

[3] G. Fox, M. Johnson, G. Lyzenga, S. Otto, J. Salmon and D. Walker. *Solving Problems on Concurrent Processors - Volume 1: General Techniques and Regular Problems.* Prentice-Hall International, Inc., Englewood Cliffs, 1988.

[4] G.H. Golub and C.F. Van Loan. *Matrix Computations - 2nd Edition.* The John Hopkins University Press, 1989.

EIGENVALUE ROULETTE AND RANDOM TEST MATRICES

A. EDELMAN
Department of Mathematics
University of California
Berkeley, CA 94720
edelman@math.berkeley.edu

KEYWORDS. CM5, Eigenvalues, Numerical Linear Algebra, Random Matrix, Roulette, Test Matrix.

1 Random matrices in numerical linear algebra

Random matrices may not be mentioned in research abstracts, but second only to Matlab, the most widely used tool of numerical linear algebraists is the "random test matrix." Algorithmic developers in need of guinea pigs nearly always take random matrices with standard normal entries or perhaps close cousins, such as the uniform distribution on $[-1, 1]$. The choice is highly reasonable: these matrices are generated effortlessly and might very well catch programming errors. What is a mistake is to psychologically link a random matrix with the intuitive notion of a "typical" matrix or the vague concept of "any old matrix."

In contrast, we argue that "random matrices" are very *special* matrices. The larger the size of the matrix the more predictable they are because of the central limit theorem. This is the beauty of a random matrix; it has more structure than a fixed matrix. For example, an n by n matrix with normally distributed entries has 2-norm very nearly $2\sqrt{n}$, spectral radius $\sqrt{n}$, and $\sqrt{2n/\pi}$ real eigenvalues. If you sort and plot the singular values, you will get nearly the same picture each time. These statements are either crude versions of rigorous theorems or empirical evidence (see Edelman [1] for one survey); however, if you perform the experiments on matrices of size $n > 50$, you will see for yourself that a random matrix has structure. Experiments with the uniform distribution or even the discrete distribution $\{-1, 1\}$ will yield essentially the same results scaled by the variance.

Recently, we took a closer look at the real eigenvalues of a matrix of standard normals [2, 3]. Since such is matrix is unlikely to be symmetric, one might expect that it would generally have some real and some complex eigenvalues. Plotting normalized

M. S. Moonen et al. (eds.), Linear Algebra for Large Scale and Real-Time Applications, 365–368.
© 1993 *Kluwer Academic Publishers.*

366

eigenvalues $\lambda/\sqrt{n}$ in the complex plane yields a curious picture. To the right are 2500 dots representing $\lambda/\sqrt{n}$ for fifty random matrices of size $n = 50$. What may appear to be a horizontal line segment is in fact, many closely huddled eigenvalues on the real axis. Furthermore, according to Girko's circular law [4], the normalized complex eigenvalues fall uniformly in the complex unit disk (in the limit as $n \to \infty$). We concentrated our studies on

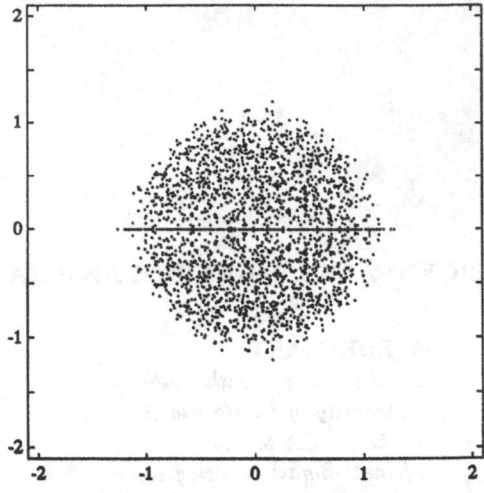

- $p_{n,k}$, the probability that an n by n matrix has exactly k real eigenvalues,

- $E_n = \sum_k k p_{n,k}$, the expected number of real eigenvalues and,

- the distribution of the real eigenvalues.

The tables below contain a wealth of information regarding the expected number of real eigenvalues of a matrix (from [3]). The reader will notice that if n is even, the expected number is a rational multiple of $\sqrt{2}$, while if n is odd, it is one more than such a multiple. We consider the CM-5 experiments a simultaneous validation of our theory, the CM-5, and LAPACK. The exact formulas contain the familiar double factorial notation which is the product of odd or even numbers, depending on the argument.

Expected number of real eigenvalues

n	E_n	
1	1	1.00000
2	$\sqrt{2}$	1.41421
3	$1 + \frac{1}{2}\sqrt{2}$	1.70711
4	$\frac{11}{8}\sqrt{2}$	1.94454
5	$1 + \frac{13}{16}\sqrt{2}$	2.14905
6	$\frac{211}{128}\sqrt{2}$	2.33124
7	$1 + \frac{271}{256}\sqrt{2}$	2.49708
8	$\frac{1919}{1024}\sqrt{2}$	2.65027
9	$1 + \frac{2597}{2048}\sqrt{2}$	2.79332
10	$\frac{67843}{32768}\sqrt{2}$	2.92799

CM-5 Experiments using LAPACK on 64 processors

n	trials	experimental E_n	theoretical E_n	minutes
80	640	7.6	7.603	1
160	640	10.7	10.569	7
320	640	14.9	14.756	51
640	128	20.8	20.673	82
900	64	24.5	24.427	107

Exact Formulas:

$E_n = \sqrt{2} \displaystyle\sum_{k=0}^{n/2-1} \frac{(4k-1)!!}{(4k)!!}$	n even
$E_n = 1 + \sqrt{2} \displaystyle\sum_{k=1}^{(n-1)/2} \frac{(4k-3)!!}{(4k-2)!!}$	n odd

In [2], we prove that a not too small random normally distributed matrix virtually never has all real eigenvalues, despite doctored examples found in elementary linear algebra textbooks. The probability that all the eigenvalues are real is $p_{n,n} = 2^{-n(n-1)/4}$. We further show that $p_{n,k}$ always has the form $r + s\sqrt{2}$ where r and s are rational numbers with de-show that $p_{n,k}$ always has the form $r + s\sqrt{2}$ where r and s are rational numbers with denominators that are powers of 2 and therefore E_n has this form also. The expected number of real eigenvalues E_n, is shown in [3] to be asymptotically $\sqrt{2n/\pi}$. Finally, we show that the real eigenvalues normalized by dividing by $\sqrt{n}$ are asymptotically uniformly distributed on $[-1,1]$.

2 Eigenvalue roulette

We conclude with a game for the amusement and edification of our readers. Let A be a random n by n matrix whose elements are independent standard normals. Our proposed game of eigenvalue roulette spins a wheel with the numbers $n, n-2, n-4, \ldots$ down to 0 or 1 with the possible number of real eigenvalues that A might have. As an example the figure to the right is our "wheel" for $n = 8$. The wheel may be spun (with a tolerance) in Matlab Version 3 by typing

```
>> rand('normal');a=rand(8);sum(abs(imag(eig(a)))<.0001)
```

Payoffs for the game may be made at the reader's discretion; fair odds, however, for each value on the wheel when $n = 8$ are as follows [2]:

Probabilities for Eigenvalue Roulette

k	$p_{n,k}$	
8	$1/16384$	0.00
6	$-1/4096 + 3851/262144\,\sqrt{2}$	0.02
4	$53519/131072 - 11553/262144\,\sqrt{2}$	0.35
2	$-53487/65536 + 257185/262144\,\sqrt{2}$	0.57
0	$184551/131072 - 249483/262144\,\sqrt{2}$	0.06

A game of a similar flavor was proposed for roots of random polynomials in 1938 by Littlewood and Offord [5]. They write

> Let the reader place himself at the point of view of A in the following situation. An equation of degree 30 with real coefficients being selected by a referee, A lays B even odds that the equation has not more than r real roots. ... What smallest possible value of r will make A safe [from being ruined]?

What makes their game so remarkable is that they did not have the technology to play it!

References

[1] A. Edelman. *Eigenvalues and Condition Numbers of Random Matrices*, PhD thesis, Department of Mathematics, MIT, 1989.

[2] A. Edelman. The probability that an n by n matrix has k real eigenvalues. In preparation.

[3] A. Edelman, E. Kostlan, M. Shub. How many eigenvalues of a random matrix are real? Submitted to *J. Amer. Math. Soc.*

[4] V.L. Girko, Circular law, *Theory Prob. Appl.*, **29**, pp 694–706, 1984.

[5] J.E. Littlewood, A.C. Offord. On the number of real roots of a random algebraic equation, *J. London Math. Soc.*, **13**, pp 288–295, 1938.

ON NUMERICAL METHODS FOR UNITARY EIGENVALUE PROBLEMS

H. FAßBENDER
Department of Mathematics and Computer Science
University of Bremen
Postfach 330 440
2800 Bremen 33
Federal Republic of Germany
heike@mathematik.uni-bremen.de

KEYWORDS. Inverse eigenvalue problem, unitary matrix, Szegö polynomial, orthogonal rational function, discrete least squares approximation.

A number of signal processing problems can be seen to require numerical methods for different unitary eigenvalue problems.

One of these problems is the discrete least-square approximation of a real-valued function f given at arbitrary distinct nodes $\{\theta_k\}_{k=1}^m$ in $[0, 2\pi)$ by trigonometric polynomials t in the discrete norm $||f - t|| = (\sum_{k=1}^m |f(\theta_k) - t(\theta_k)|^2 \omega_k^2)^{\frac{1}{2}}$, where the $\{\omega_k^2\}_{k=1}^m$ are positive weights. The problem can easily be reformulated as the standard least square problem of minimizing $DAc - Dg$ over all coefficient vectors c in the Euclidian norm, where $D = diag(\omega_1, ..., \omega_m)$ and A is the transposed $m \times n$ Vandermonde matrix

$$A = \begin{pmatrix} 1 & z_1 & z_1^2 & \cdots & z_1^{n-1} \\ \vdots & \vdots & \vdots & & \vdots \\ 1 & z_m & z_m^2 & \cdots & z_m^{n-1} \end{pmatrix}$$

with $z_k = exp(i\theta_k)$.

The usual way to solve this least square problem is to compute the QR decomposition of DA. But DA is just the Krylov matrix $K(\Lambda, q_0, n) = [q_0, \Lambda q_0, ..., \Lambda^{n-1}q_0]$ (where $\Lambda = diag(z_1, ..., z_m)$ and $q_0 = (\omega_1, ..., \omega_m)^T$). We may therefore use the following consequence of the Implicit Q Theorem to compute the desired QR decomposition. If there exists a unitary matrix U such that $U^H \Lambda U = H$ is a unitary upper Hessenberg matrix with positive subdiagonal elements, then the unique QR decomposition of $K(\Lambda, q_0, m)$ is given by UR with $R = K(H, e_1, m)$. The construction of such a unitary Hessenberg matrix from spectral data, here contained in Λ, is an inverse eigenproblem. Thus the best trigonometric approximation to f can be computed via solving this inverse eigenproblem.

M. S. Moonen et al. (eds.), Linear Algebra for Large Scale and Real-Time Applications, 369–370.
© 1993 *Kluwer Academic Publishers.*

A different approach is to reformulate the approximation problem as the standard least square problem of minimizing $D\tilde{A}\tilde{t} - D\tilde{f}$ over all coefficient vectors $\tilde{t}$ in the Euclidian norm, where

$$\tilde{A} = \begin{pmatrix} 1 & \sin\theta_1 & \cos\theta_1 & \cdots & \sin l\theta_1 & \cos l\theta_1 \\ \vdots & \vdots & \vdots & & \vdots & \vdots \\ 1 & \sin\theta_m & \cos\theta_m & \cdots & \sin l\theta_m & \cos l\theta_m \end{pmatrix}$$

and l is the degree of the desired trigonometric polynomial. $D\tilde{A}$ is the product of the modified Krylov matrix $\kappa(\Lambda, q_0, l) = [q_0, \Lambda q_0, \Lambda^H q_0, \Lambda^2 q_0, \Lambda^{H^2} q_0, ..., \Lambda^l q_0, \Lambda^{H^l} q_0]$ and a block diagonal matrix $F = diag(1, B, B, ..., B)$ with $B = \begin{pmatrix} -i & 1 \\ i & 1 \end{pmatrix}$. If there exists a unitary matrix $\tilde{Q}$ such that $\tilde{Q}^H(\Lambda - \lambda I)\tilde{Q}G_e = G_o - \lambda G_e$ is a unitary matrix pencil in parametrized form, where G_o, G_e are unitary block diagonal matrices with block size at most two, then the unique QR decomposition of $\kappa(\Lambda, q_o, l)$ is given by $\tilde{Q}\tilde{R}$ with $\tilde{R} = \kappa(G_o G_e^H, e_1, l)$. From this a unique (real-valued) QR factorization of $D\tilde{A}$ is easily obtained. The construction of such a unitary matrix pencil in parametrized form from spectral data is a generalized inverse eigenproblem.

In this talk we present algorithms for discrete least-square approximations that are based on schemes for the solution of an inverse eigenproblem for unitary Hessenberg matrices $H = H(\gamma_1, ..., \gamma_n)$ and for unitary matrix pencils in parametrized form $G_o - \lambda G_e$.

Reichel, Ammar and Gragg observe in [2] that solving an inverse eigenproblem for unitary Hessenberg matrices is equivalent to computing Szegö polynomials, that is to computing polynomials that are orthogonal with respect to an inner product on the unit circle. The scheme for solving an inverse eigenproblem for unitary matrix pencils in parametrized form is developed from a backward stable algorithm given by Bunse-Gerstner and Elsner in [1] which reduces a unitary matrix pencil to parametrized form. It is shown that this is equivalent to computing rational functions that are orthonormal with respect to an inner product on the unit circle.

The algorithms require only $O(mn)$ arithmetic operations as compared with $O(mn^2)$ operations needed for algorithms that ignore the special structure of DA. We compare the presented algorithms with each other and with a general QR decomposition. We will see that the proposed algorithms produce consistently accurate results that are often better than those obtained by general QR decomposition methods for the least-squares problem.

References

[1] A. Bunse-Gerstner and L. Elsner. Schur Parameter Pencils for the Solution of the Unitary Eigenproblem. *Lin. Alg. Appl.*, **154 - 156**, pp 741-778, 1991.

[2] L. Reichel, G. S. Ammar and W. B. Gragg. Discrete Least Squares Approximation by Trigonometric Polynomials. *Math. Comp.*, **57**, pp 273-289, 1991.

ACCURATE SINGULAR VALUES AND DIFFERENTIAL QD ALGORITHMS

K.V. FERNANDO
NAG Ltd, Wilkinson House
Jordan Hill
Oxford OX2 8DR, U.K.
  and
Division of Computer Science
University of California
Berkeley, CA 94708, U.S.A.

B.N. PARLETT
Department of Mathematics
University of California
Berkeley, CA 94720, U.S.A.
parlett@math.berkeley.edu

In September 1991 J. W. Demmel and W. M. Kahan were awarded the second SIAM prize in numerical linear algebra for their paper 'Accurate Singular Values of Bidiagonal Matrices' [1], referred to as DK hereafter. Among several valuable results was the observation that the standard bidiagonal QR algorithm used in LINPACK [2], and in many other SVD programs, can be simplified when the shift is zero and, of greater importance, no subtractions occur. The last feature permits very small singular values to be found with (almost) all the accuracy permitted by the data and at no extra cost.

In this paper we show that the DK zero shift algorithm can be further simplified and this simplicity has several benefits. One is that a new algorithm can be implemented in either parallel or pipelined format as an $\mathcal{O}(\log_2 n)$ algorithm. This is pursued in a companion paper [4].

Our investigations began with the modest goal of showing that it was preferable to replace the DK zero-shift QR transform by two steps of zero-shift LR implemented in a qd (quotient-difference) format. Root-free algorithms run considerably faster than standard ones. The surprise here is that to keep the high relative accuracy property it is necessary to use a little known variant of qd (the differential form of the progressive qd algorithm

371

M. S. Moonen et al. (eds.), Linear Algebra for Large Scale and Real-Time Applications, 371–373.
© 1993 *Kluwer Academic Publishers.*

or **dqd** [7], [9]). The standard qd will not suffice. There are no subtractions in **dqd**. We suspect that Rutishauser discovered **dqd** in 1968, just two years before his death.

What we want to stress here is that, for reasons we may never know, Rutishauser did not consider the shifted version of **dqd**. Incidentally this differential qd is not to be confused with the continuous analogue of qd (see [8]) and more recent work on QR flows. The trouble with the shifted version of the ordinary qd algorithm is that it cannot recover from a shift that is too large. Consequently qd algorithms have been shackled with very conservative shift strategies, such as Newton's method, and earned the reputation of being slow compared to the QR algorithm. Had Rutishauser considered shifts with differential qd (**dqds** hereafter) he would have realized, as we soon did, that the transformation may be split into two parts. The parts depend on whether the machine is of sequential or parallel type but, in each case, a shift that is too big reveals itself before the old matrix is overwritten and so need not be invoked. An unused shift is not wasted because it gives an improved upper bound on the smallest singular value at a cost less than one qd transformation as well as contributing to an improved shift.

Our approach frees the algorithm to exploit powerful shift strategies while preserving high relative accuracy all the time. In contrast the QR algorithm delivers high relative accuracy only with a zero shift.

Even though our algorithms must find the singular values in order we can use shift strategies that are at least quadratically convergent. This is better than fourth order convergence for QR. When only the smallest few singular values are needed this ordering constraint is a great advantage. Another rather subtle feature is that it is not necessary to make an extra $\mathcal{O}(n)$ check for splitting of the matrix into a direct sum. The necessary information is provided by the auxiliary quantities.

In June 1992 we discovered that our **dqds** algorithm enjoys high relative stability for all shifts provided that they avoid underflow, overflow or divide by zero. Consequently it can be used in a variety of applications (eigenvalues of symmetric or unsymmetric tridiagonals, zeros of polynomials, poles and zeros of transfer functions and many applications involving continued fractions) where Rutishauser's **qd** has been abandoned because of its instability in the general case.

Our error bounds for singular values are significantly smaller than those in DK and the approach is quite transparent. It was this analysis that showed us the possibility of violating positivity while still maintaining maximal relative accuracy for all singular values, not just the small ones.

It gradually dawned on us as we developed the algorithm that we were breaking away from the *orthogonal paradigm* that has dominated the field of matrix computations (called numerical linear algebra by highbrows) since the 1960's. It seems to be sacrilegious to be achieving greater accuracy and on average, a five fold speed-upby simply abandoning QR something equivalent to LR. High accuracy comes from the fact that **dqds** spends most of its time transforming lower triangular 2×2s into upper triangular 2×2s by premultiplication.

Rutishauser gave no direct explanation for the way shifts are introduced into **qd**. We have supplied one in terms of matrix factorizations.

We give a unifying general result which shows that it is possible to implement the LR-Cholesky algorithm of Rutishauser [5], [6] using orthogonal transformations only. Perhaps this is the key idea exploited in the paper. Since the term LR-Cholesky over describes the algorithm we simply refer to it as the Cholesky Algorithm. Our orthogonal Cholesky algorithm is applicable to dense matrices; this more general case is studied elsewhere [3].

We want to point out the unusual historical lineage of this algorithm. The qd algorithm begat the LR algorithm which then gave rise to the QR algorithm of Francis. This in turn led to the Golub-Kahan and Golub-Reinsch algorithms for singular values of bidiagonal matrices which lead to the DK zero-shift variant. This inspired our orthogonal algorithm of which differential qd is the root-free version. We are back to qd again but with a new implementation.

Acknowledgements

K.V. Fernando is supported by NSF, under grant ASC-9005933.
K.V. Fernando and B.N. Parlett are supported by ONR, contract N000014-90-J-1372.

References

[1] J. Demmel and W. Kahan. Accurate singular values of bidiagonal matrices. *SIAM J. Sci. Sta. Comput.*, 11, pp 873-912, 1990.

[2] J.J. Dongarra, J.R. Bunch, C.B. Moler and G.W. Stewart. *LINPACK User' Guide.* SIAM, Philadelphia, 1979.

[3] K.V. Fernando and B.N. Parlett. Orthogonal Cholesky Algorithm. *Technical Report,* under preparation, Centre for Pure and Applied Mathematics, University of California at Berkeley, 1992.

[4] K.V. Fernando and B.N. Parlett. QD algorithms for advanced architectures. *Technical Report,* under preparation, Centre for Pure and Applied Mathematics, University of California at Berkeley, 1992.

[5] H. Rutishauser. Solution of eigenvalue problems with the LR-transformation. *Nat. Bur. Standards Appl. Math. Series*, 49, pp 47-81, 1958.

[6] H. Rutishauser and H.R. Schwarz. The LR transformation method for symmetric matrices. *Numer. Math.*, 5, pp 273-289, 1963.

[7] H. Rutishauser. *Lectures on Numerical Mathematics.* Birkhäuser, Boston, 1990.

[8] H. Rutishauser. Ein infinitesimales Analogon zum Quotienten-Differenzen-Algorithmus. *Arch. Math.*, 5, pp 132-137, 1954.

[9] H. Rutishauser. *Vorlesungen über numerische Mathematik.* Birkhäuser, Basel, 1976.

ORTHOGONAL PROJECTION AND TOTAL LEAST SQUARES

R.D. FIERRO
Department of Mathematics
University of California, Los Angeles
405 Hilgard Ave, Los Angeles, CA 90024-1555, U.S.A.
fierro@math.ucla.edu

J.R. BUNCH
Department of Mathematics
University of California, San Diego
9500 Gilman Dr, La Jolla, CA 92093-0112, U.S.A.

1 Introduction

When the overdetermined system of linear equations

$$AX \approx B \qquad (1)$$

has no solution, compatibility may be restored by an orthogonal projection method to estimate the relationship between A and B, where we assume $A \in \Re^{m \times n}$ has numerical rank k and $B \in \Re^{m \times d}$ $(m \geq n + d)$. The least squares (LS) and total least squares (TLS) are two common orthogonal projection methods used to solve (1). The idea is to determine an orthogonal projection matrix P by some method M such that $[\tilde{A}\ \tilde{B}] = P[A\ B]$, $\tilde{A}X \approx \tilde{B}$ is compatible, and $\text{rank}(\tilde{A}) = k$. Denote by X_M the minimum norm solution to $\tilde{A}X = \tilde{B}$ by method M. Let $Y \in \Re^{(n+d) \times (n-k+d)}$ be an orthonormal matrix such that $\|[A\ B]Y\| \approx$ *small* and $[\tilde{A}\ \tilde{B}]Y = 0$.

2 Model and perturbation bounds

We present a model for orthogonal projection methods by reformulating the parameter estimation problem as an equivalent problem of nullspace determination, i.e., determine Y such that $[\tilde{A}\ \tilde{B}]Y = 0$, $\tilde{A}$ is well conditioned, and $[A\ B] - [\tilde{A}\ \tilde{B}]$ is not too large. Under a very mild condition, the kernel of $[\tilde{A}\ \tilde{B}]$, denoted $\ker([\tilde{A}\ \tilde{B}])$, completely determines X_M. In fact, if $Y = [Y_1^T\ Y_2^T]^T$ where $Y_2 \in \Re^{d \times (n-k+d)}$, then $X_M = -Y_1 Y_2^+$. When M $\equiv$ TLS, the model specializes to the well known TLS method [3],[4],[6].

We supply lower and upper perturbation bounds for X_M in terms of the subspace angle between approximate nullspaces. Specifically, let $[\bar{A}\ \bar{B}] = [A\ B] + [\triangle A\ \triangle B]$ denote a perturbation of $[A\ B]$ and let $\bar{X}_M$ denote the minimum norm solution to

$$\bar{A}X = \bar{B}, \qquad (2)$$

M. S. Moonen et al. (eds.), Linear Algebra for Large Scale and Real-Time Applications, 375–376.
© 1993 *Kluwer Academic Publishers.*

a compatible approximation to $\bar{A}X \approx \bar{B}$. Then it can be shown

$$\sin \theta_M \le \|X_M - \bar{X}_M\| \le \sin \theta_M \sqrt{1 + \|X_M\|^2} \sqrt{1 + \|\bar{X}_M\|^2}, \tag{3}$$

where $\sin \theta_M$ denotes the subspace angle between $\ker([\bar{A} \ \bar{B}])$ and $\ker([\check{A} \ \check{B}])$. Provided the problem is not very ill-conditioned (i.e., $\|X_M\|$ is not very large), this shows the perturbation effect depends on the noise distribution in $\ker([\check{A} \ \check{B}])$. Van Huffel and Vandewalle [5, p. 215] reached the same conclusion for the perturbation effects in the M $\equiv$ TLS problem.

The general perturbation bound is then applied to LS and TLS, where upper bounds for the respective angles are found. Similarly, we also discuss the perturbation bounds for the truncated QR method as well the URV and ULV decomposition.

3 Comparing X_M to X_{TLS}

TLS has nice properties: it solves the nearest rank-k compatible system, and both the analysis and the algorithm is based on the singular value decomposition (SVD). However, the SVD is difficult to update and computationally involved. The above perturbation bound can also be used to compare two competing orthogonal projection methods, as in [2]. We discuss how one may use a rank revealing QR algorithm to compute an approximation of the TLS solution, or a TLS solution using subspace refinement; the algorithm is a hybrid of one in [1]. Our numerical results (see Table 1) confirm that when the subspace angle is small, the solutions are initially close, and usually one or two subspace iterations are needed to compute a good TLS solution. Finally, the bounds allow one to compare the solutions to any two competing orthogonal projection methods. For instance, one can use the bounds to compute the difference in the LS and TLS solutions:

$$\|X_{LS} - X_{TLS}\| \le (\sigma_{k+1}([A \ B])/\sigma_k(A))^2 \sqrt{1 + \|X_{LS}\|^2} \sqrt{1 + \|X_{TLS}\|^2},$$

where the σ's are singular values of their respective matrices.

References

[1] T.F. Chan and P.C. Hansen. Some Applications of the Rank Revealing QR Factorization. Technical Report CAM-90-09, Department of Mathematics, University of California, Los Angeles, 1990.

[2] R.D. Fierro and J.R. Bunch. Orthogonal Projection and Total Least Squares. Submitted to *J. Numer. Lin. Alg. and Applic.*, 1992

[3] G.H. Golub and C.F. Van Loan. An Analysis of the Total Least Squares Problem. *SIAM J. Numer. Anal.* **17**, pp 883-893, 1980.

[4] S. Van Huffel and J. Vandewalle. Algebraic Connections Between the Least Squares and Total Least Squares Problems. *Numer. Math.* **55**, pp 432-449, 1989.

[5] S. Van Huffel and J. Vandewalle. *The Total Least Squares Problem: Computational Aspects and Analysis.* SIAM, 1991.

[6] M.D. Zoltowski. Generalized minimum Norm and Constrained Total Least Squares with Applications to Signal Processing. *Advanced Algorithms and Architectures for Signal Processing III*, SPIE, **975**, pp 78-85, 1988.

GAUSS QUADRATURES ASSOCIATED WITH THE ARNOLDI PROCESS AND THE LANCZOS ALGORITHM

R.W. FREUND
AT&T Bell Laboratories
600 Mountain Avenue, Room 2C-420
Murray Hill
New Jersey 07974-0636, U.S.A.
freund@research.att.com

M. HOCHBRUCK
IPS
ETH-Zentrum
CH-8092 Zürich
Switzerland
na.hochbruck@na-net.ornl.gov

KEYWORDS. Non-Hermitian matrices, Krylov subspaces, Arnoldi process, Lanczos algorithm, Gauss quadrature.

1 Introduction

Many iterative methods for large matrix computations are based on *Krylov subspaces*

$$K_n(v_1, A) := \text{span}\{v_1, Av_1, \ldots, A^{n-1}v_1\}, \quad n = 1, 2, \ldots. \tag{1}$$

Here A is a given real or complex $N \times N$ matrix, and $v_1 \in \mathbb{C}^N$ is a nonzero starting vector. Of course, the basis vectors used in the definition (1) of $K_n(v_1, A)$ tend to be numerically linearly dependent, and therefore, in general, they are of little use in practice. Indeed, the main ingredient of a Krylov subspace algorithm is the computation of a suitable sequence of vectors $v_1, v_2, \ldots$, such that

$$\text{span}\{v_1, v_2, \ldots, v_n\} = K_n(v_1, A), \quad n = 1, 2, \ldots. \tag{2}$$

Evidently, the choice of such vectors depends on the matrix A, and therefore some kind of information on A is required. There are essentially two different approaches. *Parameter-free Krylov subspace methods*, such as the Arnoldi process [1] or the Lanczos algorithm [4], obtain such information in the course of the iteration, by computing, at the nth step,

377

M. S. Moonen et al. (eds.), Linear Algebra for Large Scale and Real-Time Applications, 377–380.
© 1993 *Kluwer Academic Publishers.*

inner products with vectors $v \in K_n(v_1, A)$. We note that, on modern architectures, the computation of inner products of long vectors often constitutes a bottleneck. *Parameter-free Krylov subspace methods*, such as Chebyshev iteration [5], require a-priori information on the spectrum $\lambda(A)$ of A to determine iteration coefficients beforehand. However, in contrast to parameter-free methods, Krylov subspace algorithms in this second category do not involve computations of inner products of long vectors $v \in C^N$. This is one of the reasons for the renewed interest in parameter-dependent Krylov subspace iterations. In practice, suitable a-priori information on $\lambda(A)$ is usually not available beforehand. Instead, as a first phase, one runs a few step of a parameter-free Krylov subspace algorithm from which information on $\lambda(A)$ is extracted. Then, in a second phase, one switches to a parameter-dependent method, where the parameters are chosen based on the information obtained in the first phase.

Recently, Fischer and Freund [2] studied the problem of extracting suitable information on the spectrum $\lambda(A)$ of Hermitian matrices A. They showed that a very good approximation to the eigenvalue distribution of A can be obtained after only few iterations of the Hermitian Lanczos algorithm. Their approach is based on the intimate connection of the Hermitian Lanczos process and Gauss quadrature on a real interval. It is well known [3] that, after n Lanczos steps, the knots and weights for the corresponding nth Gauss quadrature rule can be obtained by simply solving an eigenvalue problem with the tridiagonal Lanczos matrix.

Motivated by the results in [2] for Hermitian matrices, in this paper, we consider the Arnoldi process and the nonsymmetric Lanczos algorithm for non-Hermitian matrices A. We show that, with both methods, one can associate Gauss quadratures in the complex plane.

Clearly, any sequence of vectors satisfying (2) can be written in terms of polynomials, and we have

$$v_{n+1} = \varphi_n(A)v_1, \quad \text{where} \quad \varphi_n \in \mathcal{P}_n, \quad n = 0, 1, \ldots . \tag{3}$$

Here $\mathcal{P}_n$ denotes the set of all complex polynomials of degree at most n. Moreover, we will denote by $\mathcal{P}$ the set of all complex polynomials of arbitrary degree. Finally, A always denotes a general $N \times N$ matrix.

2 The Arnoldi case

Let $v_1 \in C^N$ be an arbitrary starting vector with $\|v_1\|_2 = 1$. The Arnoldi process then generates a sequence of basis vectors $v_1, v_2, \ldots$, that are orthonormal, i.e.,

$$v_j^H v_n = 0, \quad \text{if} \quad j \neq n, \quad \text{and} \quad \|v_n\|_2 = 1.$$

Hence the corresponding polynomials defined by (3), $\varphi_0, \varphi_1, \ldots$, are orthonormal with respect to the inner product $\langle \cdot, \cdot \rangle$ on $\mathcal{P}$ that is defined as follows:

$$\langle \varphi, \psi \rangle := v_1^H (\psi(A))^H (\varphi(A)) v_1 \quad \text{for all} \quad \varphi, \psi \in \mathcal{P}. \tag{4}$$

Note that $\langle \cdot, \cdot \rangle$ is a positive semidefinite inner product on $\mathcal{P}$. Furthermore, $\langle \cdot, \cdot \rangle$ is positive definite on $\mathcal{P}_{N_A-1}$. Here N_A ($\leq N$) is the degree of the minimal polynomial of A with respect to v_1, i.e., N_A is the smallest integer such that there exists a polynomial $\varphi \in \mathcal{P}_{N_A}$

with $\varphi(A)v_1 = 0$, $\varphi \neq 0$.

In the following, let $n < N_A$, and assume that we have run the Arnoldi process for n steps. Thus we have vectors $\{v_j\}_{j=1}^{n+1}$ that satisfy

$$AV_n = V_n H_n + v_{n+1}[0 \quad \cdots \quad 0 \quad h_{n+1,n}] \quad \text{and} \quad V_{n+1}^H V_{n+1} = I_{n+1}. \tag{5}$$

Here, H_n is an $n \times n$ upper Hessenberg matrix, and $V_n := [v_1 \quad v_2 \quad \cdots \quad v_n]$. Moreover, I_n denotes the $n \times n$ identity matrix.

Using the $n \times n$ Hessenberg matrix generated by n steps of the Arnoldi process, we can now define an approximation $\langle \cdot, \cdot \rangle_n$ to the original inner product $\langle \cdot, \cdot \rangle$. More precisely, we set

$$\langle \varphi, \psi \rangle_n := e_1^H (\psi(H_n))^H (\varphi(H_n)) e_1 \quad \text{for all} \quad \varphi, \psi \in \mathcal{P}, \tag{6}$$

where $e_1 := [1 \quad 0 \quad \cdots \quad 0]^T$ denotes the first unit vector. It turns out that the inner product $\langle \varphi, \psi \rangle_n$ is just the inner product corresponding to Gauss quadrature in the complex plane. Indeed, we have the following result.

Theorem 1 *Let A be a general $N \times N$ matrix, and let $n \leq N_A$. Let $\langle \cdot, \cdot \rangle$ and $\langle \cdot, \cdot \rangle_n$ be the inner products defined by (4) and (6), respectively. Then:*

$$\langle \varphi, \psi \rangle = \langle \varphi, \psi \rangle_n \quad \text{for all} \quad \varphi \in \mathcal{P}_n \quad \text{and} \quad \psi \in \mathcal{P}_{n-1}, \tag{7}$$

$$\langle \varphi, \psi \rangle = \langle \varphi, \psi \rangle_n \quad \text{for all} \quad \varphi \in \mathcal{P}_{n-1} \quad \text{and} \quad \psi \in \mathcal{P}_n. \tag{8}$$

Proof. In view of $\langle \varphi, \psi \rangle = \overline{\langle \psi, \varphi \rangle}$ and $\langle \varphi, \psi \rangle_n = \overline{\langle \psi, \varphi \rangle_n}$, we only need to show either (7) or (8). We prove (7).

Obviously, it is sufficient to consider monomials $\varphi(\lambda) \equiv \lambda^j$ and $\psi(\lambda) \equiv \lambda^k$ in (7). Then, by (4) and (6), equation (7) is equivalent to the following statement:

$$(A^k v_1)^H (A^j v_1) = (H_n^k e_1)^H (H_n^j e_1) \quad \text{for all} \quad j = 0, 1, \ldots, n, \quad k = 0, 1, \ldots, n-1. \tag{9}$$

Using (5) and induction on j, one easily verifies that

$$\begin{aligned} A^j v_1 &= V_n H_n^j e_1 \quad \text{for all} \quad j = 0, 1, \ldots, n-1, \\ A^n v_1 &= V_n H_n^n e_1 + v_{n+1}[0 \quad \cdots \quad 0 \quad h_{n+1,n}] H_n^{n-1} e_1. \end{aligned} \tag{10}$$

Using that, by (5), $V_n^H V_n = I_n$ and $V_n^H v_{n+1} = 0$, the desired relation (9) then follows immediately from (10). $\square$

It remains to describe how to evaluate the approximation (6) to the original inner product (4). For the Hermitian positive definite case, such an algorithm can be found in Golub and Welsch [3]. Algorithms for the non-Hermitian case can be based on either the Jordan canonical form or the Schur normal form of H_n; details of such procedures will be presented elsewhere.

3 The Lanczos case

Let $v_1, w_1 \in \mathbb{C}^N$ be two arbitrary nonzero vectors. Then, starting with v_1, the Lanczos process generates a sequence of basis vectors $v_1, v_2, \ldots$, such that the corresponding polynomials (3) are orthogonal with respect to the bilinear form $[\cdot, \cdot]$ defined as follows:

$$[\varphi, \psi] := w_1^T \varphi(A)\psi(A)v_1 \quad \text{for all} \quad \varphi, \psi \in \mathcal{P}. \tag{11}$$

In addition to $v_1, v_2, \ldots$, the Lanczos process also generates a second sequence $w_1, w_2, \ldots$. We assume that these vectors have been normalized such that $w_j^T v_j = 1$ for all $j = 1, 2, \ldots$. In exact arithmetic, the Lanczos algorithm terminates after a finite number of steps, say N_L, and we have $N_L \leq N_A$.

In the following, let $n < N_L$, and let $\{v_j\}_{j=1}^{n+1}$ and $\{w_j\}_{j=1}^{n+1}$ be the vectors generated by n steps of the Lanczos algorithm. The Lanczos vectors then satisfy $W_{n+1}^T V_{n+1} = I_{n+1}$ and

$$AV_n = V_n T_n + v_{n+1} [0 \ \cdots \ 0 \ t_{n+1,n}],$$
$$A^T W_n = W_n T_n^T + w_{n+1} [0 \ \cdots \ 0 \ t_{n,n+1}].$$

Here $V_n := [v_1 \ v_2 \ \cdots \ v_n]$, $W_n := [w_1 \ w_2 \ \cdots \ w_n]$, and T_n is an $n \times n$ tridiagonal matrix. Using the Lanczos matrix T_n, we can define an approximation $[\cdot, \cdot]_n$ to the original bilinear form $[\cdot, \cdot]$. More precisely, we set

$$[\varphi, \psi]_n := e_1^T \varphi(T_n)\psi(T_n)e_1 \quad \text{for all} \quad \varphi, \psi \in \mathcal{P}. \tag{12}$$

It turns out that the bilinear form $[\varphi, \psi]_n$ just corresponds to Gauss quadrature for formally orthogonal polynomials. Indeed, we have the following result.

Theorem 2 *Let A be a general $N \times N$ matrix, and let $n \leq N_L$. Let $[\cdot, \cdot]$ and $[\cdot, \cdot]_n$ be the bilinear form defined by (11) and (12), respectively. Then:*

$$[\varphi, 1] = [\varphi, 1]_n \quad \text{for all} \quad \varphi \in \mathcal{P}_{2n-1}.$$

Proof. The proof is completely analogous to the proof of Theorem 1. □

References

[1] W.E. Arnoldi. The principle of minimized iterations in the solution of the matrix eigenvalue problem. *Quart. Appl. Math.*, **9**, pp 17–29, 1951.

[2] B. Fischer and R.W. Freund. On adaptive weighted polynomial preconditioning for Hermitian positive definite matrices. Technical Report 92.09, RIACS, NASA Ames Research Center, Moffett Field, California, March 1992.

[3] G.H. Golub and J.H. Welsch. Calculation of Gauss quadrature rules. *Math. Comp.*, **23**, pp 221–230, 1969.

[4] C. Lanczos. An iteration method for the solution of the eigenvalue problem of linear differential and integral operators. *J. Res. Nat. Bur. Standards*, **45**, pp 255–281, 1950.

[5] T.A. Manteuffel. The Tchebychev iteration for nonsymmetric linear systems. *Numer. Math.*, **28**, pp 307–327, 1977.

AN IMPLEMENTATION OF THE QMR METHOD BASED ON COUPLED TWO-TERM RECURRENCES

R.W. FREUND
AT&T Bell Laboratories
600 Mountain Avenue, Room 2C-420
Murray Hill
New Jersey 07974-0636, U.S.A.
freund@research.att.com

N.M. NACHTIGAL
RIACS, Mail Stop T041-5
NASA Ames Research Center
Moffett Field
California 94035-1000, U.S.A.
santa@riacs.edu

KEYWORDS. Krylov subspace iteration, quasi-minimal residual method, non-Hermitian matrices, coupled two-term recurrences, look-ahead techniques.

1 Introduction

Recently, we proposed a new Krylov subspace iteration, the quasi-minimal residual algorithm (QMR) [1], for solving general nonsingular non-Hermitian systems of linear equations

$$Ax = b. \tag{1}$$

The QMR method is closely related to the classical biconjugate gradient algorithm (BCG) due to Lanczos [4]. The BCG method aims at generating approximate solutions that satisfy a Galerkin condition. Unfortunately, for non-Hermitian matrices A, such iterates need not always exist, and this is the source of one of the two possible breakdowns—triggered by division by 0—that can occur during each iteration step of BCG. The second breakdown is equivalent to the possible breakdown—also triggered by division by 0—of the nonsymmetric Lanczos process [3]. In finite precision arithmetic, it is unlikely that one encounters exact breakdowns in the BCG algorithm. However, near-breakdowns can occur, which can cause a build-up of round-off in successive iterations. Another problem with BCG is the lack of a

M. S. Moonen et al. (eds.), *Linear Algebra for Large Scale and Real-Time Applications*, 381–384.
© 1993 *Kluwer Academic Publishers.*

382

residual minimization property for its iterates, which leads to a typically erratic convergence behavior, with wild oscillations in the residual norm.

The QMR method offers remedies for these problems. It generates iterates that are defined by a quasi-minimization of the residual norm, rather than a Galerkin condition. This eliminates the oscillations and leads to a smooth and nearly monotone convergence behavior. In contrast to BCG, a QMR iterate always exists at each iteration step, and this excludes breakdowns caused by non-existent iterates. Moreover, possible breakdowns in the underlying Lanczos process are prevented by using look-ahead techniques. Therefore, except for the rare event of an incurable breakdown, breakdowns cannot occur in the QMR method.

In the original QMR algorithm [1], an implementation of the Lanczos method with look-ahead is used to generate basis vectors for the underlying Krylov subspaces. In the Lanczos process, these basis vectors are generated by means of three-term recurrences. Let x_0 be an arbitrary initial guess for the solution of (1). Let $r_0 := b - Ax_0$ denote the corresponding initial residual vector, and choose $v_1 = r_0/\rho_1$, with a non-zero scalar ρ_1, as the starting right Lanczos vector. Then, letting V_n denote the matrix whose columns are the first n right Lanczos vectors, one obtains a recurrence of the form

$$AV_n = V_{n+1}H_n, \tag{2}$$

where H_n is an $(n + 1) \times n$ block tridiagonal upper Hessenberg matrix containing the coefficients of the block three-term recurrence of the look-ahead Lanczos algorithm. Once the basis vectors V_n are constructed, the nth QMR iterate x_n is defined as

$$x_n = x_0 + V_n z_n, \tag{3}$$

where $z_n \in \mathbb{C}^n$ is the solution of the least-squares problem

$$\|f_{n+1} - \Omega_{n+1}H_n z_n\| = \min_{z \in \mathbb{C}^n} \|f_{n+1} - \Omega_{n+1}H_n z\|. \tag{4}$$

Here,

$$\Omega_{n+1} := \text{diag}(\omega_1, \omega_2, \cdots, \omega_{n+1}), \quad \omega_j > 0, \quad j = 1, 2, \cdots, n+1, \tag{5}$$

is a diagonal scaling matrix, and

$$f_{n+1} := \omega_1 \rho_1 \cdot [1 \quad 0 \quad \cdots \quad 0]^T \in \mathbb{R}^{n+1}.$$

Suppose now that one computes a decomposition

$$H_n = L_n R_n \tag{6}$$

of H_n, with L_n an $(n + 1) \times n$ block bidiagonal upper Hessenberg matrix, and R_n a block bidiagonal upper triangular matrix. Then, inserting (6) in (4), one obtains the equivalent least-squares problem

$$\|f_{n+1} - \Omega_{n+1}L_n y_n\| = \min_{y \in \mathbb{C}^n} \|f_{n+1} - \Omega_{n+1}L_n y\|, \tag{7}$$

where now $y_n = R_n z_n$. This means that now the QMR iterates x_n can be computed from

$$x_n = x_n + P_n y_n, \tag{8}$$

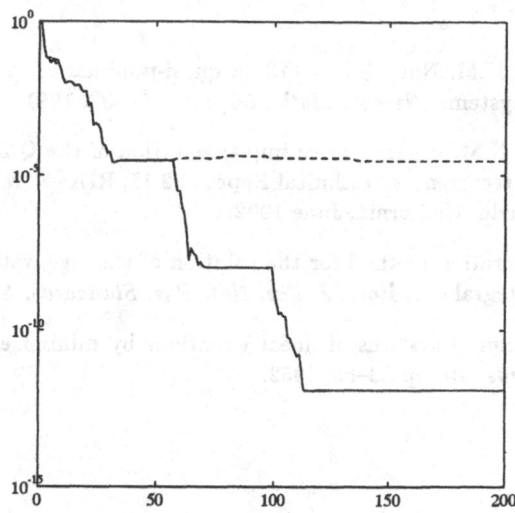

Figure 1: *Convergence curves for Petri net example.*

where, from (2) and (6), P_n is given by the coupled recurrences

$$V_n = P_n R_n \quad \text{and} \quad AP_n = V_{n+1} L_n. \tag{9}$$

Note that (3) and (8) describe two mathematically equivalent ways of computing the QMR iterates x_n. Nonetheless, it has been observed that, in finite-precision arithmetic, the QMR algorithm based on the coupled block two-term recurrence process (9) is usually more robust than the equivalent algorithm based on the block three-term recurrence (2). Figure 1 shows the computed residual norm $\|b - Ax_n\|/\|r_0\|$ versus iteration number n for the two-term and the three-term algorithms applied to a linear system arising in performance modeling of multiprocessor systems, using Petri net analysis. While the performance of the two algorithms is usually more comparable, the example does illustrate that, in extreme cases, the performance of the two algorithms can differ greatly.

Full details of the new coupled QMR algorithm, as well as more numerical examples, can be found in [2]. We give a different motivation for the decomposition (6), which enables us to characterize the vectors p_n that appear in (8). The paper also discusses the practical aspects of the new algorithm, such as the implementation of the look-ahead strategy, the choice of the scaling factors ω_i in (5), the solution of the least-squares problem (7), and the computation of the QMR iterate x_n using only short recurrences. Finally, we note that all the theoretical properties of the original QMR algorithm, discussed in [1], carry over to the new algorithm, since they are mathematically equivalent.

384

References

[1] R.W. Freund and N.M. Nachtigal. QMR: a quasi-minimal residual method for non-Hermitian linear systems. *Numer. Math.*, **60**, pp 315–339, 1991.

[2] R.W. Freund and N.M. Nachtigal. An implementation of the QMR method based on coupled two-term recurrences. Technical Report 92.15, RIACS, NASA Ames Research Center, Moffett Field, California, June 1992.

[3] C. Lanczos. An iteration method for the solution of the eigenvalue problem of linear differential and integral operators. *J. Res. Nat. Bur. Standards*, **45**, pp 255–282, 1950.

[4] C. Lanczos. Solution of systems of linear equations by minimized iterations. *J. Res. Nat. Bur. Standards*, **49**, pp 33–53, 1952.

COMPUTATIONALLY EFFICIENT HOMOTOPIES FOR THE H^2 MODEL ORDER REDUCTION PROBLEM

Y. GE and L. T. WATSON
Comp. Sc. & Math. Dept.
Virginia Polytechnic Institute
and State University
Blacksburg, VA 24061-0106, U.S.A.
ltw@vtopus.cs.vt.edu

E. G. COLLINS, JR. and L. D. DAVIS
Harris Corporation
Government Aerospace Systems
Division
Melbourne, FL 32902, U.S.A.
ecollins@x102a.ess.harris.com

KEYWORDS. Homotopy, model order reduction, optimal control.

1. Introduction

The H^2 optimal model reduction problem, i.e., the problem of approximating a higher order dynamical system by a lower order one so that a model reduction criterion is minimized, is of significant importance and is under intense study. Several earlier attempts to apply homotopy methods to the H^2 optimal model order reduction problem were not entirely satisfactory. Richter devised a homotopy approach which only estimated certain crucial partial derivatives and employed relatively crude curve tracking techniques. Žigić, Bernstein, Collins, Richter, and Watson formulated the problem so that numerical linear algebra techniques could be used to explicitly calculate partial derivatives, and employed sophisticated homotopy curve tracking algorithms, but the number of variables made large problems intractable. We propose here several ways to reduce the dimension of the homotopy map so that large problems are computationally feasible.

The problem can be formulated as: given the asymptotically stable, controllable, and observable time invariant continuous time system

$$\dot{x}(t) = A\,x(t) + B\,u(t), \qquad y(t) = C\,x(t), \tag{1}$$

where $A \in \mathbf{R}^{n \times n}$, $B \in \mathbf{R}^{n \times m}$, $C \in \mathbf{R}^{l \times n}$, the goal is to find a reduced order model

$$\dot{x}_m(t) = A_m\,x_m(t) + B_m\,u(t), \qquad y_m(t) = C_m\,x_m(t), \tag{2}$$

where $A_m \in \mathbf{R}^{n_m \times n_m}$, $B_m \in \mathbf{R}^{n_m \times m}$, $C_m \in \mathbf{R}^{l \times n_m}$, $n_m < n$ which minimizes the cost function

$$J(A_m, B_m, C_m) \equiv \lim_{t \to \infty} E\left[(y - y_m)^t R (y - y_m)\right], \tag{3}$$

where the input $u(t)$ is white noise with symmetric and positive definite intensity V and R is a symmetric and positive definite weighting matrix.

The optimal projection equations of Hyland and Bernstein [2] are basis independent and correspond to the maximum number of degrees of freedom one could plausibly use. At the other extreme, the minimal number of unknowns corresponds to the input normal form.

M. S. Moonen et al. (eds.), *Linear Algebra for Large Scale and Real-Time Applications*, 385–386.
© 1993 *Kluwer Academic Publishers*.

There are subtle differences between the optimal projection equations and input normal form formulations. Assuming a particular Jordan form for A_m leads to the minimal parameter formulation of Ly, Bryson, and Cannon. Homotopy algorithms are developed based on the input normal and Ly forms, and tested on a collection of problems from the literature. Both the input normal form and Ly parameterization use the minimum possible number of degrees of freedom, but make assumptions about the structure of A_m, and therefore may not exist. Even worse, they may exist but be arbitrarily badly ill conditioned, resulting in unstable numerical algorithms. We explore some alternative formulations using more than the minimal number of degrees of freedom, and compare to the minimal formulations.

2. Numerical results

Homotopy algorithms for the optimal projection equations, input normal form, Ly form, and a nonminimal formulation are described in detail in [1] and [3]. Table 1 gives the CPU times (seconds on a DECstation 5000/200) and number of homotopy curve tracking steps for some test problems from [3] (dash means failure).

Table 1. *Algorithm measures for three formulations.*

Ex	n_m	Input normal form		Ly's form		Over-parametrization	
		steps	time	steps	time	steps	time
1	1	7	0.06	8	0.08	12	0.08
2	1	25	0.13	40	0.25	31	0.20
3	1	23	0.10	25	0.14	30	0.16
4	2	11	0.20	16	0.37	18	0.47
5	2	14	0.22	12	0.27	10	0.29
6	2	-	-	8	0.32	35	1.3
6	3	-	-	114	6.9	125*	13.
7	3	10	0.42	12	0.70	18	1.5
8	2	22	0.50	35	1.1	35	1.1
8	3	16	0.65	17	0.96	14	1.3
9	2	127	8.0	-	-	168*	13.
9	4	8	1.9	59	15.	17*	7.5
10	8	-	-	19	35.	7	49.
11	2	6	0.13	-	-	16	0.42

[1] Y. Ge, E.G. Collins, Jr., L.T. Watson, and L.D. Davis. Minimal parameter homotopies for the L^2 optimal model order reduction problem. *Tech. Rep.* **92-36**, Dept. of Computer Science, VPI&SU, Blacksburg, VA, 1992.

[2] D.C. Hyland and D.S. Bernstein. The optimal projection equations for model reduction and the relationships among the methods of Wilson, Skelton, and Moore. *IEEE Trans. Autom. Contr.*, **AC-30**, pp 1201–1211, 1985.

[3] D. Žigić. *Homotopy methods for solving the optimal projection equations for the reduced order model problem.* M.S. thesis, Dept. of Computer Science, VPI&SU, Blacksburg, VA, 1991.

A FAST ALGORITHM FOR $\mathcal{QR}$ DECOMPOSITION OF TOEPLITZ MATRICES

G. O. GLENTIS
University of Athens
Informatics Dept.
Panepistimiopolis
TYPA Buildings 157 71, Athens
Greece
othon@grathun1.bitnet

KEYWORDS. Toeplitz matrices, LS solution, QR decomposition, fast algorithms.

In this paper new fast algorithms for the $\mathcal{QR}$ factorization of a Toeplitz matrix and the determination of the least squares solution are developed. Both topics are of general importance in a wide range of applications. Typical examples include the design of optimal FIR filters, linear prediction of speech, time series analysis, othogonal polynomials, Pade approximation.

We adopt the signal processing nomenclature. Given an input singal $x(n) \in \Re$ and a desired responce singal $y(n) \in \Re$, over an observation interval $n \in [M, N]$, $M < N$, we seek to determine the optimal filter of *order* p, $\mathbf{c}_p = [c(i)]_{i=1,\ldots,p}$, that minimizes the error norm

$$\min_{\mathbf{c}_p} \& \Psi(M, N) + \mathcal{X}(M, N)\mathbf{c}_p \& \tag{1}$$

$\mathcal{X}(M, N)$ is the input data matrix of dimensions $(N - M - p + 1) \times (N - M - p + 1)$

$$\mathcal{X}(M, N) = \begin{bmatrix} x(M+p-1) & x(M+p-2) & \ldots & x(M+1) & x(M) \\ x(M+p) & x(M+p-1) & \ldots & x(M+2) & x(M+1) \\ x(M+p+1) & x(M+p) & \ldots & x(M+3) & x(M+2) \\ \vdots & \vdots & \ddots & \vdots & \vdots \\ x(N-1) & x(N-2) & \ldots & x(N-p+1) & x(N-p) \\ x(N) & x(N-1) & \ldots & x(N-p+2) & x(N-p+1) \end{bmatrix} \tag{2}$$

and $\Psi(M, N)$ is the desired responce data vector of dimensions $(N - M - p + 1) \times 1$

$$\Psi(M, N) = [y(M+p-1) \; y(M+p) \; \ldots \; y(N-1) \; y(N)]^t \tag{3}$$

M. S. Moonen et al. (eds.), *Linear Algebra for Large Scale and Real-Time Applications*, 387–388.
© 1993 *Kluwer Academic Publishers.*

Matrix $\mathcal{X}(M,N)$ is Toeplitz. Minimization of (1) with respect to $\mathbf{c}_p$ leads to the linear system, the so called normal equations

$$\mathcal{X}(M,N)^t\mathcal{X}(M,N)\mathbf{c}_p(M,N) = -\mathcal{X}(M,N)\Psi(M,N) \tag{4}$$

The optimal filter of order p is obtain as the solution of (4)

$$\mathbf{c}_p(M,N) = -(\mathcal{X}(M,N)^t\mathcal{X}(M,N))^{-1}\mathcal{X}(M,N)\Psi(M,N) \tag{5}$$

and the residual error is given by the formula

$$E^c(M,N) = \left(\mathbf{I}_p - \mathcal{X}(M,N)(\mathcal{X}(M,N)^t\mathcal{X}(M,N))^{-1}\mathcal{X}(M,N)\right)\Psi(M,N) \tag{6}$$

Eq. (5) provides an algorithm for computating of the optimal parameter vector $\mathbf{c}_p$. The Toeplitz structure of $\mathcal{X}(M,N)$ gives rise to efficient methods for the solution of the normal equations. Several fast, order recursive algorithms exist, that solve (5) using $2Lp + O(p^2)$ MADS (Multiplications and Divisions).

An alternative methodology for determining $\mathbf{c}_p$ is based on the $\mathcal{QR}$ factorization of the data matrix $\mathcal{X}(M,N)$. $\mathcal{QR}$ methods are discussed for their superior performance over the normal equations approach, especially in badly conditioned problems. Fast $\mathcal{QR}$ algorithms have been derived by taking into accound the Toeplitz structure of $\mathcal{X}(M,N)$. Algorithms [1]- [3] are based on fast Cholesky factorization methods. Algorithm [5] utilizes a Toeplitz embeding technique ans [4], and results to a generalized lattice type structure. It computes the $\mathcal{Q}$ factor and the inverse $\mathcal{R}^{-1}$ using $9Lp + 13.5p^2$ MADS, $L = N - M + p - 1$ being the leading dimension of $\mathcal{X}(M,N)$.

The proposed method can be viewed as a fast Modified Gram- Schmidt algorithm which recursively computes the orthonormal columns $\mathcal{Q}_i$, $i = 1, 2 \ldots p$ of $\mathcal{Q}$ as well as the elements of $\mathcal{R}^{-1}$. When the determination of the optimal filter is the desired task it can be utilized to compute the least squares filter in an order recursive way. The computational complexity of the proposed method is $8Lp + 3p^2$ MADS. In the special case when $x(n) = 0, n < M + p - 1$ and $n > N - p$ the algorithm reduces to the classical lattice stucture.

References

[1] D.R. Sweet. Fast Toeplitz orthogonalization. *Numer. Math.*, **43**, pp 1-21, 1984.

[2] A.W. Bojanczyk, R.P. Brent and F.R. de Hoog. QR factorization of Toeplitz matrices. *Numer. Math.*, **49**, pp 81-94, 1986.

[3] J. Chun, T. Kailath and H. Lev-Ari. Fast parallel algorithms for QR and triangular factorization. *SIAM J. Sci. Stat. Comput.*, pp 899-913, 1987.

[4] G. Cybenko. A general orthogonalization technique with applications to time series analysis and signal processing. *Math. Computation*, pp 323-336, 1983.

[5] G. Cybenko. Fast Toeplitz orthogonalization using inner products. *SIAM J. Stat. Comput.*, pp 734- 739, 1987.

UNITARY HESSENBERG METHODS FOR TOEPLITZ APPROXIMATIONS AND APPLICATIONS

Ch. HE
Universität Bielefeld
Fakultät für Mathematik
Postfach 8640, 4800 Bielefeld 1
Deutschland

A. BUNSE-GERSTNER
Universität Bremen
Fachbereich Mathematik und Informatik
Postfach 33 04 40, 2800 Bremen 33
Deutschland
angelika@mathematik.uni-bremen.de

KEYWORDS. Toeplitz matrix, unitary Hessenberg matrix, Pisarenko frequency model, composite sinusoidal model, Toeplitz approximation model.

We consider three basic models arising in signal processing, the Pisarenko frequency model (PFM) proposed by Pisarenko in 1973, the composite sinusoidal model (CSM) given by Sagayama and Itakura in 1979, and the Toeplitz approximation model (TAM) proposed by Kung in 1981. They deal with a fundamental problem in signal processing: a complex signal $\ldots, s_{-1}, s_0, s_1, s_2, \ldots$, is approximated by an exponential sum:

$$s_k \approx \sum_{j=1}^{n} \omega_j e^{i\theta_j k}, k = 0, 1, 2, \ldots,$$

for some integer n, some frequencies $\theta_j, j = 1, \ldots, n$ and amplitudes $\omega_j, j = 1, \ldots, n$. This problem is known as the "retrieval of harmonics" problem. The frequencies and amplitudes are often determined such that the terms

$$\sum_{j=1}^{n} |\omega_j|^2 e^{i\theta_j k}, k = 0, 1, \ldots, N - 1.$$

389

M. S. Moonen et al. (eds.), Linear Algebra for Large Scale and Real-Time Applications, 389–390.
© 1993 Kluwer Academic Publishers.

match an approximation of the first N autocorrelation lags

$$t_k = \sum_{-\infty}^{\infty} \bar{s}_j s_{j-k}, k = 0, 1, \ldots, N-1$$

The Toeplitz matrix $T_N = T(t_0, t_1, \ldots, t_{N-1})$ built with these t_k is called the autocorrelation matrix of order N and the problems can thus be viewed as the Toeplitz approximation problems of approximating the entries of the autocorrelation matrix T_N in a suitable way.

The mathematical background of the "retrieval of harmonics" problem is a trigonometric approximation problem with unknown exponents. A unifying framework for the PFM, CSM and other linear prediction methods based on the classical trigonometric moment problem were developed by Delsarte, Genin, Kamp and Van Dooren.

Ammar, Gragg and Reichel showed that the PFM as well as the CSM can be treated as an eigenvalue problem for unitary matrices. Compared with the original methods which were based on computing the zeros of a related polynomial, these matrix methods are numerically more reliable. Finally for the TAM Cybenko gave a first characterization of the problem in terms of related unitary Hessenberg matrices.

Here we treat the slightly more general case of s_k being a complex signal. To the Toeplitz matrix T_N we can associate a unitary Hessenberg matrix in the following way:

let $H(\rho) = H(\gamma_1, \ldots, \gamma_{N-1}, \rho)$ be a unitary Hessenberg matrix parameterized by a sequence $\{\gamma_k\}_1^{N-1}$ and a free number ρ on the unit circle, where the $\gamma_k, k = 1, 2, \ldots, N-1$, are derived by the Levinson algorithm for obtaining the inverse Cholesky decomposition of T_N. Then the Cholesky decomposition of T_N can be given in terms of $H(\rho)$ as

$$T_N = t_0(e_1, H(\rho)e_1, \ldots, H^{N-1}(\rho)e_1)^H (e_1, H(\rho)e_1, \ldots, H^{N-1}(\rho)e_1),$$

where e_1 denotes the first unit vector. The purpose of this paper is to exploit this relation for the solutions of the Toeplitz approximation problems.

By solving the eigenproblem of $H(\rho)$ and finding the spectral decomposition of $H(\rho) = X(\rho)\Lambda(\rho)X(\rho)^H$ with $\Lambda(\rho) = diag(e^{i\theta_1(\rho)}, \ldots, e^{i\theta_N(\rho)})$, we obtain

$$t_k = \sum_{j=1}^{N} |\omega_j|^2(\rho)e^{i\theta_j(\rho)}, k = 0, 1, 2 \ldots, N-1$$

where $\omega_j(\rho) = \sqrt{t_0}e_1^T X(\rho)e_j$. The PFM and the CSM can then both be treated as eigenvalue problems of certain unitary Hessenberg matrices. We refer to these methods as unitary Hessenberg methods for solving the PFM and CSM. It is worth mentioning that the decomposition also highlights a basic algebraic background of the Levinson algorithm and the lattice algorithm as well as the Schur algorithm in matrix terms.

Based on this basic decomposition we can also solve the TAM under an assumption which is reasonable in the practical signal processing problem. We successfully reduce the TAM to an optimization problem which is solved numerically. An eigenvalue problem for a unitary Hessenberg matrix with an unknown variable is involved in this optimization problem.

PERTURBATION THEORY AND BACKWARD ERROR
FOR $AX - XB = C$

(To appear in BIT)

N.J. HIGHAM
Nuffield Science Research Fellow
Department of Mathematics
University of Manchester
Manchester, M13 9PL
England
na.nhigham@na-net.ornl.gov

KEYWORDS. Sylvester equation, Lyapunov equation, backward error, perturbation bound, condition number, error estimate, LAPACK.

Because of the special structure of the equations $AX - XB = C$ the usual relation for linear equations "backward error = relative residual" does not hold, and application of the standard perturbation result for $Ax = b$ yields a perturbation bound involving $\text{sep}(A,B)I{-}1$ that is not always attainable. An expression is derived for the backward error of an approximate solution Y; it shows that the backward error can exceed the relative residual by an arbitrary factor. A sharp perturbation bound is derived and it is shown that the condition number it defines can be arbitrarily smaller than the $\text{sep}(A,B)I{-}1$-based quantity that is usually used to measure sensitivity. For practical error estimation using the residual of a computed solution an "LAPACK-style" bound is shown to be efficiently computable and potentially much smaller than a sep-based bound. A Fortran 77 code has been written that solves the Sylvester equation and computes this bound, making use of LAPACK routines.

Acknowledgements

This work was carried out while the author was a visitor at the Institute for Mathematics and Its Applications, University of Minnesota.

M. S. Moonen et al. (eds.), Linear Algebra for Large Scale and Real-Time Applications, 391.
© 1993 *Kluwer Academic Publishers.*

FAST TRANSFORMS AND ELLIPTIC PROBLEMS

T. HUCKLE
Institut für Angewandte Mathematik und Statistik
Universität Würzburg
D-8700 Würzburg
FRG
huckle@vax.rz.uni-wuerzburg.dbp.de

KEYWORDS. Fourier Transform, sine Transform, preconditioned conjugate gradient method, elliptic differential equations, finite difference method.

We study the use of the Fourier, Sine and Cosine Transform for solving or preconditioning linear systems $Eu = f$ which arise from the discretization of elliptic problems. In the one-dimensional case the corresponding matrices are of the form

$$E = \begin{pmatrix} a_0 + a_1 & -a_1 & & & \\ -a_1 & a_1 + a_2 & -a_2 & & \\ & \ddots & \ddots & \ddots & \\ & & -a_{n-2} & a_{n-2} + a_{n-1} & -a_{n-1} \\ & & & -a_{n-1} & a_{n-1} + a_n \end{pmatrix},$$

related to the elliptic problem $-(a(x)u_x)_x = f(x)$, $a(x) > 0$ on the interval $[0,1]$ with Dirichlet boundary conditions.

In a recent paper [1] R. Chan and T. Chan analysed the preconditioned conjugate gradient method with circulant preconditioners for solving such problems. Thereby, Hermitian Toeplitz matrices are of the form $T_n = (t_{j-i})_{i,j=1}^n =: T_n(t_0, t_1, ..., t_{n-1})$, and with this notation, a Hermitian circulant matrix is given by (see [2]) $C_n = T_n(c_0, c_1, c_2, ..., \bar{c}_2, \bar{c}_1) = F_n^H \Lambda F_n$, whereby Λ is a diagonal matrix and $F_n = (exp(-2\pi ikj/n))_{k,j=0}^{n-1}$ is the Fourier matrix related to the Fast Fourier Transform. Similarly one can use skewcirculant matrices (see [4]) $S_n = T_n(s_0, s_1, s_2, ..., -\bar{s}_2, -\bar{s}_1) = \Omega F_n^H \Delta F_n \bar{\Omega}$ with $\Omega = diag(1, exp(\pi i/n), ..., exp((n-1)\pi i/n))$ for preconditioning.

Preconditioning the linear system $Eu = f$ with a skewcirculant matrix is equivalent to preconditioning the transformed equations $F_n \bar{\Omega} E \Omega F_n^H \hat{u} = \hat{f}$ by a diagonal matrix. For this transformed matrix we get the following result:

393

M. S. Moonen et al. (eds.), Linear Algebra for Large Scale and Real-Time Applications, 393–394.
© 1993 *Kluwer Academic Publishers.*

Theorem 1. *With R a rank-2 matrix, C circulant with eigenvalues $b_1, ..., b_n$, and a diagonal matrix $\Delta = I - exp(\pi i/n)\Omega^2$ it holds*

$$F_n \tilde{\Omega} E \Omega F_n^H = R + \Delta C \Delta^H.$$

For the special case $a(x) \equiv 1$ the matrix E reduces to a tridiagonal Toeplitz matrix $E_0 = tridiag(-1, 2, -1)$. The eigenvectors of E_0 are given by the Sine Transform matrix (see [3]) $S_n = (sin(\pi jk/(n+1))_{j,k=1}^n$. This connection is used by preconditioning the linear system for general $a(x)$ by E_0. Therefore, it is natural to analyse the structure of the matrix $S_n E S_n$.

Theorem 2. *With $D = diag(sin(\pi/(2(n+1))), ..., sin(n\pi/(2(n+1))))$ it holds*

$$S_n E S_n = D(T+H)D,$$

whereby T and H are Toeplitz and Hankel matrices of the form

$$T = \begin{pmatrix} t_0 & t_1 & \cdots & \cdots & t_{n-1} \\ t_1 & t_0 & t_1 & & \vdots \\ \vdots & \ddots & \ddots & \ddots & \vdots \\ \vdots & & \ddots & \ddots & t_1 \\ t_{n-1} & \cdots & \cdots & t_1 & t_0 \end{pmatrix}, \text{ and } H = \begin{pmatrix} t_2 & t_3 & \cdot & \cdot & t_n & 0 \\ t_3 & \cdot & & t_n & 0 & -t_n \\ \cdot & & \cdot & & \cdot & \cdot \\ \cdot & \cdot & & \cdot & & \cdot \\ t_n & 0 & -t_n & & \cdot & -t_3 \\ 0 & -t_n & \cdot & & -t_3 & -t_2 \end{pmatrix},$$

and the vector $(t_k)_{k=0}^n$ is given by the Cosine Transform of the vector $(a_k)_{k=0}^n$.

By a further diagonal transformation the matrix $T + H$ can be reduced to an $n \times n$ leading principal submatrix of an $n+1 \times n+1$ circulant matrix. Therefore, theorem 1 and 2 are of interest for solving such tridiagonal linear systems on a parallel computer using the Sherman-Morrison-Woodbury formula and fast transforms. Based on 2D Fourier, Sine, and Cosine Transform we get similar relations for the 2D elliptic problem which give preconditioners for solving such equations with the preconditioned conjugate gradient method.

References

[1] R.H. Chan and T.F. Chan. Circulant preconditioners for elliptic problems, to appear in *J. Num. Linear Alg. Appl.*

[2] P.J. Davis. *Circulant Matrices*. John Wiley, New York, 1979.

[3] D.F. Elliot and K.R. Rao. *Fast Transforms*. Academic Press, New York, 1982.

[4] T.K. Huckle. A note on skewcirculant preconditioners for elliptic problems, to appear in *Num. Alg.*

AN INTERIOR-POINT METHOD FOR MINIMIZING THE MAXIMUM EIGENVALUE OF A LINEAR COMBINATION OF MATRICES

F. JARRE
Inst. für Ang. Mathematik
University of Würzburg
8700 Würzburg, Germany
jarre@vax.rz.uni-wuerzburg.dbp.de

KEYWORDS. Convex program, eigenvalue, implementation.

We consider the problem of finding

$$\lambda^{\mathrm{opt}} := \inf_{x \in R^m} \left\{ \lambda_{\max}\left(A^{(0)} + \sum_{i=1}^{m} x_i A^{(i)}\right) \right\} \tag{1}$$

and an optimal solution x^{opt} if it exists. Here, the $A^{(i)}$ are given symmetric matrices in $R^{n \times n}$ for $0 \le i \le m$, and $\lambda_{\max}(A)$ is the largest eigenvalue of a matrix A. Problem (1) can be rewritten as a convex differentiable problem with a positive definite constraint. Positive definiteness can easily be verified numerically (via Cholesky decomposition). Moreover, the cone of positive definite matrices A is characterized by the domain of the smooth convex barrier function

$$-\log \det_{+}(A).$$

Here, $\det_{+}$ of a symmetric matrix is simply defined as

$$\det_{+}(A) := \begin{cases} \det(A) & \text{if } A \text{ is positive definite,} \\ 0 & \text{otherwise.} \end{cases} \tag{2}$$

Observe that problem (1) is equivalent to

$$\inf_{x \in R^{m+1}} \left\{ x_{m+1} \mid \det_{+}\left(x_{m+1}I - A^{(0)} - \sum_{i=1}^{m} x_i A^{(i)}\right) > 0 \right\}. \tag{3}$$

The domain of problem (3) will be denoted by the open set S°. The efficiency of a barrier method for solving (3) strongly depends on the properties of the barrier function

$$\phi(x) := -\log \det_{+}\left(x_{m+1}I - A^{(0)} - \sum_{i=1}^{m} x_i A^{(i)}\right) \tag{4}$$

for S°. If the matrices I, A^i are linearly independent one can show that this function is strictly convex. Another important property of this barrier function is the property of

M. S. Moonen et al. (eds.), *Linear Algebra for Large Scale and Real-Time Applications*, 395–396.
© 1993 *Kluwer Academic Publishers.*

self-concordance, which is defined and analyzed by Nesterov and Nemirovsky in [5]. For the derivation of an equivalent condition, a "relative Lipschitz condition" on the Hessian of ϕ, and for a simplified analysis we refer to [2, 3].

In our talk we describe a globally linearly convergent predictor-corrector method based on the barrier function ϕ and present some numerical experiments indicating that the method efficiently exploits the structure (convexity and differentiability) of the problem. For a detailed description and analysis of this algorithm we refer to [4]. Tow closely related algorithms are given in [1, 6]. In particular, the one in [1] is tailored to large scale applications. Finally we also would like to mention the algorithms by Overton [7, 8] which solve the same problem by sequential linear or quadratic programming.

Acknowledgements

This work was done at the Dept. of Operations Research at Stanford University, Stanford, CA, while the author was on leave from the Inst. für Ang. Mathematik, University of Würzburg.

References

[1] F. Alizadeh. Optimization over the positive definite cone: Interior-point methods and combinatorial applications. In : P. Pardalos (Ed.), *Advances in Optimization and Parallel Computing*, North Holland, Amsterdam, pp 1-25, 1992.

[2] F. Jarre. The method of analytic centers for solving smooth convex programs. In : S. Dolecki (Ed.), *Lecture Notes in Mathematics*, Vol. **1405**, Optimization, Proceedings Varetz 1988, Springer, pp 69-85, 1989.

[3] F. Jarre. Interior-point methods for convex programming. Report SOL 90-16, Dept. of Operations Research, Stanford University, Stanford, CA, 1990, to appear in *Appl. Math. and Opt.*

[4] F. Jarre. An interior-point method for minimizing the maximum eigenvalue of a linear combination of symmetric matrices. Report SOL 91-08, Dept. of Operations Research, Stanford University, Stanford, CA, 1991, to appear in *SIAM J. Contr. Opt.*

[5] J.E. Nesterov and A.S. Nemirovsky. Self-concordant functions and polynomial-time methods in convex programming. Report, Central Economical and Mathematical Institute, USSR Acad. Sci., Moscow, USSR, 1989.

[6] J.E. Nesterov and A.S. Nemirovsky. Optimization over positive semidefinite matrices. Manual, Central Economical and Mathematical Institute, USSR Acad. Sci., Moscow, USSR, 1990.

[7] M.L. Overton. On minimizing the maximum eigenvalue of a symmetric matrix. *SIAM J. Matrix Anal. Appl.*, Vol. **9**, No. 2, pp 256-268, 1988.

[8] M.L. Overton. Large-scale optimization of eigenvalues. NYU Computer Science Dept. Report No. 505, Courant Institute of Mathematical Sciences, New York Univerity, 1990.

THE LATTICE-LADDER WITH GENERALIZED FORGETTING

J. KADLEC
Department of Electrical and Electronic Engineering
The Queen's University of Belfast
Stranmillis Road, Belfast, BT9 5AH
United Kingdom
eeg0020@v2.qub.ac.uk

KEYWORDS. Estimator, systolic array, forgetting, lattice.

A novel unified algorithm is presented for the adaptive recursive least-square Lattice-Ladder prediction and filtering. The design of the filter is based on formulating the Lattice section [3, 5] as a special implementation of the factorised recursive least square algorithms [9, 11], where the data and parameter tracking reproduces the nesting and shifting property of the lower order models.

Unlike the standard RLS Lattice Ladder algorithm [10] the proposed estimator utilizes a version of the generalised exponential forgetting [7, 8] which can be on the cell level implemented as a set of specially tuned forgetting factors [5, 6]. The weights are in fact tuned to control character of the adaptation for different variances of uncorrelated backward prediction errors without affecting current estimates of the nested, lower order problems. The proposed method implemented on Ladder-Lattice pipelined array [5] provides (for the prewindowed-data case) identical results with an equal method implemented on triangular array [6]. This is possible, because here (unlike in the square root RLS Lattice-Ladder) the necessary shifting properties of matrices can be valid exactly from the first measurement even in the case of the nonzero initial conditions.

The Ladder-Lattice estimator provides on a linear array of p processors the data throughput comparable with the triangular systolic implementation of the recursive QR [2, 11] or RMGS array [9] on $pI2/2$ processors.

The proposed systolic implementation [5] provides the output prediction errors associated with the nested, lower order regression parameter models and reduces considerably the numerical problems related to the over-parametrisation or poor informative data.

The filter is particularly suitable for systolic implementation of numerically robust predictors or filters in signal processing applications with a variable or unknown number of the parameters.

M. S. Moonen et al. (eds.), *Linear Algebra for Large Scale and Real-Time Applications*, 397–398.
© 1993 *Kluwer Academic Publishers.*

Acknowledgements

The author gratefully acknowledges support by the The International Fund for Ireland through the Institute for Advanced Microelectronics.

References

[1] F.M.F. Gaston and G. W. Irwin. The systolic approach to information Kalman filtering. *Int. Jour. Control*, Vol. **15**, No. 1, pp 225-228, 1989.

[2] W.M. Gentleman amd H.T.Kung. Matrix Triangularisation by Systolic Arrays. *Real Time Signal Processing IV*, Proc. SPIE, Vol. **298**, pp 19-26, 1981.

[3] J. Kadlec. Fast Ladder-Lattice Identification Architecture with Numerically Robust Tracking of Parameters. In : P.J. Fleming and D.I. Jones (Eds.), *Algorithms and Architectures for Real-Time Control*, Proc. IFAC Workshop, Bangor, North Wales, UK, Pergamon Press, IFAC Workshop Series, No. 4, pp 105-110, 1992.

[4] J. Kadlec. A Joint Criterion for Exponential Directional and Mixed Parameter Tracking. *Preprints IFAC Int. Symp. on Adapt. Syst. in Control and Sig. Processing (ACASP 92)*, Grenoble, France, pp 687-692, 1992.

[5] J. Kadlec, F.M.F. Gaston and G. W. Irwin. Regularised Lattice-Ladder Adaptive Filter. In : *Mutual Impact of Computing Power and Control Theory*, IFAC Workshop Proceeding Series, 1992.

[6] J. Kadlec, F.M.F. Gaston and G. W. Irwin. Implementation of the Regularised Parameter Estimator. Accepted for *Proceedings of the 1992 IEEE Workshop on VLSI and Signal Processing*, Napa, CA, October 1992.

[7] R. Kulhavý. Restricted exponential forgetting in real-time identification. *Automatica*, **23**, pp 589-600, 1987.

[8] R. Kulhavý and M.B. Zarrop. On a general concept of forgetting. Accepted for *Int. J. of Control*.

[9] F. Ling, D. Manolakis, and J.G. Proakis. A recursive modified Gram-Schmidt algorithm for least squares estimation. *IEEE Trans. ASSP*, **34**, pp 829-835, 1986.

[10] L. Ljung and T. Söderström. *Theory and Practice of Recursive Identification*. Cambridge, MA, MIT Press. 1983.

[11] J.G. Mc Whirter. Algorithmic engineering - an emerging discipline. *Advanced Algorithms and Architectures for Signal Processing IV*, Proc. SPIE, Vol. **1152**, pp 2-15, 1989.

[12] V. Peterka. Bayesian approach to system identification. In : P. Eykhoff (Ed.), *Trends and Progress in System Identification*, Pergamon Press, Eindhoven, The Netherlands, 1981.

SOLVING A LEAST SQUARES PROBLEM WITH BOUNDARY CONSTRAINTS

L. KAUFMAN
Bell Laboratories
Rm. 2c-461
Murray Hill, N.J. 07974, U.S.A.
lck@research.att.com

KEYWORDS. Least squares, tomography, conjugate gradient.

Positron emission tomography(PET) is used to study blood flow and metabolism of a particular organ. The patient is given a tagged substance(such as glucose for brain study) which emits positrons. Each positron annihilates with an electron and emits two photons in opposite directions. The patient is surrounded by a ring of detectors which are wired so that whenever any pair sense a photon within a very small time interval, the count for that pair is incremented. The reconstruction problem in PET is to determine a memory map of the annihilations, and hence a map of the blood flow, given the data gathered by the ring of detectors.

In the discrete reconstruction problem let η_t represent the number of photon pairs detected in tube t. Assume that the area of interest has been divided into B boxes and that there are T detector tubes. Let P be a $B \times T$ matrix such that $p_{b,t}$ represents the probability that a photon pair emitted in box b will be detected in tube t. One wishes to determine x_b, the number of photon pairs emitted in box b. The $x's$ should be nonnegative.

One could find x by minimizing the least squares function

$$f(x) = \|P^T x - \eta\|_2$$

subject to nonnegativity constraints. For a rather small problem one may impose a 128×128 grid leading to 16384 unknowns and that there might by 256 detectors Thus P might have 500 million elements of which only about 3 million are nonzero. Because of its size and density, using a factorization of the P matrix is ill advised. The P matrix is composed of repeated submatrices which allows one to use multiprocessing when computing a matrix vector product with P and decreases the amount of storage required to reconstruct the matrix. Often there are hundreds of variables that are initially nonzero which should be zero.

399

M. S. Moonen et al. (eds.), Linear Algebra for Large Scale and Real-Time Applications, 399–400.
© 1993 *Kluwer Academic Publishers.*

400

Moreover, the determination whether a variable is small or zero is rather inconsequential and the focus should be on the large values of x.

Most efforts for minimizing the least squares function have produced ions for problems with more than 16000 variables. The most time consuming portion of each iteration are 2 matrix by vector multiplications with P. However, one can adapt the EM algorithm to the least squares problem and come up with an EM-LS algorithm of the form

$$x^{(new)} = x^{(old)} + \alpha X^{(old)} \nabla f(\mathbf{x}^{(old)})$$

where X is a diagonal matrix with $x_{ii} = x_i$ and α is a line search parameter that minimizes f. in the interval that permits nonnegativity. Experimental evidence indicates that this algorithm yields the same quality image as the EM algorithm applied to the maximum likelihood(ML) function. With ML in most instances the largest α which insures feasibility also maximizes the function. This fact has not been observed with the least squares function, which opens the way to introducing a conjugate gradient subalgorithm.

We suggest coupling the conjugate gradient algorithm with the EM algorithm to obtain a preconditioned conjugate gradient(PCG) algorithm. The algorithm below is an inner-outer algorithm where the inner algorithm is a PCG algorithm while the outer algorithm changes the preconditioner whenever the inner algorithm suggests an infeasible step.

1. Determine an initial step

2. Set $x_{ii} = x_i$.

3. Apply the conjugate gradient algorithm to the problem $X^{1/2}PP^T X^{1/2}y = X^{1/2}P\eta$ with a bounded line search to insure nonnegativity and terminate the algorithm whenever an unbounded line search would violate the nonnegativity constraint.

4. If step (3) is terminated prematurely, return to step (2) with the last feasible point, else set $x = X^{1/2}y$.

The first step of each new PCG subproblem is just the EM-LS step. In our experience about 5 iterations of the PCG algorithm are taken before a restart and after about 3 restarts a satisfatory picture is obtained. A uniform initial guess has been very successful. If one starts with a random initial guess, then values which are initially small, which should be increased, increase very slowly because of the form of the preconditioner. This problem can be overcome by changing the preconditioner so that if $\frac{\partial f}{\partial x_i} < 0$, set $x_{ii} = max(x_j)$

References

[1] I.I. Dikin. Iterative solution of problems in linear and quadratic programming. *Soviet Mathematics Doklady*, 8, pp 54-60, 1974.

[2] L. Shepp and Y. Vardi. Maximum Likelihood Reconstruction in Positron Emission Tomography. *IEEE Transactions on Medical Imaging*, pp. 113-122, 1982.

ESTIMATING THE EXTREMAL EIGENVALUES AND CONDITION NUMBER BY THE LANCZOS ALGORITHM WITH A RANDOM START

J. KUCZYŃSKI
Institute of Computer Science
Polish Academy of Sciences
Ordona 21
01-237 Warsaw
Poland

H. WOŹNIAKOWSKI
Institute of Applied Mathematics
University of Warsaw
Banacha 2
00-913 Warsaw
Poland
    and
Department of Computer Science
Columbia University
New York, N.Y. 10027, U.S.A.
henryk@c.s.columbia.edu

KEYWORDS. Eigenvalue problem, Lanczos algorithm, random start.

We analyze the problem of approximating the extremal eigenvalues of an $n \times n$ symmetric positive definite matrix A by the Lanczos algorithm. The Lanczos algorithm uses the Krylov information $[b, Ab, \ldots, A^k b]$ for some unit vector b.

If the vector b is chosen deterministically then the problem cannot be solved for some matrices A. If, however, the vector b is chosen randomly with respect to the uniform distribution over the unit sphere, then the problem can be solved for all matrices A.

We prove that $O((\ln(n)/k)^2)$ is an upper bound on the average relative error of the Lanczos algorithm. We show that the measure of the set of vectors b, for which the relative error of the Lanczos algorithm is greater than ε, goes to zero at most as $\sqrt{n}e^{-(2k-1)\sqrt{\varepsilon}}$. These bounds are independent on the distribution of eigenvalues of A.

M. S. Moonen et al. (eds.), *Linear Algebra for Large Scale and Real-Time Applications*, 401–402.
© 1993 *Kluwer Academic Publishers.*

We also provide some theoretical bounds dependent on the distribution of the eigenvalues of A. These bounds are more complicated and one can use them only when good estimates of the largest (smallest) and the second largest (second smallest) eigenvalues and their multiplicities are known.

We obtain estimates for the condition number of a matrix in the two norm by combining bounds on the smallest and largest eigenvalue. More precisely, we give some lower bounds on the measure (probability) of the set of these starting vectors b for which the k-th step of the Lanczos algorithm produces satisfactory approximation to the condition number of a matrix. One of these bounds depends on the computed approximation to the condition number at the k-th step, while the other does not. However, the latter estimate requires knowledge of an upper bound of the condition number of the matrix. Obviously, the better a priori bound on the condition number is given, the better estimate is obtained.

We performed many numerical tests to verify the sharpness of the obtained estimates. For a couple of matrices we chose randomly many vectors. For all matrices, the average value of the corresponding errors satisfied the theoretical bounds, however for some of them the bounds were poor. This was the case especially for small k.

This talk is based on two papers by Kuczyński and Woźniakowski: "Estimating the Largest Eigenvalue by the Power and Lanczos Algorithm with a Random Start" and "Estimating the Extremal Eigenvalues and Condition Number by the Lanczos Algorithm with a Random Start" to be published in SIMAX.

A GENERALIZED ADI ITERATIVE METHOD

N. LEVENBERG
Department of Mathematics and Statistics
University of Auckland
Auckland
New Zealand
levenber@mat.aukuni.ac.nz

L. REICHEL
Department of Mathematics and Computer Science
Kent State University
Kent, OH 44242
U.S.A.
reichel@mcs.kent.edu

KEYWORDS. ADI, iterative method, Sylvester's equation.

The discretization of certain separable partial differential equations as well as some problems in control theory give rise to Sylvester's equations

$$AX - XB = C, \tag{1}$$

where the matrices $A \in \mathbb{R}^{m \times m}$, $B \in \mathbb{R}^{n \times n}$ and $C \in \mathbb{R}^{m \times n}$ are given, and the solution matrix $X \in \mathbb{R}^{m \times n}$ is to be determined. Equation (1) has a unique solution if the spectra of A and B, denoted by $\lambda(A)$ and $\lambda(B)$, respectively, satisfy $\lambda(A) \cap \lambda(B) = \emptyset$. We assume that this condition holds.

If the matrices A and B are dense, then direct solution methods such as the Bartels-Stewart and Golub-Nash-Van Loan algorithms can be applied. When the matrices A and B are large and sparse, iterative solution techniques have to be employed. The ADI iterative method is an attractive solution technique when the matrices A and B are large and sparse, and some information about the spectra $\lambda(A)$ and $\lambda(B)$ is available. The iterations can be written

$$j := j + 1, \quad X_{j+k} := (A - \delta_j I)^{-1} (X_{j+k-1}(B - \delta_j I) + C), \tag{2}$$

$$k := k + 1, \quad X_{j+k} := ((A - \tau_k I)X_{j+k-1} - C)(B - \tau_k I)^{-1}, \tag{3}$$

403

M. S. Moonen et al. (eds.), *Linear Algebra for Large Scale and Real-Time Applications*, 403–404.
© 1993 *Kluwer Academic Publishers.*

where we start with $j = k = 0$. The matrix X_0 is a given initial approximate solution of (1). Strict alternation between (2) and (3) is implied. Thus, when $j + k$ is even, we have $j + k = 2j = 2k$. We refer to this scheme as the classical ADI iterative method, or briefly as the ADI method.

In applications of the ADI method one generally requires sets F and G, such that

$$\lambda(A) \subset F, \quad \lambda(B) \subset G, \quad F \cap G = \emptyset, \tag{4}$$

be explicitly known. The rate of convergence of the ADI method depends on the choice of iteration parameters δ_j and τ_k. These parameters are typically chosen to be optimal with respect to the sets F and G. The problem of how to choose optimal parameters has received considerable attention.

The ADI iterative method is particularly attractive when the matrices A and B have a structure which permits rapid robust direct solution of linear systems of equations with matrices $A - \delta I$ and $B - \delta I$ for certain values of the parameter δ. Banded matrices A and B with small bandwidth provide an example. We note that the ADI method has been used successfully for the solution of separable partial differential equations as well as for preconditioning.

The paper [1] investigates the iterative method obtained when all we require is that $j + k \to \infty$ and j and k are nondecreasing, i.e., when strict alternation between (2) and (3) is not demanded. We refer to the scheme so obtained as the generalized ADI iterative method. Roughly, when the sets F and G are of different size, with a suitable potential-theoretic definition of size, the generalized ADI method can yield a higher asymptotic rate of convergence than the classical ADI method. In particular, if F and G are intervals on the real axis of different length and the iteration parameters are chosen to be asymptotically optimal with respect to F and G, then the asymptotic rate of convergence of the generalized ADI method is higher than of the classical ADI method. Our investigation uses potential-theoretic methods. We also present some computed examples that illustrate and complement the theory.

References

[1] N. Levenberg and L. Reichel. A generalized ADI iterative method. Report ICM-9209-40, Institute for Computational Mathematics, Kent State University, Kent, OH 44242, 1992.

QUATERNIONS AND THE SYMMETRIC EIGENVALUE PROBLEM

N. MACKEY
Department of Computer Science
SUNY at Buffalo
NY-14260, U.S.A.
mackey@cs.buffalo.edu

KEYWORDS. Eigenvalue, Jacobi method, quaternion, tensor product.

In 1846, C.G.J. Jacobi [4] introduced an iterative method for solving the eigenproblem of an $n \times n$ symmetric matrix. Variations of the method are enjoying renewed interest because they are rich in vector operations, can be parallelized with ease (for example, see [6]), and can compute small eigenvalues accurately [1].

Jacobi's method diagonalizes a symmetric matrix by performing a sequence of orthogonal similarity transformations. Each similarity transformation diagonalizes some 2×2 principal submatrix, moving the weight of the annihilated elements onto the diagonal.

Using quaternions, Derek Hacon [2] recently specified a 4×4 orthogonal matrix that transforms a 4×4 skew-symmetric matrix into its real Schur form, thereby obtaining a Jacobi-like method for $n \times n$ skew-symmetric matrices.

By exploiting the algebra isomorphism between $\mathcal{M}_4(\mathcal{R})$, the algebra of 4×4 matrices, and the tensor product $\mathcal{H} \otimes \mathcal{H}$, where $\mathcal{H}$ is the algebra of quaternions, Hacon's ideas can be extended to symmetric matrices. We show how to find a 4×4 orthogonal matrix that can (2×2)-block diagonalize a 4×4 symmetric matrix via a similarity transformation. The quaternion-Jacobi method thus obtained produces four times as many zeroes at each iteration as the traditional Jacobi method. The price we pay for this abundance of zeroes is the cost of computing one right-left singular vector pair of a 3×3 matrix whose entries are easily specifiable in terms of the 4×4 principal submatrix being block diagonalized.

Choosing the singular vector pair corresponding to the largest singular value is analogous to choosing the sorting angle in the traditional Jacobi method: the eigenvalues of the 4×4 matrix that is being block diagonalized are sorted, i.e., the eigenvalues the upper block are larger than those of the lower block.

When partitioning the matrix into 2×2 contiguous blocks, complete diagonalization of the pivot submatrix is unnecessary: block diagonalization incurs no penalty in the form of

M. S. Moonen et al. (eds.), *Linear Algebra for Large Scale and Real-Time Applications*, 405–406.
© 1993 *Kluwer Academic Publishers*.

406

additional sweeps or additional iterations per sweep. Empirical evidence strongly suggests that row-cyclic quaternion-Jacobi is at least quadratically convergent, and requires one less sweep than does the Jacobi method. Hacon [2] has proved asymptotic quadratic convergence in the skew-symmetric case, but the proof does not easily transfer to the symmetric case.

Extending a theorem of Mascarenhas [5], we have proved convergence of the diagonal of the iterates of a "universal" Jacobi-type algorithm. An essential ingredient in the proof is an elegant result due to Schur [3], which states that the vector of diagonal elements of a Hermitian matrix majorizes the vector of its eigenvalues.

Theorem *Given an $n \times n$ Hermitian matrix A_0, let $A_{i+1} = J_i A_i J_i^t$ be the sequence of unitary similarity transformations of a Jacobi-type algorithm. If each J_i is a sorting transformation that diagonalizes some principal submatrix of A_i, without affecting the other diagonal elements of A_i, then the sequence of diagonal vectors $\{(a_{11}^i, a_{22}^i, \ldots, a_{nn}^i)\}_{i=1}^{\infty}$ converges to a point in $\mathcal{R}^n$.*

Showing that the off-diagonal elements converge to zero is one of the tasks at hand. We are also investigating the best way to determine one singular vector pair of a 3×3 matrix, and how the choice of singular vector pair and pivot ordering affects performance. Finally, we hope to extend the ideas described here to the non-symmetric case.

Acknowledgements

I would like to thank my thesis advisor, Dr.P.J. Eberlein, for first suggesting the problem to me, and for her unstinting help and encouragement. Thanks are also due to the organizers of the conference, NATO and NSF for making my attendance possible.

References

[1] J. Demmel and K. Veselić. Jacobi's method is more accurate than QR. *SIAM J. Matrix Anal. Appl.*, **13**, 1992.

[2] D. Hacon. Jacobi's method for skew-symmetric matrices. *SIAM J. Matrix Anal. Appl.*, to appear.

[3] R. Horn and C. Johnson. *Matrix Analysis*. Cambridge University Press, New York, pp 192-193, 1985.

[4] C.G.J. Jacobi. Über ein leichtes Verfahren die in der Theorie der Säcularstörungen vorkommenden Gleichungen numerisch aufzulösen. *Journal für die Reine und Angewandte Mathematik*, **30**, pp 51-94, 1846.

[5] W. Mascarenhas. *On the convergence of the Jacobi method for arbitrary orderings.* PhD thesis, Dept. of Math, MIT, 1991.

[6] A. Sameh. On Jacobi and Jacobi-type algorithms for a parallel computer. *Math. Comp.*, **25**, pp 579-590, 1971.

APPLICATION OF GAUSS-SEIDEL ITERATION TO THE RLS ALGORITHM

Ö. MORGÜL and A. MALAŞ
Dept. of Electrical Engineering
Bilkent University
06533, Bilkent
Ankara, Turkey
morgul@trbilun.bitnet

KEYWORDS. Gauss-Seidel, Parameter Identification, RLS, stability.

We consider the systems given by the following equation :

$$y(n) = \phi^T(n)\theta \qquad n \in \mathbf{N} \tag{1}$$

where $y(n) \in \mathbf{R}$, $\phi(n), \theta \in \mathbf{R}^N$ for some $N \in \mathbf{N}$, T stands for transpose. Here $\phi(n)$, which is called the regressor vector, and $y(n)$ are assumed to be available from measurements, and θ is called the parameter vector, which is a constant by assumption.

Given $y(n)$ and $\phi(n)$, let $\hat{\theta}(n)$ denote an estimate of the parameter vector θ which satisfies (1). Almost all of the parameter identification algorithms seek a recursive formula for $\hat{\theta}(n)$ such that $\lim_{n \to \infty} \hat{\theta}(n) = \theta$. There are various such algorithms, and among these the recursive least squares (RLS) type algorithms are widely used in practice, see [2]. The standard RLS algorithm is given by the following set of equations :

$$\hat{\theta}(n) = \hat{\theta}(n-1) + K(n)(y(n) - \phi^T(n)\hat{\theta}(n-1)) \tag{2}$$

$$R^{-1}(n) = R^{-1}(n-1) + \phi(n)\phi^T(n) \qquad , \qquad K(n) = R(n)\phi(n) \tag{3}$$

where $R(n) \in \mathbf{R}^{N \times N}$ is called the covariance matrix, and $R(0) = R^T(0) > 0$.

In this paper we propose a modification on the parameter update equation (2). Basically we apply the well-known Gauss-Seidel (GS) type iteration to (2) in which the recent updates of the components of the parameter vector are used to update the other components, see [1]. To be specific, let a subindex, as in $\hat{\theta}_i(n)$, denote the corresponding component of that vector. Then, GS applied parameter update equation becomes : $i = 1, 2, \dots, N$

$$\hat{\theta}_1(n) = \hat{\theta}_1(n-1) + K_1(n)(y(n) - \phi_1(n)\hat{\theta}_1(n-1) - \cdots - \phi_N(n)\hat{\theta}_N(n-1))$$

$$\hat{\theta}_i(n) = \hat{\theta}_i(n-1) + K_i(n)(y(n) - \phi_1(n)\hat{\theta}_1(n) - \cdots \phi_{i-1}(n)\hat{\theta}_{i-1}(n)$$
$$-\phi_i(n)\hat{\theta}_i(n-1) - \cdots - \phi_N(n)\hat{\theta}_N(n-1)) \tag{4}$$

Various modifications can be made in the update given by (4). For example, the update sequence can be changed, or block update can be made. The procedure is similar to (4),

407

M. S. Moonen et al. (eds.), *Linear Algebra for Large Scale and Real-Time Applications*, 407–408.
© 1993 *Kluwer Academic Publishers*.

and will not be pursued here. For definiteness, we choose the update equation given by (4), and call the algorithm given by (3)-(4) as the Gauss-Seidel Recursive Least-Squares Algorithm (GSRLS). By using the first equation in the second, and so on, (4) can be put into the following compact form :

$$\hat{\theta}(n) = \hat{\theta}(n-1) + (I - \Gamma(n))K(n)(y(n) - \phi^T(n)\hat{\theta}(n-1)) \tag{5}$$

where $\Gamma(n)$ is a diagonal matrix given as :

$$\Gamma(n) = diag\{0, \ K_1(n)\phi_1(n), \ldots, \ (1 - \prod_{j=1}^{N-1}(1 - K_j(n)\phi_j(n)))\} \tag{6}$$

For the convergence analysis, we assume that there exist positive constants $0 < \alpha < \beta$, and an integer S such that for all $j \in N$, the regressors satisfy the following growth condition

$$\beta I > \frac{1}{S}\sum_{i=j+1}^{j+S} \phi(i)\phi^T(i) > \alpha I \tag{7}$$

The condition given by (7) is called a *persistency of excitation* condition and is typically assumed to hold in many adaptive identification algorithms. Note that (3) and (7) imply that the maximum and the minimum eigenvalues of $R^{-1}(n)$ grow as $O(n)$ for large n, hence the maximum and the minimum eigenvalues of $R(n)$ decay as $O(1/n)$.

For the convergence analysis, we define the parameter error $\tilde{\theta}(n)$ as $\tilde{\theta}(n) = \hat{\theta}(n) - \theta$. Then the error dynamics can be found as :

$$\tilde{\theta}(n) = (I - K(n)\phi^T(n))\tilde{\theta}(n-1) + \Gamma(n)K(n)\phi^T(n)\tilde{\theta}(n-1) \tag{8}$$

Let us define the following change of variables :

$$\tilde{\theta}(n) = R(n)R^{-1}(0)s(n) \tag{9}$$

Let us assume that the regressors $\phi(n)$ are bounded for all $n \in N$. By taking norms, and by using (7), it can be shown that $s(n)$ is bounded as well. Then, from (9) we conclude that $\tilde{\theta}(n)$ approaches to zero as $O(1/n)$ for large n, hence $\lim_{n\to\infty} \tilde{\theta}(n) = \theta$. Simulation results suggest that the performance of RLS and GSRLS are similar, with GSRLS showing better transients, especially at low signal levels. We note that by using different update sequences, or by using block update, presumably we may improve the convergence rate. Intuitively, a good way of block updating is to divide the components of the parameter vector into subgroups depending on our confidence in their estimated values, and then update the parameters which have the highest confidence first and lowest confidence last. We also note that the GS iteration can be applied to other recursive identification algorithms.

References

[1] D. Bertsekas and J. N. Tsitsiklis, *Parallel and Distributed Computations : Numerical Methods*, Prentice-Hall, Englewood Cliffs, New Jersey, 1989.

[2] G. C. Goodwin and K. S. Sin, *Adaptive Filtering, Prediction and Control*, Prentice-Hall, Englewood Cliffs, New Jersey, 1984.

RANKS OF SUBMATRICES OF A MATRIX AND ITS INVERSE

C.C. PAIGE
McGill University
School of Computer Science
3480 University Street
Montreal, PQ, Canada H3A 2A7
chris@cs.mcgill.ca

M. WEI
East China Normal University
Department of Mathematics
Shanghai 200062, China

KEYWORDS. CS-decomposition, unitary matrices, rank relations, nonsingular matrices.

ABSTRACT. Although the CS decomposition (CSD) is usually proven in texts (*e.g.* [4, 6]) for a 2×2 block partitioning of a unitary matrix having square blocks on the diagonal, the CSD for a general 2×2 partitioning (*e.g.* [5]) is far more interesting and useful. As well as its better known applications, it instantly reveals some nice rank properties of these submatrices, and these can then be used almost trivially to prove some important but little known relationships between the ranks of submatrices of a nonsingular matrix and its inverse.

The CS-decomposition (CSD) of a 2-block by 2-block partitioned unitary matrix Q reveals the relationships between the singular value decompositions (SVDs) of each of the 4 subblocks of Q. The CSD shows each SVD has the form $Q_{ij} = U_i D_{ij} V_j^H$, for $i, j = 1, 2$, where each U_i and V_j is unitary, and each D_{ij} is essentially diagonal. The talk gives a simple proof of this with no restrictions on the dimensions of Q_{11}.

The CSD was originally proposed by C. Davis and W. Kahan in [1, 2], and is important in finding the principal angles between subspaces. It also arises in, for example, the analysis of the Total Least Squares (TLS) problem [7]. The relationships between the 4 subblock SVDs (in particular the way that each unitary matrix U_i or V_j appears in 2 different SVDs) has made the CSD a powerful tool for providing simple and elegant proofs of many useful results involving partitioned unitary matrices or orthogonal projectors.

In the talk we also show the CSD makes several nice rank relations obvious. For example,

M. S. Moonen et al. (eds.), Linear Algebra for Large Scale and Real-Time Applications, 409–410.
© 1993 *Kluwer Academic Publishers.*

it can be seen directly from the CSD that if $Q \in C^{n \times n}$ is a unitary matrix partitioned as

$$Q = \begin{pmatrix} Q_{11} & Q_{12} \\ Q_{21} & Q_{22} \end{pmatrix} \begin{matrix} r_1 \\ r_2 \end{matrix}, \qquad r_1 + r_2 = c_1 + c_2 = n,$$

with column labels c_1, c_2 then

$$\min\{r_1, c_1\} - \operatorname{rank}(Q_{11}) = \min\{r_2, c_2\} - \operatorname{rank}(Q_{22}),$$
$$\min\{r_2, c_1\} - \operatorname{rank}(Q_{21}) = \min\{r_1, c_2\} - \operatorname{rank}(Q_{12}).$$

The above result can be used in a very simple way to prove interesting results involving general nonsingular matrices. For example if $Z \in C^{n \times n}$ is a nonsingular matrix and

$$Z = \begin{pmatrix} Z_{11} & Z_{12} \\ Z_{21} & Z_{22} \end{pmatrix} \begin{matrix} r_1 \\ r_2 \end{matrix}, \quad Z^{-1} = \begin{pmatrix} G_{11} & G_{12} \\ G_{21} & G_{22} \end{pmatrix} \begin{matrix} c_1 \\ c_2 \end{matrix}, \quad r_1 + r_2 = c_1 + c_2 = n,$$

then

$$\min\{r_1, c_1\} - \operatorname{rank}(Z_{11}) = \min\{r_2, c_2\} - \operatorname{rank}(G_{22}),$$
$$\min\{r_2, c_1\} - \operatorname{rank}(Z_{21}) = \min\{r_1, c_2\} - \operatorname{rank}(G_{21}).$$

Acknowledgements

Several references, including [3], to earlier proofs of results on nullities of submatrices of a matrix and its inverse were supplied to us by Charles R. Johnson.

References

[1] C. Davis and W.M. Kahan. Some new bounds on perturbations of subspaces. *Bull. AMS*, **75**, pp 863-868, 1969.

[2] C. Davis and W.M. Kahan. The rotation of eigenvectors by a perturbation III. *SIAM J. Numer. Anal.*, **7**, pp 1-46, 1970.

[3] M. Fiedler and T. Markham. Completing a matrix when certain of its entries are specified, *Lin. Alg. Appl.*, **74**, pp 225-237, 1986.

[4] G.H. Golub, C.F. Van Loan. *Matrix Computations*, 2nd Edn. The Johns Hopkins Univ. Press, Baltimore, MD, 1989.

[5] C.C. Paige and M.A. Saunders. Towards a generalized singular value decomposition. *SIAM J. Numer. Anal.*, **18**, pp 398-405, 1981.

[6] G.W. Stewart, J.-G. Sun. *Matrix Perturbation Theory*. Academic Press, New York NY, 1990.

[7] M. Wei. The analysis for the total least squares problem with more than one solution. *SIAM J. Matrix Anal. Appl.*, **13**, No.3, 1992.

A QRD-BASED LEAST-SQUARES ALGORITHM FOR MULTIPULSE ANTENNA ARRAY SIGNAL PROCESSING

I.K. PROUDLER
Defence Research Agency
St. Andrews Road, Malvern
Worcestershire, WR14 3PS
UK
proudler@hermes.mod.uk

KEYWORDS. Multi-pulse radar, adaptive beamforming, RLS, QR decomposition.

The concept of adaptive antenna arrays for radar is well known [1]. One problem of interest is that of interference suppression i.e. observing signals from some directions whilst rejecting those from other directions. In a changing environment, the location of the interfering sources are not known apriori and hence the set of weights that produces a zero response (or null) in these directions needs to be determined in an adaptive manner. One method for solving this adaptive null-steering problem relies on the fact that it can be cast as a linearly constrained least-squares problem:

$$\min_{\underline{w}} \sum_{n=1}^{N} \left| \underline{x}^T(n)\underline{w} \right|^2 \qquad st. \quad \underline{c}^T \underline{w} = \mu$$

where $\underline{x}(n)$ is a p-dimensional vector of antenna outputs at time n, $\underline{w}$ is a p-dimensional vector of weights, $\underline{c}$ is a vector that specifies a given source direction and $\mu \neq 0$ is the desired magnitude of the antenna response function in this direction. The least-squares solution will correspond to that $\underline{w}$ which results in an output time series $y(n) = \underline{x}^T(n)\underline{w}$ which has minimum energy. This means that nulls will be "steered" in the directions of any sources present. The constraint ensures that a null cannot be formed in the direction specified by $\underline{c}$. Solutions to this constrained least squares problem are well known [1]. In particular, there exists an $O(p^2)$, parallel, numerically robust algorithm based on the QR decomposition (QRD) technique [2].

In order to be able to discriminate between sources on the basis of their velocity as well as their bearing from the antenna array it proves necessary to process the echo signal produced

411

M. S. Moonen et al. (eds.), Linear Algebra for Large Scale and Real-Time Applications, 411–412.

by several coherent pulses. The beamformer output is now given by

$$y(n) = \sum_{j=0}^{M-1} \underline{x}^T(n - j\tau)\underline{w}_j$$

where $\underline{w}_j$ is a p-dimensional vector, M is the number of pulses used and τ is the time, in samples, between each pulse. Adaptive null-steering in both space and frequency could then be achieved using the algorithm mentioned above; however it would require $O(p^2 M^2)$ operation per unit time. Fortunately, the data matrix has a certain amount of time-shift redundancy that can be used to reduce the computational load to $O(p^2 M)$ operations. "Fast" least-squares algorithms are well known for the case that the data matrix is Toeplitz (i.e. the ij-th element is a function of $(i - j)$). In this case, however, the data matrix has the property that the ij-th element is a function of $(i - j\tau)$. A lattice-type algorithm for the efficient solution of this adaptive null-steering problem can, nevertheless, be developed. In particular, it is possible to use the new technique of "algorithmic engineering" to derive this novel fast algorithm.

The concept of algorithmic engineering as proposed by McWhirter [3] provides a rigorous framework for describing and manipulating the type of building blocks commonly used to define parallel algorithms and architectures for digital signal processing. Each building block is treated as a mathematical operator with its parallel structure and interconnections represented in terms of a signal flow graph (SFG). It is possible to manipulate the blocks in a rigorous manner using, for example, the associative, commutative and orthogonality properties determined by the matrix (or other) algebra associated with the corresponding operators. This provides the basis of a formal design method which could be used to discover novel processing structures. In particular, a fast algorithm for the space-time adaptive null-steering problem can be easily derived using a series of these simple SFG manipulations. The resultant algorithm is similar to a conventional QRD-based lattice except that the interstage delays are now τ units long. Computer simulations comparing the performance of this algorithm to the optimum (Wiener) solution show that QRD-based least squares algorithms, and the fast versions in particular, are numerically robust.

Acknowledgements

This work was done in collaboration with L. Timmoneri, A. Farina (Alenia, Italy) and J.G. McWhirter (DRA). ©British Crown Copyright 1992 / MOD.

References

[1] S. Haykin. *Adaptive Filter Theory*, 2nd Edition. Prentice-Hall, Englewood Cliffs, New Jersey, USA, 1991.

[2] J.G. McWhirter and T.J. Shepherd. Systolic Array Processor for MVDR Beamforming. *IEE Proceedings Pt. F*, **136(2)**, pp 75-80, 1989.

[3] J.G. McWhirter. Algorithmic Engineering in Adaptive Signal Processing. *IEE Proceedings Pt. F*, **139(3)**, 1992.

ON DISPLACEMENT STRUCTURES FOR COVARIANCE MATRICES AND LOSSLESS FUNCTIONS

P. A. REGALIA and F. DESBOUVRIES
Institut National de Télécommunications
Département Electronique et Communications
9, rue Charles Fourier
91011 Evry cedex, France
regalia@galaxie.int_evry.fr

KEYWORDS. Displacement ranks, covariance matrices, lossless functions, Schur reduction, fast algorithms.

Displacement ranks play a fundamental role in the development of fast computational algorithms for structured covariance matrices [3]–[2], and share a rich connection with modern analytical results in inverse scattering and interpolation [1]. This paper derives an equivalence class between the set of $M \times M$ postive definite matrices of displacement inertia (p, q) on the one hand, and the set of $p \times q$ lossless (transfer) matrices of McMillan degree M on the other. The importance of this result is that it leads to inherently consistent parametrizations of displacement structures of covariance matrices via orthogonal realization theory, paving the way towards numerically stable fast algorithm design.

The connection with lossless systems for the case of displacement structures of positive operators has been developed in [1], and this paper may be understand as a specialization to the matrix case, which is of greater interest in algorithm design. To achieve this, we must first redefine what is meant by the displacement residue of a matrix; the modified residue treated here allows one to derive connections with lossless systems using basic polynomial algebra and matrix manipulation.

1 Summary of results

Let $\mathbf{P}$ be a symmetric $M \times M$ matrix. Most works treating displacement structures use a displacement residue of the form

$$\mathbf{P} - \mathbf{Z}\mathbf{P}\mathbf{Z}^t = \sum_{k=1}^{p} \mathbf{a}_k \mathbf{a}_k^t - \sum_{k=1}^{q} \mathbf{b}_k \mathbf{b}_k^t \tag{1}$$

with $\mathbf{Z}$ the shift matrix with ones along the subdiagonal, and where the generator vectors $\{\mathbf{a}_k\}$ and $\{\mathbf{b}_k\}$ are reduced to a linearly independent set; $\mathbf{P}$ then has displacement inertia (p, q), and displacement rank $p + q$. Many signal processing applications involve structured covariance matrices of low displacement rank, and many fast algorithms are algebraically

413

M. S. Moonen et al. (eds.), Linear Algebra for Large Scale and Real-Time Applications, 413–414.
© 1993 *Kluwer Academic Publishers.*

414

equivalent to manipulating the displacement residue of a matrix rather than the matrix itself. Some of these algorithms prove numerically unstable, and it is our claim that instability problems derive from a poor parametrization of the displacement structure, namely via the generator vectors $\{\mathbf{a}_k\}$ and $\{\mathbf{b}_k\}$ themselves.

Greater analytical insight is achieved by treating a modified displacement residue of the form

$$\begin{bmatrix} \mathbf{P} & \mathbf{0} \\ \mathbf{0}^t & 0 \end{bmatrix} - \begin{bmatrix} 0 & \mathbf{0}^t \\ \mathbf{0} & \mathbf{P} \end{bmatrix} = \sum_{k=1}^{p} \mathbf{a}_k\,\mathbf{a}_k^t - \sum_{k=1}^{q} \mathbf{b}_k\,\mathbf{b}_k^t \tag{2}$$

with $\mathbf{0}$ the null column vector of M elements, and with the generator vectors now of length $M + 1$. Associated to the generator vectors are the polynomial functions

$$\begin{aligned} A_k(z) &= [1 \; z \; \cdots \; z^M]\,\mathbf{a}_k, & k &= 1,\ldots,p; \\ B_k(z) &= [1 \; z \; \cdots \; z^M]\,\mathbf{a}_k, & k &= 1,\ldots,q. \end{aligned}$$

We show that $\mathbf{P}$ can be recovered from its displacement residue in (2) if and only if

$$\sum_{k=1}^{p} A_k(z)\,A_k(z^{-1}) = \sum_{k=1}^{q} B_k(z)\,B_k(z^{-1}), \qquad \text{for all } z, \tag{3}$$

which shows "redundancy" among the generator vectors. If the generator vectors are numerically manipulated, then locally independent errors in the their elements will lead to a violation of (3); such errors clearly cannot be backwards stable.

More enlightening is interpret (3) as saying that the vectors $\mathbf{A}(z) \triangleq [A_1(z),\ldots,A_p(z)]^t$ and $\mathbf{B}(z) \triangleq [B_1(z),\ldots,B_q(z)]^t$ are different spectral factors of the same function. Our principal result shows that $\mathbf{P}$ in (2) is positive definite if and only if the generator polynomials can be reconciled as $\mathbf{A}(z) = \mathbf{U}(z)\,\mathbf{B}(z)$ with $\mathbf{U}(z)$ a $p \times q$ lossless matrix of McMillan degree M. "Lossless" here means that (i) $\mathbf{U}(z)$ is para-unitary: $\mathbf{U}^t(z^{-1})\,\mathbf{U}(z) = \mathbf{I}_q$, and (ii) $\mathbf{U}(z)$ is analytic in $|z| \geq 1$.

Any lossless function $\mathbf{U}(z)$ can be parametrized by a sequence of rotation angles (say, $\{\theta_k\}$) such that, for all $\{\theta_k\}$ in some well defined region (e.g., $|\theta_k| < \pi/2$), $\mathbf{U}(z)$ remains lossless of McMillan degree M. Hence any computational algorithm rephrased in terms of the parameter set $\{\theta_k\}$ will, in finite precision arithmetic, remain inherently consisent with the exact solution obtained from a perturbed positive definite matrix of the same displacement inertia. The importance of this result in numerically stable fast algorithm design will be illustrated by examples.

References

[1] D. Alpay, P. Dewilde, and H. Dym. On the existence and construction of solutions to the partial lossless inverse scattering problem, with applications to estimation theory. *IEEE Trans. Information Theory*, **IT-35**, pp 1184–1205, 1989.

[2] F. Desbouvries. *Rangs de Déplacement et Algorithmes Rapides*, doctoral thesis, Ecole Nationale Supérieure des Télécommunications, Paris, Jan. 1991.

[3] T. Kailath, S. Y. Kung, and M. Morf. Displacement ranks of matrices and linear equations. *J. Math. Anal. Applic.*, **68**, pp 395–407, 1979.

[4] H. Lev-Ari and T. Kailath. Lattice filter parametrization and modelling of nonstationary processes. *IEEE Trans. Information Theory*, **IT-30** pp 2–16, p 878, 1984.

APPLICATION OF VECTOR EXTRAPOLATION AND CONJUGATE GRADIENT TYPE METHODS TO THE SEMICONDUCTOR DEVICE PROBLEM

W.H.A. SCHILDERS
Philips Research Laboratories
P.O. Box 80000
5600 JA Eindhoven
The Netherlands
schildr@prl.philips.nl

KEYWORDS. Vector extrapolation, RRE, GMRES, CGS, Bi-CGSTAB, semiconductor device simulation.

Semiconductor device simulation is an area which has received much attention over the past fifteen years. The rapid development and miniaturisation of devices has urged the need for robust and efficient algorithms to be able to perform such simulations (cf. [1]). From a mathematical point of view, it is a very tempting problem: the system of three partial differential equations describing the behaviour of semiconductor devices is singularly perturbed and extremely nonlinear. Furthermore, the resulting linear systems are badly conditioned. Therefore, sophisticated discretisation and solution techniques have to be employed to solve the problem. In this paper we discuss two situations in which recently developed iterative techniques have been employed to improve the convergence in the area of semiconductor device simulation.

1 Reduced rank extrapolation

The nonlinear discretised system of equations describing the behaviour of semiconductor devices is often solved using Gummel's method, which is essentially a nonlinear block Gauss-Seidel method. Despite the good convergence properties of this method, its convergence may be very slow: convergence factors of 0.95 are no exception. Because of this, we have investigated accelleration techniques for Gummel's method based on the reduced rank extrapolation method (RRE). Recently, this method has gained much attention; Smith, Ford and Sidi ([5]) presented an excellent survey of extrapolation methods. The application of RRE to the accelleration of Gummel's method is described in several papers ([2, 3]). The method has turned out to be very succesful; modifications developed by the author of this paper have shown even faster convergence.

An interesting property of RRE is its equivalence to GMRES when applied to vector sequences generated with a constant linear operator ([4]). The main advantage of RRE over

415

M. S. Moonen et al. (eds.), Linear Algebra for Large Scale and Real-Time Applications, 415–416.
© 1993 *Kluwer Academic Publishers.*

GMRES is that it is defined in terms of the iterates, making it ideally suited for situations in which the operator is not known explicitly. Modifications of RRE immediately lead to modifications of GMRES, and vice versa. For example, recently developed nested GMRES methods (Van der Vorst) are very similar to some of the RRE-modifications suggested in [3]. It would be very interesting to investigate whether QMR can be cast into an RRE-like form, and investigate its convergence properties when applied to the accelleration of vector sequences.

2   Bi-CGSTAB applied to the semiconductor problem

In the case of the semiconductor problem, the linear systems arising when solving the discretised equations using Newton's method are extremely badly conditioned. We have employed many iterative solution methods for solving the nonsymmetric systems (cf. [1]) and found ILU-CGS to be the most robust method. However, for advanced semiconductor problems, CGS sometimes fails to converge. The convergence problems were significantly less when applying ILU-bi-CGSTAB. Our experiences with this method are summarized as follows (notation as in [6]):

- $\omega_i = (t,s)/(t,t)$ is used instead of $\omega_i = (K_L^{-1}t, K_L^{-1}s)/(K_L^{-1}t, K_L^{-1}s)$ (cf. [6])
- the parameter $\rho_j = (\bar{r}, r_j)$ can be evaluated in single or full precision, or in the approximate form $\rho_j = -\omega_j(\bar{r}, t)$; the latter form is not recommended
- a different choice of $\bar{r}$ may lead to faster convergence; for the moment, the default choice still is $\bar{r} = r_0$
- current conservation (essential property in device simulation) is not as good with bi-CGSTAB as compared to CGS

References

[1] S.J. Polak, C. den Heijer, W.H.A. Schilders and P.A. Markowich. Semiconductor device modelling from the numerical point of view. *Int. J. Numer. Meth. Engng.*, **24**, pp 763-838, 1987.

[2] W.H.A. Schilders, P.A. Gough and K. Whight. Extrapolation techniques for improved convergence in semiconductor device simulation. In : J.J.H. Miller (Ed.), *Proc. NASECODE VIII Conf.*, Boole Press, Dublin, pp 94-95, 1992.

[3] W.H.A. Schilders. Modified reduced rank extrapolation methods with applications. To be published.

[4] A. Sidi. Convergence and stability properties of minimal polynomial and reduced rank extrapolation algorithms. *SIAM J. Numer. Anal.*, **23**, pp 197-209, 1986.

[5] D.A. Smith, W.F. Ford and A. Sidi. Extrapolation methods for vector sequences. *SIAM Review*, **29**, pp 199-233, 1987.

[6] H.A. van der Vorst. Bi-CGSTAB: a more smoothly converging variant of CG-S for the solution of nonsymmetric linear systems. *SIAM J. Sci. Stat. Comp.*, **13**, pp 631-644, 1992.

ACCURATE SYMMETRIC EIGENREDUCTION BY A JACOBI METHOD

I. SLAPNIČAR
LG Mathematische Physik
Fernuniversität Hagen
P.O. Box 940
5800 Hagen 1
Germany
MA703@DHAFEU11.BITNET

KEYWORDS. Symmetric eigenvalue problem, error analysis, relative accuracy.

A real symmetric matrix H is called "well-behaved" if the small relative changes in its elements cause small relative changes in its eigenvalues. We show that an algorithm which implements decomposition followed by Jacobi-type iteration computes the eigenvalues with nearly the same accuracy. The positive definite case was treated in [2], our results cover general indefinite matrices.

We use the following two-step algorithm originally proposed in [5] (see also [3]):

1. Decompose H as $H = GJG^T$, where J is diagonal with ± 1's on the diagonal, and G has full column rank. This is a modification of the symmetric indefinite decomposition [1] with complete pivoting.

2. Apply the implicit (one-sided) J-orthogonal Jacobi method to the pair G, J [5] to find the non-zero eigenvalues and the corresponding eigenvectors of H. Here both trigonometric and hyperbolic plane rotations are used.

For most well-behaved matrices we have proved good relative error bounds for the computed eigenvalues and the norm error bounds for the computed eigenvectors [3]. Numerical experiments [3] also give strong evidence that our algorithm is never much worse, and can be much better than QR or the standard Jacobi algorithm applied directly to H.

The preturbation theory [6] for non-singular H is the following: let $|H| = \sqrt{H^2}$, $D = (diag|H|)^{1/2}$, and set $|H| = DAD$. Let δH be a symmetric perturbation such that $|\delta H_{ij}| \leq \varepsilon |H_{ij}|$. Let λ and $\lambda + \delta\lambda$ denote the i-th eigenvalue of H and $H + \delta H$, respectively. Then

$$\frac{|\delta\lambda|}{|\lambda|} \leq \varepsilon n \|A^{-1}\|_2. \tag{1}$$

417

M. S. Moonen et al. (eds.), *Linear Algebra for Large Scale and Real-Time Applications*, 417–418.
© 1993 *Kluwer Academic Publishers*.

418

By [4] $\kappa(A) \leq n\kappa(|H|) = n\kappa(H)$, where $\kappa(\cdot)$ is the spectral condition number. Thus, (1) is never much worse than the standard perturbation bound $|\delta\lambda|/|\lambda| \leq \varepsilon\sqrt{n}\kappa(H)$, and it is possible that $\kappa(A) \ll \kappa(H)$, in which case (1) is much better.

We also need the perturbation bound in the case when only the factor G is perturbed. Let δG be the perturbation such that $|\delta G_{ij}| \leq \varepsilon|G_{ij}|$. Set $G = BD$, where D is diagonal positive definite such that the columns of B have unit norms. Then [6]

$$(1 - \eta)^2 \leq \frac{\lambda + \delta\lambda}{\lambda} \leq (1 + \eta)^2, \quad \eta = \varepsilon\sqrt{n}\|B^{-1}\|_2. \tag{2}$$

Let $\lambda + \delta\lambda$ now denote the i-th eigenvalue computed in the floating-point arithmetic with precision ε. Set $G = \bar{D}\bar{B}$, where $\bar{D}$ is diagonal positive definite such that the rows of $\bar{B}$ have unit norms. Then

$$\frac{|\delta\lambda|}{|\lambda|} \leq 272n^2\varepsilon\|(\bar{D}^{-1}|GJG^T|\bar{D}^{-1})^{-1}\|_2 + O(\|B^{-1}\|_2\varepsilon). \tag{3}$$

The first term of the right hand-side of (3) comes from the error analysis of the decomposition [3] and (1), and the rest comes from the error analysis of the implicit Jacobi method [3] and (2). Numerical experiments [3] give strong numerical evidence that $\|(\bar{D}^{-1}|GJG^T|\bar{D}^{-1})^{-1}\|_2 \approx \|A^{-1}\|_2$. Further, $\|B^{-1}\|_2$ is bounded by $O(n^2 3.781^{2n})$ irrespectively of $\kappa(H)$ [3]. In practice $\|B^{-1}\|_2$ is generally much smaller (≤ 30) and it does not grow much during the Jacobi process, which means that the error induced by the Jacobi part is usually negligible to the error induced by the decomposition. Altogether, our algorithm computes the eigenvalues with nearly the accuracy predicted by (1), and (3) is uniformly better than the relative error bounds for QR or the standard Jacobi algorithm.

References

[1] J. R. Bunch and B. N. Parlett. Direct Methods for Solving Symmetric Indefinite Systems of Linear Equations. *SIAM J. Numer. Anal.*, Vol. 8, No. 4, pp 639-655, 1971.

[2] J. Demmel and K. Veselić. Jacobi's method is more accurate than QR. To appear in *SIAM J. Mat. Anal. Appl.*

[3] I. Slapničar. *Accurate Symmetric Eigenreduction by a Jacobi Method*. Ph.D. Thesis, FB Mathematik, Fernuniversität, Hagen, 1992.

[4] A. van der Sluis. Condition Numbers and Equilibration of Matrices. *Numer. Math.*, Vol. 14, pp 14-23, 1969.

[5] K. Veselić. An Eigenreduction Algorithm for Definite Matrix Pairs. To appear in *Numer. Math.*

[6] K. Veselić and I. Slapničar. Floating-point perturbations of Hermitian matrices. *Preprint*, FB Mathematik, Fernuniversität, Hagen 1991, submitted to *Lin. Alg. Appl.*

THE ORDER-RECURSIVE CHANDRASEKHAR EQUATIONS FOR FAST SQUARE-ROOT KALMAN FILTERING

D.T.M. SLOCK
Eurecom Institute
2229 route des Crêtes, Sophia Antipolis
F-06560 Valbonne
France
slock@eurecom.fr

KEYWORDS. Kalman filtering, square-root, Chandrasekhar equations, order-recursive.

It has been three decades now that Kalman filtering has been the tool of choice for least-squares estimation in linear state-space models. The Kalman filter makes no distinction between time-varying and time-invariant models. About two decades ago, Kailath introduced the Chandrasekhar equations, which generally reduce the computational complexity of the Kalman filter when the linear state-space model is time-invariant. Along another venue, the so-called square-root forms of the Kalman filter (SRCF, SRIF) were introduced, which propagate a matrix square-root of the covariance matrix of the state estimation error. In this way, the covariance matrix itself, being the square of its square-root, is inherently non-negative definite and symmetric at all times. The square-root algorithms offer various numerical advantages [2] and have a similar algorithmic complexity as their Riccati equation based counterpart(s).

Recursive least-squares (RLS) parameter estimation can be formulated as a special case of Kalman filtering, namely as a fixed-point smoothing problem for the case of no process noise. In the seventies, fast algorithms have been developed for the solution of systems of equations appearing in least-squares filtering problems. The underlying concept for these fast algorithms is the *displacement structure* of the covariance matrix involved. These ideas have been further developed into fast RLS algorithms for estimating the impulse response of a FIR filter. These fast RLS algorithms come into two groups: the Fast Transversal Filter (FTF) algorithms and the Fast Lattice/ Fast QR (FLA/FQR) algorithms, corresponding respectively to a transversal and a lattice realization of the estimated filter. The FLA/FQR algorithms are fast algorithms for updating the QR factorization of the Toeplitz data matrix involved in the least-squares problem, and hence can be viewed as fast versions of the corresponding square-root RLS algorithm.

Hence we get the following picture for recursive least-squares estimation:

419

M. S. Moonen et al. (eds.), Linear Algebra for Large Scale and Real-Time Applications, 419–420.
© 1993 *Kluwer Academic Publishers.*

	Kalman	SQRT Kalman	RLS	SQRT RLS
regular algorithm	Kalman filter	SRCF, SRIF	RLS alg.	square-root RLS
fast algorithm	Chandrasekhar eq.	?	FTF	FLA/FQR

We see that for the special case of FIR filter estimation, the regular algorithm exists in its two versions, propagating the covariance matrix or its square-root, and both versions have a fast alternative. For the general Kalman filtering problem though, a fast alternative exists only for the classical Riccati equation based solution. No fast version of the square-root Kalman filtering algorithms has been known thus far. We present such a fast alternative, which offers the following advantages (see [1] also):

• *order-recursiveness*: the filtering problem is solved for all model orders up to some maximum order. This offers interesting perspectives for model reduction.

• *numerical robustness*: the order-recursive Chandrasekhar equations being a fast form of the numerically robust Square-Root Kalman filters, one can expect excellent numerical behavior (whereas the numerical behavior of the original Chandrasekhar equations is unclear). Specialized to the FIR filter estimation problem, the Chandrasekhar equations are knwon to lead to the FTF algorithm, which has a tricky numerical behavior (basically unstable, but can be stabilized). The order-recursive Chandrasekhar equations specialize to the FLA/FQR algorithms though, of which the good numerical properties are well-known. Using the framework introduced in [2], we may actually be able to demonstrate the numerical properties of the order-recursive Chandrasekhar equations.

• *parallel implementations*: the inner products in the original Chandrasekhar algorithm make it little amenable to parallel implementation. The different lattice sections of the order-recursive version however can be pipelined and mapped to parallel processors.

When considering fast RLS parameter estimation applications, we omit the process noise, and consider a fixed-point smoothing formulation (the parameters are the unknown initial state). Many parameter estimation problems can be formulated in this way [3, chapter 7]. The following ones deserve special attention:

• FLA/FQR algorithms, as already mentioned above.

• δ-parameterization: we can easily derive FLA/FQR algorithms for this recently popularized alternative parameterization of discrete-time systems.

• Signal modeling by fitting a sum of exponentials (with known exponent). Especially when the exponents are on the unit circle, this provides a fast RLS alternative to sliding FFT, with complete flexibility in the choice of frequencies though.

References

[1] D.T.M. Slock. The Order-Recursive Chandrasekhar Equations for Fast Square-Root Kalman Filtering. In : *Proc. 26th Information Sciences and Systems Conference*, Princeton University, NJ, USA, March 18-20 1992. See ICASSP-93 also.

[2] D.T.M. Slock. Backward Consistency Concept and Round-Off Error Propagation Dynamics in Recursive Least-Squares Algorithms. *Optical Engineering*, **31(6)**, 1992.

[3] D.T.M. Slock. *Fast algorithms for Fixed-Order Recursive Least-Squares Parameter Estimation*. PhD thesis, Stanford University, Stanford, CA, USA, September 1989.

HYBRID ITERATIVE METHODS BASED ON FABER POLYNOMIALS

G. STARKE
Institut für Praktische Mathematik
Universität Karlsruhe
Englerstrasse 2
7500 Karlsruhe 1
Germany
starke@ipmsun1.mathematik.uni-karlsruhe.de

KEYWORDS. Iterative methods for nonsymmetric systems, hybrid methods, Arnoldi process, Faber polynomials.

In recent years, a lot of progress has been made in the field of iterative methods for large nonsymmetric systems of linear equations. Most of this work dealt with generalizations of the classical conjugate gradient algorithm to problems which are not positive definite (see [2] for an extensive survey of these developments). These CG-like methods belong to the larger class of polynomial iterative methods, i.e., starting from an initial guess $x_0 \in \mathbf{R}^N$, the iterates can be expressed as

$$x_m = x_0 + q_{m-1}(A)r_0$$

where $q_{m-1}(A)$ is a polynomial of degree $m - 1$ in the coefficient matrix $A \in \mathbf{R}^{N \times N}$ and $r_0 = b - Ax_0$ denotes the initial residual. For CG-like methods, the iteration polynomial q_{m-1} is constructed during the iteration process based on certain orthogonality relations, therefore requiring the computation of inner products in each step. A second type of polynomial methods consists of the so-called Chebyshev-like or semi-iterative methods. For this class, the iteration polynomial is constructed beforehand based on some information about the underlying linear system. Here, the goal is to choose the iteration polynomial q_m such that the norm of the residual

$$r_m = (I - q_{m-1}(A)A)r_0 = p_m(A)r_0$$

is reduced. In other words, we have to choose the residual polynomial p_m (normalized by $p_m(0) = 1$) in such a way that $\|p_m(A)\|$ becomes small. One motivation for the study of Chebyshev-like methods is that they avoid inner products which can be quite expensive (compared to matrix-vector multiplications) on certain parallel architectures. Moreover, CG-like methods for nonsymmetric systems are susceptible to breakdowns or stagnation.

M. S. Moonen et al. (eds.), Linear Algebra for Large Scale and Real-Time Applications, 421–422.
© 1993 *Kluwer Academic Publishers.*

In practice, Chebyshev-like methods are implemented as hybrid schemes which consist of a beginning phase where information about the matrix A is acquired, and a second phase where a polynomial iteration designed with respect to this information is carried out. A number of such hybrid methods have been proposed in recent years; for an overview see [3].

Phase I: Arnoldi vs. Field of Values. Most of these hybrid algorithms are based on eigenvalue estimates obtained by the Arnoldi process, i.e., eigenvalues of a Hessenberg matrix $H_m = V_m^T A V_m \in \mathbf{R}^{m \times m}$ where the columns of V_m form an orthonormal basis for the Krylov subspace span$\{\mathbf{v}, A\mathbf{v}, \ldots, A^{m-1}\mathbf{v}\}$. However, these eigenvalue estimates, the so-called Ritz values associated with the Arnoldi process, do not necessarily give a good representation of the spectrum of A for $m \ll N$. In contrast to the situation when A has real spectrum, it can happen that part of $\sigma(A)$ is not approximated well by $\sigma(H_m)$ (see the examples in [4]). Another shortcoming of the Arnoldi process is the fact that the Ritz values can only be guaranteed to lie in the field of values of A but not necessarily close to its spectrum. In particular, Arnoldi eigenvalue estimates could be located close to the origin while $\sigma(A)$ is actually well-separated from 0. In both of these cases, any iterative method based on Ritz values will converge poorly. To overcome this problem, we approximate the set $W(A) \cap 1/W(A^{-1})$ in the Krylov subspace spanned by the columns of V_m and use the resulting polygonal domain for optimizing the polynomial iteration in the second phase.

Phase II: Faber Polynomials. The use of Faber polynomials for polynomial iterative methods was studied by Eiermann [1]. Faber polynomials (properly normalized) have the remarkable property that they are almost optimal as residual polynomials with respect to a given set Ω (see [1], [5]). In addition, they can be constructed for a variety of regions, for example, if Ω is bounded by a polygon. Then, the recursion coefficients for the Faber polynomials can be extracted from the Schwarz-Christoffel representation of the corresponding conformal mapping (see [5] for details).

Final Remarks. Experimental and theoretical results concerning the above mentioned techniques are currently written up in [4] which will soon be available as a technical report.

References

[1] M. Eiermann. On semiiterative methods generated by Faber polynomials. *Numer. Math.*, **56**, pp 139–156, 1989.

[2] R.W. Freund, G.H. Golub and N.M. Nachtigal. Iterative solution of linear systems. *Acta Numerica*, **1**, pp 57–100, 1992.

[3] N.M. Nachtigal, L.Reichel and L.N. Trefethen. A hybrid GMRES algorithm for non-symmetric matrix iterations. *SIAM J. Matrix Anal. Appl.*, **13**, pp 796–825, 1992.

[4] G. Starke. On hybrid iterative methods for nonsymmetric systems of linear equations. *Numer. Math.*, 1992. In preparation.

[5] G. Starke and R.S. Varga. A hybrid Arnoldi-Faber method for nonsymmetric systems of linear equations. *Numer. Math.*, 1992. To appear.

ASPECTS OF IMPLEMENTING A 'C' MATRIX LIBRARY

D.E. STEWART
Program in Advanced Computation
School of Mathematical Sciences
Australian National University
GPO Box 4, Canberra, ACT 2601
Australia
des@thrain.anu.edu.au

ABSTRACT. A library for numerical linear algebra called "Meschach" is described. This library is written in the 'C' programming language and uses the features of the language (data structures, dynamic memory allocation, error handling) to provide a clean and efficient way of programming algorithms. This library is now part of *netlib*.

KEYWORDS. Data structures, 'C', error handling, memory management, sparse matrices.

Introduction

The 'C' programming language provides a number of features that can ease the burden of programming, but which seem to be under-utilised by the scientific computing community. A matrix library called "Meschach" developed by the author is used to illustrate how 'C' can be used to obtain both good performance and ease the programmer's lot. This library uses self-contained data structures to implement matrices, vectors, permutations and sparse matrices. Dynamic memory allocation and deallocation is used to create, destroy and resize these data structures. Input/output routines for the data structures are part of Meschach; output can be read back in, and comments can be part of the input. MATLAB compatible input/output is also part of Meschach. Error/exception handling is built in and uses a "catch/throw" model[2], which enables the user to gain control after an error has occurred. In addition, the library implements a wide range of numerical linear algebra operations including:

- basic operations (inner products, linear combinations, matrix-vector and matrix-matrix products).

423

M. S. Moonen et al. (eds.), Linear Algebra for Large Scale and Real-Time Applications, 423–424.
© 1993 *Kluwer Academic Publishers.*

- a wide range of dense matrix factorizations (Cholesky, LU, QR, LDL^T, BKP) and sparse matrix factorization (Cholesky and LU) and solve routines.

- incomplete, symbolic and modified factorizations (Cholesky and LU).

- eigenvalue/vector and singular value/vector decompositions (real Schur form, SVD, eigenvalue/vector extraction).

- iterative routines for sparse and procedurally-defined matrices (pre-conditioned conjugate gradients, CGS, LSQR, Lanczos, Arnoldi).

Meschach has been compiled and used on a variety of machines: Firstly on a Zilog Zeus Z8000, then on a Pyramid 9810, an IBM RT network, Sun 3's, SPARCstations, and RS/6000 and on Silicon Graphics as well as on an IBM PC. Meschach is compilable using either "Kernighan and Ritchie C"[1] or ANSI C compilers. Meschach has also been used with C++.

Portability is achieved partly through having three files which deal with machine dependent characteristics. Currently a library of various versions of these files for different operating systems and compilers is being developed.

Meschach can be obtained from the AT&T Netlib site using the Unix commands

```
mail netlib@research.att.com
send all from c/meschach
```

More extensive documentation than is provided with the source code can be obtained on request from the author at des@thrain.anu.edu.au until further arrangements can be made.

References

[1] B. Kernighan and D. Ritchie, *The 'C' Programming Language*. Prentice Hall, Englewood Cliffs, NJ, 2nd edition 1988 (1st edition 1978).

[2] G. Steele, *Common Lisp: The Language*. Digital Press, Burlington, MA, 1984.

INTERMEDIATE FILL-IN IN SPARSE QR DECOMPOSITION

M. TUMA
Institute of Computer Science
Czechoslovak Academy of Sciences
Pod vodárenskou věží 2
182 07 Praha 8 - Libeň
Czechoslovakia
tuma@cspgcs11.bitnet

KEYWORDS. QR decomposition, fill-in, elimination tree.

1 Introduction

We consider QR factorization of the large sparse matrix $A \in R^{m,n}$, $m > n$. We concentrate on methods using Householder reflections (HR) and Givens rotations (GR). Type of the algorithm we consider is the scheme where dense working matrices correspond to the nodes of the elimination tree of $A^T A$ (see [4], [5]). In each node corresponding to an iteration step we assume using of one or more HR and/or GR.

The feature differing sparse decomposition from the dense case is the necessity of keeping small size of fill-in (number of new nonzero positions appearing throughout the algorithm in A). Small size of fill-in implies short time of computation. Possible means to reduce fill-in are permutations of columns and rows of A.

We distinguish two different forms of fill-in: fill-in in the factor R (related to the structure of the upper triangle of A) which is not influenced by row permutations of A and intermediate fill-in in working matrices used throughout the decomposition, which is finally removed. We will not touch the problem of fill-in in R. For problems connected with column ordering which minimizes fill-in in R and estimation of structure of R see [1], [3], [6]. We are interested mainly in the intermediate fill-in.

2 Intermediate fill-in in a multifrontal-like method

Structural properties of GR and HR are similar : both the algorithms can be interpreted as a row merge where the resulting rows have the structure corresponding to a merge of row

M. S. Moonen et al. (eds.), Linear Algebra for Large Scale and Real-Time Applications, 425–426.

426

patterns of all the rows involved. Only positions which were eliminated are forced to be zeros. The serious problem is to decide how to give rows of working matrices into groups to be eliminated together. To spare arithmetics and symbolic manipulations, groups should be preferably larger, but, on the other side, forming of these groups of rows using some local search rule can be costly. The other problem is to find the additional column ordering to minimize intermediate fill-in keeping the structure of R possibly unchanged.

If we can easily find column reordering of A in order to minimize intermediate fill-in such that the fill-in in R is kept the same or only "slightly" changed, we can also use postorder of processed columns resulting in an effective implementation (see [4]).

Theorem 2.1 : *Postordering of the elimination tree does not change overall size of the intermediate fill-in created throughout the decomposition.*

The following observation shows how to obtain a case of A for which we can easily reorder its columns. The same reformulation with even better results can be done for a slightly changed width-2 clique ordering of [2] and some other orderings.

Observation 2.2 : *Let us have width-2 node ordering of A as in [2]. Its rows can be reordered such that all its column blocks of nonzeros are continuous (1-block shape).*

Using the structure of a 1-block matrix we can effectively permute its (block) columns in order to minimize the intermediate fill-in. Additional advantage of the 1-block shape is that the beginnings and ends of the column blocks of A imply an *automatic* rule to determine actual strategy of merges of groups of rows in the working matrices corresponding to the nodes of elimination tree. This rule can be implemented with small symbolical overhead.

As the general conclusion we state the following. Matrices having the 1-block shape can be easily obtained from A using row and column reorderings. Their use implies very effective multifrontal-like algorithms as it has been verified by numerical experiments.

References

[1] T.F. Coleman, A. Edenbrandt and J.R. Gilbert. Predicting fill for sparse orthogonal factorization. *JACM*, **33**, pp 517-532, 1986.

[2] A. George, J. Liu and E. Ng. Row-ordering schemes for sparse Givens Transformations. I. Bipartite graph model. *Linear Algebra and its Appl.*, **61**, pp 55-81, 1984.

[3] D.R. Hare, C.R. Johnson, D.D. Olesky and P. van den Driessche. Sparsity Analysis of the QR Factorization. *preprint*, 1991.

[4] J.W.H. Liu. The role of elimination trees in sparse factorization. *SIAM J. Matrix Anal. Appl.*, **11**, pp 134-172, 1990.

[5] P.Matstoms. *The multifrontal solution of sparse linear least squares problems.* Licentiat thesis No. 293, University of Linköping, 1991.

[6] A.Pothen. Predicting the structure of sparse orthogonal factors, *preprint*, revised October 1991.

SHIFTING STRATEGIES FOR THE PARALLEL QR ALGORITHM

D.S. WATKINS
Department of Pure and Applied Mathematics
Washington State University
Pullman, WA 99164-3113
U.S.A.
na.watkins@na-net.ornl.gov

KEYWORDS. QR algorithm, parallel implementation, shifts.

Consider the problem of finding all eigenvalues of a real square matrix A that has no special structure to exploit. The QR algorithm has been the method of choice for solving such problems for the past 30 years. However, in recent years it has been suggested that the days of the QR algorithm may be numbered, for the QR algorithm has proven difficult to parallelize. Such predictions are at best premature. A simple modification of the shifting strategy promises to improve the parallel performance of the QR algorithm significantly and keep the competition at bay for years to come.

The QR algorithm is an iterative process that produces a sequence of unitarily similar upper-Hessenberg matrices (A_k) that (usually) converges to quasi-triangular form, thereby revealing the eigenvalues. In order to accelerate convergence, we shift the matrix at each step by an amount that approximates an eigenvalue (or so we hope).

The standard implementations of the QR algorithm for real, nonsymmetric matrices perform double steps, that is, two steps at a time. A double QR step on a Hessenberg matrix proceeds roughly as follows: Two shifts are chosen. These (and other information) determine an initial similarity transformation that sets the step in motion by creating a small bulge in the Hessenberg form at the upper left hand corner of the matrix. The subsequent transformations return the matrix to upper Hessenberg form by chasing the bulge down the diagonal until it disappears off of the edge of the matrix.

This is a highly sequential process. One fairly obvious way to make it parallel is to pipeline it: Once the first bulge has been chased just a small distance down the diagonal, there is enough room to start the second bulge. Then we can chase two bulges simultanously. Shortly thereafter we can start a third bulge, and so on. In this way we can chase a large number of bulges simultaneously in pipeline fashion. This scheme generates enough

M. S. Moonen et al. (eds.), Linear Algebra for Large Scale and Real-Time Applications, 427–428.
© 1993 *Kluwer Academic Publishers.*

arithmetic to keep many processors busy. It should be possible to implement the scheme on parallel computers of both shared memory and distributed memory types.

This idea is not new. It has been suggested by several authors, but so far it has not caught on in a big way. An important reason for this is that until now nobody has thought to change the shifting strategy. It has been assumed that one should continue to use the standard shifts, which are the eigenvalues of the lower right-hand 2×2 submatrix. The problem with this is that if one is chasing, say, 20 bulges in pipeline and one chooses shifts from the lower right-hand corner of the matrix, then the shifts that one gets are not really the right shifts for the next step; they are the shifts that would have been right 20 iterations ago. The effect of using such out-of-date shifts is to degrade the convergence rate severely [2].

A way around this problem is to change the shifting strategy. Instead of getting just two shifts at a time, get (say) 40 shifts by calculating the eigenvalues of the lower right-hand 40×40 submatrix. The cost of doing so is insignificant if the matrix is sufficiently large, and it provides enough fresh shifts for a batch of 20 double steps. Preliminary experiments indicate that strategies of this type are competitive with the standard shifting strategies, even in a sequential environment. It follows that they will be far superior when run in pipeline fashion on a parallel machine, since they allow many bulges to be chased at once without any of the shifts being out of date.

There is, however, one drawback. If we wish to obtain quadratic convergence, we must finish each batch, and thereby empty the pipeline, before computing the shifts for the next batch. This is probably unsatisfactory for parallel processing. If we wish to keep the pipeline full at all times, we cannot avoid using out-of-date shifts. However, by taking large enough batches or spacing the bulges appropriately, we can guarantee that our shifts are never out of date by more than one batch. The analysis of van de Geijn [2] shows that under these circumstances the convergence rate is degraded from 2 (quadratic) to 1.62, which is still not bad.

This brief note is a partial summary of [3].

Acknowledgment

The author thought to try this new class of shifting strategies after reading about some interesting experiments with the multishift QR algorithm by A. A. Dubrulle [1].

References

[1] A. A. Dubrulle. The multishift QR algorithm—is it worth the trouble? Manuscript, 1991.

[2] R. A. van de Geijn. Deferred shifting schemes for parallel QR methods. *SIAM J. Matrix Anal. Appl.*, to appear.

[3] D. S. Watkins. Shifting strategies for the parallel QR algorithm. Manuscript submitted for publication, 1992.

LIST OF PARTICIPANTS

Gregory S. Ammar
Northern Illinois University, Mathematics Department, DeKalb, IL, USA
ammar@math.niu.edu

Miguel Anjos
Stanford University, Department of Computer Science, Stanford, CA, USA
anjos@sccm.stanford.edu

Peter Arbenz
ETH Zentrum, Institute for Scientific Computing, Zuerich, Switzerland
arbenz@inf.ethz.ch

Suzanne Balle
Technical University of Denmark, Institute for Numerical Analysis, Lyngby, Denmark
unismb@wuli.uni-c.dk

Valeriu Beiu
Katholieke Universiteit Leuven, ESAT Laboratory, Heverlee, Belgium
beiu@esat.kuleuven.ac.be

Michael W. Berry
University of Tennessee, Department of Computer Science, Knoxville, TN, USA
berry@cs.utk.edu

Laurens Bierens
TNO Physics and Electronic Laboratory, The Hague, The Netherlands
laurens.bierens@fel.tno.nl

Petter E. Bjørstad
University of Bergen, Institutt for Informatikk, Bergen, Norway
petter@eik.ii.uib.no

Adam W. Bojanczyk
Cornell University, School of Electrical Engineering, Ithaca, NY, USA
adamb@toeplitz.ee.cornell.edu

Daniel Boley
University of Minnesota, Computer Science Department, Minneapolis, MN 55455, USA
boley@janus.cs.umn.edu

Stephen Boyd
Stanford University, Information Systems Laboratory, Stanford, CA, USA
boyd@isl.stanford.edu

Adhemar Bultheel
Katholieke Universiteit Leuven, Computer Science Department, Heverlee, Belgium
Adhemar.Bultheel@cs.kuleuven.ac.be

429

James R. Bunch
University of California San Diego, Mathematics Department, La Jolla, CA, USA
jrb@sdna3.ucsd.edu

Angelika Bunse-Gerstner
Universität Bremen, Fachbereich Mathematik und Informatik, Bremen, Germany
angelika@mathematik.uni-bremen.de

Daniela Calvetti
Stevens Inst. of Technology, Dept. of Pure and Applied Mathematics, Hoboken, NJ, USA
calvetti@sitult.stevens-tech.edu

Moody T. Chu
North Carolina State University, Department of Mathematics, Raleigh, NC, USA
chu@gauss.math.ncsu.edu

George Cybenko
Dartmouth College, Thayer School of Engineering, Hanover, NH, USA
george_cybenko@dartmouth.edu

Pieter de Groen
Vrije Universiteit Brussel, Department of Mathematics, Brussel, Belgium
pieter@tena2.vub.ac.be

Thomas Dehn
Universität Karlsruhe, Institut fur Praktische Mathematik, Karlsruhe, Germany
af06@dkauni2.bitnet

James W. Demmel
University of California Berkeley, Division of Computer Science, Berkeley, CA, USA
demmel@arpd.berkeley.edu

Bart De Moor
Katholieke Universiteit Leuven, ESAT Laboratory, Heverlee, Belgium
demoor@esat.kuleuven.ac.be

Ed Deprettere
Delft University of Technology, Dept. of Electrical Engineering, Delft, The Netherlands
ed@dutentb.et.tudelft.nl

Patrick Dewilde
Delft University of Technology, Dept. of Electrical Engineering, Delft, The Netherlands
dewilde@dutentb.et.tudelft.nl

Zlatko Drmac
FernUniversität Hagen, Lehrgebiet Mathematische Physik, Hagen, Germany
in708@dhafeu11.bitnet

Frits Dumortier
Universiteit Gent, Laboratorium Regeltecniek, Zwijnaarde, Belgium
fd@autoctrl.rug.ac.be

Alan Edelman
University of California Berkeley, Department of Mathematics, Berkeley, CA, USA
edelman@robalo.Berkeley.edu

Heike Fassbender
Universität Bremen, Fachbereich Mathematik und Informatik, Bremen, Germany
heike@mathematik.uni-bremen.de

Vince Fernando
University of California Berkeley, Division of Computer Science, Berkeley, CA, USA
fernando@math.berkeley.edu

Ricardo Fierro
University of California Los Angeles, Department of Mathematics, Los Angeles, CA, USA
fierro@math.ucla.edu

Roland W. Freund
AT&T Bell Laboratories, Murray Hill, NJ, USA
freund@research.att.com

Yves Genin
Universite Catholique de Louvain, Inst. Math. Pure et Appl., Louvain-la-Neuve, Belgium
genin@anma.ucl.ac.be

Morven Gentleman
Natl Research Council Canada, Inst of Information Technology, Ottawa, Ontario, Canada
gentleman@iit.nrc.ca

George Glentis
University of Athens, Department of Informatics, Athens, Greece
othon@grathun1.bitnet

Gene H. Golub
Stanford University, Department of Computer Science, Stanford, CA, USA.
na.golub@na-net.stanford.edu

Sven Hammarling
The Numerical Algorithms Group Ltd, Oxford, United Kingdom
sven@vax.nag.co.uk

Simon Haykin
McMaster University, Communications Research Laboratory, Hamilton, Ontario, Canada
haykin@mcbuff.eng.mcmaster.ca

Chunyang He
University of Bielefeld, Department of Mathematics, Bielefeld, Germany
he@math4.mathematik.uni-bielefeld.de

Nick J. Higham
University of Manchester, Department of Mathematics, Manchester, United Kingdom
higham@vortex.mathematics.manchester.ac.uk

432

Marlis Hochbruck
ETH-Zentrum, IPS Supercomputing, Zuerich, Switzerland
marlis@ips.id.ethz.ch

Thomas Huckle
Universität Wuerzburg, Institut Angew. Mathematik, Wuerzburg, Germany
huckle@vax.rz.uni-wuerzburg.dbp.de

Khakim D. Ikramov
Moscow State University, Faculty of Numerical Maths & Cybernetics, Moscow, Russia

Florian Jarre
Universität Wurzburg, Institut fur Angewandte Mathematik, Wuerzburg, Germany
angm026@vax.rz.uni-wuerzburg.dbp.de

Søren Holdt Jensen
Technical University of Denmark, Electronics Institute, Lyngby, Denmark
soeren@eiffel.ei.dth.dk

Jiri Kadlec
Queen's University, Dept. of Electrical and Electronic Eng., Belfast, Northern Ireland
eeg0020@vax2.queens-belfast.ac.uk

Bo Kagstrom
University of Umea, Institute of Information Processing, Umea, Sweden.
bokg@cs.umu.se

Nick Kalouptsidis
University of Athens, Department of Informatics, Athens, Greece
spa64@grathun1.bitnet

Linda Kaufman
AT&T Bell Laboratories, Murray Hill, NJ, USA
lck@research.att.com

Jacek Kuczynski
Polish Academy of Sciences, Institute of Computer Science, Warsaw, Poland

Wang Liang
Faculte Polytechnique, Service d'Informatique, Mons, Belgium
wang@pip.umh.ac.be

Franklin Luk
Rensselaer Polytechnic Institute, Computer Science Department, Troy, NY, USA
luk@cs.rpi.edu

Niloufer Mackey
SUNY at Buffalo, Department of Computer Science, Amherst, NY, USA
mackey@cs.Buffalo.edu

Pierre Manneback
Faculte Polytechnique, Service d'Informatique, Mons, Belgium
pierre@pip.umh.ac.be

433

Roy Mathias
University of Minnesota, Institute for Maths and its Applications, Minneapolis, MN, USA
mathias@imafs.ima.umn.edu

Volker Mehrmann
Universität Bielefeld, Fakultät fuer Mathematik, Bielefeld, Germany
mehrmann@math1.mathematik.uni-bielefeld.de

Jean Meinguet
Université Catholique de Louvain, Inst. Math. Pure et Appl., Louvain-la-Neuve, Belgium
meinguet@anma.ucl.ac.be

Carl Meyer
North Carolina State University, Mathematics Department, Raleigh, NC, USA
meyer@math.ncsu.edu

Marc Moonen
Katholieke Universiteit Leuven, ESAT Laboratory, Heverlee, Belgium
moonen@esat.kuleuven.ac.be

Omer Morgul
Bilkent University, Dept. of Electrical and Electronics Engineering, Bilkent, Ankara, Turkey
morgul@trbilun.bitnet

Bart Motmans
Katholieke Universiteit Leuven, ESAT Laboratory, Heverlee, Belgium
motmans@esat.kuleuven.ac.be

Noel M. Nachtigal
RIACS, NASA Ames Research Center, Moffett Field, CA, USA
santa@riacs.edu

Nancy Nichols
University of Reading, Department of Mathematics, Reading, United Kingdom
smsnicho@cms.am.cc.reading.ac.uk

Chris Paige
McGill University, Computer Science Department, Montreal, Canada
chris@opus.cs.mcgill.ca

Pythagoras Papadimitriou
University of Manchester, Department of Mathematics, Manchester, England
mbbgppa@cms.mcc.ac.uk

Beresford Parlett
University of California Berkeley, Mathematics Department, Berkeley, CA, USA
parlett@math.berkeley.edu

Olli-Pekka Piirila
Helsinki University of Technology, Institute of Mathematics, Espoo, Finland
opp@dopey.hut.fi

Ian K. Proudler
Defence Research Agency, RSRE, Malvern, Worcestershire, United Kingdom
proudler@hermes.mod.uk

Sanzheng Qiao
McMaster University, Communications Research Laboratory, Hamilton, Ontario, Canada
qiao@maccs.dcss.mcmaster.ca

Phil A. Regalia
Institut National des Telecommunications, Dept. Electronique et Communications,
Evry, France
regalia@frint51.bitnet

Lothar Reichel
Kent State University, Department of Mathematics and Computer Science, Kent, OH, USA
reichel@mcs.kent.edu

Acar Savaci
Istanbul Technical University, Electrical-Electronics Engineering, Maslak, Istanbul, Turkey
eecuneyt@tritu.bitnet

Geert Schelfhout
Katholieke Universiteit Leuven, ESAT Laboratory, Heverlee, Belgium
schelfhout@esat.kuleuven.ac.be

Willy H.A. Schilders
Philips Research Laboratory, Eindhoven, The Netherlands
schildr@prl.philips.nl

Ivan Slapnicar
Fernuniversität Hagen, LG Mathematische Physik, Hagen, Germany
ma703@dhafeu11.bitnet

Dirk Slock
Institut EURECOM, Sophia Antipolis, Valbonne, France
slock@eurecom.cica.fr

Gerhard Starke
Universität Karlsruhe, Institut fuer Praktische Mathematik, Karlsruhe, Germany
na.starke@na-net.ornl.gov

David Stewart
Australian National University, School of Mathematical Sciences, Canberra, Australia
des@thrain.anu.edu.au

G.W. Stewart
University of Maryland, Department of Computer Science and Institute for Advanced
Computer Studies, College Park, MD, USA
stewart@cs.umd.edu

Shilpa Talwar
Stanford University, Computer Science Department, Stanford, CA, USA
talwar@sccm.stanford.edu

Philippe Toint
Facultés Universitaires ND de la Paix, Department of Mathematics, Namur, Belgium
phtoint@bnandp51.bitnet

Miroslav Tuma
Institute of Computer Sciience CSAV, Prague, Czechoslovakia
cvs80@cspgcs11.bitnet

Marc Vanbarel
Katholieke Universiteit Leuven, Department of Computer Science, Heverlee, Belgium
marc@cs.kuleuven.ac.be

Joos Vandewalle
Katholieke Universiteit Leuven, ESAT Laboratory, Heverlee, Belgium
vandewalle@esat.kuleuven.ac.be

Charles Van Loan
Cornell University, Department of Computer Science, Ithaca, NY, USA
cv@gvax.cs.cornell.edu

Peter Van Overschee
Katholieke Universiteit Leuven, ESAT Laboratory, Heverlee, Belgium
vanovers@esat.kuleuven.ac.be

Filiep Vanpoucke
Katholieke Universiteit Leuven, ESAT Laboratory, Heverlee, Belgium
vpoucke@esat.kuleuven.ac.be

Paul Vanvuchelen
Katholieke Universiteit Leuven, ESAT Laboratory, Heverlee, Belgium
vanvuche@esat.kuleuven.ac.be

Johan Verbeke
Katholieke Universiteit Leuven, Department of Computer Science, Heverlee, Belgium
Johan.Verbeke@cs.kuleuven.ac.be

Kresimir Veselic
Fernuniversität Hagen, Lehrgebiet Mathematische Physik, Hagen, Germany
ma704@dhafeu11.bitnet

David Watkins
University of Washington, Applied Mathematics, Seattle, WA, USA
watkins@amath.washington.edu

Layne T. Watson
Virginia Polytechnic Institute and State University, Dept. Computer Science,
Blacksburg, VA, USA
ltw@cayuga.cs.vt.edu

Margaret Wright
AT&T Bell Laboratories, Murray Hill, NJ, USA
mhw@research.att.com

AUTHORS INDEX

437